# Lecture Notes in Control and Information Sciences

Edited by M. Thoma and A. Wyner

Vol. 62: Analysis and Optimization of Systems
Proceedings of the Sixth International Conference on Analysis and Optimization of Systems
Nice, June 19–22, 1984
Edited by A. Bensoussan, J. L. Lions
XIX, 591 pages. 1984.

Vol. 63: Analysis and Optimization of Systems
Proceedings of the Sixth International Conference on Analysis and Optimization of Systems
Nice, June 19–22, 1984
Edited by A. Bensoussan, J. L. Lions
XIX, 700 pages. 1984.

Vol. 64: Arunabha Bagchi
Stackelberg Differential Games in Economic Models
VIII, 203 pages, 1984

Vol. 65: Yaakov Yavin
Numerical Studies in Nonlinear Filtering
VIII, 273 pages, 1985.

Vol. 66: Systems and Optimization
Proceedings of the Twente Workshop
Enschede, The Netherlands, April 16–18, 1984
Edited by A. Bagchi, H. Th. Jongen
X, 206 pages, 1985.

Vol. 67: Real Time Control of Large Scale Systems
Proceedings of the First European Workshop
University of Patras, Greece, Juli 9–12, 1984
Edited by G. Schmidt, M. Singh, A. Titli, S. Tzafestas
XI, 650 pages, 1985.

Vol. 68: T. Kaczorek
Two-Dimensional Linear Systems
IX, 397 pages, 1985.

Vol. 69: Stochastic Differential Systems – Filtering and Control
Proceedings of the IFIP-WG 7/1 Working Conference
Marseille-Luminy, France, March 12-17, 1984
Edited by M. Metivier, E. Pardoux
X, 310 pages, 1985.

Vol. 70: Uncertainty and Control
Proceedings of a DFVLR International Colloquium
Bonn, Germany, March, 1985
Edited by J. Ackermann
IV, 236 pages, 1985.

Vol. 71: N. Baba
New Topics in Learning Automata Theory and Applications
VII, 231 pages, 1985.

Vol. 72: A. Isidori
Nonlinear Control Systems: An Introduction
VI, 297 pages, 1985.

Vol. 73: J. Zarzycki
Nonlinear Prediction Ladder-Filters for Higher-Order Stochastic Sequences
V, 132 pages, 1985.

Vol. 74: K. Ichikawa
Control System Design based on Exact Model Matching Techniques
VII, 129 pages, 1985.

Vol. 75: Distributed Parameter Systems
Proceedings of the 2nd International Conference, Vorau, Austria 1984
Edited by F. Kappel, K. Kunisch, W. Schappacher
VIII, 460 pages, 1985.

Vol. 76: Stochastic Programming
Edited by F. Archetti, G. Di Pillo, M. Lucertini
V, 285 pages, 1986.

Vol. 77: Detection of Abrupt Changes in Signals and Dynamical Systems
Edited by M. Basseville, A. Benveniste
X, 373 pages, 1986.

Vol. 78: Stochastic Differential Systems
Proceedings of the 3rd Bad Honnef Conference, June 3–7, 1985
Edited by N. Christopeit, K. Helmes, M. Kohlmann
V, 372 pages, 1986.

Vol. 79: Signal Processing for Control
Edited by K. Godfrey, P. Jones
XVIII, 413 pages, 1986.

Vol. 80: Artificial Intelligence and Man-Machine Systems
Edited by H. Winter
IV, 211 pages, 1986.

For information about Vols. 1–61 please contact your bookseller or Springer-Verlag.

# Lecture Notes in Control and Information Sciences

Edited by M. Thoma and A. Wyner

122

J. Descusse, M. Fliess, A. Isidori, D. Leborgne (Editors)

# New Trends in Nonlinear Control Theory

Proceedings of an International Conference on Nonlinear Systems, Nantes, France, June 13–17, 1988

Springer-Verlag Berlin
Heidelberg GmbH

**Editors**

J. Descusse
Laboratoire d'Automatique
de Nantes, ENSM
1, rue de la Noë
44072 Nantes Cedex 03
France

A. Isidori
Dipartimento di Informatica
e Sistemistica
Università di Roma »La Sapienza«
18, via Eudossiana
00184 Rome
Italy

Michel Fliess
Laboratoire des Signaux
& Systèmes, CNRS-ESE
Plateau du Moulon
91192 Gif-Sur-Yvette Cedex
France

D. Leborgne
UER de Mathématiques
Université de Nantes
2 chemin de la Houssinière
44072 Nantes Cedex 03
France

ISBN 978-3-540-51075-8 ISBN 978-3-540-46143-2 (eBook)
DOI 10.1007/978-3-540-46143-2


Originally published by Springer-Verlag Berlin Heidelberg New York in 1989.


Offsetprinting: Mercedes-Druck, Berlin

2161/3020-543210

# FOREWORD

This Conference on nonlinear control theory was organized within a special "Nonlinear Year" of the French "Centre National de la Recherche Scientifique". It was held in Nantes from June 13th to June 17th 1988. There were only invited speakers and we hope they gave a correct sample of the new trends in research from all over the world. We tried to bring together theoretical talks by pure mathematicians and more applied communications which were dedicated to robotics, electrical engines, biology and computer science.

When comparing this with an analogous meeting which was organized in Paris in June 1985 [1], it should first be noted that techniques from differential geometry are still developing, as demonstrated by the table of contents. The birth of methods from differential algebra, which has permitted the understanding of some specific problems such as input-output inversion and dynamic feedback decoupling, should also be pointed out. Stabilization, which is certainly one of the most important questions in nonlinear control, is now being attacked by several new and promising approaches.

Finally, we wish to acknowledge the financial support of the following French institutions: Centre National de la Recherche Scientifique, Direction des Recherches et Etudes Techniques de la Délégation Générale pour l'Armement, Ecole Nationale Supérieure de Mécanique de Nantes, Chambre de Commerce et de l'Industrie de Nantes.

Reference

[1] M. Fliess and M. Hazewinkel, eds, Algebraic and Geometric Methods in Nonlinear Control Theory, Reidel, Dordrecht, 1986.

Nantes, Paris, Rome,
December 1988

J. Descusse
M. Fliess
A. Isidori
D. Leborgne

## AUTHORS

AEYELS (BELGIUM)
AKSAS (FRANCE)
ANDREINI (ITALY)
ARAPOSTATHIS (U.S.A.)
BACCIOTTI (ITALY)
BASTIN (BELGIUM)
BEJCZY (U.S.A.)
BENOIT (FRANCE)
BENVENISTE(FRANCE)
BOIERI (ITALY)
BONNARD (FRANCE)
BORNARD (FRANCE)
BYRNES (U.S.A.)
CAMPILLO (FRANCE)
CANDELPERGHER (FRANCE)
CAPOLINO (FRANCE)
CELLE (FRANCE)
CHALEYAT-MAUREL (FRANCE)
CLAUDE (FRANCE)
COUENNE (FRANCE)
CROUCH (U.S.A.)
DAUPHIN-TANGUY (FRANCE)
DESCUSSE (FRANCE)
DI BENEDETTO (ITALY)
DING (U.S.A.)
DIOP (FRANCE)
EL ASSOUDI (FRANCE)
FLIESS (FRANCE)
GAUTHIER (FRANCE)
GLAD (SWEDEN)
GRIZZLE (U.S.A.)
HAMMER (U.S.A.)
HILL (AUSTRALIA)
HU (FRANCE)
HUILLET (FRANCE)
IGHNEIWA (U.S.A.)
ISIDORI (ITALY)
JAKUBCZYK (POLAND)
JURDJEVIC (U.S.A.)
KNOBLOCH (GERMANY)
KUPKA (CANADA)
LAMNABHI (FRANCE)
LE GLAND (FRANCE)
LE GUERNIC (FRANCE)
LEE (U.S.A.)
LEVINE (FRANCE)
LOBRY (FRANCE)
MARCUS (U.S.A.)
MEIZEL (FRANCE)
MICHEL (FRANCE)
MOKKADEM (FRANCE)
MONIN (FRANCE)
MONTSENY (FRANCE
MOOG (FRANCE)
OUSTALOUP (FRANCE)
PARDOUX (FRANCE)
POMET (FRANCE)
PRALY (FRANCE)
RAMIS (FRANCE)
ROTELLA (FRANCE)
SALLET (FRANCE)
SALUT (FRANCE)
SLOTINE (FRANCE)
STEFANI (ITALY)
SUSSMAN (U.S.A.)
TARN (U.S.A.)
TCHON (POLAND)
VAN DER SHAFT (THE NETERLANDS)
WEN (AUSTRALIA)

# TABLE OF CONTENTS / TABLE DES MATIERES

## I - DIFFERENTIAL GEOMETRIC SYSTEM THEORY

### 1 - STRUCTURAL PROPERTIES



### 2 - FEEDBACK SYNTHESIS



### 3 - OBSERVERS

## II - ALGEBRAIC SYSTEM THEORY



## III - OPTIMAL CONTROL



## IV - STABILITY, ROBUSTNESS AND SINGULAR PERTURBATIONS TECHNIQUES

### 1 - STABILIZATION TECHNIQUES



### 2 - ROBUSTNESS AND SINGULAR PERTURBATIONS TECHNIQUES

## 2 - MISCELLANEOUS

# I - DIFFERENTIAL GEOMETRIC SYSTEM THEORY

## STRUCTURAL PROPERTIES

# CHARACTERIZATIONS OF HAMILTONIAN CONTROL SYSTEMS

B. Jakubczyk

Institute of Mathematics
Polish Academy of Sciences
00-950 Warsaw, Śniadeckich 8, Poland

## ABSTRACT

Given a control system with an output, we give explicit conditions on the system which guarantee that it admits a Hamiltonian structure. The conditions are stated directly in terms of vector fields and observation functions defined by the system.

## §1. INTRODUCTION.

In recent years, several papers were devoted to extending a linear theory of Hamiltonian control systems (cf.Brockett, Rahimi [2]) to nonlinear case. A definition of Hamiltonian input-output system was proposed by Van der Schaft [10] and characterizations of Hamiltonian systems in this sense were given by Crouch and Irving [4] and Crouch and Van der Schaft [11] and [6]. Their characterizations are stated directly in terms of the input-output map of the system or in terms of its Volterra kernels (see [5] for a review) and refer to realization theory of nonlinear systems. Therefore, we may call them external characterizations.

Other external characterizations of Hamiltonian systems were given by this author in [7],[8] and [9]. Our characterizations, similarly as those of [6], concerned a completely general class of systems. The method was based on an internal characterization of Hamiltonian input-output systems given in terms of vector fields and observation functions of the system. The definition of Hamiltonian system used for this internal characterization was different, however.

In this paper we extend our method and present internal characterizations of Hamiltonian control systems in the Van der Schaft's sense. We show that these characterizations are equivalent, in the case of linear systems

$$\dot{x} = Ax + Bu, \quad y = Cx, \qquad x \in \mathbb{R}^n, \; u, y \in \mathbb{R}^m,$$

to the familiar criterion

$$(CA^kB)^T = C(-A)^kB, \quad k \geq 0.$$

The latter criterion is usually written in terms of the transfer function $G$ of the system as the equality $G^T(s) = G(-s)$ (cf. [3]).

In this paper we give sketches of proofs of the main results, only. Detailed proofs, together with an extended version of these results, will appear in another paper.

## §2. PRELIMINARIES.

We shall first recall some basic definitions. Let $X$ be a differential paracompact manifold of class $C^\infty$. A symplectic form on $X$ is a closed nondegenerate 2-form on $X$ of class $C^\infty$. A differential manifold $X$ together with a symplectic form $\omega$ on it is called a symplectic manifold. Given a symplectic manifold $(X, \omega)$ and a $C^\infty$ function $\phi$ on $X$, there is a unique vector field $f_\phi$ on $X$ given by the equality

$$d\phi = \omega(\cdot, f_\phi).$$

Such a vector field is called a Hamiltonian vector field and the function $\phi$ is called the Hamiltonian of this vector field. A standard symplectic manifold is $X = \mathbb{R}^{2k}$ with the symplectic form

$$\omega = \sum_{i=1}^{k} dq_i \wedge dp_i,$$

where $x = (p_1, ..., p_k, q_1, ..., q_k)$ are standard coordinates on $\mathbb{R}^{2k}$. A Darboux theorem says that given any symplectic form $\omega$, there always exists a local coordinate system in which this form takes the above canonical form. Given the canonical symplectic form and a function $\phi$, the corresponding Hamiltonian vector field takes the familiar form

$$f_\phi = (-\partial\phi/\partial q_1, ..., -\partial\phi/\partial q_k, \partial\phi/\partial p_1, ..., \partial\phi/\partial p_k).$$

On a symplectic manifold $(X, \omega)$ one defines Poisson bracket of functions on $X$ as

$$\{\phi, \psi\} = \omega(f_\phi, f_\psi) = f_\phi \psi,$$

where the most right expression denotes the directional derivative of $\psi$ along the vector field $f_\phi$. The map $\phi \longrightarrow f_\phi$ is the homomorphism of Lie algebras, i.e.

$$f_{\{\phi,\psi\}} = [f_\phi, f_\psi],$$

where $[\cdot,\cdot]$ denotes Lie bracket of vector fields. The standard Poisson structure on $\mathbb{R}^{2k}$ gives the standard Poisson bracket

$$\{\phi, \psi\} = \sum_{i=1}^{k} \left(\frac{\partial\phi}{\partial p_i}\frac{\partial\psi}{\partial q_i} - \frac{\partial\phi}{\partial q_i}\frac{\partial\psi}{\partial p_i}\right).$$

Consider the following input-output system

$$\dot{x} = f(x, u), \quad y = h(x, u), \tag{2.1}$$

where the state $x(t)$ lives in $\mathbb{R}^n$ or, more generally, in an n-dimensional differentiable manifold $X$ of class $C^\infty$, the input $u(t) = (u_1(t), ..., u_m(t))$ takes values in a connected, simply connected subset $U$ of $\mathbb{R}^m$ and the output $y(t) = (y_1(t), ..., y_m(t))$ takes values in $\mathbb{R}^m$. To avoid some degeneracy of the set $U$ we assume that for any two points $u,\ v \in U$ there is a smooth curve joining $u$ and $v$ whose interior lies entirely in the interior of $U$ (in particular, this means that the interior of $U$ is nonempty).

Following [10], we shall say that the above system is ***Hamiltonian locally around*** $x_0 \in X$ if $n = 2k$ and there exists a function $H : V \times U \longrightarrow \mathbb{R}$, where $V$ is a neighborhood of $x_0$, and a system of local coordinates $(p_1, ..., p_k, q_1, ..., q_k)$ on $V$ such that the system (2.1) takes the following form in these coordinates

$$\dot{p}_i = -\frac{\partial}{\partial q_i}H, \quad \dot{q}_i = \frac{\partial}{\partial p_i}H, \quad i = 1, ..., k$$
$$y_j = \frac{\partial}{\partial u_j}H, \quad j = 1, ...m. \tag{2.2}$$

The function $H$ is called the Hamiltonian of the system. We shall call system (2.1) *globally Hamiltonian* if there are a smooth function $H : X \times U \longrightarrow \mathbb{R}$ and a symplectic form $\omega$ on $X$ such that

$$dH = \omega(\cdot, f), \quad \text{and} \quad h = \frac{\partial}{\partial u}H. \tag{2.3}$$

As a special case, we will consider the following class of systems

$$\dot{x} = g_0(x) + \sum_{i=1}^{m} u_i g_i(x), \quad y = h(x). \tag{2.4}$$

In this case the output function is independent of $u$, therefore, a Hamiltonian system should have the Hamiltonian $H$ affine with respect to $u$ and of the form

$$H(x, u) = h_0(x) + \sum_{j=1}^{m} u_j h_j(x). \tag{2.5}$$

Thus, an unknown function to find is $h_0$ instead of $H$, in this case. It follows then that the definitions of locally and globally Hamiltonian systems can be rephrased as follows. System (2.4) is called *locally Hamiltonian around* $x_0 \in X$ if there exists a local function $h_0$ and local coordinates $p_1, ..., p_k, q_1, ..., q_k$ around $x_0$ such that in these coordinates system (2.4) takes the form

$$\dot{p}_i = -\frac{\partial}{\partial q_i}h_0 - \sum_{i=1}^{m} u_i \frac{\partial}{\partial q_i}h_i, \quad \dot{q}_i = \frac{\partial}{\partial p_i}h_0 + \sum_{i=1}^{m} u_i \frac{\partial}{\partial p_i}h_i, \quad i = 1, ..., k,$$
$$y_j = h_j, \quad j = 1, ..., m.$$

System (2.4) is called *globally Hamiltonian* if there is a function $h_0$ on $X$ and a symplectic form $\omega$ on $X$ such that

$$dh_i = \omega(\cdot, g_i), \quad i = 0, ..., m.$$

## §3. CHARACTERIZATIONS.

Both problems of deciding whether a given system in form (2.1) or in form (2.4) can be given a Hamiltonian structure (local or global) can be reduced to the following problems.

*Problem 1.* Let $H = \{h_\alpha\}_{\alpha\in A}$ be a family of $C^\infty$ functions on $X$ and let $F = \{f_\alpha\}_{\alpha\in A}$ be a family of $C^\infty$ vector fields on $X$. We ask whether there exists a symplectic form $\omega$ on $X$ such that the vector fields $f_\alpha$, $\alpha \in A$, are Hamiltonian vector fields with Hamiltonians $h_\alpha$, i.e.

$$dh_\alpha = \omega(\cdot, f_\alpha), \quad \alpha \in A. \tag{3.1}$$

*Problem 2.* Let $H = \{h_\alpha\}_{\alpha\in A}$ be a family of $C^\infty$ functions on $X$ and let $F = \{f_\alpha\}_{\alpha\in A}$ be a family of $C^\infty$ vector fields on $X$. Given another $C^\infty$ vector field $f_0$ on $X$, we ask whether there exists a symplectic form $\omega$ on $X$ and a $C^\infty$ function $h_0$ such that the vector fields $f_0$ and $f_\alpha$, $\alpha \in A$, are Hamiltonian vector fields with Hamiltonians $h_0$ and $h_\alpha$, i.e.

$$dh_0 = \omega(\cdot, f_0), \quad dh_\alpha = \omega(\cdot, f_\alpha), \quad \alpha \in A. \tag{3.2}$$

The case of the special system (2.4) clearly reduces to Problem 2 when we take $A = \{1, ..., m\}$, $f_0 = g_0$, and $f_\alpha = g_\alpha$ for $\alpha \in \{1, ..., m\}$.

The general case of system (2.1) can be reduced to Problem 2 by defining $A = \{\alpha = (i, u) \mid i = 1, ..., m,\ u \in U\}$, and

$$h_\alpha = h_i(\cdot, u), \quad f_\alpha = \frac{\partial}{\partial u_i} f(\cdot, u), \quad \text{and} \quad f_0 = f(\cdot, u_0),$$

where $u_0$ is a fixed element of $U$. In order that the two problems be equivalent we have to assume, additionally, that

$$\frac{\partial}{\partial u_i} h_j = \frac{\partial}{\partial u_j} h_i, \quad i, j = 1, ..., m. \tag{3.3}$$

The equivalence of Problem 2 to the original problem of existence of the Hamiltonian structure of system (2.1) can be seen as follows. If our system (2.1) is Hamiltonian, then differentiating the first equality in (2.3) with respect to $u_i$ we obtain that Problem 2 is solvable. Vice versa, if Problem 2 is solvable, then integrating the second equality in (3.2) along any curve $\gamma$ in $U$ joining $u_0$ with $u$ and defining

$$H(x, u) = h_0(x) + \int_\gamma \sum_{j=1}^{m} h_{(j,u)}(x) du_j$$

we obtain the equality (2.3). Note that in the above integration the result depends on the ends of the curve only which follows from our assumption (3.3).

Now, we will concentrate our attention on solving Problem 2. We will reduce it to Problem 1 which was already solved in [8]. In [8] several characterizations of existence of the Hamiltonian structure for the families $H$ and $F$ were given. We shall use one of them to solve Problem 2.

In order to state a solution of Problem 1 we need the following notation. Given a vector field $f$ and a function $h$ on $X$, we denote by $fh$ the directional derivative of $h$ along $f$. In particular, if $X = \mathbb{R}^n$ and $f = (f_1, ..., f_n)^T$, then

$$fh = \sum_{i=1}^{n} f_i \frac{\partial}{\partial x_i} h.$$

Let us define the following family of functions

$$h_{\alpha_q \cdots \alpha_1} = f_{\alpha_q} \cdots f_{\alpha_2} h_{\alpha_1}, \quad \alpha_1, ..., \alpha_q \in A,$$

where on the right hand side we take the iterated directional derivative of the function $h_{\alpha_1}$ along the vector fields $f_{\alpha_2}, ..., f_{\alpha_q}$. We shall also use the following notation for iterated Lie brackets of vector fields

$$f_{[\alpha_q \cdots \alpha_1]} = [f_{\alpha_q}, [f_{\alpha_{q-1}}, ..., [f_{\alpha_2}, f_{\alpha_1}]...]].$$

For completeness we also denote $f_{[\alpha_1]} = f_{\alpha_1}$.

Given families $H = \{h_\alpha | \alpha \in A\}$, $F = \{f_\alpha | \alpha \in A\}$ and the extended families

$$\hat{H} = \{h_{\alpha_q \cdots \alpha_1} \mid q \geq 1, \alpha_1, ..., \alpha_q \in A\}, \quad \hat{F} = \{f_{[\alpha_q \cdots \alpha_1]} \mid q \geq 1, \alpha_1, ..., \alpha_q \in A\},$$

we say that the system $(H, F)$ is *minimal* if, at each point $x \in X$, the differentials of the functions in $\hat{H}$ span the whole cotangent space, and the vector fields in $\hat{F}$ span the whole tangent space. The definition can be extended to any families of functions $H$ and vector fields $F$ on $X$, not necessarily indexed by the same parameter. In this case we define $\hat{H}$ as all iterated directional derivatives of functions from $H$ along vector fields in $F$, and $\hat{F}$ as all iterated Lie brackets of vector fields from $F$. The following theorem was proved in [8].

*Theorem 3.1* If the system $(H, F)$ is minimal, then Problem 1 is solvable if and only if the following equalities hold

$$f_{\alpha_0} h_{\alpha_q \cdots \alpha_1} + f_{[\alpha_q \cdots \alpha_1]} h_{\alpha_0} = 0, \quad \forall k \geq 1, \ \forall \alpha_0, ..., \alpha_q \in A. \tag{3.4}$$

In order to give some insight into the above result we shall sketch the simple proof of necessity of condition (3.4). The first fact we should mention is that for a Hamiltonian system $(H, F)$ on a symplectic manifold the functions $h_{\alpha_q \cdots \alpha_1}$ can be expressed in terms of Poisson bracket as

$$h_{\alpha_q \cdots \alpha_1} = \{h_{\alpha_q}, ..., \{h_{\alpha_2}, h_{\alpha_1}\}...\}.$$

This follows by iterative application of the formula $\{\phi, \psi\} = f_\phi \psi$. As the Lie bracket of Hamiltonian vector fields corresponds to the Poisson bracket of their Hamiltonians it follows that the iterated Lie brackets $f_{[\alpha_q \cdots \alpha_1]}$ are Hamiltonian vector fields with the Hamiltonians $h_{\alpha_q \cdots \alpha_1}$. Now, formula (3.4) can be equivalently written as

$$\{h_{\alpha_0}, h_{\alpha_q \cdots \alpha_1}\} + \{h_{\alpha_q \cdots \alpha_1}, h_{\alpha_0}\} = 0$$

and it simply follows from the antisymmetry of Poisson bracket.

To formulate another criterion of solvability of Problem 1, we shall use the following functions. Define

$$h_{[\alpha_1]} = h_{\alpha_1}$$

and, by induction,

$$h_{[\alpha_q \cdots \alpha_1]} = f_{\alpha_q} h_{[\alpha_{q-1} \cdots \alpha_1]} - f_{[\alpha_{q-1} \cdots \alpha_1]} h_{\alpha_q}. \tag{3.5}$$

We have the following result ([8]).

*Theorem 3.2* If the system (H,F) is minimal, then Problem 1 is solvable if and only if

$$h_{[\alpha_q \cdots \alpha_1]} = q\ h_{\alpha_q \cdots \alpha_1}, \quad \forall\, q \geq 1,\ \forall\, \alpha_1, \ldots, \alpha_q \in A. \tag{3.6}$$

The main additional difficulty in solving Problem 2 lies in finding the function $h_0$. If this function was known we would immediately reduce our problem to Problem 1 by extending the families $H$ and $F$ to $\tilde{H} = H \cup \{h_0\}$ and $\tilde{F} = F \cup \{f_0\}$. Not having the function $h_0$ given we will construct it using the Frobenius theorem.

Let us define the following vector fields on $X \times \mathbb{R}$

$$\tilde{f}_{[\alpha_q \cdots \alpha_1]} = (f_{[\alpha_q \cdots \alpha_1]}, -f_0 h_{\alpha_q \cdots \alpha_1})^T, \quad q \geq 1,\ \alpha_1 \in A,\ \alpha_2, .., \alpha_q \in A \cup \{0\}. \tag{3.7}$$

and, additionally,

$$\tilde{f}_{[\alpha_q \cdots \alpha_2 \alpha_1]} = -\tilde{f}_{[\alpha_q \cdots \alpha_1 \alpha_2]}, \quad \text{if} \quad \alpha_1 = 0, \quad \text{and} \quad \tilde{f}_0 = (f_0, 0). \tag{3.8}$$

In particular, this means that $\tilde{f}_{[\alpha_q \cdots 00]} = 0$. We denote the distribution spanned by these vector fields by

$$\Delta(x) = \text{span}\ \ \{\tilde{f}_{[\alpha_q \cdots \alpha_1]}(x) \mid \alpha_1, ..., \alpha_q \in A \cup \{0\},\ q \geq 1\} = n.$$

A solution to Problem 2 is given by the following theorem.

*Theorem 3.3* If the system $(H, \tilde{F})$, where $\tilde{F} = F \cup \{f_0\}$, is minimal then Problem 2 is locally solvable if and only if the following conditions are satisfied:

$(A)$ $\qquad f_{\alpha_0} h_{\alpha_q \cdots \alpha_1} + f_{[\alpha_q \cdots \alpha_1]} h_{\alpha_0} = 0, \quad \alpha_0, \alpha_1 \in A,\ \alpha_2, ..., \alpha_q \in A \cup \{0\}, q \geq 1,$

(B) The distribution $\Delta$ is involutive and of dimension $n$.

If, additionally, the manifold $X$ is simply connected, then Problem 2 is (globally) solvable if and only if conditions (A) and (B) are satisfied.

We note here that the dimension of the distribution $\Delta$ in condition (B) in the above theorem is at least $n$ which follows from the minimality assumption. On the other hand, this dimension can not exceed the dimension of the space which is equal to $n+1$. It is this condition which, together with the Frobenius theorem, will enable us to construct the function $h_0$.

In a similar way, Theorem 3.2 can be used to prove the following criterion. However, since $h_0$ is not given, we can not use the previous definition (3.5) when $\alpha_q = 0$. Thus, we use the following modification of this definition. For $q = 1$, we define

$$h_{[\alpha_1]} = h_{\alpha_1}, \quad \alpha_1 \in A,$$

and leave $h_{[0]}$ undefined. Next, for $q = 2$, we define

$$h_{[00]} = 0, \quad h_{[\alpha 0]} = -h_{[0\alpha]} = -2f_0 h_\alpha, \quad \alpha \in A.$$

Finally, for $q \geq 3$, we define inductively

$$h_{[\alpha_q \cdots \alpha_1]} = f_{\alpha_q} h_{[\alpha_{q-1} \cdots \alpha_1]} - f_{[\alpha_{q-1} \cdots \alpha_1]} h_{\alpha_q}, \quad \text{if} \quad \alpha_q \neq 0, \tag{3.9}$$

and

$$h_{[\alpha_q \cdots \alpha_1]} = f_{\alpha_q} h_{[\alpha_{q-1} \cdots \alpha_1]} + f_{\alpha_q} h_{\alpha_{q-1} \cdots \alpha_1}, \quad \text{if} \quad \alpha_q = 0. \tag{3.10}$$

*Theorem 3.4* If the system $(H, \tilde{F})$, where $\tilde{F} = F \cup \{f_0\}$, is minimal then Problem 2 is locally solvable if and only if the following conditions are satisfied:

(*A*) $$h_{[\alpha_q \cdots \alpha_1]} = q h_{\alpha_q \cdots \alpha_1}, \quad \alpha_1, ..., \alpha_q \in A \cup \{0\}, q \geq 2,$$

(B) The distribution $\Delta$ is involutive and of dimension $n$.

If, additionally, the manifold $X$ is simply connected, then Problem 2 is (globally) solvable if and only if conditions (A) and (B) are satisfied.

As solvability of Problem 2 is equivalent to existence of a Hamiltonian structure for our systems (2.1) and (2.4), from Theorems 3.3 and 3.4 we immediately obtain criteria for existence of a Hamiltonian structure.

*Corollary 3.1.* Define $A = \{1, ..., m\}$, $f_\alpha = g_\alpha$, $\alpha \in A$, $f_0 = g_0$, and assume that the system $(\{h_\alpha \mid \alpha \in A\}, \{f_\alpha \mid \alpha \in A \cup \{0\}\})$ is minimal. Then the system (2.4) is locally Hamiltonian if and only if conditions (A) and (B) of Theorems 3.3 or 3.4 are satisfied. If $X$ is simply connected, then (A) and (B) are equivalent to the fact that (2.4) is globally Hamiltonian.

*Corollary 3.2.* Let $u_0$ be an element of $U$ and let us define $A = \{1, ..., m\} \times U$, and

$$f_\alpha = \frac{\partial}{\partial u_i} f(\cdot, u), \quad h_\alpha = h_i(\cdot, u), \quad \alpha = (i, u) \in A, \quad \text{and} \quad f_0 = f(\cdot, u_0).$$

Assume that the system $(\{h_\alpha \mid \alpha \in A\}, \{f_\alpha \mid \alpha \in A \cup \{0\}\})$ is minimal. Then the system (2.1) is locally Hamiltonian if and only if condition (3.3) and conditions (A) and (B) of Theorems 3.3 or 3.4 are satisfied. If $X$ is simply connected, then (3.3), (A) and (B) are equivalent to the fact that (2.1) is globally Hamiltonian.

Let us interpret our conditions for the linear system

$$\dot{x} = Ax + Bu = Ax + \sum_{i=1}^{m} u_i b_i, \quad y_j = c_j x, \quad 1 \leq j \leq m,$$

where $c_1, \ldots, c_m$ are row vectors which constitute an $m \times n$ matrix $C$. We have that $f_\alpha = g_\alpha$, $h_\alpha = C_\alpha x$, for $\alpha = 1, \ldots, m$, and $f_0 = Ax$. It is easy to see that

$$h_{0\ldots 0j} = c_j A^k x, \qquad h_{i0\ldots 0j} = c_j A^k b_i,$$

and the other functions $h_{\alpha_q \ldots \alpha_1}$, with $\alpha_1 \neq 0$, are zero. We also have that

$$f_{[0\ldots 00i]} = -f_{[0\ldots 0i0]} = (-A)^k b_i,$$

and the other vector fields $f_{[\alpha_q \ldots \alpha_1]}$ are zero. The only nontrival conditions in condition (A) of Theorem 3.3 are

$$f_i h_{0\ldots 0j} + f_{[0\ldots 0j]} h_i = 0,$$

which, explicitely, take the form

$$c_j A^k b_i + c_i (-A)^k b_j = 0, \qquad i, j = 1, \ldots, m, \; k = q - 1 \geq 0. \tag{3.11}$$

Now, we shall interpret condition (B) of Theorem 3.3 for the linear system. The only nontrivial vector fields among the fields (3.7) defining the distribution $\Delta$ are

$$\tilde{f}_{[0\cdots 0i]} = ((-A)^k b_i, -c_i A^{k+1} x)^T = g_i^k.$$

The Lie bracket of such two vector fields is

$$[g_i^k, g_j^s] = (0, -c_j A^{s+1}(-A)^k b_i + c_i A^{k+1}(-A)^s b_j)^T.$$

As our systems are assumed to be transitive, the components along $X$ of the vector fields in $\Delta$ span the whole space. Thus, from our assumption that $\dim \Delta = n$ it follows that the obove vector fields are zero (otherwise, $\dim \Delta = n + 1$). We obtain that

$$c_j (-A)^{s+k+1} b_i + c_i A^{s+k+1} b_j = 0,$$

which is the same as the condition (3.11). In terms of the transfer matrix $G(s) = \sum s^{-k-1} C A^k B$ the condition (3.11) takes the form of the familiar criterion for Hamiltonian linear systems $G(s)^T = G(-s)$ ([2], [3]).

Analogously, one can interpret the condition (A) of Theorem 3.4.

## §4. SKETCHES OF PROOFS.

As we have already said, Theorem 3.1 was proved in [8]. We shall use this theorem to prove Theorem 3.3.

*Proof of Theorem 3.3. Sufficiency.* We shall first construct the function $h_0$. Let us denote points in $X \times \mathbb{R}$ by $(x,y)$, where $x \in X$ and $y \in \mathbb{R}$. From condition (B) it follows that the distribution $\Delta$ satisfies the assumptions of the Frobenius theorem, (cf. [1], p.159). Therefore, this distribution is integrable. Take any maximal connected integral manifold $S$ of this distribution. From the minimality assumption and the definition of $\Delta$ it follows that the projection of $\Delta$ onto $X$ gives the distribution of full dimension. This means that $S$ is transversal to the lines $\{x\} \times \mathbb{R}$. Therefore, locally this submanifold is the graph of a function $y = h_0(x)$. In general, the manifold $S$ is a covering space of the manifold $X$ (note that $\Delta$ does not depend on $y$, so the effect of "escaping to infinity" does not appear). If $X$ is simply connected it follows that this covering is trivial (one fold) and so the manifold $S$ is the graph of a global function $h_0 : X \longrightarrow \mathbb{R}$.

The submanifold $S$ is given by the equation $H_0 = h_0(x) - y = 0$. Since the distribution $\Delta$ does not depend on $y$, the other integral manifolds of this distribution can be obtained by shifting $S$ along $y$, i.e. they are of the form $H_0 = h_0(x) - y =$ const. It follows that the directional derivative of $H_0$ along any vector field $\tilde{f}_{[\alpha_q \cdots \alpha_1]}$ vanishes. This, together with the definition (3.7) of the vector fields $\tilde{f}_{[\alpha_q \cdots \alpha_1]}$ gives that

$$f_{[\alpha_q \cdots \alpha_1]} h_0 + f_0 h_{\alpha_q \cdots \alpha_1} = 0, \quad \alpha_1, ..., \alpha_q \in A \cup \{0\}. \tag{4.1}$$

More precisely, the above equalities follow directly for $\alpha_1 \neq 0$ only. By an additional argument one can show that they also hold for $\alpha_1 = 0$.

Having constructed a local or global function $h_0$ we can reduce Problem 2 to Problem 1. Define the following extended families

$$\tilde{H} = H \cup \{h_0\}, \quad \text{and} \quad \tilde{F} = F \cup \{f_0\}.$$

Now it is enough to check that the condition (3.4) in Theorem 1 holds for the families $\tilde{H}$ and $\tilde{F}$. From condition (A) it follows that (3.4) holds for $\tilde{H}$ and $\tilde{F}$ when $\alpha_0 \neq 0$ and $\alpha_1 \neq 0$. Together with condition (4.1) this means that (3.4) holds for $\tilde{H}$ and $\tilde{F}$ except of the case when $\alpha_0 \neq 0$ and $\alpha_1 = 0$. This case follows from the other cases (we leave the details to the reader).

If the equalities (3.4) hold, we can apply Theorem 1 to the families $\tilde{H}$ and $\tilde{F}$. In conclusion we obtain that Problem 2 is solvable.

*Necessity.* Necessity of condition (A) follows from antisymmetry of Poisson bracket in a similar way to the argument following Theorem 4.1. To prove necessity of condition (B) it is enough to show that the distribution $\Delta$ anihilates a nondegenerate function on $X \times \mathbb{R}$. We take $H_0 = h_0(x) - y$ as this function. Then, it easily follows from the definition of the vector fields in $\Delta$ that $\tilde{f} H_0 = 0$ for any vector field $f$ in $\Delta$. Thererefore, the distribution $\Delta$ is integrable with the level sets of $H_0$ as integral manifolds of $\Delta$, and so this distribution is involutive and of dimension $n$. ■

*Proof of Theorem 3.4.* The proof of this theorem is similar to the proof of Theorem 3.3 and is omited.

*Proof of Corollary 3.1.* The corollary follows from Theorem 3.2 as the problem of existence of Hamiltonian structure for the system (2.4) can be reduced to Problem 2 as described in Section 2. To get the local version one should additionally use the Darboux theorem. ∎

*Proof of Corollary 3.2.* As described in Section 2, existence of Hamiltonian structure for system (2.1) can be reduced to Problem 2 and so the corollary follows from Theorem 3.2. ∎

## REFERENCES.


[1] W.M. Boothby, "An Introduction to Differentiable Manifolds and Riemannian Geometry", Academic Press, New York 1975.

[2] R.W. Brockett, A. Rahimi, Lie algebras and linear differential equations, in "Ordinary Differential Equations", ed. L. Weiss, Academic Press, New York 1972.

[3] C.I. Byrnes, T.E. Duncan, A note on the topology of spaces of Hamiltonian transfer functions, in Lectures in Applied Mathematics, Vol.18, American Mathematical Society 1980.

[4] P.E. Crouch, M. Irving, On finite Volterra series which admit Hamiltonian realizations, *Math. Systems Theory* 17 (1984),293-318.

[5] P.E. Crouch, Hamiltonian realizations of finite Volterra series, in "Geometric Theory of Nonlinear Control Systems", eds. B. Jakubczyk, W. Respondek, K. Tchon, Technical University of Wroclaw 1985.

[6] P.E. Crouch, A.J. Van der Schaft, "Variational Characterization of Hamiltonian Systems", Lecture Notes in Control and Information Sciences, Springer 1987.

[7] B. Jakubczyk, Hamiltonian Realizations of Nonlinear Systems, Proc. MTNS-85, C. Byrnes, A. Lindquist eds. North-Holland 1986.

[8] B. Jakubczyk, Poisson structures and relations on vector fields and their Hamiltonians, *Bull. Pol. Acad. Sci. Ser. Math.* 34, (1986), 713-721.

[9] B. Jakubczyk, Existence of Hamiltonian realizations of nonlinear causal operators, *Bull. Pol. Acad. Sci. Ser. Math* 34. (1986), 737-747.

[10] A.J. Van der Schaft, "System Theoretic Descriptions of Physical Systems", CWI Tract No. 3, Amsterdam 1984.

[11] A.J. Van der Schaft, P.E. Crouch, Hamiltonian and self-adjoint control systems, *Systems and Control Letters* 8 (1987), 289-295.

# INVARIANTS IN THE FEEDBACK CLASSIFICATION OF NONLINEAR SYSTEMS

Bernard Bonnard

LAG, ENSIEG, UA CNRS 228
BP 46
38402 Saint-Martin-D'Hères FRANCE

RESUME Dans cette note heuristique on montre le lien entre le problème de classification des systèmes non linéaires à contrôle affine relativement à l'action du groupe feedback et la classification différentielle d'une famille d'équations différentielles hamiltoniennes qui représentes les singularités de l'application entrée-sortie. Cela à pour conséquence de situer le problème de feedback classification dans le contexte classique de classification de tenseurs et de fournir une approche efficace pour le calcul des invariants.

ABSTRACT In this heuristic article we connect the classification of affine control systems under the action of the feedback group and the differential classification of hamiltonian vector fields which are a representation of the singularities of the input-output mapping. This relates the feedback classification with a problem of tensors classification and gives a powerful tool to compute the invariants.

1. INTRODUCTION The main difficulty in the feedback classification of non linear systems is to understand the interaction between the change of state coordinates and the feedback.

To a system we can associate a map which is the input-output mapping. Such a map has singularities which are called in optimal control the singular trajectories [3]. A singularity is feedback invariant.

If we substract to the set of systems a bad set, which is in our approach systems where the singularities are poor, like linear systems or systems with too much inputs, one can show that *the classification of those singularities is equivalent to the feedback classification.*

Using the Pontryagin's maximum principle those singularities can be parametrized by a constained hamiltonian vector field. And the feedback classification is related to the classification of those vector fields under the action of diagonal symplectic transformations.

In others words we have connected the feedback classification to the *classification of tensors, on which the feedback doesn't act anymore.* As the classification of tensors is the main part of the classical invariant theory, one can used all the machinery of this theory , particularly to compute invariants [6].

Our approach is applied to describe invariants in the feedback classification of quadratic control systems and to connect the feedback equivalence with quadratic control system and the differential classification of homogeneous polynomial differential equations extending Poincaré and Dulac's theory [8,9].

## 2. PRELIMINARIES

### 2.1 FEEDBACK CLASSIFICATION

Consider the set $\mathcal{A}$ of $C^\omega$ affine control systems in $R^n$ of the form :

(1) $\dot{x}(t) = X(x(t)) + Y(x(t))u(t)$,

$u(t) \in R^p$, $Y = (Y_1,\dots,Y_p)$.

Let $(X,Y)$, $(X',Y')$ be two affine control systems. They are called feedback equivalent , if there exists a $C^\omega$ diffeomorphism $\xi$ of $R^n$, a feedback $\alpha \in C^\omega(R^n,R^p)$ and a state dependent change of input coordinates $\beta \in C^\omega(R^n,GL(p,R))$ such that :

$$X' = \frac{\partial \xi}{\partial x}^{-1} (X + Y\alpha) \circ \xi \quad , \quad Y' = \frac{\partial \xi}{\partial x}^{-1} (Y\beta) \circ \xi \ .$$

This action defines on the set of triplets $(\xi, \alpha, \beta)$ a group structure denoted by $G_f$.

Denote by $G_d$ the subgroup of the diffeomorphisms of $G_f$ and by $G_d^Y$ the subgroup of $G_d$ leaving invariant the distribution generated by the vector fields $Y_1,\dots,Y_p$.

### 2.2 SINGULAR TRAJECTORIES

2.2.1 DEFINITION Let $x_0 \in R^n$, $T > 0$ be fixed . Denote by $x(t,x_0,u)$ the response of system (1) to the control $u$ such that $x(0) = x_0$. The input-output mapping is the map $E : u(.) \longrightarrow x(t,x_0,u)$. The response $x(t,x_0,u)$ is called singular on $[0,T]$ if the Fréchet derivative of $E$, $L^\infty$ being the norm on the inputs space , is not of full rank $n$, when evaluated at $u(.)$.

This derivative can be computed and we can parametrize the singular trajectories using the Pontryagin's maximum principle, see [3] for the details.

2.2.2 PROPOSITION The trajectory $x(t,x_0,u)$ is singular on $[0,T]$ iff there exists $p(t) \in R^n$ (the adjoint vector), never vanishing and solution a.e. of :

(2) $\dot{p}(t) = - p(t) \left( \frac{\partial X}{\partial x}(x(t)) + \frac{\partial Y}{\partial x}(x(t))\, u(t) \right)$

such that :

(3) $\langle p(t),Y(x(t))\rangle = 0, \ \forall t \in [0,T]$.

The pair $(x(t),p(t))$ is called a singular extremal.

It is convenient to introduce the Hamiltonian formalism.

2.2.3 ASSUMPTION From now we will assume that the singular control $u$ is almost everywhere $C^\omega$ with respect to $(x,p)$.

2.2.4 NOTATIONS Let $\Sigma = \langle p,Y(x)\rangle$, the set $\Sigma = 0$ is called the switching surface and Let $H = \langle p,X + Yu\rangle$. Then the equations (1),(2),(3) can be written as the constrained hamiltonian differential equation :

$$(1)\quad \dot{x} = \frac{\partial H}{\partial p}, \qquad (2)\quad \dot{p} = -\frac{\partial H}{\partial x} \qquad (3)\quad \Sigma = 0.$$

2.2.5 NOTATION Denote by $\omega$ the map which associates to a system (X,Y) the constrained hamiltonian equation defined by the equations (1),(2) and (3).

## 2.3 FEEDBACK GROUP $G_f$ AND HAMILTONIAN DIFFERENTIAL EQUATIONS

Consider the space $\mathscr{B}$ of hamiltonian differential equations :

$$(4)\quad \dot{x} = \frac{\partial H}{\partial p}, \qquad \dot{p} = -\frac{\partial H}{\partial x}$$

One define the action of the feedback group $G_f$ on the equations (4) by :

(i) Let $x = \xi(y)$ be a diffeomorphism on $R^n$. Then $\xi$ is lifted into the diagonal symplectic diffeomorphism $\xi_s : x = \xi(y)$, $p = q\,\partial\xi^{-1}/\partial y$, which acts on (4) by : $\xi_s.H = H \circ \xi_s$.

(ii) Let $u = \alpha(x) + \beta(x)v$ be a feedback. Define the action of $(\alpha,\beta)$ on (4) by : $(\alpha,\beta).H = H$.

This action has the following geometric meaning. The equation (4) is interprated as a system without input , this explains the trivial feedback action. A diffeomorphism of $R^n$ is lifted into a symplectic diffeomorphism which preserves the canonical lift of a vector field into an hamiltonian one.

## 3. FEEDBACK CLASSIFICATION AND SINGULAR TRAJECTORIES

3.1 THEOREM The following diagram is commutative :

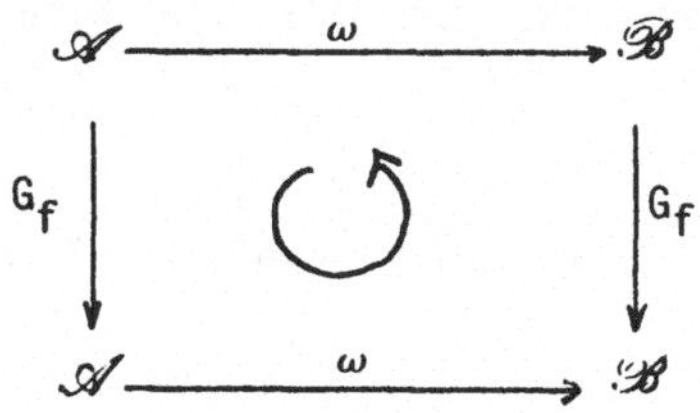


PROOF

Let $\xi$ be a diffeomorphism : $x = \xi(y)$, then $\xi_s$ is the symplectic map : $x = \xi(y)$, $p = q\,\partial\xi^{-1}/\partial y$. The hamiltonian equation (1),(2) whose hamiltonian is denoted by H is transformed into an hamiltonian system , whose hamiltonian denoted by H' is given by : $H'(y,q) = H(x,p)$.

By definition $H(x,p) = \langle p, X + Yu\rangle = \langle q\,\partial\xi^{-1}/\partial y, (X + Yu)(\xi(y))\rangle = \langle q, (\partial\xi^{-1}/\partial y)(X + Yu)(\xi(y))\rangle$.

Similarly the constrains $\langle p, Y(x)\rangle = 0$ is equivalent to $\langle q, (\partial\xi^{-1}/\partial y)(Y(\xi(y)))\rangle = 0$, and $\xi$ preserves the switching surfaces.

Let $u = \alpha(x) + \beta(x)v$ be a feedback. The constrains $\langle p, Y(x)\rangle = 0$ and $\langle p, Y\beta(x)\rangle = 0$

are equivalent and define the same switching surface. Using the relation $\langle p,Y(x) = 0$, we check that the constrained systems whose hamiltonian are respectively $H = \langle p,X + Yu\rangle$ and $H' = \langle p,X + Y\alpha + Y\beta v\rangle$ are identical.

3.2 SINGULAR CONTROLS We have to prove the existence of singular trajectories by computing the singular control, deriving the constrain $\Sigma = 0$ along the singular extremals.

Deriving in t the relations :

(3) $\langle p(t),Y_i(x(t))\rangle = 0, \; i = 1,\dots,p$

we get :

(5) $\langle p,[X,Y_i](x)\rangle + \sum_{j=1}^{p} u_j \langle p,[Y_j,Y_i](x)\rangle = 0, \; i = 1,\dots,p$, where the Lie bracket is computed with the convention :

$$[Z_1,Z_2](x) = \frac{\partial Z_2}{\partial x}(x)Z_1(x) - \frac{\partial Z_1}{\partial x}(x)Z_2(x) .$$

Let $\theta(x,p)$ be the $p \times p$ matrix $\langle p,[Y_j,Y_i](x)\rangle$ and $L(x,p)$ be the $p \times 1$ matrix : $\langle p,[X,Y_i](x)\rangle$.

The equation (5) can be written :

(5) $L(x,p) + \theta(x,p)u = 0.$

Since the Lie bracket is antisymmetrical , the matrix $\theta$ is antisymmetrical.

Let M be the kernel of $\theta(x,p)$, then M is a vector subspace whose (generic) dimension is denoted by k.

The singular control u(t) is computed for generic (x,p) as a dynamic feedback $u(x(t),p(t))$ whose role is to make $\Sigma = 0$ and and additional variety of dimension $2n-k$, related to M, a.e. invariant for the solutions of the equations (1),(2).

Let us explain the procedure with two examples , the generalization being straightforward.

Case 1 : the number p of inputs is even

Then for (X,Y) generic, the kernel of $\theta(x,p)$ is for almost every (x,p) trivial and $u(t) = u(x(t),p(t))$ a.e., with :

(6) $u(x,p) = -\theta^{-1}(x,p)L(x,p).$

Case 2 : the number p of inputs is odd

Then the control is computed into two steps. This is illustrated by the single input case :

we derive in t the relation :

(3) $\langle p(t),Y(x(t))\rangle = 0.$

We get :

(3') $\langle p(t),[X,Y](x(t))\rangle = 0.$

Deriving in t the relation (3') we get :

(7) $\langle p(t),[X,[X,Y]](x(t))\rangle + u(t)\langle p(t),[Y,[X,Y]](x(t))\rangle = 0$.

And the singular control is computed a.e. using (7) as $u(t)=u(x(t),p(t))$ with :

(8) $$u(x,p) = -\frac{\langle p,[X,[X,Y]](x)\rangle}{\langle p,[Y,[X,Y]](x)\rangle}$$

Let us denote by $\Sigma'$ the constrains (3) in the even case and (3),(3') in the odd case , the dimension of $\Sigma'$ being even. We have :

3.3 PROPOSITION The singular extremals are for a.e. (x,p) the solutions of the constrained hamiltonian differential equation :

(1) $\dot{x} = \frac{\partial H}{\partial p}$, (2) $\dot{p} = -\frac{\partial H}{\partial x}$, $(x,p) \in \Sigma'$,

with $H = \langle p,X(x) + u(x,p)Y(x)\rangle$. The singular control $u(x,p)$ being a.e. $C^{\omega}$ and $\Sigma'$ is a subvariety of even dimension.

By theorem 3.1, $\omega$ is a covariant. The second step in our analysis is to investigate when this covariant is complete .

This is indicated heuristically by the following result .

3.4 MAIN THEOREM

Assumptions and notations One assume that the dimension of the state space is $n \geq 3$, the number of inputs is 1. We suppose that Y never vanishes and for almost every x, $[Y,[X,Y]](x)$ doesn't belong to the vector space spanned by $Y(x)$ and $[X,Y](x)$.

Let U be an open neighboorhood such that Y is diffeomorphic to $\partial/\partial x_n$. Let X be a vector field, on U,X can be decomposed into $X = X_1 + X_2$, where $X_1$ is orthogonal to Y and $X_2$ is colinear to Y. Let $\pi_1$ be the projection : $Y^{\perp} \oplus Y \longrightarrow Y^{\perp}$. One may define a group action of $G_d^Y$ on the vector fields by : if $\xi \in G_d^Y$, $X \longrightarrow \xi \circ X = \pi_1(\xi . X) = \pi_1(\frac{\partial \xi}{\partial x}^{-1} X_1 \circ \xi)$.

THEOREM The following assertions are equivalent.

(i) (X,Y) and (X',Y') are $G_f$-equivalent.

(ii) The constrained hamiltonian differential equations $\omega.(X,Y)$ and $\omega.(X',Y')$ are $G_f$ (or $G_d$) -equivalent.

(iii) X and X' are $G_d^Y$-equivalent.

SKETCH OF THE PROOF

Let U be an open neighborhood such that Y is diffeomorphic to $\partial/\partial x_n$.

If (X,Y) and (X',Y') are $G_f$-equivalent, using a diffeomorphism , one may assume that Y' is colinear to $Y = \partial/\partial x_n$.

In those coordinates , let $(\xi,\alpha,\beta) \in G_f$ such that $(\xi,\alpha,\beta).(X,Y) = (X',Y')$.

Then $\xi \in G_d^Y$ and the action can be decomposed into : $(\xi, \alpha, \beta).(X,Y) = ((\xi \circ X, *), \lambda(x)Y)$, where $\xi \circ X$ is orthogonal to Y and * colinear to Y.

Since the feedback doesn't act on the component orthogonal to Y we have proved the equivalence between (i) and (iii).

To prove the equivalence between (i) and (ii) we must check that the constrained differential hamiltonian equation defining the singular extremals roughly contains all the information about the orbit of (X,Y).

First notice that the direction Y can be recovered from the switching surface $\Sigma = \{\langle p, Y(x)\rangle = 0\}$ ,since $\Sigma = \Sigma'$ iff Y and Y' are colinear.

Therefore we can choose coordinates such that Y' is colinear to $Y = \partial/\partial x_n$. Now we have to study the equivalence between $\omega.(X,Y)$ and $\omega.(X',Y')$ for the changes of coordinates leaving invariant the direction Y,i.e., for the subgroup action $G_d^Y$.

Using the assumptions , the equation whose solutions are the singular trajectories can be written locally as :

$$\dot{x} = X_1(x) + u(x,p)Y,$$

where the singular control u is a.e. defined and depends upon $p \in R^{n-3}$ parameters and $X_1$ belongs to the orthogonal of Y.

Therefore $X_1$ can be recovered from the above equation and a change of coordinates $\xi$ of $G_d^Y$ acts on the equation as : $\xi.(X_1, u(x,p)Y) = (\xi \circ X, *)$.

This proves the equivalence between (i) and (ii).

## 4 QUADRATIC CONTROL SYSTEMS

4.1 DEFINITIONS Consider the set $\mathscr{C}$ of quadratic control systems in $R^n$ of the form :

$$(9) \qquad \dot{x}(t) = Q(x(t)) + Bu(t),$$

$Q = (Q_1, \dots, Q_n)$, $Q_i$ = quadratic form and B = p x p matrix.

Consider the set $G_f^{'}$ of triplets $(P, \alpha, \beta)$, where $P \in GL(n,R)$, $\alpha : R^n \longrightarrow R^p$, $\alpha = (\alpha_1, \dots, \alpha_p)$, $\alpha_i$ quadratic form and $\beta \in GL(p,R)$. Let $g_1 = (P_1, \alpha_1, \beta_1)$ , $g_2 = (P_2, \alpha_2, \beta_2)$ and endow $G_f^{'}$ with the Lie group stucture given by :

$$g_2.g_1 = (P_1P_2, \alpha_1 + \beta_1\alpha_2 \circ P_1^{-1}, \beta_1\beta_2).$$

The group $G_f^{'}$ acts on $\mathscr{C}$ with the law :

$$(P, \alpha, \beta).(Q,B) = (P^{-1}(Q + B\alpha) \circ P, P^{-1}B\beta),$$

and is a transformation subgroup of $G_f$ preserving the class $\mathscr{C}$.

4.2 DEFINITIONS Let E,F be two space vectors on a field K and let $\Gamma$ be a Lie group acting linearly on E and F. A homorphism $\chi : \Gamma \rightarrow K^* = K - \{0\}$ is called a character. A relative invariant of $\Gamma$ of weight $\chi$ is a mapping $\omega: E \rightarrow K$ such that $\forall g \in \Gamma$, $\forall x \in E$, $\omega(g.x) = \chi(g).\omega(x)$. It is called absolute if $\chi = 1$. A mapping $\omega : E \longrightarrow F$ is a relative concomitant (covariant) of weight $\chi$ if $\forall g \in \Gamma$, $\forall x \in E$, $\omega(g.x) = \chi(g)$

g. $\omega(x)$, it is called <u>absolute</u> if $\chi = 1$.

Using the section 3, we have :

4.3 PROPOSITION The singular extremals $(x(t),p(t))$ for the quadratic system (9) are a.e. on [0,T] the solutions of :

$$(9) \quad \dot{x}(t) = Q(x(t)) + Bu(t)$$

$$(10) \quad \dot{p}(t) = - p(t) \left( \frac{\partial Q}{\partial x}(x(t)) \right),$$

satisfying the 2p constrains :

$$(11) \quad \langle p,B\rangle = \langle p,[Q,B]\rangle = 0,$$

the singular control being obtained by solving :

$$(12) \quad L'(x,p) + \Theta'(x,p)u = 0,$$

where $L'(x,p)$ is the p x 1 matrix $(\langle p,[Q,[Q,b_i]]\rangle)$ and $\Theta'(x,p)$ is the p x p symmetrical matrix $(\langle p,[b_j,[Q,b_i]]\rangle)$

4.4 NOTATION Denote by $\mathscr{D}$ the set of constrained equations defined by (9),(10),(11),(12), i.e., $\mathscr{D} = \omega(\mathscr{C})$. And consider the linear symplectic action of $G_f^! \subset G_f$ on $\mathscr{D}$ induced by the action of $G_f$ on $\mathscr{B}$ defined in 2.3.

From theorem 3.1, we have :

4.5 PROPOSITION For the action of $G_f^!$ on $\mathscr{C}$ and $\mathscr{D}$ , the mapping $\omega$ is an absolute concomitant .

From theorem 3.4, we have :

4.6 PROPOSITION With the assumptions of theorem 3.4, two quadratic control systems (Q,b) and (Q',b') are $G_f^!$ equivalent iff $\omega$.(Q,b) and $\omega$.(Q',b') are $G_f^!$ equivalent(i.e., up to the diagonal linear changes of coordinates : $x = Py$, $p = q\,P^{-1}$).

## 5 QUADRATIC SYSTEMS IN $R^3$

Consider a quadratic control system in $R^3$ :

$$(13) \quad \dot{v}(t) = Q(v(t)) + u(t)b$$

Our approach will be used to investigate the following problems :

(i) Describe the algebra of covariants in the $G_f^!$ classification of quadratic systtems.

(ii) The feedback equivalence of a non linear system with a quadratic system.

5.1 PROPOSITION Let (X,Y) be a single input system in $R^3$. Denote by D = det(Y,[X,Y],

[Y,[X,Y]]), D' = det (Y,[X,Y],[X,[X,Y]]). The singular trajectories are the solutions of the differential equation in $R^3$ :

$$(14) \qquad \dot{v}(t) = X(v(t)) - \frac{D'(v(t))}{D(v(t))} Y(v(t))$$

PROOF Using the relations : $\langle p,Y\rangle = \langle p,[X,Y]\rangle = \langle p,[X,[X,Y]] + u[Y,[X,Y]]\rangle = 0$, $p \in R^3$, $p \neq 0$, the vector p can be eliminated and the singular control is the static feed-back satisfying : $D' + uD = 0$.

5.2 NOTATION To the equation (14) one may associate the $C^{\omega}$ differential equation :

$$(15) \qquad \dot{v}(t) = D(v(t))X(v(t)) - D'(v(t))Y(v(t)) = (\text{Notation})\ \tilde{X}_Y.$$

Both equations describe the same distribution.

5.3 PROPOSITION Let (Q,b) be a single input control system in $R^3$ such that $D \neq 0$. Let $g = (P,\alpha,\beta) \in G_f^1$. Then the action induced on (15) is equivalent to the linear change of coordinates $x = Py$, up to a weight related to det P and $\beta$ . Therefore making $G_f^1$ acts on the tensors as GL(3,R), we have the following relative covariants :

(i) $\pi_1 : (Q,b) \longrightarrow D$

(ii) $\omega' : (Q,b) \longrightarrow \tilde{Q}_b$.

The proof which is a trivial computation is omitted.

The classification of systems (Q,b) by feedback equivalence is then generically reduced to the linear classification of the (1,3) tensor $\tilde{Q}_b$. A non trivial application is to give a set of invariants for the feedback classification problem, see [3] for the details.

5.4 PROPOSITION Denote by $D'' = \det(b,[Q,b],Q)$ and decompose $\tilde{Q}_b$ as the sum of a conservative vector field $\tilde{Q}_b^c$ and $1/5 \operatorname{div} \tilde{Q}_b \cdot v$.

For the action of $G_f^1$ on the tensors , we have the following relative covariants :

(i) $\pi_1 : (Q,b) \longrightarrow D$

(ii) $\pi_2 : (Q,b) \longrightarrow D''$

(iii) $\pi_3 : (Q,b) \longrightarrow D'$ restricted to $D = 0$

(iv) $\pi_4 : (Q,b) \longrightarrow \operatorname{div} \tilde{Q}_b$

(v) $\pi_5 : (Q,b) \longrightarrow \tilde{Q}_b^c$ .

Geometric interpretation

The plane $D = 0$ is an invariant for the solutions of $\dot{v} = \tilde{Q}_b$.
Similarly the cubic D" is an invariant for the solutions . It plays a fundamental

role in the time minimal control problem : D" separates the domains of slow and fast singular trajectories, given by the strong Legendre condition.

The covariant $\pi_3$ is connected to the behaviors of the singular trajectories near the set $D = 0$.

The covariant $\pi_4$ describes the invariance of the divergence in the classification of differential equations.

The covariant $\pi_5$ describes the projected differential equation on the shere $S^2$.

Conclusion We gave an insight into the structure of the algebra of invariants in the feedback classification problem. One can used the classical theory and the symbolic calculus [6] to compute this algebra. This is a non trivial and interesting problem which shall be didcussed in a forthcoming paper.

## 5.5 TOWARDS A DICTIONARY BETWEEN FEEDBACK CLASSIFICATION AND LINEAR CLASSIFICATION OF SINGULAR FLOWS

It is interesting to relate the properties of systems preserved by feedback equivalence with the properties of (15). We give an example.

5.5.1 DEFINITIONS AND NOTATIONS Denote by w the direction such that $[Q,b](w)$ is colinear to b. Define the vectors $v_i$, $i= 1,2,3,4$ by : $v_1=b$, $v_2=[[Q,b],v_1]$, $v_3=[[Q,v_2],b]$ and $v_4=[[Q,v_2],v_2]$. Denote by $L_0$ the vector space generated by $\{v_i\}$.

A system (1) is called weakly controllable if the Lie algebra L generated by $\{X,Y_i\}$ is of full rank : $\dim L(x) = n\ \forall x \in R^n$. From [2] the quadratic control system is weakly controllable iff $\dim L_0 = 3$.

We have the following results , the computations being omitted, see [3] for the details.

5.5.2 LEMMA If b and w are linearly independent then $D = 0$ is the plane generated by b and w; $D = 0$ is $R^3$ iff b and w are colinear.

5.5.3 PROPOSITION Assume b and w linearly independent . The system is not weakly controllable iff D is a factor of D', i.e., $D' = 0$ on $D = 0$.

5.5.4 PROPOSITION The solutions of $D' = D = 0$ are w and two (real or complex) lines denoted by $L_1$ and $L_2$. The system is weakly controllable iff $v_1,v_2,v_3$ or $v_1,v_2,v_4$ are independent. The vectors $v_1,v_2,v_3$ are dependent iff $L_1=L_2$. In this case there exist(high order) singular trajectories contained in $D = 0$ and on the line $L_1$

Conclusion The equation $\dot{v} = \tilde{Q}_b(v)$ contains generically all the information about the orbit of system(Q,b) under $G_f^{!}$. For instance b is a ray solution of the equation contained in $D = 0$ which causes the blowing up of the solutions [3].

Other properties like a cascade decomposition of (Q,b) will be related to structural properties of (15) in a forthcoming paper.

5.6 FEEDBACK EQUIVALENCE WITH A QUADRATIC SYSTEM

Let (X,Y) be a single input system in $R^3$, one may ask the question : when is this system feedback equivalent to a quadratic system (Q,b) ?

If (X,Y) is equivalent to (Q,b) then the differential equations $\dot{v} = \tilde{X}_Y$ and $\dot{v} = \tilde{Q}_b$ are up to a weight , $C^\omega$equivalent, this by a slight generalization of the proposition 5.3.

The feedback equivalence with a quadratic system is then connected with the $C^\omega$equivalence of a differential equation with a homogeneous polynomial one. This problem is a classical problem solved by Poincaré and Dulac in the linear case [9] and generalized in the non linear case in [8].

This can be generalized for systems (X,Y) in $R^n$ the singular equation preserving the homogeneity properties.

6 CONCLUSION

The feedback equivalence problem is tightly connected with the classification of differential equations and more precisely with the classification of the problems of the calculus of variations.

The analysis of singular trajectories is in my opinion the main problem to solve to understand the control properties of a system and the time minimal synthesis for the (unbounded) control system :

$$(1) \qquad \dot{x}(t) = X(x(t)) + Y(x(t))u(t), \quad u(t) \in R^p,$$

is one major control problem.

REFERENCES


[1] V. ARNOLD , "Chap. supplémentaires de la théorie des équations diff. ordinaires" Ed. Mir 1980.

[2] B. BONNARD, "Quadratic control systems", to appear in MCSS.

[3] B. BONNARD, "Contribution à l'étude du problème singulier" Note interne LAG, Novembre 1987.

[4] E CARTAN, "Oeuvres complètes" Part II, Vol. 2, Gauthier-Villars, 1953

[6] G.B. GUREVITCH, "Foundations of the theory of algebraic invariants", Noordhoff Ed. , 1964.

[7] A. LICHNEROWICKZ, "Les variétés de Poisson et leurs algèbres de Lie associées", in Choix d'oeuvres math., Hermann Ed. 1982.

[8] F. TAKENS, "Singularities of vector fields", Pub. IHES, N° 43. 1973.

[9] V. ARNOLD, "Méthodes math. de la Mécanique classique", Ed. Mir 1976.

# ON NORMAL FORMS OF AFFINE SYSTEMS UNDER FEEDBACK

Krzysztof Tchon
Institute of Engineering Cybernetics, Technical University of Wroclaw,
ul.Janiszewskiego 11/17, 50-372 Wroclaw, Poland.

Summary: The paper is concerned with the static state feedback theory for affine control systems. The problem is studied of the existence of structurally stable normal forms of affine systems under the feedback. Two such forms have been found directly and the existence of other structurally stable normal forms has been excluded, except for the case when the dimension of the state space exceeds the number of inputs by one. Approximate feedback groups are introduced as a basic working instrument. The action of the approximate feedback group of order 1 has been examined in very detail, and consequences derived for the class of bilinear systems.
Key words: Affine control system, feedback, normal form, structural stability, Lie group, codimension, bilinear system.

## 1. Introduction.

The concept of feedback, undoubtedly, lies at the heart of modern control and systems theory. Static feedback, often accompanied by a dynamic compensation, seems to be a paradigmatic approach to solving many control synthesis problems, both for linear and non-linear control systems. The static feedback theory of linear systems culminated with the celebrated Brunovsky classification theorem, and established a deep relationship between the structure of a linear system and sequences of integers called controllability indices. However, a similar theory of static state feedback for non-linear systems, even local, still remains to be developed. This paper may be regarded as a step towards such a theory.

More specifically, we concentrate our attention on smooth affine control systems subject to the non-linear, static state feedback, and study the problem of existence of structurally stable normal forms of such systems. Intuitively, a normal form is called structurally stable, if any system, sufficiently close to a given one, can be transformed to the same normal form as the given system.In order that handle the problem mathematically we need a topological structure within the class of affine systems. An appropriate topology is well-known and widely used in differential topology and singularity theory. This is $C^\infty$-Whitney topology. Having imposed this topology, we are able to precisely define structural stability of normal forms.

Within the class of affine systems it is easy to find two structurally stable normal forms: one global, for systems whose number of inputs equals the dimensionality of the state space, and one local, for systems with one input and 2-dimensional state space. To treat the other cases we have introduced a family of approximate feedback groups acting on approximate affine systems. Using a simple Lie-theoretic argument we show that, except for the case when the dimension of the state space exceeds the number of inputs by one, there are no structurally stable normal forms whatever. Obstacles to the existence of the forms lie in the action of some approximate feedback groups, in most cases just of the group of order 1. This result seems to be of independent interest, with a clear reference to normal forms of bilinear systems. Indeed, we have called the approximate feedback group of order 1 the bilinear feedback group, and found some structurally stable, local, approximate normal forms of bilinear systems under this group.

The paper is composed as follows. In Section 2 we introduce basic concepts and define the systemic subject of the paper. Section 3 presents approximate feedback groups. Main results are contained in Section 4. They are concerned with the existence of structurally stable normal forms of affine systems and a detailed investigation of the approximate feedback group of order 1. The paper is concluded with Sectioon 5 devoted to inhomogeneous bilinear systems.

## 2. Basic concepts.

We shall study smooth *affine control systems* of the form

$$\dot{x} = f(x) + g(x)u = f(x) + \sum_{i=1}^{m} u_i g_i(x) \quad . \tag{2.1}$$

defined on $\mathbb{R}^n$ such that $f(0)=0$. The controls are bounded measurable and take values in $\mathbb{R}^m$, $m \le n$ . Maps $f, g_1, \ldots, g_m$ are assumed to be smooth, i.e. of class $C^\infty$. Given a system (2.1), it is natural to identify the system with a tuple of smooth maps, $\sigma = (f, g_1, \ldots, g_m) = (f,g) \in \Sigma \cong C_0^\infty(\mathbb{R}^n, \mathbb{R}^{(m+1)n})$ (subscript 0 serves to note that $f(0)=0$). The last space of smooth maps will always be considered along with $C^\infty$-Whitney topology, [1],[2],[3].

Consider two affine systems: $\sigma=(f,g)$, $\sigma'=(f',g')\in\Sigma$ They will be called *(globally) feedback equivalent*, if there exists a diffeomorphism $\varphi \in \mathrm{Diff}(\mathbb{R}^n)$ of the state space, $\varphi(0)=0$, a feedback map $\eta \in C^\infty(\mathbb{R}^n, \mathbb{R}^m)$, $\eta(0)=0$, and a state-dependent change of input coordinates $\psi \in C^\infty(\mathbb{R}^n, GL_m(\mathbb{R}))$ such that

$$f' = (\frac{\partial \varphi}{\partial x})^{-1}(f \circ \varphi + g \circ \varphi \cdot \eta) \quad , \quad g' = (\frac{\partial \varphi}{\partial x})^{-1} g \circ \varphi \cdot \psi \quad . \tag{2.2}$$

Expression (2.2) can be interpreted as a result of the action of the so-called *global feedback group*. he group, denoted by G, consists of triples $(\varphi, \eta, \psi)$ defined above, so $G \cong \mathrm{Diff}(\mathbb{R}^n) \times C^\infty(\mathbb{R}^n, \mathbb{R}^m) \times C^\infty(\mathbb{R}^n, GL_m(\mathbb{R}))$. The group multiplication is described by the formula

$$(\varphi', \eta', \psi') \cdot (\varphi, \eta, \eta) = (\varphi \circ \varphi', \eta \circ \varphi' + \psi \circ \varphi' \cdot \eta', \psi \circ \varphi' \cdot \psi'), \tag{2.3}$$

The identity element $e=(id, 0, I_m)$, the inverse $(\varphi, \eta, \psi)^{-1} = (\varphi^{-1}, -\psi^{-1} \circ \varphi^{-1} \cdot \eta \circ \varphi^{-1}, \psi^{-1}_{\circ} \varphi^{-1})$. The action of $(\varphi, \eta, \psi)$ on an affine system $\sigma=(f,g)$

produces the system $\sigma'=(f',g')$ given by (2.2). It can be proved that w.r.t. the product $C^{\infty}$-Whitney topology G is a topological group, and the action (2.2) is continuous.

In many cases it is sufficient to deal with a local feedback equivalence instead of the global one. A suitable definition will be given in terms of map-germs, [4].. Thus let $\Sigma_0$ denote the set of germs of affine systems (2.1) at $0\in\mathbb{R}^n$, similarly let $G_0$ be the set of germs of feedbacks at the origin. Two affine systems $\sigma=(f,g), \sigma'=(f',g')\in\Sigma$ will be called *(locally) feedback equivalent*, if there exists a triple of germs $(\varphi,\eta,\psi)\in G_0$ such that formula (2.2) holds for map-germs. Equivalently, one can introduce the *local feedback group* $G_0$ of germs, with the multiplication inherited from (2.3), acting on system germs $\Sigma_0$ in accordance with the following rule

$$\gamma:(\varphi,\eta,\psi)(f,g)\mapsto((\frac{\partial\varphi}{\partial x})^{-1}(f\circ\varphi + g\circ\varphi\cdot\eta),\quad (\frac{\partial\varphi}{\partial x})^{-1}g\circ\varphi\cdot\psi\ ). \qquad (2.4)$$

As a consequence, systems $\sigma,\sigma'$ are then said to be locally feedback equivalent, if their germs at $0\in\mathbb{R}^n$ belong to the same orbit of the local feedback group. From now on, by the *feedback group* we shall mean the local feedback group $G_0$ acting on germs $\Sigma_0$ of affine systems in agreement with (2.4).

Having introduced the local feedback equivalence, it is of natural interest to look for some nice, simple, *local normal forms* of affine systems w.r.t. the feedback equivalence. The local normal forms classify the orbits of $G_0$ in $\Sigma_0$ . From the systemic point of view, there is of considerable importance to know if there exist any normal forms which would be insensitive to small changes in underlying systems. Such normal forms will be called *structurally stable*. More specifically, an affine system $\sigma\in\Sigma$ is called a *structurally stable local normal form* of affine systems) whenever there exists an open neighbourhood V of $\sigma$ (w.r.t. $C^{\infty}$-Whitney topology) such that, for any $\sigma'\in V$ the germs at $0\in\mathbb{R}^n$ of $\sigma,\sigma'$ lie in the same orbit of the feedback group, [5],[6]. In what follows we shall focus our attention on the existence problem of structurally stable normal forms of affine control systems.

## 3. Approximate feedback groups.

In order to solve the problem mentioned above we prefer to deal, instead of the feedback group $G_0$, with a collection of finite dimensional objects called approximate feedback groups, [5],[6]. They will be introduced as follows.

Given the set of system germs $\Sigma_0$, we define for every k=0,1,... the set of *k-th approximates* at $0\in\mathbb{R}^n$ of affine systems $\Sigma_0^k$ by setting

$\Sigma_o^k=\{j^k\sigma : \sigma\in\Sigma_o\}$, where $j^k\sigma=\sigma^k=(j^kf(0),j^kg(0))$ is a tuple of k jets of f,g at $0\in\mathbb{R}^n$. It is easily observed that $\Sigma_o^k$ is a real vector space, and $\dim\Sigma_o^k = (m+1)n\binom{n+k}{n} -n$ , the last subtraction due to $f(0)=0$. Analogous construction for the feedback group $G_o$ should be made with some care, taking into account the form of (2.2) and (2.4).

As a matter of fact, if $(\varphi,\eta,\psi)\in G$ then we define the *k-th approximate feedback group* $G_o^k$ (at $0\in\mathbb{R}^n$) as consisting of triples $(j^{k+1}\varphi(0),j^k\eta(0),j^k\psi(0))$. It is not hard to notice that $G_o^k$ is an open (and dense) subset of a real vector space $\mathbb{R}^N$ , where $N= n\binom{n+k+1}{n}+ m\binom{n+k}{n} + m^2\binom{n+k}{n} - n - m$ (because $\varphi(0)=0,\eta(0)=0$). Moreover, two elements of $G_o^k$ can be multiplied,

$$(j^{k+1}\varphi',j^k\eta',j^k\psi')\cdot(j^{k+1}\varphi,j^k\eta,j^k\psi) = (j^{k+1}(\varphi\circ\varphi'),j^k(\eta\circ\varphi'+\psi\circ\varphi'\cdot\eta'), j^k(\psi\circ\varphi'\cdot\psi')), \tag{3.1}$$

there is in $G_o^k$ the identity element $e^k=(0,I_n,0,\dots,0;0;I_m,0,\dots,0)$, and the inverse of $(j^{k+1}\varphi,j^k\eta,j^k\psi)$ is equal to $(j^{k+1}\varphi^{-1},-j^k(\psi^{-1}\circ\varphi^{-1}\cdot\eta\circ\varphi^{-1}), j^k(\psi^{-1}\circ\varphi^{-1}))$, all the jets taken at $0\in\mathbb{R}^n$). Now a glance at (3.1) shows that the group operations in $G_o^k$ are smooth, so $G_o^k$ is a Lie group. Furthermore, the action (2.4) of the feedback group induces the following natural action of $G_o^k$ on $\Sigma_o^k$:

$$\gamma_k:(j^{k+1}\varphi,j^k\eta,j^k\psi)(j^kf,j^kg)\mapsto(j^k(\tfrac{\partial\varphi}{\partial x})^{-1}(f\circ\varphi + g\circ\varphi\cdot\eta),j^k(\tfrac{\partial\varphi}{\partial x})^{-1}g\circ\varphi\cdot\psi). \tag{3.2}$$

Thus (3.2) is a Lie group action on k-th approximates of afine systems.

The usefulness of the concepts just introduced for the existence problem of structurally stable normal forms flows from the following characterization of the structural stability, [6],[7],cf. also [1].

<u>Lemma 3.1.</u> Let $\sigma\in\Sigma$ be a structurally stable local normal form of affine systems. Then the orbit $G_o^k(j^k\sigma)$ of the group $G_o^k$ passing through the k-th approximate $j^k\sigma$ of $\sigma$ has non-void interior in $\Sigma_o^k$ for every $k\geq0$.

## 4. Existence of structurally stable normal forms.

We begin with two particular cases when the existemce of structurally stable normal forms can be demonstrated easily and directly, [6]. These are the following situations. Let m=n (the number of controls equals the dimension of the state space), and define in $\Sigma$ a subset $\Sigma_{nn}= \{(f,g)\in\Sigma: \forall x \text{ rank } g(x)=n\}$. It results immediately from the definition of $C^\infty$-Whitney topology that $\Sigma_{nn}$ is open. On the other hand, any $\sigma=(f,g)\in\Sigma_{nn}$ can be given the *normal form* $\bar{\sigma}=(0,I_n)$ by the feedback $\varphi(x)=x, \eta(x)=-g(x)^{-1}f(x), \psi(x)=g(x)^{-1}$. Observe that the form is *global*. The set $\Sigma_{nn}$ is open in $\Sigma$ but not dense, so $\bar{\sigma}$ is structurally stable, but not *generic*, [6],[8].

The other straightforward case is that of m=1, n=2. Assume this and define in $\Sigma$ a subset $\Sigma_{12}=\{(f,g)\in\Sigma: \forall x \ \text{rank}\{g,[f,g](x)=2\}\}$, where the Lie bracket [f,g] is equal to $\frac{\partial g}{\partial x}f - \frac{\partial f}{\partial x}g$ . Clearly $\Sigma_{12}$ is open in $\Sigma$ w.r.t. $C^{\infty}$-Whitney topology. Now, by an argument used in [6], any $\sigma=(f,g)\in\Sigma_{12}$ can be transformed to *normal form* $\bar{\sigma}_1=(\binom{0}{x},\binom{1}{0})$. This form is *local* and *structurally stable*, but not generic,cf.[8].

The cases m=n and m=1,n=2 are the only ones when we can find structurally stable normal forms directly. Now we are going to describe those ranges of dimensions (m,n) in which definitely there are no structurally stable forms. To this purpose we shall use Lemma 3.1. Our strategy will be based upon showing that for some k≥0 the action (3.3) of the Lie group $G_o^k$ on $\Sigma_o^k$ produces orbits of positive codimensions, cf. [4],[7],[9],[10].We offer a two step approach to the problem. At first we find a rough estimate of the number k, then we shall improve the estimate considerably. The rough result is as follows, [6].

Proposition 4.1. Let $m\le n-2$ and $m\ge 1$. Then, for $k>\frac{n^2}{m(n-m-1)}-1$, the orbits of $G_o^k$ have positive codimensions in $\Sigma_o^k$.

Proof: Clearly, the orbits have positive codimensions, if $\dim G_o^k < \dim \Sigma_o^k$, [11]. Both the dimensions have been found in Section 3, so $\dim G_o^k = n\binom{n+k+1}{n} + m\binom{n+k}{n} + m^2\binom{n+k}{n} - n - m$, $\dim \Sigma_o^k = (m+1)n\binom{n+k}{n} - n$, and the result follows.

QED

By combining Proposition 4.1 with Lemma 3.1 we obtain the following theorem which generalizes a result by Jakubczyk-Przytycki proved for symmetric control systems, [10],p.21.

Theorem 4.2. There are no structurally stable normal forms of affine systems within the range of dimensions $m\ge 1$, $m\le n-2$.

In order that improve the estimate for k found in Proposition 4.1, we first recall that given a k-th approximate $\sigma^k\in\Sigma_o^k$, the orbit of $G_o^k$ through $\sigma^k$ has dimension given by

$$\dim \mathcal{O}_{\sigma^k} = \dim G_o^k - \dim \text{stab}\ \sigma^k = \text{codim stab}\ \sigma^k, \tag{4.1}$$

where $\text{stab}\ \sigma^k = \{g\in G_o^k : g\sigma^k = \sigma^k\}$ denotes the stabilizer of $\sigma^k$ w.r.t. the action (3.3), [11]. Thus, to calculate the dimension of $\mathcal{O}_{\sigma^k}$ it is enough to count independent equations defining stab $\sigma^k$. Below we analyze in detail the orbits of $G_0^k$ for $k=0$ and $k=1$, [7]. The first result is immediate.

<u>Lemma 4.3.</u> The orbit space of $G_0^0$ acting on $\Sigma_0^0$ is finite and contains an open-dense orbit whose normal form is $\begin{bmatrix} I_m \\ 0 \end{bmatrix}$.

Now consider the action $\gamma_1$ of $G_0^1$ on $\Sigma_0^1$. It is not hard to see that $\Sigma_0^1 \cong \mathrm{Mat}(n)\times\mathrm{Mat}(n,m)\times\mathrm{Mat}(n,m)^n$, whereas $G_0^1 \cong GL_n(\mathbb{R})\times\mathrm{Sym}(n)^n\times\mathrm{Mat}(m,n)\times GL_m(\mathbb{R})\times\mathrm{Mat}(m)^n$. Hereabove Mat(n), Sym(n) denote, respectively, square and square symmetric matrices $n\times n$, superscripts stand for Cartesian powers. The action $\gamma_1$, after some developments, can be represented as follows:

$$\gamma_1 : \left(\frac{\partial\varphi}{\partial x}, \frac{\partial^2\varphi_1}{\partial x^2}, \dots, \frac{\partial^2\varphi_n}{\partial x^2}, \frac{\partial\eta}{\partial x}, \psi, \frac{\partial\psi}{\partial x_1}, \dots, \frac{\partial\psi}{\partial x_n}\right)\left(\frac{\partial f}{\partial x}, g, \frac{\partial g^1}{\partial x}, \dots, \frac{\partial g^n}{\partial x}\right) \mapsto$$

$$\Bigl(\left(\frac{\partial\varphi}{\partial x}\right)^{-1}\frac{\partial f}{\partial x}\frac{\partial\varphi}{\partial x} + \left(\frac{\partial\varphi}{\partial x}\right)^{-1} g\,\frac{\partial\eta}{\partial x},\ \left(\frac{\partial\varphi}{\partial x}\right)^{-1} g\psi,\ \left(\frac{\partial\varphi}{\partial x}\right)^{T}\frac{\partial^2\varphi_i}{\partial x^2}^{-1} g\psi + \sum_{s=1}^{n}\frac{\partial\varphi_i}{\partial x_s}^{-1}\left(\frac{\partial\varphi}{\partial x}\right)^{T}\frac{\partial g^s}{\partial x}\psi +$$

$$\sum_{s=1}^{n}\frac{\partial\varphi_i}{\partial x_i}^{-1}\left(g\,\frac{\partial\psi}{\partial x}\right)^{s},\ i=1,2,\dots,n\Bigr), \tag{4.2}$$

where $\left(\frac{\partial g^s}{\partial x}\right)_{ij} = \frac{\partial g_{sj}}{\partial x_i}$, $\left(g\frac{\partial\psi}{\partial x_i}\right)^s_{ij} = \left(g\frac{\partial\psi}{\partial x_i}\right)_{sj}$, $\left(\frac{\partial\psi}{\partial x_s}\right)_{ij} = \frac{\partial\psi_{ij}}{\partial x_s}$.

$i,s = 1,2,\dots,n$; $j = 1,2,\dots,m$; $g = [g_1, g_2, \dots, g_m]$ .

Looking at (4.2) one easily recognizes that the first two terms on r.h.s. represent the action of the *linear feedback group* on $(\frac{\partial f}{\partial x}, g)$. Therefore, $G_0^1$ acts with open orbits only in the case when $(\frac{\partial f}{\partial x}, g)$ has choose an element $\sigma^1 \in \Sigma_0^1$ in the form $\sigma^1 = (\frac{\partial f}{\partial x}, g, \frac{\partial g^1}{\partial x}, \dots, \frac{\partial g^n}{\partial x})$ with $(\frac{\partial f}{\partial x}, g)$ in the generic Brunovsky form and $\frac{\partial g^i}{\partial x}$ arbitrary, and study the orbit $\mathcal{O}_{\sigma^1}$ of $\gamma_1$ through $\sigma^1$. Knowing that the projection of $\mathcal{O}_{\sigma^1}$ onto $\mathrm{Mat}(n)\times\mathrm{Mat}(n,m)$ is and of agents like $\frac{\partial^2\varphi_i}{\partial x^2}, \frac{\partial\psi}{\partial x}$ is able to produce an open orbit $\mathcal{O}_{\sigma^1}$.

<u>Proposition 4.4.</u> With $\sigma^1$ as above, $\operatorname{codim} \mathcal{O}_{\sigma^1} \geq \max(0, \frac{m}{2}(m-1)(m-n)-m^2)$.

<u>Proof</u>: We have $\dim \mathcal{O}_{\sigma^1} = \operatorname{codim} \operatorname{stab} \sigma^1$, so it suffices to find equations determining the stabilizer of $\sigma^1$ w.r.t. $\gamma_1$. Taking into account (4.2), the equations are as follows:

$$\left(\frac{\partial\varphi}{\partial x}\right)^{-1}\frac{\partial f}{\partial x}\frac{\partial\varphi}{\partial x} + \left(\frac{\partial\varphi}{\partial x}\right)^{-1} g\,\frac{\partial\eta}{\partial x} = \frac{\partial f}{\partial x}, \quad \left(\frac{\partial\varphi}{\partial x}\right)^{-1} g\psi = g, \tag{4.3a}$$

$$\left(\frac{\partial\varphi}{\partial x}\right)^{T}\frac{\partial^2\varphi_i}{\partial x^2}^{-1} g\psi + \sum_{s=1}^{n}\frac{\partial\varphi_i}{\partial x_s}^{-1}\left(\frac{\partial\varphi}{\partial x}\right)^{T}\frac{\partial g^s}{\partial x}\psi + \sum_{s=1}^{n}\frac{\partial\varphi_i}{\partial x_s}^{-1}\left(g\,\frac{\partial\psi}{\partial x}\right)^{s} = \frac{\partial g^i}{\partial x}, \tag{4.3b}$$

$i=1,2,\dots,n$.

The solution of (4.3a) for $(\frac{\partial f}{\partial x}, g)$ in the generic form is well-known, [13],[14], and can be written down as follows:

$$\psi^{-1} = \begin{bmatrix} A & 0 \\ D & B \end{bmatrix} \qquad (\frac{\partial \varphi}{\partial x})^{-1} = \begin{bmatrix} P & 0 \\ Q & R \end{bmatrix} , \tag{4.4}$$

where $P=A\otimes I_{k+1}$, $Q=C\otimes J_{k,k+1} + D\otimes J^{*}_{k+1}$, $R=B\otimes I_k$ for non-singular matrices A,B of respective sizes $r\times r$, $(m-r)\times(m-r)$, and arbitrary matrices C,D of size $(m-r)\times r$. Hereabove $n=km+r$, $0\le r<m$, $J_{k,k+1}=[I_k, 0]_{k\times(k+1)}$, $J^{*}_{k,k+1}=[0,I_k]_{k\times(k+1)}$, $I_k, I_{k+1}$ are identity matrices and $\otimes$ denotes the Kronecker product of matrices. Now, after substituting (4.4) into (4.3b), one arrives at the following four groups of equations first introduced in [7]:

1. $i=l(k+1)$, $l=1,2,\dots,r$:

$$(\frac{\partial \varphi}{\partial x})^T \frac{\partial^2 \varphi_i}{\partial x^2} g\psi + \sum_{t=1}^{r} a_{lt} (\frac{\partial \varphi}{\partial x})^T \frac{\partial g}{\partial x}^{t(k+1)} \psi + \sum_{t=1}^{r} a_{lt} \frac{\partial \psi_t}{\partial x} = \frac{\partial g^i}{\partial x} ,$$

2. $i=r(k+1)+pk$, $p=1,2,\dots,m-r$:

$$(\frac{\partial \varphi}{\partial x})^T \frac{\partial^2 \varphi_i^{-1}}{\partial x^2} g\psi + \sum_{t=1}^{r} c_{pt} (\frac{\partial \varphi}{\partial x})^T \frac{\partial g}{\partial x}^{t(k+1)} \psi + \sum_{t=1}^{r} d_{pt} (\frac{\partial \varphi}{\partial x})^T \frac{\partial g}{\partial x}^{t(k+1)} \psi$$

$$+ \sum_{s=1}^{m-r} b_{ps} (\frac{\partial \varphi}{\partial x})^T \frac{\partial g}{\partial x}^{r(k+1)+sk} \psi + \sum_{t=1}^{r} d_{pt} \frac{\partial \psi_t}{\partial x} + \sum_{s=1}^{m-r} b_{ps} \frac{\partial \psi_{r+s}}{\partial x} = \frac{\partial g}{\partial x}^{i} ,$$

3. $i=(l-1)(k+1)+t$, $l=1,2,\dots,r$, $t=1,2,\dots,k$: (4.5)

$$(\frac{\partial \varphi}{\partial x})^T \frac{\partial^2 \varphi_i^{-1}}{\partial x^2} g\psi + \sum_{j=1}^{r} a_{lj} (\frac{\partial \varphi}{\partial x})^T \frac{\partial g}{\partial x}^{(j-1)(k+1)+t} \psi = \frac{\partial g^i}{\partial x} ,$$

4. $i=r(k+1)+(s-1)k+w$, $w=1,2,\dots,k-1$, $s=1,2,\dots,m-r$:

$$(\frac{\partial \varphi}{\partial x})^T \frac{\partial^2 \varphi_i}{\partial x^2} g\psi + \sum_{j=1}^{r} c_{sj} (\frac{\partial \varphi}{\partial x})^T \frac{\partial g}{\partial x}^{(j-1)(k+1)+w} \psi + \sum_{j=1}^{r} d_{js} (\frac{\partial \varphi}{\partial x})^T \frac{\partial g}{\partial x}^{(j-1)(k+1)+w+1} \psi +$$

$$+ \sum_{t=1}^{m-r} b_{st} (\frac{\partial \varphi}{\partial x})^T \frac{\partial g}{\partial x}^{r(k+1)+(t-1)k+w} \psi = \frac{\partial g^i}{\partial x} ,$$

where $(\frac{\partial \psi_k}{\partial x})_{ij} = \frac{\partial \psi_{kj}}{\partial x_i}$ .

In equations (4.5) the unknowns to be determined are $\frac{\partial^2 \varphi^{-1}}{\partial x^2}$, $\frac{\partial \psi}{\partial x}$, and $\frac{\partial \varphi}{\partial x}$, $\psi$. The data g are specified completely (the Brunovsky generic form), while $\frac{\partial g^i}{\partial x}$ are fixed, but not given any definite value. On solving

(4.5;1,2) one immediately observes that, due to non-singularity of matrices A,B in (4.4), these equations determine all the derivatives $\frac{\partial\psi}{\partial x}$, i.e. their contribution to codim stab $\sigma^1$ amounts to $n\cdot m^2$. Next, taking into account the form of g, premultiplying both sides of (4.5;3,4) by $(\frac{\partial\varphi}{\partial x})^{T^{-1}}$ and postmultiplying by $\psi^{-1}$, one arrives at $(n-m)\cdot m+\frac{m(m+1)}{2}$ equations for entries of every $\frac{\partial^2}{\partial x^2}\varphi_i^{-1}$ appearing in (4.5,3,4). This gives a total of $(n-m)\cdot((n-m)\cdot m+\frac{m(m+1)}{2})$ equations for $\frac{\partial^2\varphi^{-1}}{\partial x^2}$. But, due to symmetry of $\frac{\partial^2\varphi^{-1}_i}{\partial x^2}$, (4.5;3,4) contain additional $\frac{m}{2}(m-1)(n-m)$ equations which should be solved for still undetermined entries of $\frac{\partial\varphi}{\partial x}, \psi$. For general $\frac{\partial g^i}{\partial x}$ standing in (4.5;3,4), the additional equations determine $\min(\frac{m}{2}\cdot(m-1)(n-m),\ m^2)$ elements of matrices A,B,C,D in (4.4). Eventually, codim stab $\sigma^1\le n^2+mn+nm^2+(n-m)((n-m)\cdot m+\frac{m(m+1)}{2})+\min(\frac{m}{2}\cdot(m-1)(n-m),m^2)$, so codim $\mathcal{O}_{\sigma^1}=\dim\Sigma^1_0-$ codim stab $\sigma^1\ge\max(0,\frac{m}{2}\cdot(m-1)(n-m)-m^2)$.

QED

It is easily checked that the Proposition allows for the existence of open orbits of the action (4.2), among other, for m=n-2, n-arbitrary. This case, however, can be examinated further. What is observed is that only entries of $\psi^{-1}$ (i.e. matrices A,B,D) are present in the equations of stabilizer in this case. This observation leads to the following result.

<u>Corollary 4.5.</u> For $\sigma^1$ as in Proposition 4.4, and m=n-2, codim $\mathcal{O}_{\sigma^1}\ge\max(0,m-4)$.

<u>Remark 4.6.</u> Proposition 4.4 and Corollary 4.5 imply that open orbits of the action (4.2) can be looked for exclusively within the range of (m,n) exhibited below:

1. m=1, m=n, m=n-1, n - arbitrary.
2. m=2, n=4,5,6. 3. m=3, n=5,6. 4. m=4, n=6.

In some cases mentioned above open orbits of (4.2) do exist and can be described by simple normal forms. Below we give an example.

<u>Example 4.7.</u> Let $\sigma^1\in\Sigma^1_0$ be of the form used in Proposition 4.4. Then, for (m,n) given below there exist open-dense orbits of the action (4.2), characterized by the following normal forms:

1. m=1, n - arbitrary : $\frac{\partial g^1}{\partial x_1}=\frac{\partial g^2}{\partial x_2}=\ldots=\frac{\partial g^n}{\partial x_n}=0$ .

2. m=n, n - arbitrary : $\frac{\partial g}{\partial x}=\frac{\partial g}{\partial x}=\ldots=\frac{\partial g}{\partial x}=0$ .

3. m=2, n=3 : $\frac{\partial g^1}{\partial x} = \begin{bmatrix} 0 & 0 \\ 0 & 0 \\ 1 & 0 \end{bmatrix}$ , $\frac{\partial g^2}{\partial x} = \frac{\partial g^3}{\partial x} = 0$

4. m=2, n=4 : $\frac{\partial g^1}{\partial x} = \begin{bmatrix} 0 & 0 \\ 0 & 0 \\ 0 & 0 \\ 1 & 0 \end{bmatrix}$ , $\frac{\partial g^2}{\partial x} = \frac{\partial g^3}{\partial x} = \frac{\partial g^4}{\partial x} = 0$

5. m=3, n=4 : $\frac{\partial g^1}{\partial x} = \begin{bmatrix} 0 & 0 & 0 \\ 0 & 0 & 0 \\ 0 & 1 & 0 \end{bmatrix}$ , $\frac{\partial g^2}{\partial x} = \frac{\partial g^3}{\partial x} = \frac{\partial g^4}{\partial x} = 0$ .

Remark 4.8. It should be observed that Proposition 4.4 strenghtens remarkably the results of Proposition 4.1 by asserting that an obstacle to the structural stability in most cases lies just in the action of $G_o^1$ on $\Sigma_o^1$.

## 5. Conclusion: an application to bilinear systems.

Consider the class B of inhomogeneous bilinear systems described by the following formula, [15]:

$$\dot{x} = Ax + Bu + \sum_{i=1}^{m} u_i B_i x \qquad (5.1)$$

where $x \in \mathbb{R}^n$, $u \in \mathbb{R}^m$, $A, B, B_1, \ldots, B_m$ are appropriately dimensioned real matrices. A bilinear system $\beta \in B$ will be identified with a tuple of matrices, $\beta = (A, B, B_1, \ldots, B_m) \in B \cong \mathbb{R}^N$, $N = n^2 + nm + mn^2$. Clearly, (5.1) can be viewed as an affine system (2.1) with $f(x) = Ax$, $g_i(x) = b_i + B_i x$, $B = [b_1, \ldots, b_m]$.

It is well-known that the feedback group (2.2) does not preserve the bilinear structure of (5.1), while the approximate group $G_o^1$ does and seems to be a proper feedback group for bilinear system. We shall call this group the *bilinear feedback group*. Its action can be interpreted as an effect of the truncation of the feedback group action (2.2) around $0 \in \mathbb{R}^n$ form quadratic terms in x onwards.

Now we are in a position to establish a clear relevance of results obtained in Section 4 to the bilinear case. The conclusion is as follows.

Theorem 5.1. Given the class of bilinear systems (5.1) acted on by the bilinear feedback group (4.2). Then the action produces no open orbits except possibly for the cases referred to in Remark 4.6. Remaining orbits have codimensions greater than or equal to either $\frac{m}{2}\cdot(m-1)(n-m)-m^2$, if $n>m+2$, or $m-4$, if $n=m+2$.

It is easy to write down some structurally stable normal forms of bilinear systems under the bilinear feedabck group within the range of dimensions indicated in Example 4.7. Below we present a sample:

1. m=2, n=3 : $\dot{x}_1=x_2+u_1x_3$ , $\dot{x}_2=u_1$, $\dot{x}_3=u_2$.

2. m=2, n=4 : $\dot{x}_1=x_2+u_1x_4$, $\dot{x}_2=u_1$, $\dot{x}_3=x_4$, $\dot{x}_4=u_2$.

3. m=3, n=4 : $\dot{x}_1=x_2+u_2x_4$, $\dot{x}_2=u_1$, $\dot{x}_3=u_2$, $\dot{x}_4=u_3$.

References:

[1] M.Golubitsky, V.Guillemin, *Stable Mappings and Their Singularities* Springer-Verlag, Berlin, 1973.
[2] M.W.Hirsch, *Differential Topology* Springer-Verlag, Berlin, 1976.
[3] P.W.Michor, *Manifolds of Differentiable Mappings* Shiva Publ. Ltd.,Orpington, 1980.
[4] R.Thom, H.I.Levine, Singularities of differentiable mappings. In:*Proc of Liverpool Singularities I* Springer-Verlag, Berlin, 1971, 1-89.
[5] K.Tchon',On approximate equivalence of affine control systems. *Internat. J. Control* 44(1986), 259-266.
[6] K.Tchon', The only stable normal forms of affine systems under feedback are linear. *Systems and Control Letters* 8(1987),359-365.
[7] K.Tchon', On structural instability of normal forms of affine systems subject to static state feedback. *Linear Algebra and Applications*, to appear.
[8] C.I.Byrnes, Remarks on nonlinear planar control systems which are not linearizable by feedback. *Systems and Control Letters* 5(1985), 363-367.
[9] M.Golubitsky, D.Tischler, A survey of the singularities and stability of differential forms. *Asterisque* 59-60(1978), 43-62.
[10] B.Jakubczyk, F.Przytycki, *Singularities of k-tuples of Vector Fields* . Polish Sci. Publ., Warsaw, 1984.
[11] F.W.Warner, *Foundations of Differentiable Manifolds and Lie Groups*. Scott-Foresman, Glenview, 1971.
[12] P.Brunovsky', A classification of linear controllable systems.*Kybernetika* 6(1970), 173-187.
[13] K.Tchon', Generic properties of linear systems: An overview. *Kybernetika* 19(1983), 467-474.
[14] R.W.Brockett, The geometry of the set of controllable linear systems. *Res Reports of Aut Cont Lab of Nagoya University* 24(1977).
[15] A.Isidori, A.Ruberti, Realization theory of bilinear systems. In: *Geometric Methods in System Theory*, ed. by D.Q.Mayne and R.W.Brockett. D.Reidel, Dordrecht, 1973, 84-129.

# TRANSFORMATIONS OF NONLINEAR SYSTEMS UNDER EXTERNAL EQUIVALENCE

A.J. van der Schaft
Department of Applied Mathematics, University of Twente
P.O. Box 217, 7500 AE Enschede, The Netherlands

## 1. Introduction

It is well-known that linear (finite-dimensional) systems can be represented in different ways. Most common are the transfer matrix representation $y(s) = G(s)u(s)$ (or, transformed to the time-domain, the impulse response matrix representation), the state-space representation $\dot{x} = Ax + Bu$, $y = Cx + Du$, with $G(s)$ equal to $C(Is-A)^{-1}B + D$, as well as matrix polynomial representations such as $D(s)y(s) = N(s)u(s)$, with $D^{-1}(s)N(s)$ a polynomial matrix fraction of $G(s)$. These representations all have their own advantages, and part of the power of linear systems theory lies in the ability to readily switch over from one representation to another.

In the nonlinear systems literature the emphasis so far has been mainly on the nonlinear generalization of the transfer matrix and impulse response matrix representation, and on nonlinear state space representations. Only recently some attention has been paid to the nonlinear generalization of polynomial descriptions. Motivated by work of Willems [W3,W1], in [S1,S2] nonlinear systems were considered described by sets of higher-order implicit nonlinear differential equations in the inputs and outputs

$$(1.1) \qquad P_i\big(y,\dot{y},\ldots,y^{(k)},u,\dot{u},\ldots,u^{(k)}\big) = 0 \qquad i = 1,\ldots,\ell$$

which can be regarded as the nonlinear generalization of the polynomial matrix description (in time-domain)

$$(1.2) \qquad D\big(\tfrac{d}{dt}\big)y(t) = N\big(\tfrac{d}{dt}\big)u(t),$$

and conditions for transforming these systems into state space form were deduced. Recently, in a series of papers Fliess, e.g. [F1,F2,F3], has underlined the importance of system representations as (1.1) in modeling and control problems such as invertibility and decoupling (see also [DM], [P], and very recently [CL], [CMP], [Re], [S3,4], [So]).The present paper is mainly concerned with the notion of <u>equivalence</u> <u>transformation</u> for systems like (1.1), and is partly inspired by a recent paper of Schumacher [Sc] on equivalence transformations for linear systems. Motivated by the discussions in Willems [W1,W2,W3] we start from a general system description of the form

$$(1.3) \qquad P_i\big(w,\dot{w},\ldots,w^{(k)},\xi,\dot{\xi},\ldots,\xi^{(k)}\big) = 0 \qquad i = 1,\ldots,\ell$$

where $w \in \mathbb{R}^q$ denote the external variables, and $\xi \in \mathbb{R}^s$ the auxiliary or latent variables. We call (1.3) an external differential system. The set of external variables consists of both the input and the output variables; the idea however is that the distinction between inputs and outputs may not be given a priori, and/or that the splitting of w into input and output components may be itself a modeling question. The nature of the latent variables in (1.3) can range from states, partial states (such as in the Rosenbrock framework [R]) to hidden inputs or driving variables.

Two external differential systems (1.3) are called equivalent (see Willems [W1,W2]) if the sets of trajectories that they allow for the external variables are the same. To avoid (difficult) technical problems we shall only consider trajectories that are smooth ($C^\infty$) functions of time. Formally we have

Definition 1.1. Consider an external differential system (1.3). Its behaviour is defined as

$$\mathcal{B} = \{ w: \mathbb{R} \to \mathbb{R}^q,\ w \text{ is } C^\infty \mid \exists \xi: \mathbb{R} \to \mathbb{R}^s,\ \xi \text{ is } C^\infty, \text{ such that } (w(t),\xi(t)) \text{ is solution of (1.3) for any } t \}$$

Two external differential systems with behaviours $\mathcal{B}_1$, respectively $\mathcal{B}_2$, are called equivalent if $\mathcal{B}_1 = \mathcal{B}_2$.

An equivalence transformation for a system (1.3) then is a transformation of the system to another system of the form (1.3) which is equivalent. Note that two equivalent systems may have a different number of latent variables and/or a different number of defining equations. In this terminology the problem of state space representation (or realization) consists in finding an equivalence transformation from (1.3) to a system (1.3) of the special form (an input-state-output system)

$$\text{(1.4)} \qquad \begin{aligned} \dot{x} &= f(x,u) \\ y &= h(x,u) \end{aligned}$$

with x the latent variables, and the vector (y,u) being some permutation of the vector w ([W1],[S1],[Sc]).

The main result of this paper is that, under constant rank assumptions, we can always locally transform a general system (1.3) into an equivalent system not involving any latent variables. Furthermore we will show how locally this equivalent system without latent variables can be transformed into a reduced form which is more amenable for analysis, in particular for realization purposes.

A particular example of eliminating the latent variables in a general external system (1.3) is the problem of transforming an input-state-output system (1.4) into a system (1.1) by elimination of the state variables x. In [S3], see also [CMP,Re], it has been shown that in this case the general procedure given in the next section simplifies considerably, basically because of the simple dependence of equations (1.4) on the time-derivative $\dot{x}$. Finally, during the present conference algorithms have been presented ([G],[D]) for converting *polynomial* state space systems (1.4) into systems (1.1), using tools from differential algebra (cf. [F1,F2,F3,P]). It seems to be of interest to extend these last algorithms to the general problem considered here, in case (1.3) consists of polynomial equations.

## 2. Equivalence transformations on external differential systems

The basic idea for the equivalence transformation employed in this section is already contained in [S1], and is inspired by the following kind of equivalence transformation for linear systems, cf. [W1,Sc,Ka]. Consider a linear external differential system

$$(2.1)\qquad R\left(\frac{d}{dt}\right)w(t) = Q\left(\frac{d}{dt}\right)\xi(t)$$

with R(s) and Q(s) polynomial matrices in the indeterminate s. Now let U(s) be a unimodular polynomial q×q matrix (i.e. det U(s) = constant ≠ 0), and define $\bar{R}(s) = U(s)R(s)$, $\bar{Q}(s) = U(s)Q(s)$. Then

$$(2.2)\qquad \bar{R}\left(\frac{d}{dt}\right)w(t) = \bar{Q}\left(\frac{d}{dt}\right)\xi(t)$$

is easily seen to be an equivalent system, and so U(s) defines an equivalence transformation. In particular we may take U(s) to be the q×q identity matrix with a particular j-th row replaced by $(\alpha_1 s^{\nu_1}, \alpha_2 s^{\nu_2}, \ldots, \alpha_q s^{\nu_q})$, with $\nu_j = 0$ and $\alpha_j \neq 0$. In this last case U(s) is said to define an elementary equivalence transformation; the j-th equation in (2.1) is replaced by a linear combination of the j-th equation and the time- derivatives (up to some order) of the other equations (note that multiplication by s corresponds to derivation). This idea generalizes to the nonlinear case as follows.

Lemma 2.1. Let $f_1(x,z),\ldots,f_m(x,z)$ be smooth functions of $x \in \mathbb{R}^n$, $z \in \mathbb{R}^k$. Let $(\bar{x},\bar{z}) \in \mathbb{R}^n \times \mathbb{R}^k$ be such that $f_1(\bar{x},\bar{z}) = \ldots = f_m(\bar{x},\bar{z}) = 0$. Suppose that the Jacobian matrix of $f = (f_1,\ldots,f_m)^T$ with respect to x

$$(2.1)\qquad D_x f(x,z) = \left[\frac{\partial f_i}{\partial x_j}(x,z)\right]_{\substack{i=1,\ldots,m\\ j=1,\ldots,n}}$$

has constant rank m-p, p > 0, in a neighbourhood of $(\bar{x},\bar{z})$. Then there exist smooth functions $\varphi_1(y,z),\ldots,\varphi_p(y,z)$, $y \in \mathbb{R}^m$, $z \in \mathbb{R}^k$, defined on a neighbourhood of $(0,\bar{z}) \in \mathbb{R}^m \times \mathbb{R}^k$ such that $\varphi_1,\ldots,\varphi_p$ are independent as functions of y, and the functions $\psi_1(x,z),\ldots,\psi_p(x,z)$ defined as

(2.2) $\psi_i(x,z) = \varphi_i\big(f_1(x,z),\ldots,f_m(x,z),z_1,\ldots,z_k\big)$, $\quad i = 1,\ldots,p$

do not depend on x, i.e.

(2.3) $\frac{\partial\psi_i}{\partial x_j}(x,z) = 0, \quad i = 1,\ldots,p, \; j = 1,\ldots,n$

Additionally $\varphi_1,\ldots,\varphi_p$ can be chosen in such a way that

(2.4) $\varphi_i(0,z) = 0, \quad i = 1,\ldots,p$

for z in a neighbourhood of $\bar{z}$. Moreover, permute the functions $f_1,\ldots,f_m$ in such a way that

(2.5) $$\operatorname{rank}\left[\frac{\partial f_{p+i}}{\partial x_j}(\bar{x},\bar{z})\right]_{\substack{i=1,\ldots,m-p\\ j=1,\ldots,m}} = m-p$$

then

(2.6) $$\operatorname{rank}\left[\frac{\partial \varphi_i}{\partial x_j}(0,\bar{z})\right]_{\substack{i=1,\ldots,p\\ j=1,\ldots,p}} = p$$

Finally, (2.6) together with (2.4) implies that for (x,z) close to $(\bar{x},\bar{z})$

(2.7) $\varphi_i\big(f_1(x,z),\ldots,f_p(x,z),0,\ldots,0,z_1,\ldots,z_k\big) = 0 \quad , \quad i = 1,\ldots,p,$

if and only if $\quad f_i(x,z) = 0 \quad , \quad i = 1,\ldots,p.$

Proof. For $k = 0$ the lemma is a simple consequence of the well-known rank theorem for smooth mappings (see for example [Sp]) applied to the map $f\colon \mathbb{R}^n \to \mathbb{R}^m$, which yields the existence of local coordinates $\tilde{x}$ for $\mathbb{R}^n$ and $\tilde{y}$ for $\mathbb{R}^m$ such that f is locally described as $(\tilde{x}_1,\ldots,\tilde{x}_n) \longmapsto (0,\ldots,0,\tilde{x}_1,\ldots,\tilde{x}_{m-p})$. The above parameter dependent version (with $z \in \mathbb{R}^k$ the parameter) follows completely analogously. □

The importance of Lemma 2.1 for our purposes lies in the following corollary, which immediately follows from (2.7).

Corollary 2.2. Around $(\bar{x},\bar{z})$ the solution set (x,z) of

(2.8) $f_1(x,z) = \ldots = f_m(x,z) = 0$

equals the solution set of

(2.9a) $\varphi_i\big(f_1(x,z),\ldots,f_p(x,z),f_{p+1}(x,z),\ldots,f_m(x,z),z\big) = 0, \quad i = 1,\ldots,p$

(2.9b) $f_{p+1}(x,z) = \ldots = f_m(x,z) = 0$

where by (2.3) the equations (2.9a) do not depend on x.

Before applying Lemma 2.1 and Corollary 2.2 to external differential systems we first introduce some terminology. Consider a higher-order differential equation

$$P(w,\dot{w},\ldots,w^{(k)},\xi,\dot{\xi},\ldots,\xi^{(k)}) = 0. \tag{2.10}$$

Let $(\bar{w}(t),\bar{\xi}(t))$, $t \in (-\epsilon,\epsilon)$, be a solution of (2.10). We will denote the point in the k-jet space defined by $(\bar{w}(t),\bar{\xi}(t))$ by taking the derivatives up to order k in $t = 0$ by $(\bar{w},\bar{\xi})$. Clearly $(\bar{w},\bar{\xi})$ is a solution point of the equations (2.10) in the indeterminates $w,\dot{w},\ldots,w^{(k)},\xi,\dot{\xi},\ldots,\xi^{(k)}$. The degree $\sigma$ of (2.10) with respect to $\xi$ is defined as the largest integer ($\leq k$) such that

$$\frac{\partial P}{\partial \xi^{(\sigma)}}(w,\ldots,w^{(k)},\xi,\ldots,\xi^{(k)}) \neq 0 \tag{2.11}$$

for some $(w,\xi)$ close to $(\bar{w},\bar{\xi})$ (in the k-jet space). If $\sigma$ is not defined (i.e. P does not depend on $\xi$) then set $\sigma = 0$. The first-order <u>prolongation</u> of an equation (2.10) is defined as

$$\dot{P}(w,\ldots,w^{(k+1)},\xi,\ldots,\xi^{(k+1)}) = \frac{\partial P}{\partial w}\dot{w} + \ldots + \frac{\partial P}{\partial w^{(k)}} w^{(k+1)} + \frac{\partial P}{\partial \xi}\dot{\xi} + \ldots + \frac{\partial P}{\partial \xi^{(k)}}\xi^{(k+1)} = 0 \tag{2.12}$$

Since $\frac{d}{dt} P(w(t),..,w^{(k)}(t),\xi(t),..,\xi^{(k)}(t)) = \dot{P}(w(t),..,w^{(k+1)}(t),\xi(t),..,\xi^{(k+1)}(t))$ it follows that if $(w(t),\xi(t))$, $t \in (-\epsilon,\epsilon)$, is a solution of (2.10), then it is also a solution of (2.12). Furthermore if the degree with respect to $\xi$ of (2.10) is $\sigma$ then

$$\frac{\partial \dot{P}}{\partial \xi^{(\sigma+1)}}(w,\ldots,w^{(k+1)},\xi,\ldots,\xi^{(k+1)}) = \frac{\partial P}{\partial \xi^{(\sigma)}}(w,\ldots,w^{(k)},\xi,\ldots,\xi^{(k)}) \tag{2.13}$$

and the degree w.r.t. $\xi$ of $\dot{P}$ equals $\sigma+1$. In general, we denote by $P^{(j)}$ the j-th order prolongation of P. It follows that the degree of $P^{(j)}$ is $\sigma+j$.
Now let us return to the external differential system (1.3). We will give an algorithm to locally transform (1.3) into an equivalent external differential system <u>not</u> involving any latent variables. Let $\sigma_i$ be the degree w.r.t. $\xi$ of $P_i$, $i = 1,\ldots,\ell$ (around $(\bar{w},\bar{\xi})$). The <u>leading</u> <u>row coefficient</u> (l.r.c) matrix with respect to $\xi$ of (1.3) is defined as the $\ell \times s$ matrix

$$\left[\frac{\partial P_i}{\partial \xi_j^{(\sigma_i)}}(w,\dot{w},\ldots,w^{(k)},\xi,\dot{\xi},\ldots,\xi^{(k)})\right]_{\substack{i=1,\ldots,\ell\\ j=1,\ldots,s}} \tag{2.14}$$

which we denote as $A(w,\xi)$. If rank $A(w,\xi) = \ell$ for all $(w,\xi)$ close to $(\bar{w},\bar{\xi})$ then the algorithm terminates. Now consider the case that rank $A(\bar{w},\bar{\xi}) < \ell$.

<u>Assumption 1</u>. rank $A(w,\xi) < \ell$ for any $(w,\xi)$ close to $(\bar{w},\bar{\xi})$.

It follows that around $(\bar{w},\bar{\xi})$ there exist smooth functions $\alpha_1(w,\xi),\ldots,\alpha_\ell(w,\xi)$

(here $(w,\xi)$ denotes $\left(w,\dots,w^{(k)},\xi,\dots,\xi^{(k)}\right)$), not all identically zero such that

(2.15) $$\sum_{i=1}^{\ell} \alpha_i(w,\xi)A_i(w,\xi) = 0$$

with $A_i$ the i-th row of A. Now reorder the equations $P_1,\dots,P_\ell$ in such a way that $\alpha_i \neq 0$ for $i = 1,\dots,\ell'$ and $\alpha_i = 0$ for $i = \ell'+1,\dots,\ell$ $(\ell' \leq \ell)$, and furthermore that $\sigma_1 \geq \sigma_i$, $i = 1,\dots,\ell'$ (everything locally around $(\bar{w},\bar{\xi})$).

Assumption 2. Around $(\bar{w},\bar{\xi})$

(2.16) $$\operatorname{rank}\left(\frac{\partial P_i}{\partial \xi_j^{(\sigma_i)}}\right)_{\substack{i=2,\dots,\ell'\\ j=1,\dots,s}} = \operatorname{rank}\left(\frac{\partial P_i}{\partial \xi_j^{(\sigma_i)}}\right)_{\substack{i=1,\dots,\ell'\\ j=1,\dots,s}} = \text{constant}$$

Let us now consider the prolonged equations

(2.17) $$P_i^{(\sigma_1-\sigma_i)}\left(w,\dot{w},\dots,w^{(k')},\xi,\dot{\xi},\dots,\xi^{(\sigma_1)}\right) = 0, \qquad i = 1,2,\dots,\ell'$$

with $k' \leq k + \sigma_1$, which all have degree $\sigma_1$ (w.r.t. $\xi$). By (2.13) and (2.16) it follows that

(2.18) $$\operatorname{rank}\left(\frac{\partial P_i^{(\sigma_1-\sigma_i)}}{\partial \xi_j^{(\sigma_1)}}\right)_{\substack{i=2,\dots,\ell'\\ j=1,\dots,s}} = \operatorname{rank}\left(\frac{\partial P_i^{(\sigma_1-\sigma_i)}}{\partial \xi_j^{(\sigma_1)}}\right)_{\substack{i=1,\dots,\ell'\\ j=1,\dots,s}} = \text{constant}$$

Hence we can apply Lemma 2.1 and Corollary 2.2 to the prolonged equations (2.17), regarded as equations in the independent variables $\xi^{(\sigma_1)}$, and parameters $w,\dot{w},\dots,w^{(k')},\xi,\dot{\xi},\dots,\xi^{(\sigma_1-1)}$.
For simplicity we will not use Lemma 2.1 in its full generality; we only deduce from it the local existence of a single smooth function

(2.19) $$\varphi\left(y_1,\dots,y_{\ell'},w,\dot{w},\dots,w^{(k')},\xi,\dot{\xi},\dots,\xi^{(\sigma_1-1)}\right)$$

satisfying

(2.20) $$\varphi\left(0,\dots,0,w,\dots,w^{(k')},\xi,\dots,\xi^{(\sigma_1-1)}\right) = 0, \quad \frac{\partial\varphi}{\partial y_1}(0,\dots,0,\bar{w},\bar{\xi}) \neq 0$$

and such that the function

(2.21) $$\psi\left(w,\dots,w^{(k')},\xi,\dots,\xi^{(\sigma_1)}\right) := \varphi\left(P_1^{(\sigma_1-\sigma_1)}(w,\xi),\dots,P_{\ell'}^{(\sigma_1-\sigma_{\ell'})}(w,\xi),w,\dots,w^{(k')},\xi,\dots,\xi^{(\sigma_1-1)}\right)$$

does not depend on $\xi^{(\sigma_1)}$. It follows from Corollary 2.2 that if we replace $P_1(w,\xi) = 0$ in the reordered equations (1.3) by $\psi(w,\xi) = 0$ then we have obtained an equivalent external differential system. In analogy with the linear case we will call this an elementary equivalence transformation.
The crucial thing is that by the above elementary equivalence transformation the degree (w.r.t. $\xi$) of the first equation has been strictly diminished (while the other equations are unchanged). Hence the leading row coefficient matrix of the newly defined equivalent external differential system will be in general different from (2.14), and we can perform the same procedure once

again. If the rank of the new l.r.c. matrix equals $\ell$ then the algorithm terminates. If the rank is less than $\ell$ then under Assumptions 1 and 2 for the newly defined external differential system there exists an elementary equivalence transformation which will strictly diminish the degree of one of the equations. It is clear that if at every step the constant rank assumptions 1 and 2 are met then the algorithm will terminate after a finite number steps because of one of the following reasons:

(i) the rank of the l.r.c. matrix has become equal to $\ell$. (Note that at every step the rank of the l.r.c. matrix does not decrease.)

(ii) the degrees (w.r.t. $\xi$) of the dependent rows in the l.r.c. matrix have become zero.

This is summarized in

<u>Theorem 2.3</u>. Consider an external differential system (1.3). If at every step of the algorithm the constant rank assumptions 1 and 2 are satisfied, then after a finite number of steps we obtain locally around $(\bar{w},\bar{\xi})$ an equivalent external differential system of the form

(2.22a) $\quad P_i(w,\dot{w},\dots,w^{(\bar{k})}) = 0 \qquad i = 1,\dots,\bar{\ell} \qquad (\bar{\ell} \le \ell)$

(2.22b) $\quad P_i(w,\dot{w},\dots,w^{(\bar{k})},\xi,\dot{\xi},\dots,\xi^{(k)}) = 0 \qquad i = \bar{\ell}+1,\dots,\ell$

where the last equations satisfy

(2.23) $$\operatorname{rank}\left[\frac{\partial P_i}{\partial \xi_j^{(\sigma_i)}}\right]_{\substack{i=\bar{\ell}+1,\dots,\ell \\ j=1,\dots,s}} = \ell-\bar{\ell} \quad , \quad \text{around } (\bar{w},\bar{\xi}).$$

Furthermore we have the following lemma (compare [W1, Prop. 3.3])

<u>Lemma 2.4.</u> Consider an external differential system (1.3) such that

(2.24) $$\operatorname{rank}\left[\frac{\partial P_i}{\partial \xi_j^{(\sigma_i)}}\right]_{\substack{i=1,\dots,\ell \\ j=1,\dots,s}} = \ell \quad , \quad \text{around } (\bar{w},\bar{\xi}).$$

By permutation of the latent variables $\xi_1,\dots,\xi_s$, and by elementary equivalence transformations and permutations on the equations $P_1 = .. = P_\ell = 0$ we can transform locally around $(\bar{w},\bar{\xi})$ the system into an equivalent external differential system $\tilde{P}_1 = \dots = \tilde{P}_\ell = 0$, with the same set of degrees $\sigma_1,\dots,\sigma_\ell$ (with respect to $\xi$) as the original system satisfying

(2.25a) $\quad \sigma_1 \le \sigma_2 \le \dots \le \sigma_\ell$

(2.25b) $$\frac{\partial \tilde{P}_i}{\partial \xi_j^{(\sigma_i)}} = 0 \qquad j < i \qquad i,j = 1,\dots,\ell$$

$$(2.25c) \qquad \frac{\partial \tilde{P}_i}{\partial \xi_i^{(\sigma_i)}} \neq 0 \qquad\qquad i = 1,\ldots,\ell$$

Moreover it follows from (2.25) that for <u>any</u> smooth function $w(t)$ close to $\bar{w}$ there exists a smooth function $\xi(t)$ close to $\bar{\xi}$ such that $(w(t),\xi(t))$ satisfies (1.3).

Before giving the proof, we note that application of this lemma to the system (2.22b) satisfying (2.23) implies that for any smooth $w(t)$ close to $\bar{w}$ there exists $\xi(t)$ close to $\bar{\xi}$ such that $(w(t),\xi(t))$ satisfies (2.22b). Hence (2.22b) are <u>ineffective</u> constraint equations (cf. [Sc,Prop.2.1]), and we have the following corollary to Theorem 2.3:

<u>Corollary 2.5.</u> Around $(\bar{w},\bar{\xi})$ the external differential system (2.22) is equivalent to the external differential system (2.22a) <u>without</u> latent variables.

<u>Proof of Lemma 2.4.</u> Reorder the equations $P_1,\ldots,P_\ell$ in such a way that $\sigma_1 \le \sigma_2 \le \ldots \le \sigma_\ell$. By (2.24) we can reorder the variables $\xi_1,\ldots,\xi_s$ in such a way that

$$(2.26) \qquad \operatorname{rank} \left( \frac{\partial P_i}{\partial \xi_j^{(\sigma_i)}} \right)_{\substack{i=1,\ldots,\ell \\ j=1,\ldots,\ell}} = \ell \quad , \quad \text{around } (\bar{w},\bar{\xi}).$$

Furthermore we can ensure that $\frac{\partial P_1}{\partial \xi_1^{(\sigma_1)}}$ does not vanish around $(\bar{w},\bar{\xi})$. Denote $\tilde{P}_1 = P_1$. Now consider the two functions

$$(2.27) \qquad \begin{array}{l} \tilde{P}_1^{(\sigma_2-\sigma_1)}(w,\dot{w},\ldots,\xi,\dot{\xi},\ldots,\xi^{(\sigma_2)}) \\ \\ P_2(w,\dot{w},\ldots,\xi,\dot{\xi},\ldots,\xi^{(\sigma_2)}) \end{array}$$

Regard $\xi_1^{(\sigma_2)}$ as the independent variable, and all other variables as parameters. Then by Lemma 2.1 and Corollary 2.2 we may locally replace the equation $P_2 = 0$ in (1.3) by an equation $\tilde{P}_2 = 0$, where the function $\tilde{P}_2$, which is defined through $P_2$ and $\tilde{P}_1^{(\sigma_2-\sigma_1)}$, does not depend anymore on $\xi_1^{(\sigma_2)}$. In the same way, by considering pairs of functions

$$(2.28) \qquad \begin{array}{l} \tilde{P}_1^{(\sigma_i-\sigma_1)}(w,\dot{w},\ldots,\xi,\dot{\xi},\ldots,\xi^{(\sigma_i)}) \\ \\ P_i(w,\dot{w},\ldots,\xi,\dot{\xi},\ldots,\xi^{(\sigma_i)}) \end{array}$$

we can replace the equations $P_i = 0$ by equations $\tilde{P}_i = 0$, where $\tilde{P}_i$ does not depend on $\xi_1^{(\sigma_i)}$, $i = 2,\dots,\ell$. It follows that the degrees of $\tilde{P}_2,\dots,\tilde{P}_\ell$ are still equal to $\sigma_2,\dots,\sigma_\ell$. Now rename $P_i := \tilde{P}_i$, $i = 3,\dots,\ell$. Reorder the variables $\xi_2,\dots,\xi_\ell$ in such a way that $\frac{\partial \tilde{P}_2}{\partial \xi_2^{(\sigma_2)}} \neq 0$ around $(\bar{w},\bar{\xi})$. Then, proceeding in the same way, consider the functions

$$(2.29)\qquad \begin{aligned} &\tilde{P}_2^{(\sigma_3-\sigma_2)}(w,\dot{w},\dots,\xi,\dot{\xi},\dots,\xi^{(\sigma_3)}) \\ &P_3(w,\dot{w},\dots,\xi,\dot{\xi},\dots,\xi^{(\sigma_3)}) \end{aligned}$$

where $\xi_2^{(\sigma_3)}$ is the independent variable, and the other variables are parameters. Again by Lemma 2.1 and Corollary 2.2 we may locally replace the equation $P_3 = 0$ by an equations $\tilde{P}_3 = 0$ which does not depend on $\xi_2^{(\sigma_3)}$. Furthermore since $\tilde{P}_2^{(\sigma_3-\sigma_2)}$ and $P_3$ by construction do not depend on $\xi_1^{(\sigma_3)}$ we can ensure that $\tilde{P}_3$ will not depend on $\xi_1^{(\sigma_3)}$. In the same way we can replace the equations $P_4 = 0$ up to $P_\ell = 0$ by equations $\tilde{P}_4 = 0,\dots,\tilde{P}_\ell = 0$ which do not depend respectively on $(\xi_1^{(\sigma_4)},\xi_2^{(\sigma_4)})$ up to $(\xi_1^{(\sigma_\ell)},\xi_2^{(\sigma_\ell)})$. Continuing in this way we obtain the desired equations $\tilde{P}_1 = \dots = \tilde{P}_\ell = 0$ satisfying (2.25).
Finally, take any smooth functions $w(t)$ close to $\bar{w}(t)$, and $\xi_{\ell+1}(t),\dots,\xi_s(t)$ close to $\bar{\xi}(t)$. Then by (2.25b,c) for $i = \ell$ and the implicit function theorem we can locally rewrite $\tilde{P}_\ell(w,\dot{w},\dots,w^{(k)},\xi,\dot{\xi},\dots,\xi^{(k)}) = 0$ as

$$(2.30)\qquad \frac{d}{dt}\,\xi_\ell^{(\sigma_\ell-1)} = R_\ell(w,\dot{w},\dots,w^{(k)},\xi_\ell,\dots,\xi_\ell^{(\sigma_\ell-1)},\hat{\xi})$$

where $\hat{\xi}$ denotes dependence on $\xi_{\ell+1},\dots,\xi_s$ and their time-derivatives. For $\sigma_\ell = 0$ this immediately determines $\xi_\ell$, while for $\sigma_\ell > 0$ this is a (possibly higher-order) differential equation in $\xi_\ell$ which has a unique locally defined solution $\xi_\ell(t)$ for any initial condition close to $\bar{\xi}$. After having determined $\xi_\ell(t)$, we proceed by noting that by (2.25b,c) for $i = \ell-1$ and the implicit function theorem we can locally rewrite $P_{\ell-1} = 0$ as

$$(2.31)\qquad \frac{d}{dt}\,\xi_{\ell-1}^{(\sigma_{\ell-1}-1)} = R_{\ell-1}(w,\dot{w},\dots,w^{(k)},\xi_{\ell-1},\dots,\xi_{\ell-1}^{(\sigma_{\ell-1}-1)},\hat{\xi})$$

where $\hat{\xi}$ now denotes dependence on $\xi_{\ell+1},\dots,\xi_s$ <u>and</u> $\xi_\ell$ (and their time-derivatives). Again (2.30) yields for any initial condition a unique solution $\xi_{\ell-1}(t)$. In this way we determine successively $\xi_\ell(t),\xi_{\ell-1}(t)$ up to $\xi_1(t)$, such that $(w(t),\xi(t))$ satisfies (1.3). □

Let us now continue with an external differential system

$$(2.32)\qquad P_i(w,w,\dots,w^{(k)}) = 0 \qquad i = 1,\dots,\ell, \quad w \in \mathbb{R}^q$$

only involving the external variables w. By using the same algorithm as above, but now with respect to the w-variables instead of the $\xi$-variables, we immediately obtain from Theorem 2.3 and Lemma 2.4

Proposition 2.6. Consider an external differential system (2.32). Let $\sigma_1,\dots,\sigma_\ell$ be the degrees of $P_1,\dots,P_\ell$ with respect to w, and apply the same algorithm as above but now with respect to w. If at every step of this algorithm Assumptions 1 and 2 are satisfied around a point $\bar{w}$ (in k-jet space), then after a finite number of steps we obtain locally around $\bar{w}$ an equivalent external differential system

$$(2.33)\qquad P_i\big(w,\dot{w},\dots,w^{(\bar{k})}\big) = 0 \qquad i = 1,\dots,\bar{\ell} \qquad (\bar{\ell} \le \ell)$$

satisfying

$$(2.34)\qquad \operatorname{rank}\left[\frac{\partial P_i}{\partial \xi_j^{(\sigma_i)}}\right]_{\substack{i=1,\dots,\bar{\ell}\\ j=1,\dots,q}} = \bar{\ell}$$

together with $\ell-\bar{\ell}$ equations of the form $0 = 0$, which can be deleted. Furthermore by permutation of $w_1,\dots,w_q$ and by elementary equivalence transformations and permutations on $P_1,\dots,P_{\bar{\ell}}$ we can transform (2.33) locally around $\bar{w}$ into an equivalent system $\tilde{P}_1 = \dots = \tilde{P}_{\bar{\ell}} = 0$ satisfying

$$(2.35a)\qquad \sigma_1 \le \sigma_2 \le \dots \le \sigma_{\bar{\ell}}$$

$$(2.35b)\qquad \frac{\partial \tilde{P}_i}{\partial w_j^{(\sigma_i)}} = 0 \qquad j < i \qquad i,j = 1,\dots,\bar{\ell}$$

$$(2.35c)\qquad \frac{\partial \tilde{P}_i}{\partial w_i^{(\sigma_i)}} \neq 0 \qquad i = 1,\dots,\bar{\ell}$$

In analogy with the linear case (cf. [Ka],[Sc],[Wl]) an external system (2.33) satisfying (2.34) is called row-reduced, and if it satisfies (2.35) also column-reduced. In [S4] it has been shown that once an external differential system has been converted into such a doubly reduced form (i.e. satisfying (2.34) and (2.35)) then the conditions for the existence of a state space representation (originally deduced in [S1]), and the state space representation itself are easily obtained. More applications of these reduced forms are currently under investigation.

## References


[CL] P.E. Crouch, F. Lamnabhi-Lagarrigue, "State space realizations of nonlinear systems defined by input-output differential equations", Analysis and Optimization of Systems (eds. A. Bensoussan, J.L. Lions), INRIA, June 1988, LNCIS 111, Springer, Berlin, 1988, pp. 138-149.

[CMP] G. Conte, C.M. Moog, A. Perdon, "Un theorème sur la représentation entrée-sortie d'un système nonlinéaire", to appear in C.R. Acad. Sci. Paris.

[D] S. Diop, "A state elimination procedure for nonlinear systems", preprint 1988, presented at the Nantes Nonlinear Control Conf. June 1988.

[DM] J. Descusse, C.H. Moog, "Dynamic decoupling for right-invertible nonlinear systems", Syst. & Contr. Lett., 8 (1987), pp. 345-349.

[F1] M. Fliess, "A note on the invertibility of nonlinear input-output systems", Systems & Control Lett. 8 (1986), pp. 147-151.

[F2] M. Fliess, "Nonlinear control theory and differential algebra", Modelling and Adaptive Control (C.I. Byrnes, A. Kurzhanski, eds.), LNCIS 105, Springer, Berlin, 1988, pp. 134-145.

[F3] M. Fliess, "Automatique et corps differentiels", to appear Forum Math. 1, 1989.

[G] S.T. Glad, Nonlinear state space and input output descriptions using differential polynomials, preprint 1988, presented at the Nantes Nonlinear Control Conf., June 1988.

[Ka] T. Kailath, Linear Systems, Prentice-Hall, Englewood Cliffs, 1980.

[P] J.F. Pommaret, "Géométrie différentielle algébrique et théorie du contrôle", C.R. Acad. Sci. Paris, t. 302, Série I, no 15, 1986, pp. 547-550.

[R] H.H. Rosenbrock, State Space and Multivariable Theory, Wiley, New York, 1970.

[Re] W. Respondek, "Transforming a nonlinear control system into a differential equation in the inputs and outputs", in preparation.

[S1] A.J. van der Schaft, "On realization of nonlinear systems described by higher-order differential equations", Math. Systems Theory 19 (1987), pp. 239-275; Correction, Math. Systems Theory, 20 (1987), pp. 305-306.

[S2] A.J. van der Schaft, System Theoretic Descriptions of Physical Systems, CWI Tract 3, CWI, Amsterdam, 1984.

[S3] A.J. van der Schaft, "Representing a nonlinear state space system as a set of higher-order differential equations in the inputs and outputs", Memo nr. 698, University of Twente, March 1988, to appear in Syst. Contr. Letters.

[S4] A.J. van der Schaft, "Transformations of nonlinear systems under external equivalent, Memo nr. 700, University of Twente, March 1988.

[Sc] J.M. Schumacher, "Transformations of linear systems under external equivalence", Lin. Alg. and its Appl., 102 (1988), pp. 1-33.

[So] E.D. Sontag, "Bilinear realizability is equivalent to existence of a singular affine differential I/O equation", to appear in Syst. Contr. Lett.

[Sp] M. Spivak, Differential Geometry, Vol. 1, Publish or Perish, Boston, 1970.

[W1] J.C. Willems, "Input-output and state-space representations of finite dimensional linear time-invariant systems", Linear Algebra and its Appl., 50 (1983), pp. 581-608.

[W2] J.C. Willems, "From time series to linear systems", Part I: Finite dimensional linear time invariant systems", Automatica, 22 (1986), pp. 261-580.

[W3] J.C. Willems, "System theoretic models for the analysis of physical systems", Ricerche di Automatica, 10, 1979, pp. 71-106.

# TRACKING THROUGH SINGULARITIES

F. LAMNABHI-LAGARRIGUE(*), P.E. CROUCH(**) and I. IGHNEIWA(**)

(*)Laboratoire des Signaux et Systèmes
CNRS - ESE
Plateau du Moulon
91190 GIF-SUR-YVETTE, FRANCE

(**)Department of Electrical and Computer Engineering
Arizona State University
TEMPE, AZ 85287, U.S.A.

Results on the invertibility of nonlinear systems and their implication in the study of the output tracking problem, i.e., the determination of whether the output of the system can be made to follow a preassigned trajectory over a specific interval of time, are now well known (see Hirschorn [4] and the references therein or the recent paper by Grasse [3]). In these studies, nonlinear systems are either defined by a state-space representation or by a set of input-output differential equations. However, these approaches can break down when singularity is present. Recently Hirschorn and Davis [5] studied output tracking with "singular points": if the singularity is assumed to be at the origin, the authors give necessary and sufficient conditions involving, as in the regular case, only derivatives at the origin of the output function to be tracked, in terms of an integer $\beta$ called the degree of singularity and depending on the singular point. However, their paper gives little indication of the complexity of the problem and does not, for example, raise the existence issue.

In this paper we give a new formulation of the problem which directly involves existence, uniqueness and regularity issues. An input-output formulation of the singular output tracking problem is proposed, although the details of this new concept will be discussed in a future paper. The main results are then illustrated with some worked examples.

## 1. TRACKING PROBLEM

Consider the single-input, single output affine nonlinear system

$$\Sigma \qquad \begin{aligned} \dot{x} &= f(x) + u(t)g(x) \\ y &= h(x), \end{aligned}$$

$x \in \mathbf{R}^n$, $x(0) = x_0$. All data is assumed to be real analytic.

Assuming the accessibility property at $x_0$, the output tracking problem is the identification of functions $y_d(t)$ (desired output) which can appear as output for the system $\Sigma$, $y_d(.) = y(., u, x_0)$ for some admissible control u, and the construction of a control $u_d$ for which $y_d(.) = y(., u_d, x_0)$.

If $\alpha$ denotes the relative degree, i.e., the least non-negative integer k such that $y^{(k)}$ depends explicitly on the control u, we have

$$y^{(\alpha)} = f^{\alpha}h(x) + u(t)\, gf^{\alpha-1}h(x)$$

Let $S = \{x;\ gf^{\alpha-1}h(x) = 0\}$.It can be shown that

**Theorem 1**: Hirschorn [4]. If $x_0 \notin S$, $y_d$ is analytic on $[0, \infty)$ and

$$y_d^{(j)}(0) = f^{j}h(x_0)\,, \qquad 0 \le j \le \alpha\text{-}1,$$

then there exists an admissible control u such that $y_d$ can be tracked on some interval $[0, t_s)$, $t_s > 0$. One can use

$$u(t) = \frac{y_d^{(\alpha)}(t) - f^{\alpha}h(x(t))}{gf^{\alpha-1}h(x(t))} := u_D(x(t), y_d^{(\alpha)}(t), t)$$

The set S is called the set of singular points for the output tracking. In the following we denote by $x_d(t)$ the closed loop response to $u_D$ and by $u_d(t)$ the corresponding open loop control

$$u_d(t) := u_D(x_d(t), y_d^{(\alpha)}(t), t).$$

Note that at time $t_s$ either x(t) "blows up" or $x(t_s) = x_s \in S$. In the following we are interested only in the case where $x_s \in S$.

**Definition 1.** $y_d$ can be *tracked through the singularity* $x_s \in S$ at time $t_s$ if $x_d$ and $u_d$ can be extended to $[0, t')$, $t' > t_s$ such that $x_d(t) \notin S$, $t \neq t_s$ and $x_d(t_s) = x_s$ so that $y_d$ is tracked on $[0, t')$.

As in the regular case, let us substitute $u_D$ into our system. We obtain a *singular* system of differential equations

$$gf^{\alpha-1}h(x)\,\dot{x} = gf^{\alpha-1}h(x)\, f(x) + (y_d^{(\alpha)}(t) - f^{\alpha}h(x))\, g(x).$$

Unfortunately the literature on singular equations is complex even in the linear case. They are mostly studied in the complex domain:

$$e(x,t)\,\dot{x} = F(x,t)\ , \qquad x \in \mathbf{C}^n,\ t \in \mathbf{C} \text{ when e and F are analytic.}$$

Let $S = \{(x,t);\ e(x,t) = 0\}$. If $(x(0),0) \notin S$ the existence, the uniqueness and the analyticity of the solution can be shown, but when $(x(0), 0) \in S$ none of these properties are true in general.

What does this mean and what can we say for the tracking problem? This problem was first studied by Hirschorn and Davis [5] who introduced the "degree of singularity" $\beta(x_s)$, depending on the singular point $x_s$. In the following we give a new formulation of the problem by introducing a new integer $\beta(x_s, y_d)$ depending not only on the singular point $x_s$ but also on the output to be tracked.

## 2. SINGULAR TRACKING PROBLEM

Let us first introduce some notation. Let us denote by $a_j(x)$ the functions $f^j h(x)$, $0 \leq j \leq \alpha$, $b(x)$ the function $gf^{\alpha-1}h(x)$, and $a_{\alpha+k}(x)$ the functions

$$a_\alpha^{(k)}(x) + \sum_{j=0}^{k-1} \binom{k}{j} u^{(j)} b^{(k-j)}(x) , \quad k > 0$$

so that one can write

$$y^{(\alpha)} = a_\alpha(x) + u\, b(x)$$

and

$$y^{(\alpha+k)} = a_{\alpha+k}(x, u, \ldots, u^{(k-1)}) + u^{(k)} b(x), \quad k > 0.$$

Moreover, $\bar{u}^k := (u, \dot{u}, \ldots, u^{(k)})'$, $\bar{y}^k := (y, \dot{y}, \ldots, y^{(k)})'$ and $\bar{a}_k := (a_0, \ldots, a_k)'$. In the following we assume $x_0 \in S$. Therefore we need to track *$y_d$ through $x_0$ at time $t_s = 0$.*

**Definition 2.** Let $\beta(x_0, y_d) \in Z^+ \cup \{+\infty\}$ be the *maximum* integer (if it exists) such that there exists $\bar{u}^{\beta-1} \in R^\beta$ satisfying

$$\bar{y}_d^{\alpha+\beta}(0) = \bar{a}_{\alpha+\beta}(x_0, \bar{u}^{\beta-1}) \tag{1}$$

We call $\beta = \beta(x_0, y_d)$ the *rank of singularity at $x_0$ with $y_d$* and we denote the set of solutions $\bar{u}^{\beta-1}$ satisfying (1) by $E(x_0, y_d)$.

**Remarks:** i) (1) includes the identities

$$y_d^j(0) = f^j h(x_0) \quad , \quad 0 \leq j \leq \alpha - 1 ,$$

as in the case $x_0 \notin S$.

ii) This definition *does not* coincide with the definition of the degree of singularity given by Hirschorn and Davis [5]. Indeed, consider

$$\Sigma_{x_0, y_d} \quad \begin{cases} \dot{x}_1 = 1 \\ \dot{x}_2 = -2u \\ \dot{x}_3 = x_2 + u x_1 \end{cases}$$
$$y = x_3$$

where $\alpha = 1$ and $S = \{(x_1, x_2, x_3);\ x_1 = 0\}$. Let $x_0 = (0, 0, 1)' \in S$. The degree of singularity is $\beta(x_0) = 1$;

but in general if $\dot{y}_d(0) = \ddot{y}_d(0) = 0$ and $y_d(0) = 1$, then from definition 2, $\beta(x_0, y_d) = \infty$. However, if $\ddot{y}_d(0) \neq 0$, $\dot{y}_d(0) = 0$ and $y_d(0) = 1$, then $\beta(x_0, y_d) = 1$. In general the rank of singularity $\beta(x_0, y_d)$ is greater than $\beta(x_0)$.

**Definition 3.** The following system derived from $\Sigma$ and denoted $\Sigma_{x_0, y_d}$

$$\begin{cases} \dot{x} = f(x) + u_0 g(x) \\ \dot{u}_0 = u_1 \\ \vdots \\ \dot{u}_{\beta-2} = u_{\beta-1} \\ \dot{u}_{\beta-1} = \dfrac{y_d^{(\alpha+\beta)}(t) - a_\beta(x, u_0, \ldots, u_{\beta-1})}{b(x)} \end{cases}$$
$$y = h(x)$$

where $u_0(t) = u(t)$, $x(0) = x_0$ and $\bar{u}^{\beta-1}(0) = (v_0, v_1, \ldots, v_{\beta-1})' = \bar{v}^\beta \in E(x_0, y_d)$ is called the *singular system associated with $x_0 \in S$ and $y_d$.*

We can now state the necessary conditions for singular output tracking:

**Theorem 2.** Assuming that the singular system associated with $x_0 \in S$ and $y_d$ has *a* solution passing through $(x_0, \bar{v}^\beta)'$ then the output $y(t)$ equals $y_d(t)$ on some interval $(-\varepsilon, \varepsilon)$ containing the origin.The *maximum regularity* associated to the solutions $( u_d, x_d )$of the singular tracking problem with $x_0$ *and* $y_d$ given is $u_d \in C^{\beta-1}$ and $x_d \in C^\beta$. The solutions are not unique in general.

**Proof.** The necessary conditions are easily checked using Definition 1 for the extended system $\Sigma_{x_0 y_d}$ and differentiating $y(t) = h(x(t))$, $\alpha + \beta$ times.

**Remarks:** i) $\beta(x_0, y_d) = \infty$ is admissible, in which case we may obtain $C^\infty$ solutions, if any exist.

ii) We may obtain not only discrete numbers of solutions. Indeed, assume

$$\bar{y}_d^{\alpha+k}(0) = \bar{a}_{\alpha+k}(x_0, \bar{u}^{k-1}), \quad k \geq 0.$$

Actually this identity involves the derivatives

$$u^{(k_0)}, u^{(k_1)}, \ldots, u^{(k_\gamma)} \text{ with } 0 \leq k_0 < k_1 < \ldots < k_\gamma \leq \ell - 1, \gamma \leq \ell - 1$$

and the Jacobian of the mapping

$$(u^{(k_0)}, u^{(k_1)}, \ldots, u^{(k_\gamma)}) \to \bar{a}_{\alpha+\ell}(x_0, \bar{u}^{\ell-1})$$

has maximal rank $r(\ell)$. Then if a solution $C^\infty$ exists there will be a $\max_\ell [\ell - r(\ell)]$ *parameter* family of solutions. In our examples this number of parameters turns out to be equal to $\beta - r(\beta)$ even when solutions are not $C^\infty$ because of the linear (time-varying) nature of the systems.

## 3. INPUT-OUTPUT FORMULATION

Recently several authors have underlined the importance of representing nonlinear systems by sets of differential equations in u and y, e.g. $y^{(n)} = F(\bar{y}^{n-1}, \bar{u}^{m})$. Before illustrating Theorem 2 with some examples, we will consider some aspects of the singular tracking problem in this context. These ideas are briefly outlined, as a more detailed discussion will be presented in a future paper. In order to understand how the singularities appear in the input-output representation, we must first clarify the method of obtaining such a representation from the state-space representation of $\Sigma$. Van der Schaft recently proposed [7] an algorithm for solving this question from a general state space representation. But in our simpler context (single-input, single-output) [1], this transformation is much easier and can be solved directly (see also Glad [2]). Let us reconsider the system $\Sigma$. As previously, let us write

$$\bar{y}^{(\alpha+k)} := a_{\alpha+k}(x, \bar{u}^{k-1}) + b(x)u^{(k)}, \qquad k \geq 0,$$

where $\alpha$ is the relative degree. This can also be written

$$\bar{y}^{\alpha+k} := \hbar_{\alpha+k}(x, \bar{u}^{k}) \text{ and in particular , if } m = n - \alpha,\ \bar{y}^{n-1} := \hbar_{n-1}(x, \bar{u}^{m-1}).$$

**Definition 4.** $\hat{x}$ is said to be an input-output singularity if

$$\text{Rank } D_x\hbar_{n-1}(\hat{x}, \bar{u}^{m-1}) < n, \text{ for all } \bar{u}^{m-1} \in \mathbf{R}^m.$$

Let IS be the set of all input-output singularities.

If $\hat{x} \notin$ IS, let $\hat{x} = d(\bar{y}^{n-1}, \bar{u}^{m-1})$ be the unique solution of $\bar{y}^{n-1} = \hbar_{n-1}(\hat{x}, \bar{u}^{m-1})$ where it exists, $(\bar{u}^{m-1}, \bar{y}^{n-1}) \in D_n \subset \mathbf{R}^m \times \mathbf{R}^n$.

Since

$$y^{(n)} = a_n(x, \bar{u}^{m-1}) + b(x)u^{(m)}$$

or

$$y^{(n)} = a_n(d(\bar{y}^{n-1}, \bar{u}^{m-1}), \bar{u}^{m-1}) + b(d(\bar{y}^{n-1}, \bar{u}^{m-1}))u^{(m)}.$$

We obtain an input-output equation of the form

$$f(\bar{y}^{n-1}, \bar{u}^{m-1})u^{(m)} = y^{(n)} - e(\bar{y}^{n-1}, \bar{u}^{m-1}) \qquad (2)$$

with $(\bar{u}^{m-1}, \bar{y}^{n-1}) \in D_n$, and where e and f are analytic functions.

Note that

$$f(\bar{y}^{n-1}, \bar{u}^{m-1}) = 0 \text{ when } \hat{x} = d(\bar{y}^{n-1}, \bar{u}^{m-1}) \in S.$$

A similar input-output equation may be obtained when $x \in IS$ by considering higher order derivations of y. In both cases we again obtain singular differential equations:

$$f^*(\bar{y}^{N-1}, \bar{u}^{M-1})u^{(M)} = y^{(N)} - e^*(\bar{y}^{N-1}, \bar{u}^{M-1}),$$

where $(\bar{y}^{N-1}, \bar{u}^{M-1}) \in \mathbf{R}^M \times \mathbf{R}^N$, $N \geq n$, $M \geq m$, $N-M=\alpha$, $f^*$ and $e^*$ are analytic functions, and

$$S = \{(\bar{y}^{N-1}, \bar{u}^{N-1}) / f^*(\bar{y}^{N-1}, \bar{u}^{M-1}) = 0\}.$$

To study the tracking problem we first substitute $y = y_d$ into (2) to find *candidate* open loop controls $u = u_d$. However, the relation

$$y^{(k+\alpha)} = a_{\alpha+k}(x, \bar{u}^{k-1}) + b(x)u^{(k)}$$

shows that the initial states are uniquely defined only through $x_0 \notin S$.

A general scheme will therefore be the following:

i) Find general solutions for open loop controls $u_d$

ii) Solve the state equations of $u_d$ to obtain candidate trajectories $x_d$.

iii) Find the solutions to singular tracking by testing the candidate solutions against identity, i.e.

$$u_d(t) = u_D(x_d(t), y_d^{(\alpha)}(t), t),$$

when $x_d(t) \notin S$.

iv) Investigate the behaviour at singular points.

**Example 1.** Consider the system $\Sigma$ with $x \in \mathbf{R}^2$,

$$f(x) = (0, x_1^2),\ g(x) = (1, 0),\ h(x) = x_2.$$

By direct computations, $S = IS = \{(x_1, x_2);\ x_1 = 0\}$ and $u_D = \ddot{y}/_{2x_1}$.

For $x_0 \notin IS$ we obtain the input-output equations

$$\ddot{y} = 2u\sqrt{|\dot{y}|}\ ,\ \ddot{y} = -2u\sqrt{|\dot{y}|}\ ,\ \dot{y} \neq 0.$$

For $x_0 \in IS$ the input-output equation is

$$\dddot{y} = 2u^2 + \frac{\dot{u}\ddot{y}}{u}\ ;\ u \neq 0.$$

Let $x_0 = (0, 0)' \in S$.

To compute $\beta(x_0, y_d)$, note that $y = x_2$, $\dot{y} = x_1^2$, $\ddot{y} = 2x_1u$, $\dddot{y} = 2u^2 + 2x_1\dot{u}$, $y^{(4)} = 6u\dot{u} + 2x_1\ddot{u}$, $y^{(5)} = 8u\ddot{u} + 6(\dot{u})^2 + 2x_1\dddot{u}$ .... Therefore in general,

i) if $\dddot{y}_d(0) > 0$, $\beta(x_0, y_d) = \infty$

ii) if $\dddot{y}_d(0) < 0$, $\beta(x_0, y_d) = 0$

iii) if $\dddot{y}_d(0) = 0$, we need further information.

Let $y_d = t^3/3$, from Theorem 1, we expect to be able to construct $C^\infty$ solutions (case i)). Let us consider the closed loop system

$$\begin{cases} \dot{x}_1 = t/x_1 \\ \dot{x}_2 = x_1^2 \end{cases} \qquad x_0 = (0, 0)'$$

Solving this system yields $C^\infty$ solutions; $x_{1d}(t) = \pm t$, $x_{2d}(t) = t^3/3$ and $u_d(t) = \pm 1$.

Now, let $y_d(t) = t - \sin t$, $u_D = \sin t/2x_1$. In this case we may also expect $C^\infty$ solutions. In fact the solutions of the closed loop system are

$$x_{1d}(t) = \pm\sqrt{2}\sin t/2,\ x_{2d}(t) = t - \sin t, \text{ and } u_d(t) = \pm(\cos t/2)/\sqrt{2}.$$

Clearly $x_d(t) \in S$ for $t = 2n\pi$, $n \in Z^+$. Therefore we conclude here that the system tracks through an infinite number of singularity points.

**Example 2** [5]. Consider the system $\Sigma$ with $x \in \mathbf{R}^3$,

$$f(x) = (1, 0, x_2),\ g(x) = (0, -2, x_1) \text{ and } h(x) = x_3.$$

In that case $\alpha = 1$, $u_D = (\dot{y} - x_2)/x_1$, $S = \{(x_1, x_2, x_3);\ x_1 = 0\}$, $IS = \emptyset$ and an input-output equation is $\dddot{y} = \frac{(\ddot{y}+u)}{\dot{u}}\ddot{u}$, $\dot{u} \neq 0$. Let $x_0 = (0, 0, 1)' \in S$. In general, if $\dot{y}_d(0) = \ddot{y}_d(0) = 0$, $y_d(0) = 1$, then $\beta(x_0, y_d) = \infty$ and we expect to be able to construct $C^\infty$ solutions.

However, let $y_d(t) = 1 + t^2 + t^3/3$. Here $\ddot{y}_d(0) \neq 0$ but $y_d(0) = 1$ and $\dot{y}_d(0) = 0$ so $\beta(x_0, y_d) = 1$. Therefore we expect solutions $x_d \in C^1$ and $u_d \in C^0$ to exist but not *more*. In this case the closed loop equations have the form

$$\dot{x}_2 - \frac{2x_2}{t} = -4 - 2t \quad, x_2(0) = 0 \tag{3}$$

with the solutions

$$\begin{cases} x_{2d}(t) = 4t - 2t^2(\ln t + c_1), & u_d(t) = -2 + t + 2t(\ln t + c_1),\ t \geq 0 \\ x_{2d}(t) = 4t + 2t^2(\ln\text{-}t + c_2), & u_d(t) = -2 + t\text{-}2t(\ln\text{-}t + c_2),\ t \leq 0 \end{cases}$$

$c_1, c_2 \in \mathbf{R}$.

**Remarks:** Note that equation (3) is a linear time-varying singular equation. In general when the closed set of equations is a linear time-varying set of singular equations we may have recourse to classical results such as the Fuchsian theory and/or the Frobënius method (see Ince [6], for instance).

Consider now $y_d(t) = 1 + t^2/4$. In this case we expect $C^\infty$ solutions. Indeed, the closed loop equations become

$$\dot{x}_2 - \frac{2x_2}{t} = -1, \qquad x_2(0) = 0.$$

Thus $x_{2d}(t) = t + bt^2$, $u_d(t) = -\frac{1}{2} - bt$, $b \in \mathbf{R}$. The arbitrary constant b is due to the fact that (setting $\beta$=2)

$$\bar{a}_{\alpha+\beta} = (x_3, x_2, -u, x_1)' \text{ and}$$
$$\bar{a}_3(0, \bar{u}^1) = (1, 0, -u, 0)'.$$

Thus the map $\bar{u}^1 \to \bar{a}_3(0, \bar{u}^1)$ has rank r=1 and the number of free parameters is $\beta - r = 1$. In this case we may also obtain the solutions using the input-output differential equation

$$\ddot{u} - \frac{\dot{u}\dddot{y}}{\ddot{y}+u} = 0.$$

If $y_d(t) = 1 + t^2/4$, we see that the candidate open loop controls are of the form

$$u_d(t) = c + dt.$$

The corresponding state equation $\dot{x}_2 = -2u$, $x_2(0) = 0$ gives

$$x_{2d}(t) = -2ct - dt^2.$$

Now the constraint $u_d(t) = u_D(x_d(t), y_d^{(\alpha)}(t), t)$ becomes

$$c + dt = \frac{t/2 - (-2ct - dt^2)}{t} = \frac{1}{2} + 2c + dt.$$

thus $c = -\frac{1}{2}$ and d is arbitrary as before.

**Example 3.** Consider now the system $\Sigma$ with

$$f(x) = (1, 0, x_2)', \quad g(x) = (0, -1, x_1)', \quad h(x) = x_3.$$

In this example,

$$\alpha = 1; \quad u_D = \frac{(\dot{y} - x_2)}{x_1}; \quad S = \{(x_1, x_2, x_3); x_1 = 0\} \text{ and } IS = \emptyset.$$

An input-output representation can be written $\dddot{y} = \frac{\ddot{u}\ddot{y}}{\dot{u}} + \dot{u}$, $\dot{u} \neq 0$. Let $x_0 = (0, 0, 0)' \in S$. In general, if $y_d(0) = \dot{y}_d(0) = \ddot{y}_d(0) = 0$, we have $\beta(x_0, y_d) = \infty$. Let $y_d(t) = t^4$. In this case we expect $C^\infty$ solutions. The closed loop equations give

$$\dot{x}_2 - \frac{x_2}{t} = -4t^2, \qquad x_2(0) = 0$$

whose solutions can be written

$$x_{2d}(t) = -2t^3 + ct, \quad c \in \mathbf{R}$$
$$u_d(t) = 6t^2 - c.$$

Note that $u_d(0)$ is completely arbitrary!

As previously, if $\beta=2$, $\bar{a}_{\alpha+\beta}(x, \bar{u}^{\beta-1}) = (x_3, x_2, 0, \dot{u})'$ and $\bar{a}_3(0, \bar{u}^1) = (0, 0, 0, \dot{u})'$. Moreover, Rank $(\bar{u}^1 \rightarrow (0, 0, 0, \dot{u}))$ is $r=1$. Thus again $\beta - r = 2 - 1 = 1$.

Consider now the function $y_d(t) = t^2$, then $\beta(x_0, y_d) = 0$. Thus from Theorem 2 the maximum regularity we expect for $x_d$ is $C^0$ and not even boundedness for $u_d$! (Note that this is not technically an **admissible** solution of the tracking problem, although $x_d(t)$ remains bounded). Here the closed loop equation $\dot{x}_2 - \frac{x_2}{t} = -2$, $x_2(0) = 0$ yields the solutions

$$\begin{cases} x_{2d}(t) = at - 2t^2 \ln t, & u_d(t) = -a + 2 + 2\ln t, \ t > 0 \\ x_{2d}(t) = at + 2t \ln -t, & u_d(t) = -a - 2 - 2\ln -t, \ t < 0, \ a \in \mathbf{R}. \end{cases}$$

**Example 4.** Consider the system $\Sigma$ with

$$f(x) = (1, 0, x_2), \quad g(x) = (0, (x_1 - 1), x_1)', \quad h(x) = x_3.$$

Easy computations show that $\dot{y} = x_2 + x_1 u$; $\ddot{y} = x_1(u + \dot{u})$; $\dddot{y} = (u + \dot{u}) + x_1(\dot{u} + \ddot{u})$ and in general

$$y^{(k)} = ((k-2)\,(u^{(k-3)} + u^{(k-2)}) + x_1(u^{(k-2)} + u^{(k-1)}),\ \ k \geq 2.$$

We also have $\alpha = 1$, $u_D = \dfrac{\dot{y} - x_2}{x_1}$; $S = \{(x_1, x_2, x_3)';\ x_1 = 0\}$ and $IS = \emptyset$. Let $x_0 = (0, 0, 0)' \in S$. In general, if $y_d(0) = \dot{y}_d(0) = \ddot{y}_d(0) = 0$, we obtain $\beta(x_0, y_d) = \infty$. Thus we expect $C^\infty$ solutions.

Let $y_d(t) = t^3$. We then have

$$x_{2d}(t) = 3t^2 - t\,(6 + ae^{-t}),\quad u_d(t) = 6 + ae^{-t}.$$

In this case we may compute the number of free parameters for an arbitrary choice of $\beta$. Let $\beta = k - 1$,

$$\bar{a}_{\alpha+\beta}(x_0, \bar{u}^{\beta-1}) = \bar{a}_k(0, \bar{u}^{k-2}) = (0, \dots, 0, u + \dot{u}, \dots, (k-2)(u^{(k-3)} + u^{(k-2)})).$$

Thus Rank $\bar{u}^{k-2} \rightarrow \bar{a}_k(0, \bar{u}^{k-2})$ is $r=k-2$. Therefore we expect $\beta - r = (k-1) - (k-2) = 1$ free parameter.

## REFERENCES


[1] P.E. CROUCH and F. LAMNABHI-LAGARRIGUE, State-space realization of nonlinear systems defined by input-output differential equations, Proc. INRIA Conf., Antibes, Lect. Notes Contr. Inform. Sc., 111, 1988, pp.138-149.

[2] S.T. GLAD, Nonlinear state-space and input-output descriptions using differential polynomials. These Conference Proceedings.

[3] K.A. GRASSE, Sufficient conditions for the functional reproducibility of time-varying input-output systems, SIAM J. Contr. Optimiz., 26, pp.230-249, 1988.

[4] R.M. HIRSCHORN, Output tracking in multivariable nonlinear systems, IEEE AC, 26, pp.593-595, 1981.

[5] R.M. HIRSCHORN and J. DAVIS, Output tracking for nonlinear systems with singular points, SIAM J. Contr. Optimiz., 25, pp.547-557, 1987.

[6] E.L. INCE, Ordinary differential equations. Dover Pubs., New York, 1956.

[7] A.J. VAN DER SHAFT, Representing a nonlinear state space system as a set of higher-order differential equations in the inputs and outputs, Proc. Nantes Conf., These Conference Proceedings.

# CONTROLLABILITY OF RIGHT-INVARIANT SYSTEMS ON SEMI-SIMPLE LIE GROUPS

by: **R. EL ASSOUDI**
Laboratoire d'Automatique de Grenoble
BP 46
38402 SAINT-MARTIN-D'HERES (France)

and: **J.P. GAUTHIER**
Université Claude Bernard - LYON 1
Laboratoire d'Automatique - ISIDT
43 boulevard du 11 Novembre 1918
69022 VILLEURBANNE (France)

**Abstract**: We deal with controllability of right-invariant control systems on Lie groups. Following the ideas of a series of papers by JURDJEVIC-KUPKA, GAUTHIER-BORNARD, GAUTHIER-KUPKA-SALLET, LEITE-CROUCH, EL ASSOUDI-GAUTHIER, we obtain a generalization of these results in the general case of real semi-simple Lie groups.

**Keywords**: Semi-simple Lie algebras, Lie groups, Controllability, Invariant vector fields, Root systems.

**AMS subject classification**: Primary : 93 B05, 17B20.
Secondary : 93C10, 22E46.

## I - INTRODUCTION

For the statement of the problem, we refer to the notations defined in the appendix below.

We deal with controllability of right-invariant control affine families of vector fields $\Gamma = \{A + uB \mid u \in R\}$ on a real semi-simple Lie group G, with finite center, whose Lie algebra is denoted L . $((A,B)) \in L^2$.

We assume that :

($H_1$) B is strongly regular.

($H_2$) Let s be maximal in Sp (B) (in the sense of the appendix) :

- if rel (s) $\neq 0$, $A(s)$ and $A(-s) \neq 0$.
- if Im (s) $= 0$, $\mathrm{Kil}(A(s), A(-s)) < 0$.

($H_3$) If $\alpha$ is primitive, $A(\alpha) \neq 0$.

The origin of this work is the following theorem, from [J.K.] :

**Theorem 1** : If ($H_1$), ($H_2$), ($H_3$) hold, $\Gamma$ is controllable (that is to say the semi group $\Gamma^+$ generated by $\{ \exp t x \mid t \geq 0, x \in \Gamma \}$ is all of G).

This statement is not exactly the one that can be found in [J.K.]. It is the statement given in a preprint. A refinement of this statement is given in the published version).

Our aim in this paper is to prove that :

**Theorem 2** : If ($H_1$) and ($H_2$) hold, then a necessary and sufficient condition for controllability is Lie $(A, B) = L$ (Lie $(A, B)$ is the Lie subalgebra of L generated by A and B).

That is to say, the so called "controllability rank condition" is necessary and sufficent for controllability under assumptions ($H_1$), ($H_2$).

Notice that ($H_1$), ($H_3$) imply Lie $(A, B) = L$ (straightforward).

In the paper [G.B.], theorem 2 is proved in case of $S\ell(n, R)$, with adB real diagonalizable. In [G.K.S.], it is proved in case of split real forms of $A_r$, $D_r$, $E_6$, $E_7$, $E_8$ with the same assumption on B. In [L.C.] assumption ($H_1$) is slightly released in the cases $A_r$, $D_r$. In [EA.G], theorem 2 is proved in the case of any remaining split real form. Our goal in this paper is to prove theorem 2 for any simple, *split or not*, Lie group with finite center. The reader will easily devine the proof in the semi-simple case.

It is clear that the rank condition Lie $(A, B) = L$ is necessary. For sufficiency, we shall need some lemmas that either can be found in [J.K.] or are proved in section III herein.

## II - PROOF OF THE RESULTS

As in [EA.G], we split the root system S relative to Ker $(\mathrm{ad}_{\mathbb{C}} B) = h$ in subsets, in the following way. Let $S_1$ be the set of roots $\hat{\alpha}$ such that $\hat{\alpha} + \hat{s}$ or $\hat{\alpha} - \hat{s}$ is also a root. Since $S = -S$, $S_1 = -S_1$, and $S_1^+$ will refer the set of positive elements in $S_1$.

Let $S_2 = S \setminus S_1 \cup \{\hat{s}, -\hat{s}\}$, $\hat{s}$ the maximal root (simple case).

Clearly $S_2 = -S_2$.

Let LS $(\Gamma)$ be the "Lie-Satured Cone" of $\Gamma$. That is the set of $x \in$ Lie $(A, B)$ such that $\{\exp tx \mid t \geq 0\} \subset C\ell\,(\Gamma^+)$ ($C\ell$ = Closure).

Let $C_1$ be the Lie subalgebra of L generated by L (s), L (-s) and the L $(\alpha)$, L $(\alpha) \subset$ LS $(\Gamma)$ and $\hat{\alpha} \in S_1$.

Let $C_2$ be the vector space generated by the L $(\gamma)$, $\hat{\gamma} \in S_2$ and A $(\gamma) \neq 0$.

Let $C_0$ = span {A (0), B}.

Let $\tilde{L}$ be the Lie algebra generated by $C_0$, $C_1$ and $C_2$.

We want to prove :

**Proposition** :

a) $\tilde{L}$ = Lie (A, B).
b) $C_1$ is an ideal in Lie (A, B).

**Proof** :

a) holds with a similar proof to that given in [EA.G.].

To prove b), we shall need the lemmas of section III.
We shall also need some properties of root spaces, that we list just below, with the notations of the appendix :

$L^{\mathbb{C}} = L_0^c \oplus \bigoplus_{\alpha \in Sp(\beta)} L_\alpha^c$, $L = L(0) \oplus \bigoplus_{\alpha \in Sp^+(\beta)} L(\alpha)$, where $L(\alpha) = (L_\alpha^c + L_{\bar\alpha}^c) \cap L$.

- $[L(\alpha), L(\beta)] = L(\alpha+\beta) + L(\alpha+\bar\beta)$, if $\hat\alpha + \hat\beta$ and $\hat\alpha + \hat{\bar\beta}$ are roots ( if $\hat\alpha + \hat\beta \notin S$ and $\hat\alpha + \hat\beta \neq 0$ (resp. $\hat\alpha + \hat{\bar\beta}$ ), $L(\alpha+\beta) = 0$ (resp. $L(\alpha + \bar\beta) = 0$ ) ).

We will often use the following relation :

- $[L(\alpha), [L(\alpha), L(\beta)]] = L(2\alpha+\beta) + L(2\alpha+\bar\beta) + L(\alpha+\bar\alpha+\beta)$.

Since if $\hat\alpha + \hat{\bar\alpha} + \hat\beta$ and $\hat\alpha + \hat{\bar\alpha} + \hat{\bar\beta}$ are both roots, one has :

$\sigma(\hat\alpha + \hat{\bar\alpha} + \hat{\bar\beta}) = \hat{\bar\alpha} + \hat\alpha + \hat{\bar\beta} = \hat\alpha + \hat{\bar\alpha} + \hat{\bar\beta}$.

__Proof of b)__ :

It is sufficient to prove that $[C_1, C_2] \subset C_1$. We verify this property on the generators of $C_1$ and $C_2$.

Let $L(\alpha) \subset C_1$, then either $\hat\alpha \in S_1$ or $\hat\alpha \in \{-\hat s, +\hat s\}$ and $L(\alpha) \subset LS(\Gamma)$.

Let $L(\gamma) \subset C_2$ ; hence $\hat\gamma \in S_2$ with $A(\gamma) \neq 0$.

- If $\hat\alpha \in S_1$, by lemma 4 we get : $[L(\alpha), L(\gamma)] \subset LS(\Gamma)$.

  By lemma 1, b), $\hat\alpha + \hat\gamma$, either is not a root or is in $S_1$.

  If $\hat\alpha + \hat\gamma$ is not a root, since $S_1 = -S_1$ and $S_1 \cap S_2 = \emptyset$, $\hat\alpha + \hat\gamma$ is also non zero, then $L(\alpha+\gamma) = 0 \subset C_1$.

  If $\hat\alpha + \hat\gamma$ is a root, since it is in $S_1$, $L(\alpha+\gamma)$ is in $C_1$.

  Now, for $\hat\alpha + \hat{\bar\gamma}$, one has $\hat{\bar\gamma} \in S = S_1 \cup \{\hat s, -\hat s\} \cup S_2$. If $\hat{\bar\gamma}$ is in $S_2$, the same applies and $L(\alpha + \bar\gamma)$ is in $C_1$.

  If $\hat{\bar\gamma}$ is in $S_1$, $A(\bar\gamma) = A(\gamma)$, $A(\bar\gamma) \neq 0$, by lemma 3, $L(\bar\gamma) \subset LS(\Gamma)$.

  Then $L(\gamma) = L(\bar\gamma) \subset C_1$, but $C_1$ is an algebra, $[L(\alpha), L(\gamma)] \subset C_1$.

  If $\hat{\bar\gamma} = \pm \hat s$, by definition of $C_1$, $L(\bar\gamma) \subset C_1$ and clearly $[L(\alpha), L(\gamma)] \subset C_1$.

  At the end, $[L(\alpha), L(\gamma)] = L(\alpha+\gamma) + L(\alpha+\bar\gamma) \subset C_1$.

- If $\hat{\alpha} \in \{\hat{s}, -\hat{s}\}$, by definition of $S_2$, $\hat{\alpha} + \hat{\gamma}$ is not a root and $\hat{\alpha} + \hat{\gamma}$ is non zero, then $L(\alpha + \gamma) = 0 \subset C_1$. For $\hat{\alpha} + \hat{\bar{\gamma}}$, if $\hat{\bar{\gamma}} \in S_2$, similarly $L(\alpha + \bar{\gamma}) = 0$. If $\hat{\bar{\gamma}} \in S_1 \cup \{\hat{s}, -\hat{s}\}$, we have $L(\gamma) = L(\bar{\gamma}) \subset C_1$, then $[L(\alpha), L(\gamma)] \subset C_1$. □

**Proof of theorem 2 (sufficiency) :**

If Lie (A, B) = L, $C_1$ is an ideal of $\tilde{L}$ (by the proposition, part b), a)), hence of L. Otherwise, $C_1$ is non trivial (it contains at least L (s) (*)). Then $C_1 = L$ (L is simple). $L \subset LS(\Gamma)$, $\Gamma$ is controllable. □

## III - LEMMAS

**Lemma 1 :**

a) Let $\hat{\alpha}$ and $\hat{\beta}$ be roots in $S_1$, $\hat{\alpha}, \hat{\beta} > 0$ (resp. < 0). If $\hat{\alpha} + \hat{\beta}$ is also a root, it is equal to $\hat{s}$ (resp. $-\hat{s}$).

b) Let $\hat{\gamma} \in S_2$ and $\hat{\alpha} \in S_1^+$ (resp. $\hat{\alpha} \in S_1^-$). If $\hat{\alpha} + \hat{\gamma}$ is a root, it is in $S_1^+$ (resp. $S_1^-$).

**Proof :**

a) If $\hat{\alpha} \in S_1^+$, since $\hat{s}$ is the maximal root, $\hat{\alpha} + \hat{s}$ is not a root. Hence $\hat{\alpha} - \hat{s} = \hat{\alpha}'$ is a root. Similarly, if $\hat{\beta} \in S_1^+$, $\hat{\beta}' = \hat{\beta} - \hat{s}$ is a root. We have : $\hat{\alpha} + \hat{\beta} = \hat{s} + \hat{\alpha}' + \hat{s} + \hat{\beta}' = 2\hat{s} + \hat{\alpha}' + \hat{\beta}'$. But $2\hat{s} + \hat{\alpha}' + \hat{\beta}'$ is a root only if $\hat{\alpha}' + \hat{\beta}' = -\hat{s}$, hence if $\hat{\alpha} + \hat{\beta}$ is a root therefore $\hat{\alpha} + \hat{\beta} = 2\hat{s} - \hat{s} = \hat{s}$.

b) Let $\hat{\alpha} \in S_1^+$. $\hat{\alpha} - \hat{s} = \hat{\alpha}'$ is a root. If $\hat{\alpha} + \hat{\gamma} = \hat{\alpha}' + \hat{s} + \hat{\gamma}$ is a root, since $\hat{\gamma} + \hat{s}$ is not a root (by definition of $S_2$), $\hat{\alpha}' + \hat{\gamma}$ is a root : this is a particular case of the following properties of roots :

(**) If $\hat{\gamma}_1, \hat{\gamma}_2$ and $\hat{\gamma}_3$ are roots such that $\hat{\gamma}_1 + \hat{\gamma}_2$ and $\hat{\gamma}_1 + \hat{\gamma}_2 + \hat{\gamma}_3$ are also roots, then either $\hat{\gamma}_2 + \hat{\gamma}_3$ or $\hat{\gamma}_1 + \hat{\gamma}_3$ is a root.

With the notations of the appendix, by Jacobi identity :

$[[E_{\gamma_1}, E_{\gamma_2}], E_{\gamma_3}] + [[E_{\gamma_2}, E_{\gamma_3}], E_{\gamma_1}] + [[E_{\gamma_3}, E_{\gamma_1}], E_{\gamma_2}] = 0.$

If $\hat{\gamma}_2 + \hat{\gamma}_3 \notin S$ and $\hat{\gamma}_1 + \hat{\gamma}_3 \notin S$ we would have :

$[E_{\gamma_1}, E_{\gamma_3}] = [E_{\gamma_2}, E_{\gamma_3}] = 0$, which contradicts $[[E_{\gamma_1}, E_{\gamma_2}], E_{\gamma_3}] \neq 0$.

Therefore, if $\hat{\alpha}' + \hat{\gamma}$ is a root, $\hat{\alpha} + \hat{\gamma} - \hat{s} = \hat{\alpha}' + \hat{\gamma} \in S$. This implies $\hat{\alpha} + \hat{\gamma} \in S_1^+$. □

Lemma 2 was proved on p. 173, proposition 11, d) of [J.K.].

**Lemma 2** : Let $B \in L$ be strongly regular. Assume that $R\,B \subset LS\,(\Gamma)$.
If $X \in LS\,(\Gamma)$ and $-X \in LS\,(\Gamma)$ then $L\,(\alpha) \subset LS\,(\Gamma)$ $\forall\, \alpha \in Sp\,(B)$ such that $X\,(\alpha) \neq 0$.

**Lemma 3** : Let $A \in LS\,(\Gamma)$. For any $\hat{\alpha} \in S_1$, $A\,(\alpha) \neq 0$ implies $L\,(\alpha) \subset LS\,(\Gamma)$.

**Proof** :

Let $\hat{\alpha} \in S_1$. Then $\hat{\alpha} + \hat{s}$ or $\hat{\alpha} - \hat{s}$ is also a root. Assume that $\hat{\alpha} + \hat{s}$ is a root.
Pick $X\,(s) \in L\,(s)$, $X\,(s) \neq 0$, for any real v, since the edge of $LS\,(\Gamma)$ normalizes $LS\,(\Gamma)$, $\phi\,(v, X\,(s)) = \exp v \operatorname{ad} X\,(s)\,(A) = \sum_{i \geq 0} \frac{v^i}{i\,!} \operatorname{ad}^i X\,(s)\,(A) \in LS\,(\Gamma)$

But $R \operatorname{ad}^2 X\,(s)\,(A) = R \sum_{\gamma \in S^+p(B)} \operatorname{ad}^2 X\,(s)\,(A\,(\gamma)) \subset LS\,(\Gamma)$ since :

$\operatorname{Ad}^2 X\,(s)\,(A\,(\gamma)) \in [\,L\,(s), [L\,(s), L\,(\gamma)]\,] = L\,(2s + \gamma) + L\,(2s + \overline{\gamma}) + L\,(s + \overline{s} + \gamma)$

Since $2\hat{s}$ is not a root $2\hat{s} + \hat{\gamma} \neq 0$, $2\hat{s} + \hat{\overline{\gamma}} \neq 0$ and $\hat{s}$ is the maximal then $\hat{s} + \hat{\overline{s}} + \hat{\gamma} \neq 0$.

As $\hat{s}$ is the maximal root one has :

- $2\hat{s} + \hat{\gamma}$ is a root only if $\hat{\gamma} = -\hat{s}$ and then $L(\gamma) = L(-s) \subset LS(\Gamma)$
- Similarly, if $\hat{s} + \hat{\bar{s}} + \hat{\gamma}$ is a root, $rel(s + \bar{s} + \gamma) = 2\, rel(s) + rel(\gamma)$, then

  $rel(\gamma) = -\, rel(s)$, (by (*) in the appendix) $L(\gamma) \subset LS(\Gamma)$.

Therefore $R\, ad^2 X(s)(A) \subset LS(\Gamma)$. It is a straightforward consequence that

$R\, ad^i X(s)(A) \subset LS(\Gamma)$ for any $i \geq 2$.

Hence, $\phi(v, X(s)) - \sum_{i \geq 2} \frac{v^i}{i!}\, ad^i X(s)(A) \in LS(\Gamma)$.

This is equal to $A + v[X(s), A]$.

Since $LS(\Gamma)$ is a closed convex cone $\lim_{v \to \infty} \frac{A + v[X(s), A]}{|v|} = \pm [X(s), A]$ is in $LS(\Gamma)$.

By lemma 2, if $A(\alpha) \neq 0$, $L(\alpha+s) \subset LS(\Gamma)$.

But $L(\alpha) \subset [L(\alpha+s), L(-s)]$, hence $L(\alpha) \subset LS(\Gamma)$.

The case when $\hat{\alpha} - \hat{s}$ is a root is analogously obtained by considering $\phi(v, X(-s))$. □

**Lemma 4** : If $\hat{\alpha} \in S_1$ and $L(\alpha) \subset LS(\Gamma)$ then $[L(\alpha), L(\gamma)] \subset LS(\Gamma)$, for any root $\hat{\gamma} \in S_2$ such that $A(\gamma) \neq 0$.

**Proof** :

Let $\hat{\gamma} \in S_2$, $\hat{\bar{\gamma}}$ is also a root (since if $\gamma \in Sp(B)$, $\sigma(\gamma) = \bar{\gamma} \in Sp(B)$ ).

$\hat{\bar{\gamma}} \in S = S_1 \cup \{\hat{s}, -\hat{s}\} \cup S_2$.

- If $\hat{\bar{\gamma}} \in S_1$ (resp. $\hat{\bar{\gamma}} \in \{\hat{s}, -\hat{s}\}$ ) since $A(\gamma) \neq 0$, $A(\bar{\gamma}) \neq 0$, by lemma 3

  $L(\bar{\gamma}) = L(\gamma) \subset LS(\Gamma)$ (resp. $L(\bar{\gamma}) = L(\pm s) \subset LS(\Gamma)$. Clearly $[L(\alpha), L(\gamma)] \subset LS(\Gamma)$.
- Therefore we only have to consider the case where both $\hat{\gamma}$ and $\hat{\bar{\gamma}}$ are in $S_2$.

  Let $X(\alpha)$ be a non zero element of $L(\alpha)$. If we prove that $R\, ad^2 X(\alpha)(A) \subset LS(\Gamma)$, it will imply that, for any real v,

$$\phi(v, X(\alpha)) - \sum_{i\geq 2} \frac{v^i}{i!} \operatorname{ad}^i X(\alpha)(A) - A + v[X(\alpha), A] \in LS(\Gamma).$$

With lemma 2, we shall have $[L(\alpha), L(\gamma)] \subset LS(\Gamma)$ for $A(\gamma) \neq 0$.

Let us show that $R \operatorname{ad}^2 X(\alpha)(A) \subset LS(\Gamma)$.

We are only interested with the terms $R \operatorname{ad}^2 X(\alpha)(A(\gamma))$ when $\hat{\gamma}$ and $\hat{\bar{\gamma}}$ are in $S_2$.

But $\operatorname{ad}^2 X(\alpha)(A(\gamma)) \in [L(\alpha), [L(\alpha), L(\gamma)]] - L(2\alpha + \gamma) + L(2\alpha + \bar{\gamma}) + L(\alpha + \bar{\alpha} + \gamma)$.

Clearly $2\hat{\alpha} + \hat{\gamma} \neq 0$, $2\hat{\alpha} + \hat{\bar{\gamma}} \neq 0$ since $2\hat{\alpha}$ is not a root. Also $\hat{\alpha} + \hat{\bar{\alpha}} + \hat{\gamma} \neq 0$ since $\hat{\alpha} \in S_1$, $\hat{\gamma} \in S_2$, by lemma 1, b) $\hat{\alpha} + \hat{\gamma}$ and $\hat{\alpha}$ have the same sign.

Then $\hat{\alpha} + \hat{\gamma} - -\hat{\bar{\alpha}}$ is impossible.

Applying lemma 1, b), a), if $2\hat{\alpha} + \hat{\gamma}$ is a root, it must be equal either to $\hat{s}$ or $-\hat{s}$.

The same applies for $2\hat{\alpha} + \hat{\bar{\gamma}}$. Hence $L(2\alpha + \gamma) + L(2\alpha + \bar{\gamma})$ is in $LS(\Gamma)$.

It remains to prove that $L(\alpha + \bar{\alpha} + \gamma) \subset LS(\Gamma)$.

For that two cases have to be examined :

**First case** : $\operatorname{rel} \alpha \neq 0$. $\alpha$ and $\bar{\alpha}$ have the same sign.

- If $\hat{\bar{\alpha}} \in \overset{+}{S_1}$ (resp. $\hat{\bar{\alpha}} \in \overset{-}{S_1}$), by lemma 1, if $\hat{\alpha} + \hat{\bar{\alpha}} + \hat{\gamma}$ is a root, it is equal to $\hat{s}$ (resp. $-\hat{s}$). (Then $L(\alpha + \bar{\alpha} + \gamma) \subset LS(\Gamma)$.)
- If $\hat{\bar{\alpha}} - \hat{s}$ (resp. $\hat{\alpha} - -\hat{s}$), $\hat{\bar{\alpha}} + \hat{\gamma}$ is not a root (definition of $S_2$). $\hat{\alpha} + \hat{\bar{\alpha}}$ is neither a root since $\hat{\alpha}$ is positive (resp. $\hat{\alpha}$ negative) and $\hat{s}$ maximal (resp. $-\hat{s}$ minimal) and, by property of roots (**), either $\hat{\alpha} + \hat{\gamma}$ is not a root and $[L(\alpha), L(\gamma)] - 0 \subset LS(\Gamma)$ or $\hat{\alpha} + \hat{\bar{\alpha}} + \hat{\gamma}$ is not a root, hence $L(\alpha + \bar{\alpha} + \gamma) - 0 \subset LS(\Gamma)$.
- If $\hat{\bar{\alpha}} \in S_2$, by lemma 1, if $\hat{\alpha} + \hat{\bar{\alpha}} + \hat{\gamma}$ is a root, it is in $S_1$.

  But $R \operatorname{ad}^3 X(\alpha)(A) \subset LS(\Gamma)$ : $\operatorname{ad}^3 X(\alpha)(A) - \operatorname{ad} X(\alpha)(\operatorname{ad}^2 X(\alpha)(A))$ is in $LS(\Gamma)$.

  if $[L(\alpha), L(\alpha + \bar{\alpha} + \gamma)] \subset LS(\Gamma)$ since we know that $L(2\alpha + \gamma) + L(2\alpha + \bar{\gamma}) \subset LS(\Gamma)$

  and $[L(\alpha), L(\alpha + \bar{\alpha} + \gamma)] - L(\alpha + \alpha + \bar{\alpha} + \gamma) + L(\alpha + \bar{\alpha} + \alpha + \bar{\gamma})$, by lemma 1,

  if $\hat{\alpha} + \hat{\alpha} + \hat{\bar{\alpha}} + \hat{\gamma}$ and $\hat{\alpha} + \hat{\bar{\alpha}} + \hat{\alpha} + \hat{\bar{\gamma}}$ are roots, they are equal to $\hat{s}$ or $-\hat{s}$,

  hence $[L(\alpha), L(\alpha + \bar{\alpha} + \gamma)] \subset LS(\Gamma)$.

It follows that, for any real v, $\phi(v, X(\alpha)) - \sum_{i\geq 3} \frac{v^i}{i!} ad^i X(\alpha)(A) \in LS(\Gamma)$.

Dividing by $v^2$ and making $v \to \infty$, we get that $y = + ad^2 X(\alpha)(A) \in LS(\Gamma)$.

Applying lemma 3 to y, since $\hat{\alpha} + \hat{\bar{\alpha}} + \hat{\gamma} \in S_1$, if $y(\alpha + \bar{\alpha} + \gamma) \neq 0$, hence $L(\alpha + \bar{\alpha} + \gamma) \subset LS(\Gamma)$, this is what we need.

**Second case** : rel $\alpha = 0$, $\bar{\alpha} = -\alpha$, hence $\hat{\alpha}$ and $\hat{\bar{\alpha}}$ have opposite signs.

Return to the definition of $S_1$. Since $\hat{\alpha} \in S_1$, $\hat{\alpha} + \hat{s}$ or $\hat{\alpha} - \hat{s}$ is a root.

Assume that $\hat{\beta} = \hat{\alpha} + \hat{s}$ is a root. $\hat{\beta} \in S_1$ ($\hat{\beta} - \hat{s} = \hat{\alpha}$ is a root) and $L(\beta) \subset LS(\Gamma)$ since $L(\beta) \subset [L(\alpha), L(s)]$ and both $L(\alpha)$ and $L(s)$ are in $LS(\Gamma)$. Moreover rel $\beta$ = rel $(\alpha + s)$ = rel $s \neq 0$. We apply the results of the first case to $\hat{\beta}$. $[L(\beta), L(\gamma)] \subset LS(\Gamma)$ for any $\hat{\gamma} \in S_2$ such that $A(\gamma) \neq 0$.

Since $\hat{\alpha} = \hat{\beta} - \hat{s}$, $L(\alpha) \subset [L(\beta), L(-s)]$ and $[L(\alpha), L(\gamma)] \subset [[L(\beta), L(-s)], L(\gamma)]$.

By Jacobi identity, $[[L(\beta), L(-s)], L(\gamma)] = [L(\beta), [L(-s), L(\gamma)]] + [L(-s), [L(\gamma), L(\beta)]]$.

But $[L(-s), L(\gamma)] = 0 \subset LS(\Gamma)$ since $\hat{\gamma}$ and $\hat{\bar{\gamma}}$ are in $S_2$, $-\hat{s} + \hat{\gamma}$ and $-\hat{s} + \hat{\bar{\gamma}}$ are not roots. Hence $[L(\alpha), L(\gamma)] \subset [L(-s), [L(\gamma), L(\beta)]] \subset LS(\Gamma)$.

If $\hat{\alpha} - \hat{s}$ is a root instead of $\hat{\alpha} + \hat{\beta}$, we change $\hat{s}$ for $-\hat{s}$ in the above. □

# IV - APPENDIX

## IV-1. Preliminaries, notations

### IV-1.1. Semi-simple Lie algebras, root systems

For details about this section, see [W], [B].

Let L be a real semi-simple Lie algebra. $L^{\mathbb{C}} = L \otimes_{\mathbb{R}} \mathbb{C}$ is a complex simple Lie algebra. Let h be a Cartan subalgebra of $L^{\mathbb{C}}$. We denote by S a root system associated to $(L^{\mathbb{C}}, h)$.

One has :

- $L^{\mathbb{C}} = h \oplus \bigoplus_{\hat{\alpha} \in S} L^{c}_{\hat{\alpha}}$ , $\dim_{\mathbb{C}} L^{c}_{\hat{\alpha}} = 1$, with :

$$[L^{c}_{\hat{\alpha}}, L^{c}_{\hat{\beta}}] = L^{c}_{\hat{\alpha}+\hat{\beta}} \quad \text{if } \hat{\alpha} + \hat{\beta} \in S$$

$$(\text{if } \hat{\alpha} + \hat{\beta} \notin S \text{ and } \hat{\alpha} + \hat{\beta} \neq 0 \ [L^{c}_{\hat{\alpha}}, L^{c}_{\hat{\beta}}] = 0).$$

$$[L^{c}_{\hat{\alpha}}, L^{c}_{-\hat{\alpha}}] \subset h$$

$$[h, L^{c}_{\hat{\alpha}}] = L^{c}_{\hat{\alpha}}$$

- For any $\hat{\alpha} \in S$, there is a unique $H_{\alpha} \in h$ such that $Kil(H, H_{\alpha}) = \hat{\alpha}(H)$ for any $H \in h$, where $Kil(X, Y) = Trace(ad X \cdot ad Y)$, the killing form.

There is also $E_{\alpha} \in L^{c}_{\hat{\alpha}}$ such that $Kil(E_{\alpha}, E_{-\alpha}) = 1$ and $[E_{\alpha}, E_{-\alpha}] = H_{\alpha}$

## IV-1.2. Strongly regular elements :

$B \in L$ is said "strongly regular" whenever $Ker(ad_{\mathbb{C}} B)$ is a Cartan subalgebra of $L^{\mathbb{C}}$, and all the non zero eigenspaces of $ad_{\mathbb{C}} B$ are one-dimensional.
For B strongly regular, set :

$L^{c}_{0} = Ker(ad_{\mathbb{C}} B)$, $L^{c}_{\alpha} = Ker(ad_{\mathbb{C}} B - \alpha \dot{I})$, $L(0) = Ker(ad B)$,

$L(\alpha) = (L^{c}_{\alpha} + L^{c}_{\bar{\alpha}}) \cap L$ ($\bar{\alpha}$ = complex conjugate of $\alpha$). Clearly, $L(\bar{\alpha}) = L(\alpha)$.

Therefore, $L^{c} = L^{c}_{0} \oplus \bigoplus_{\alpha \in Sp(B)} L^{c}_{\alpha}$ (where $Sp(B)$ is the set of non zero eigenvalues of $ad_{\mathbb{C}} B$), any $X \in L^{\mathbb{C}}$ can be written in a unique way :

$X = X_0 + \sum_{\alpha \in Sp(B)} X_{\alpha}$, $X_{\alpha} \in L^{c}_{\alpha}$. Similarly, $L = L(0) \oplus \bigoplus_{\alpha \in Sp^{+}(B)} L(\alpha)$
where $Sp^{+}(B) = \{ \alpha \in Sp(B) \mid Im(\alpha) \geq 0 \}$. Any $X \in L$ can be uniquely written

$X = X(0) + \sum_{\alpha \in Sp^{+}(B)} X(\alpha)$, $X(\alpha) \in L(\alpha)$.

$\dim_{\mathbb{R}} L(\alpha) = 1$ if $\alpha$ is real, $\dim_{\mathbb{R}} L(\alpha) = 2$ otherwise.

In this last case, a basis of $L(\alpha)$ is $\{ u_{\alpha} + \sigma u_{\alpha}, \sqrt{-1}(u_{\alpha} - \sigma u_{\alpha})$ for some $u_{\alpha} \neq 0 \in L^{c}_{\alpha}$, where $\sigma$ is the natural involution in $L^{\mathbb{C}}$. At the end, $\sigma u_{\alpha} = u_{\bar{\alpha}}$ and $u(\alpha) = u_{\alpha} + u_{\bar{\alpha}}$ . $u_{\alpha} \neq 0$ iff $u(\alpha) \neq 0$.

Recall that Sp (B) can be identified to the root system of $L^{\mathbb{C}}$ associated to $L_0^c$ - h,

by the bijection : $S \to Sp(B)$

$$\hat{\alpha} \to \alpha - \hat{\alpha}(B)$$

### IV-1.3. Ordering roots :

We order roots via the identification of S and Sp (B). A subset of $\mathbb{C}$ is ordered with the natural lexicographic order on $\mathbb{C}$. ($Sp(B) \subset \mathbb{C}$).

- Z is said fundamental (or primitive) in S if it cannot be written as a sum of at least two positive elements in S.
- $Z \in S$ is said maximal (resp. minimal) if $Z > 0$ (resp. $Z < 0$) and, for any $w \in S$, $w > 0$ ($w < 0$), $Z + w \notin S$.

## 2. A Key property used all long the paper

Let $\Gamma = \{ A + uB \mid u \in R \}$ be a family of right invariant vector fields satisfying $(H_1)$, $(H_2)$ of the introduction. Let $\hat{s} = \sup_{\alpha \in Sp(B)} (\alpha)$.

(*) Let $\beta$ be a root, with rel ($\beta$) = rel ($\pm$ s). For any $X \in LS(\Gamma)$ such that $X(\beta) \neq 0$, then $L(\beta) \subset LS(\Gamma)$. This is proved in [J.K.], page 177. In particular, $L(s) + L(-s) \subset LS(\Gamma)$.

## V - REFERENCES

[J.K.] V. JURDJEVIC, I. KUPKA : "Controllability of right invariant systems on semi-simple Lie groups and their homogeneous spaces", Annales de l'Institut Fourier, 31 (4), 1981.

[G.B.] J.P. GAUTHIER, G. BORNARD : "Controllabilité des systèmes bilinéaires", SIAM Journal on Control and opt., Vol. 20 N° 3, May 1982.

[G.K.S.] J.P. GAUTHIER, I. KUPKA, G. SALLET : "Controllability of right invariant systems on real simple Lie groups", Systems and control letters, 5, 1984.

[L.C.] S. LEITE, P. CROUCH : "Controllability on classical Lie groups", Mathematics for control, signals and systems, Vol. 1 N° 1, 1988, pp. 31-42.

[EA.G] R. EL ASSOUDI, J.P. GAUTHIER : "Controllability of right invariant systems on real simple Lie groups of type $F_4$, $G_2$, $B_n$, $C_n$", Mathematics for Control, signals and systems, Vol. 1 N° 1, 1988, pp. 293-301.

[B] N. BOURBAKI : "Groupes et algèbres de Lie", Fasc.XXXVIII, Ch.7 et 8, Herman, Paris, 1975.

[W] G. WARNER : "Harmonic analysis on semi-simple Lie groups", Die Grundlehen der Math. Wissenschaften, Bd 188 Berlin-Heidelberg, New-York, 1972.

# FEEDBACK SYNTHESIS

# STABLE NONINTERACTING CONTROL

J.W. Grizzle[1] and A. Isidori[2]

**Abstract**

Necessary conditions and sufficient conditions for the existence of a state feedback that achieves noninteraction and internal stability are presented. This entails characterizing the set of all controllability distributions that can arise as solutions to the noninteracting control problem. This characterization is then used to sort through all possible solutions and identify a fixed internal dynamics that is common to every noninteractive closed-loop. The stability properties of this dynamics is shown to be the key factor in the problem of achieving noninteracting control with internal stability.

# 1 Preliminaries

This article summarizes some recent work of the authors on the geometric aspects of characterizing those systems that can be rendered noninteractive and internally stable; for the proofs and a much more complete bibliography, the reader is referred to [1,2]. By an internally asymptotically stable closed-loop system is meant a system in which, when the reference input v is set equal to zero, a prescribed equilibrium state is locally asymptotically stable in the sense of Liapunov. It is shown that in a system for which noninteracting control can be achieved by regular static state-feedback, the noninteraction requirement induces a well-defined subsystem whose dynamics, and hence its stability property, is independent of the particular decoupling feedback used. For left-invertible square systems, this is the precise extension of a previous result of E. Gilbert, to the nonlinear setting. A necessary condition therefore for noninteracting control with internal asymptotic stability is that this induced subsystem be asymptotically stable. Later on it is also shown that if the induced subsystem is asymptotically stable and the original system was asymptotically stabilizable by state-feedback, then noninteracting control with asymptotic stability can be achieved, even in cases of critical stability. A third, and perhaps more striking result is that the aforementioned obstruction to the achievement of stable noninteracting control via static state-feedback, namely the presence of an unstable induced subsystem, *cannot* in general be overcome by dynamic compensation. This is radically different from what is known to be possible for the class of linear systems.

Consider an affine nonlinear system whose outputs have been grouped into blocks:

$$\Sigma: \quad \begin{aligned} \dot{x} &= f(x) + \sum_{j=1}^{m} g_j(x) u_j \\ y_i &= h_i(x) \qquad 1 \leq i \leq \mu \end{aligned} \tag{1.1}$$

[1]Research supported by the National Science Foundation under Grant ECS-88-96136
Department of Electrical Engineering and Computer Science, University of Michigan, Ann Arbor, MI 48109-2122 USA

[2]Dipartimento di Informatica e Sistemistica, Università di Roma, "La Sapienza", Rome, Italy

where $x \in R^n$, $u_j \in R$, $y_i \in R^{p_i}$, and $f, g_j, h_i$ are analytic. The system is said to be *noninteractive* with respect to a given partition $u^T = (u_1, \ldots, u_\mu,\ u_{\mu+1})$ of the inputs, $u_i$ possibly vector valued, if $y_i$ is unaffected by $u_j$ for $i \neq j$. The *noninteractive control problem* is to find, if possible, a *regular static state variable feedback*, $u = \alpha(x) + \beta(x)v$, $|\beta(x)| \neq 0\ \forall x$, (regular refers to the invertibility of $\beta$) and a partitioning of the new command inputs $v$ so that the resulting closed-loop system is noninteractive. If in addition, the zero-input dynamics (or drift term) $\dot{x} = f(x) + \sum_{i=1}^{m} g_i(x)\alpha_i(x)$, where $\alpha(x) = (\alpha_1(x), \ldots, \alpha_m(x))^T$, is asymptotically stable (in the sense of Liapunov) about a given equilibrium point $x_o$, the problem is said to be solvable with *stability*.

Criteria for the achievement of noninteraction without the constraint of internal stability are now well-known , but it turns out that they are not convenient starting points for the additional analysis needed to characterize stabilizing solutions. In the following, $\mathcal{G}$ will denote the distribution $\mathcal{G}(x) := \text{span}\{g_1(x), \ldots, g_m(x)\}$ , which is assumed to have dimension $m$.

**Theorem 1.1** *([1], see also [2-4]) Suppose that the system (1.1) can be rendered noninteractive with a regular static state feedback. Then there exist controllability distributions $\mathcal{R}_1, \ldots, \mathcal{R}_m$ satisfying*

$$\mathcal{R}_i \subset \ker dh_i \tag{1.2}$$

$$\mathcal{G} \cap \mathcal{R}_i + \bigcap_{j \neq i} (\mathcal{G} \cap \mathcal{R}_j) = \mathcal{G} \tag{1.3}$$

*for each $1 \leq i \leq \mu$. Conversely, if in a neighborhood of some point $x_o$ there exist controllability distributions $\mathcal{R}_1, \ldots, \mathcal{R}_\mu$ satisfying (1.2) and (1.3), and if in addition the distributions $\mathcal{G}$, $\mathcal{R}_i$, $\mathcal{R}_i \cap \mathcal{G}$, and $\bigcap_{j \neq i} (\mathcal{G} \cap \mathcal{R}_j)$, $1 \leq i \leq \mu$ are constant dimensional, then, on a possibly smaller neighborhood of $x_o$, there exists a regular static state feedback rendering (1.1) noninteractive.*

A set of controllability distributions $\{\mathcal{R}_1, \ldots, \mathcal{R}_\mu\}$ satisfying the constant dimensionality assumptions of Theorem 1.1 will be called a *regular set*; at times the qualifier, "about the point $x_o$", will be added for reasons of clarity. If in addition $\{\mathcal{R}_1, \ldots, \mathcal{R}_\mu\}$ satisfies (1.2) and (1.3), it will be called a *regular solution to the noninteracting control problem.* Note that if $\{\mathcal{R}_1, \ldots, \mathcal{R}_\mu\}$ is any such regular solution, then

$$\mathcal{R} := \bigcap_{j=1}^{\mu} \mathcal{R}_j \tag{1.4}$$

is a controlled-invariant distribution (at least on a neighborhood of $x_o$) since it is the intersection of compatible distributions. The maximal controllability distribution contained in $\mathcal{R}$ will be denoted by $\mathcal{Q}$.

## 2 Fixed Modes of Noninteracting Control

The goal is to identify a certain internal (sub)dynamics that is imposed by the noninteraction requirement. Temporarily, let $\mathcal{R}$ be *any* involutive controlled-invariant distribution, $\mathcal{Q}$ the maximal controllability distribution contained in $\mathcal{R}$ (so that $\mathcal{R} \cap \mathcal{G} = \mathcal{Q} \cap \mathcal{G}$), *let* $x_o$ be an equilibrium point of (1.1) (i.e., $f(x_o) = 0$), and suppose that $\mathcal{R}$, $\mathcal{Q}$, $\mathcal{Q} \cap \mathcal{G}$ and $\mathcal{G}$ have constant dimension around $x_o$. Since $\mathcal{R}$ is controlled-invariant, there exists a feedback function $\alpha(x)$ such that $[f + g\alpha, \mathcal{R}] \subset \mathcal{R}$ (this entails automatically that $[f + g\alpha, \mathcal{Q}] \subset \mathcal{Q}$]). Moreover, $\alpha$ can always be chosen such that $\alpha(x_o) = 0$, i.e., it preserves the equilibrium point . Therefore $f + g\alpha$ is tangent to $M$, the leaf of $\mathcal{R}$ passing through $x_o$, and $f + g\alpha|M$ is a well-defined vector field of $M$. Consider the quotient manifold $\frac{M}{\mathcal{Q}}$ which, because of constant dimensionality assumptions, locally around $x_o$ is a smooth manifold of dimension $\dim \mathcal{R} - \dim \mathcal{Q}$. Let $\pi$ denote the canonical projection $\pi : M \to M/\mathcal{Q}$. Since $\mathcal{Q}$ is invariant under $f + g\alpha|M$, locally around $\pi(x_o)$ there is a well-defined induced vector field on $M/\mathcal{Q}$; i.e. a vector field $\hat{f}$ such that $\pi_*(f + g\alpha|M) = \hat{f}\pi$. This vector field will be denoted by $f + g\alpha|\frac{M}{\mathcal{Q}}$. Note that if $(x_1, x_2)$ is a choice of local coordinates on $M$ such that $\mathcal{Q} = \text{span}\{\frac{\partial}{\partial x_1}\}$, the submanifold $L = \{(x_1, x_2) : x_1 = 0\}$ is a local representation of $\frac{M}{\mathcal{Q}}$. In these coordinates the vector field $f + g\alpha|M$ is represented as

$$f + g\alpha|M = \begin{bmatrix} \tilde{f}_1(x_1, x_2) \\ \tilde{f}_2(x_2) \end{bmatrix}$$

and $\tilde{f}_2(x_2)$ is a local representation of $f + g\alpha|\frac{M}{\mathcal{Q}}$.

**Lemma 2.1** *[1] The vector field $f + g\alpha|\frac{M}{\mathcal{Q}}$ is independent of the particular choice of $\alpha$, so long as $\alpha$ preserves the equilibrium point and renders $\mathcal{R}$ invariant.*

Applying this to the noninteracting control problem yields

**Theorem 2.2** *[1] Let $x_o$ be an equilibrium point of (1.1). Suppose that on a neighborhood of $x_o, \{\mathcal{R}_1, \ldots, \mathcal{R}_\mu\}$ is a regular solution to the noninteracting control problem, and moreover, that $\mathcal{R} := \bigcap_{j=1}^{\mu} \mathcal{R}_j$ and $\mathcal{Q}$, the maximal controllability distribution contained in $\mathcal{R}$, are constant dimensional. Let $u = \alpha(x) + \beta(x)v$ be any regular static state feedback constructed from $\{\mathcal{R}_1, \ldots, \mathcal{R}_\mu\}$ which renders (1.1) noninteractive and preserves the equilibrium point. Then the leaf of $\mathcal{R}$ passing through $x_o$, denoted $M$, is an invariant submanifold of $\dot{x} = (f + g\alpha)(x)$ and moreover, $f + g\alpha|\frac{M}{\mathcal{Q}}$ does not depend upon the particular choice of $\alpha$.*

The vector field $f + g\alpha|\frac{M}{\mathcal{Q}}$ will be called the *fixed dynamics* (or fixed modes) *of the solution* $\{\mathcal{R}_1, \ldots, \mathcal{R}_\mu\}$. Clearly this dynamics must be stable if an asymptotically stable noninteractive closed-loop is to be constructed from $\{\mathcal{R}_1, \ldots, \mathcal{R}_\mu\}$. The next step is to search over all possible solutions $\{\mathcal{R}_1, \ldots, \mathcal{R}_\mu\}$ to find the minimal fixed dynamics and hence the minimal obstruction to attaining stability. This next result is key.

**Lemma 2.3** *[1] Suppose that the noninteractive control problem admits a regular solution $\{\mathcal{R}_1, \ldots, \mathcal{R}_\mu$ and that the set $\{\mathcal{P}_1^*, \ldots, \mathcal{P}_\mu^*\}$ is regular. Then $\{\mathcal{R}_1, \ldots, \mathcal{R}_\mu,\ \mathcal{P}_1^*, \ldots, \mathcal{P}_\mu^*\}$ is a locally compatible* *set of controllability distributions and so in particular,* $\{\mathcal{P}_1^*, \ldots, \mathcal{P}_\mu^*\}$ *is a regular solution. Moreover, there locally exists a regular static state feedback and a partition of the new command inputs,* $u = \alpha(x) + \sum_{i=1}^{\mu+1}\left(\sum_{j\in I_i} \beta_j(x) v_j\right)$, *such that*

*(a)* $[f + g\alpha,\ \mathcal{P}_i^*] \subset \mathcal{P}_i^*, \qquad [(g\beta)_k,\ \mathcal{P}_i^*] \subset \mathcal{P}_i^*, \qquad 1 \leq k \leq m$

*(b)* $[f + g\alpha,\ \mathcal{R}_i] \subset \mathcal{R}_i, \qquad [(g\beta)_k,\ \mathcal{R}_i] \subset \mathcal{R}_i, \qquad 1 \leq k \leq m$

*(c)* $(g\beta)_j \in \mathcal{R}_i \quad \text{all} \quad j \notin I_i$

Note in passing that if $\mathcal{S}_1$ and $\mathcal{S}_2$ are two compatible controllability distributions, then $\mathcal{S}_1 = \mathcal{S}_2$ if, and only if, $\mathcal{S}_1 \cap \mathcal{G} = \mathcal{S}_2 \cap \mathcal{G}$. This observation, combined with Lemma 2.3 yields the following result, relating arbitrary regular solutions to the maximal solution, when the latter is also regular.

**Theorem 2.4** *[1] Suppose that* $\{\mathcal{R}_1, \ldots, \mathcal{R}_\mu\}$ *is any regular solution to the noninteracting control problem and suppose that the maximal solution* $\{\mathcal{P}_1^*, \ldots, \mathcal{P}_\mu^*\}$ *is also regular. Then, for each* $1 \leq i \leq \mu$,

$$\mathcal{R}_i + \mathcal{Q}^* = \mathcal{P}_i^*.$$

From this theorem, one sees that whenever $\mathcal{Q}^* = \{0\}$, there is only one solution to the noninteracting control problem, and hence, in this case, Theorem 2.4 yields a general necessary condition for the existence of a noninteractive stabilizing solution. When $\mathcal{Q}^* \neq \{0\}$, one must dig a little further.

**Theorem 2.5** *[1] Suppose that* $\{\mathcal{R}_1, \ldots, \mathcal{R}_\mu\}$ *is any regular solution to the noninteracting control problem and suppose that the maximal solution* $\{\mathcal{P}_1^*, \ldots, \mathcal{P}_\mu^*\}$ *is also regular. Let* $\mathcal{R} := \bigcap_{j=1}^{\mu} \mathcal{R}_j$, $\mathcal{P}^* := \bigcap_{j=1}^{\mu} \mathcal{P}_j^*$ *and* $\mathcal{Q}^* =$ *maximal controllability distribution contained in* $\mathcal{P}^*$. *Then, if* $\mathcal{R} \cap \mathcal{Q}^*$ *is constant dimensional,*

$$\mathcal{P}^* = \mathcal{R} + \mathcal{Q}^*.$$

As an immediate corollary one has:

**Corollary 2.6** *[1] Let* $\mathcal{P}^*$, $\mathcal{Q}^*$ *and* $\mathcal{R}$ *be as above. Suppose that* $\mathcal{P}^*$, $\mathcal{Q}^*$ *and* $\mathcal{Q}^* \cap \mathcal{R}$ *all have constant dimension about some equilibrium point* $x_o$, *and that* $\alpha$ *preserves the equilibrium point and renders* $\{\mathcal{R}_1, \ldots, \mathcal{R}_\mu, \mathcal{P}_1^*, \ldots, \mathcal{P}_\mu^*\}$ *invariant. Let* $M^*$ *denote the leaf of* $\mathcal{P}^*$ *passing through* $x_o$ *and* $M$ *denote the leaf of* $\mathcal{R}$ *passing through* $x_o$. *Then*

$$f + g\alpha\Big|\frac{M^*}{\mathcal{Q}^*} = f + g\alpha\Big|\frac{M}{\mathcal{R} \cap \mathcal{Q}^*}$$

To summarize what has been established up to this point, by Lemma 2.3 there exists a regular static feedback $u = \alpha(x) + \beta(x)v$ rendering $\{\mathcal{R}_1, \ldots, \mathcal{R}_\mu, \mathcal{P}_1^*, \ldots, \mathcal{P}_\mu^*\}$ simultaneously invariant; hence the same feedback also renders $\mathcal{Q}^*$, $\mathcal{P}^*$, $\mathcal{Q}$ and $\mathcal{R}$ invariant. By Lemma 2.1, the fixed dynamics $f + g\alpha|\frac{M}{\mathcal{Q}}$ and $f + g\alpha|\frac{M^*}{\mathcal{Q}^*}$ are independent of the particular choice of $\alpha$, so long as $x_o$ is an equilibrium point preserved by $\alpha$. Therefore, they will each be part of the drift (i.e., zero input) dynamics of any noninteractive loop obtained from $\{\mathcal{R}_1, \ldots, \mathcal{R}_\mu\}$ and $\{\mathcal{P}_1^*, \ldots, \mathcal{P}_\mu^*\}$ respectively. By construction, everything in sight is invariant under $f + g\alpha$, and thus $f + g\alpha|\frac{M}{\mathcal{R}\cap\mathcal{Q}^*}$ is well-defined. Now, since $\mathcal{Q} \subset \mathcal{R}$ and $\mathcal{Q} \subset \mathcal{Q}^*$, it follows that $\mathcal{Q} \subset \mathcal{Q}^* \cap \mathcal{R}$, so the dynamics $f + g\alpha|\frac{M}{\mathcal{R}\cap\mathcal{Q}^*}$ is contained within the dynamics $f + g\alpha|\frac{M}{\mathcal{Q}}$; more precisely, $f + g\alpha|\frac{M}{\mathcal{R}\cap\mathcal{Q}^*}$ is a projection of $f + g\alpha|\frac{M}{\mathcal{Q}}$. Hence, applying Corollary 2.6, one obtains that the dynamics $f + g\alpha|\frac{M^*}{\mathcal{Q}^*}$ is a projection of the dynamics $f + g\alpha|\frac{M}{\mathcal{Q}}$. If one chooses local coordinates $(x_1, x_2, x_3, x_4)$ on $M^*$ such that $\mathcal{Q} = \text{span}\{\frac{\partial}{\partial x_1}\}$, $\mathcal{Q}^* \cap \mathcal{R} = \text{span}\left\{\frac{\partial}{\partial x_1}, \frac{\partial}{\partial x_2}\right\}$, $\mathcal{R} = \text{span}\left\{\frac{\partial}{\partial x_1}, \frac{\partial}{\partial x_2}, \frac{\partial}{\partial x_4}\right\}$ and $\mathcal{Q}^* = \text{span}\left\{\frac{\partial}{\partial x_1}, \frac{\partial}{\partial x_2}, \frac{\partial}{\partial x_3}\right\}$, then the vector field $f + g\alpha|M^*$ is represented as $f_1(x_1, x_2, x_3, x_4)\frac{\partial}{\partial x_1} + f_2(x_2, x_3, x_4)\frac{\partial}{\partial x_2} + f_3(x_3)\frac{\partial}{\partial x_3} + f_4(x_4)\frac{\partial}{\partial x_4}$, and $f + g\alpha|M = f_1(x_1, x_2, 0, x_4)\frac{\partial}{\partial x_1} + f_2(x_2, 0, x_4)\frac{\partial}{\partial x_2} + f_4(x_4)\frac{\partial}{\partial x_4}$, $f + g\alpha|\frac{M}{\mathcal{Q}} = f_2(x_2, 0, x_4)\frac{\partial}{\partial x_2} + f_4(x_4)\frac{\partial}{\partial x_4}$, $f + g\alpha|\frac{M^*}{\mathcal{Q}^*} = f + g\alpha|\frac{M}{\mathcal{R}\cap\mathcal{Q}^*} = f_4(x_4)\frac{\partial}{x_4}$.

This analysis yields the following "necessary" condition for the achievement of noninteracting control with stability.

**Theorem 2.7** *[1] Suppose in a neighborhood of an equilibrium point $x_o$ that $\{\mathcal{P}_1^*, \ldots, \mathcal{P}_\mu^*\}$ is a regular solution to the noninteracting control problem. Suppose in addition that $\mathcal{P}^*$ and $\mathcal{Q}^*$ have constant dimension near $x_o$ and that $u = \alpha(x) + \beta(x)v$ is any regular static state feedback rendering $\{\mathcal{P}_1^*, \ldots, \mathcal{P}_\mu^*\}$ invariant. Then, if the fixed dynamics $f + g\alpha|\frac{M^*}{\mathcal{Q}^*}$ is not asymptotically stable, there does not exist any other regular solution $\{\mathcal{R}_1, \ldots, \mathcal{R}_\mu\}$, satisfying $\mathcal{R}$, $\mathcal{Q}$ and $\mathcal{R} \cap \mathcal{Q}^*$ constant dimensional, that admits an asymptotically stable noninteractive closed-loop.*

The above result is rather weak because there is no effective means of testing the hypotheses, due to the assumption on $\mathcal{R} \cap \mathcal{Q}^*$. However, it does show that the only other way for a stabilizing noninteractive control law to exist is through the occurrance of singularities.

The next step is to investigate whether the stability of the fixed dynamics constitutes a sufficient condition for achieving noninteracting control with asymptotic stability via static state feedback. It can be shown that if one excludes critical asymptotic stability, then this is indeed the case. Stabilization via dynamic feedback, even in critical cases, will be addressed in the next section.

Stabilization by static state feedback will be accomplished by passing through a general decomposition result, which is of independent interest.

**Theorem 2.8** *([1]; see also [2-5]) Fix $x_o \in \mathbb{R}^n$. Suppose in a neighborhood of $x_o$, that $\{\mathcal{P}_1^*, \ldots, \mathcal{P}_\mu^*\}$ is a regular solution to the decoupling problem, that $\mathcal{P}^*$ and $\mathcal{Q}^*$ have constant dimension, and that $\Sigma$ (1.1) has already been rendered noninteractive with a feedback which is compatible with $\{\mathcal{P}_1^*, \ldots, \mathcal{P}_\mu^*\}$. Then there exist coordinates $(x_1, \ldots, x_{\mu+3})$ in which $\Sigma$ decomposes as*

$$\dot{x}_1 = f_1(x_1, x_{\mu+3}) + \sum_{k \in I_1} g_{k1}(x_1, x_{\mu+3})u_k$$

$$\begin{aligned}
&\vdots \\
\dot{x}_\mu &= f_\mu(x_\mu, x_{\mu+3}) + \sum_{k\in I_\mu} g_{k\mu}(x_\mu, x_{\mu+3})u_k \\
\dot{x}_{\mu+1} &= f_{\mu+1}(x_1, \ldots, x_{\mu+3}) + \sum_{j=1}^{m} g_{j(\mu+1)}(x_1, \ldots x_{\mu+3})u_j \\
\dot{x}_{\mu+2} &= f_{\mu+2}(x_1, \ldots x_\mu,\ x_{\mu+2},\ x_{\mu+3}) + \\
&\quad + \sum_{j=1}^{\mu}\left(\sum_{k\in I_j} g_{k(\mu+2)}(x_1, \ldots, x_\mu,\ x_{\mu+2},\ x_{\mu+3})u_k\right) \\
\dot{x}_{\mu+3} &= f_{\mu+3}\,(x_{\mu+3}) \\
y_i &= h_i(x_i,\ x_{\mu+3}) \qquad 1 \le i \le \mu
\end{aligned}$$

*where $\{I_j\}_{j=1}^{\mu+1}$ is a suitable partition of the inputs. Moreover, if $x_o$ is an equilibrium point, the coordinates can always be chosen in such a way that $x_o = 0$, and then the fixed dynamics is precisely*

$$\dot{x}_{\mu+2} = f_{\mu+2}(0, \ldots, 0,\ x_{\mu+2},\ 0). \tag{2.1}$$

This leads to:

**Theorem 2.9** *[1] Consider the system $\sum$ given in (1.1). Let $\mathcal{P}_i^*$ denote the maximal controllability distribution contained in $\ker dh_i$. Define $\mathcal{P}^* := \bigcap_{j=1}^{\mu} \mathcal{P}_j^*$, $\mathcal{Q}^*$ the maximal controllability distribution contained in $\mathcal{P}^*$, and $\mathcal{G} = \text{span}\{g_1, \ldots, g_m\}$. Assume that $\mathcal{G}$, $\mathcal{Q}^*$, $\mathcal{P}^*$, $\mathcal{P}_i^*$, $\mathcal{P}_i^* \cap \mathcal{G}$ and $\bigcap_{j\neq i}(\mathcal{P}_j^* \cap \mathcal{G})$ are all constant dimensional on a neighborhood of a given point $x_o \in \mathbb{R}^n$. Then there exists a neighborhood of $x_o$ on which is defined a regular static state feedback rendering $\sum$ noninteractive if, and only if, for $1 \le i \le \mu$*

$$\mathcal{P}_i^* \cap \mathcal{G} + \bigcap_{j\neq i}(\mathcal{P}_j^* \cap \mathcal{G}) = \mathcal{G}\ . \tag{2.2}$$

*Moreover, if in addition $x_o$ is an equilibrium point, the linearization of $\sum$ about $x_o$ is (asymptotically) stabilizable and the linearization of the fixed-dynamics (2.1) about $x_o$ is asymptotically stable, then there exists a regular static state feedback which simultaneously achieves noninteraction and asymptotically stabilizes the drift dynamics.*

## 3 On the Role of Dynamic Feedback

We have seen in the previous section that imposing noninteraction via static state-feedback imposes on the closed-loop system a certain internal behavior which must be stable for noninteracting control with stability to be possible. For the class of linear systems it is known that the fixed dynamics are no longer an obstruction when dynamic state-feedback is allowed. Indeed, by a proper choice of dynamic compensator, one can always arrive at a system which is noninteractive and, more importantly, has $\mathcal{P}^* = 0$. Unfortunately, and perhaps quite surprisingly, such a result cannot be extended to the class of nonlinear systems.

*A family of counterexamples*[2]. Consider the following class of systems, having two inputs, two outputs and a 3-dimensional state space:

$$\begin{aligned} \dot{x}_1 &= u_1 \\ \dot{x}_2 &= u_2 \\ \dot{x}_3 &= a(x_1, x_2, x_3) \\ y_1 &= x_1 \\ y_2 &= x_2. \end{aligned} \tag{3.1}$$

Suppose that x=0 is an equilibrium for this system (i.e., a(0,0,0)=0) and assume that

$$\frac{\partial a}{\partial x_1}(0) \neq 0, \quad \frac{\partial a}{\partial x_2}(0) \neq 0. \tag{A1}$$

The system in question is already noninteractive, and an easy calculation shows that, because of (A1), $\mathcal{P}^* = \text{span}\{(0\ 0\ 1)^T\}$ . Because $\mathcal{Q}^* = \{0$, the fixed dynamics associated with $\mathcal{P}^*$ is $\dot{x}_3 = a(0,0,x_3)$ . Hence, if we suppose

$$\frac{\partial a}{\partial x_3}(0) > 0 \tag{A2}$$

we see that the system itself is unstable and there is no way to find a *static* state-feedback yielding a noninteractive but stable control loop, even though, if one drops the requirement of noninteraction, the system *is* stabilizable by static feedback; indeed, it is locally feedback linearizable. It can be shown that even by means of *dynamic* state-feedback, if one in addition assumes

$$\frac{\partial^2 a}{\partial x_1 \partial x_2}(0) \neq 0 \tag{A3}$$

it is still impossible to simultaneously achieve noninteraction and stability. Indeed, consider *any* (regular) dynamic state-feedback

$$\begin{aligned} u_1 &= \alpha_1(x,z) + \beta_{11}(x,z)v_1 + \beta_{12}(x,z)v_2 \\ u_2 &= \alpha_2(x,z) + \beta_{21}(x,z)v_1 + \beta_{22}(x,z)v_2 \\ \dot{z} &= \gamma(x,z) + \delta_1(x,z)v_1 + \delta_2(x,z)v_2 \end{aligned}$$

with $z \in R^k$, producing a noninteracting closed-loop (i.e., the decoupling matrix of the closed-loop system is diagonal and nonsingular). Let the closed-loop system be denoted by

$$\begin{aligned} \dot{\xi} &= F(\xi) + G_1(\xi)v_1 + G_2(\xi)v_2 \\ y_1 &= H_1(\xi) \\ y_2 &= H_2(\xi) \end{aligned}$$

where $\xi = (x_1, x_2, x_3, z)^T$ and

$$F = \begin{bmatrix} \alpha_1 \\ \alpha_2 \\ a \\ \gamma \end{bmatrix}, \; G_1 = \begin{bmatrix} \beta_{11} \\ \beta_{21} \\ 0 \\ \delta_1 \end{bmatrix}, \; G_2 = \begin{bmatrix} \beta_{12} \\ \beta_{22} \\ 0 \\ \delta_2 \end{bmatrix}$$

$$H_1 = x_1 \, , \; H_2 = x_2;$$

assume that the dynamic feedback preserves the equilibrium point, i.e., that $F(\xi) = 0$ at $\xi = 0$. Without loss of generality, one may suppose that the decoupling matrix of the closed-loop system is the indentity matrix. .

To establish the result, one shows first that, in the closed-loop system, the distribution $\mathcal{P}^*$ contains at least one nonzero vector field. Since by assumption the system is decoupled, we have $\mathcal{P}^* = \mathcal{P}_1^* \cap \mathcal{P}_2^*$, where

$$\mathcal{P}_1^* = < F, G_1, G_2 | span\{G_2\} > \; , \; \mathcal{P}_2^* = < F, G_1, G_2 | span\{G_1\} > .$$

Let $\rho_1$ and $\rho_2$ denote the characteristic numbers associated with $y_1$ and $y_2$ (for the closed-loop system), and define the vector fields $X_1 = ad_F^{\rho_1} G_1$ and $X_2 = ad_F^{\rho_2} G_2$. Then one can show, for any i and j, $[X_i, \mathcal{P}_j^*] \subset \mathcal{P}_j^*$ . It then follows from the Jacobi identity that : $Y = [X_2, [F, X_1]] \in \mathcal{P}_1^* \cap \mathcal{P}_2^*$. Since $Y \in \mathcal{P}_1^* \cap \mathcal{P}_2^* \subset \ker\{dH_1\} \cap \ker\{dH_2\}$, it follows that the first two components of Y are zero. Finally, Y can be shown to have the form

$$Y = \begin{bmatrix} 0 \\ 0 \\ (-1)^{\rho_1+\rho_2} \frac{\partial^2 a}{\partial x_1 \partial x_2} \\ * \end{bmatrix}$$

Under Assumption (A3), this vector field is nonzero, establishing that $\mathcal{P}^*$ cannot be annihilated by any regular dynamic compensator yielding a noninteractive closed-loop. (This is a significant departure from what is possible for linear systems, where $\mathcal{P}^*$ can always be eliminated by appropriately choosing the dynamic compensator). It is then easy to use this fact to show that the matrix $(\frac{\partial F}{\partial \xi})(0)$ has $(\frac{\partial a}{\partial x_3})(0)$ among its eigenvalues and thus that the closed-loop is always unstable (because of Assumption (A2)). One concludes from this analysis that if the system (3.1) is such that (A1), (A2) and (A3) are true, there is no way to achieve noninteracting control and stability by means of any regular dynamic state-feedback whatsoever. Note that if the system were linear (in which case (A3) would not be satisfied) then noninteracting control with stability would be easily achieved by means of a simple 1-dimensional dynamic state-feedback.

A positive aspect of the role of dynamic feedback will now be exhibited by showing that the achievement of asymptotically stable noninteracting control is possible whenever the following hypotheses are satisfied:

(i) noninteracting control via static state-feedback is possible, and the solution is unique ( i.e., $\mathcal{Q}^* = \{0\}$),

(ii) the system is asymptotically stabilizable via (a possibly dynamic) state-feedback,

(iii) the induced fixed dynamics $\dot{x}_{\mu+2} = f_{\mu+2}(0, ..., 0, x_{\mu+2}, 0)$ is asymptotically stable.

Suppose the system has been already rendered noninteractive by static-feedback and consider the resulting structural decomposition shown in Theorem 2.8. Assume, in order to reduce the notational burden, $m = \mu = 2$; also since $x_3$ is the empty vector due to $\mathcal{Q}^* = \{0\}$, relabel $x_4$ as $x_3$ and $x_5$ as $x_4$ . In this case, the structural decomposition in question reduces to

$$\begin{aligned}
\dot{x}_1 &= f_1(x_1, x_4) + g_{11}(x_1, x_4)u_1 \\
\dot{x}_2 &= f_2(x_2, x_4) + g_{22}(x_2, x_4)u_2 \\
\dot{x}_3 &= f_3(x_1, x_2, x_3, x_4) + \sum_{i=1}^{2} g_{i3}(x_1, x_2, x_3, x_4)u_i \\
\dot{x}_4 &= f_4(x_4) \\
y_1 &= h_1(x_1, x_4) \\
y_2 &= h_2(x_2, x_4);
\end{aligned} \tag{3.2}$$

recall that by assumption all $f_i$'s vanish at x=0.

By condition (iii), the subsystem

$$\dot{x}_3 = f_3(0, 0, x_3, 0) \tag{3.3}$$

which in the present state-space coordinates describes the fixed dynamics induced by the noninteracting requirement, is asymptotically stable. Observe that condition (i) is invariant under regular static feedback, and therefore system (3.2) is stabilizable by a (possibly dynamic) state feedback, that we shall denote as:

$$\begin{aligned}
u_1 &= \phi_1(x_1, x_2, x_3, x_4, z) \\
u_2 &= \phi_2(x_1, x_2, x_3, x_4, z) \\
\dot{z} &= \gamma(x_1, x_2, x_3, x_4, z).
\end{aligned} \tag{3.4}$$

In other words, the composition of (3.2) and (3.4) is asymptotically stable. Without loss of generality we may assume that $\phi_1$, $\phi_2$ and $\gamma$ vanish at the origin, so that the latter is still an equilibrium of this closed-loop.

If we now impose, on the first input channel of (3.2), the dynamic state feedback

$$\begin{aligned}
u_1 &= \phi_1(x_1, \eta_2, \eta_3, x_4, \eta_5) + v_1 \\
\dot{\eta}_2 &= f_2(\eta_2, x_4) + g_{22}(\eta_2, x_4)\phi_2(x_1, \eta_2, \eta_3, x_4, \eta_5) \\
\dot{\eta}_3 &= f_3(x_1, \eta_2, \eta_3, x_4) + \sum_{i=1}^{2} g_{i3}(x_1, \eta_2, \eta_3, x_4)\phi_i(x_1, \eta_2, \eta_3, x_4, \eta_5) \\
\dot{\eta}_5 &= \gamma(x_1, \eta_2, \eta_3, x_4, \eta_5)
\end{aligned}$$

and similarly, on the second input channel of (3.2), the dynamic state feedback

$$\begin{aligned}
u_2 &= \phi_2(\xi_1, x_2, \xi_3, x_4, \xi_5) + v_2 \\
\dot{\xi}_1 &= f_1(\xi_1, x_4) + g_{11}(\xi_1, x_4)\phi_2(\xi_1, x_2, \xi_3, x_4, \xi_5) \\
\dot{\xi}_3 &= f_3(\xi_1, x_2, \xi_3, x_4) + \sum_{i=1}^{2} g_{i3}(\xi_1, x_2, \xi_3, x_4)\phi_i(\xi_1, x_2, \xi_3, x_4, \xi_5) \\
\dot{\xi}_5 &= \gamma(\xi_1, x_2, \xi_3, x_4, \xi_5)
\end{aligned}$$

The system thus obtained has the form:

$$
\begin{aligned}
\dot{x}_1 &= \hat{f}_1(x_1, x_4, \xi) + g_{11}(x_1, x_4)v_1 \\
\dot{\eta} &= \hat{\gamma}_1(x_1, x_4, \eta) \\
\dot{x}_2 &= \hat{f}_2(x_2, x_4, \xi) + g_{22}(x_2, x_4)v_2 \\
\dot{\xi} &= \hat{\gamma}_2(x_2, x_4, \xi) \\
\dot{x}_3 &= \hat{f}_3(x_1, x_2, x_3, x_4, \eta, \xi) + \sum_{i=1}^{2} g_{i3}(x_1, x_2, x_3, x_4)v_i \\
\dot{x}_4 &= f_4(x_4) \\
y_1 &= h_1(x_1, x_4) \\
y_2 &= h_2(x_2, x_4)
\end{aligned}
\qquad (3.5)
$$

where $\eta = (\eta_2, \eta_3, \eta_5)$ and $\xi = (\xi_1, \xi_3, \xi_5)$. This system is still noninteractive and is also locally asymptotically stable. To see this, set $v_1 = 0$, $v_2 = 0$ and observe that the subsystem of (3.5) involving $x_1, \eta, x_4$ is just a copy of the closed-loop system (3.2) and (3.4), which by assumption is asymptotically stable. So is the subsystem of (3.5) involving $x_2, \xi, x_4$. Consider now the remaining equation of (3.5): $\dot{x}_3 = \hat{f}_3(x_1, x_2, x_3, x_4, \eta, \xi)$. Since $\phi_1$ and $\phi_2$ vanish at the origin,

$$\hat{f}_3(0, 0, x_3, 0, 0, 0) = f_3(0, 0, x_3, 0);$$

the latter coincides with the fixed dynamics, and is aymptotically stable. In view of known stability properties of composite systems, the overall system is locally asymptotically stable [6].

## References


[1] J. W. Grizzle and A. Isidori, "Block noninteracting control with stability via static state feedback", to appear in *Mathematics of Control Signals and Systems*, pre-print February 1988.

[2] A. Isidori and J. W. Grizzle, "Fixed modes and nonlinear noninteracting control with stability," to appear in *IEEE Transactions Autom. Control.*, pre-print May, 1987.

[3] H. Nijmeijer and J. M. Schumacher, "Zeros at infinity for affine nonlinear systems," *IEEE Trans. Automat. Contr.*, Vol. AC-30, pp. 566-573, 1985.

[4] H. Nijmeijer, "The regular local noninteracting control problem for nonlinear control systems," *Proc. of the 24th CDC*, Ft. Lauderdale, pp. 388-392, December, 1985.

[5] I. J. Ha, "Nonlinear decoupling theory with applications to robotics," Ph.D. Dissertation, Univ. of Michigan, Ann Arbor, CRIM report RSD-TR-8-85, 1985.

[6] K. Wagner, Personal Communication.

TOWARDS THE SOLUTION OF THE NONLINEAR BLOCK DECOUPLING PROBLEM

J. DESCUSSE

LAN UA CNRS 823
ENSM 1 rue de La Noë 44072 Nantes cedex 03 , France

Abstract: In this paper , we tackle the dynamic block decoupling problem , the solution of which remains unknown up to now. We try to generalize the successful approach taken for solving Morgan's Problem , i.e. the row by row decoupling problem , in [2] , [3]. The preliminary results , we have obtained , are strongly connected with the structure algorithm of Hirschorn [8]. Using the latter , it is possible to exhibit particular (degenerate) controlled invariant distributions which appear to be crucial in the solution of the considered problem.

## 1)Introduction

In the linear case , the decoupling problem has received a great deal of interest (for a bibliography see [17]). In the nonlinear case , there have only been a few contributions. The first ones are probably those of [6] and [16] , which were attempts to generalize the well known result of [4] for the linear Morgan's Problem. More appealing works can be found in [10] and [11].

Solutions with dynamic compensation can be found in [14] , [15] , [ 2] , [3] , [5] , [ 12] , for the nonlinear Morgan's Problem. Up to now , no solution has been proposed for the dynamic block decoupling problem. In this paper , we give preliminary results , for solving the latter , using an approach similar to that one used in [2] , [3].

The paper is organized as follows. In §2 , we recall the structure algorithm of Hirschorn [8]. In §3 , we develop important properties of this algorithm and we introduce (degenerate) controlled invariant and controllability distributions which play a key role in the next sections. In §4 , the static state feedback block decoupling problem is introduced ; we give a new necessary and sufficient condition for achieving decoupling. It is algebraic and well suited for tackling the dynamic block decoupling problem , which is investigated in §5. The solution needs an algorithmic procedure with two steps , the first one of which is given in the present paper. It generalizes the procedure of [2] and [3]. The second one will be provided in a forthcoming and more complete version of this work.

## 2) The Structure algorithm of Hirschorn [8 ]

Let us consider affine nonlinear systems described by differential equations of the type :

$$\Sigma_o \begin{cases} \dot{x} = A(x) + \Sigma_{i\in \underline{m}} B_i(x)\, u_i & (2.1) \\ \\ y = C(x) & (2.2) \end{cases}$$

The state x belongs to an n- dimensional analytic manifold M, $u_i$ belongs to $\mathbb{R}$, the field of real numbers, the vector fields A(x) and $\{B_i(x)\}_m$ are analytic on M and C : M --> N is an analytic map from M to a p-dimensional analytic manifold N ; for shortness we often write B(x) = matrix $(B_1(x), \ldots, B_m(x))$.

Using $M_0 = I_p(d/dt)$ as an output differential transformation yields a new system representation defined by (2.1) and

$$\dot{y} = L_A C + L_B C\, u \qquad (2.3)$$

Let $q_1$ = rank $L_B C$ and let $\overline{D}_1$ the submatrix formed from the first $q_1$ independent rows of $L_B C$. There then exists an m x m nonsingular matrix $S_1$ such that :

$$S_1 L_B C = \begin{bmatrix} \overline{D}_1 \\ \cdots\cdots \\ O \end{bmatrix}.$$

Using $S_1$ as an output transformation yields, from (2.1) and (2.3), a new system representation $\Sigma_1$ defined by (2.1) and

$$y_1 = C_1 + D_1 u$$

where

$$y_1 = S_1 M_0 y$$

$$C_1 = S_1 L_A C \quad \text{and } D_1 = \begin{bmatrix} \overline{D}_1 \\ \cdots\cdots \\ 0 \end{bmatrix}$$

It will prove convenient to partition $y_1$ and $C_1$ conformably with $D_1$ as

$$y_1 = \begin{bmatrix} \overline{y}_1 \\ \cdots \\ \tilde{y}_1 \end{bmatrix} \qquad C_1 = \begin{bmatrix} \overline{C}_1 \\ \cdots\cdots \\ \tilde{C}_1 \end{bmatrix}$$

From $\Sigma_0$ and $\Sigma_1$ we can define a sequence $\Sigma_k$ inductively. Assume that $\Sigma_k$ has the form

$$\dot{x} = A(x) + B(x)\, u$$
$$y_k = C_k(x) + D_k(x)\, u$$

with the partitionning

$$y_k = \begin{bmatrix} \overline{y}_k \\ \cdots\cdots \\ \tilde{y}_k \end{bmatrix} \qquad C_k = \begin{bmatrix} \overline{C}_k \\ \cdots\cdots \\ \tilde{C}_k \end{bmatrix} \qquad D_k = \begin{bmatrix} \overline{D}_k \\ \cdots\cdots \\ 0 \end{bmatrix}$$

where $\overline{D}_k$ has $q_k$ rows and rank $q_k$, $\overline{y}_k$ and $\overline{C}_k$ have $q_k$ rows and $\tilde{y}_k$ and $\tilde{C}_k$ have p - $q_k$ rows.

Observe that if

$$M_k = \left[\begin{array}{c|c} I_{q_k} & 0 \\ \hline 0 & I_{p-q_k}(d/dt) \end{array}\right]$$

then

$$M_k y_k = \begin{bmatrix} \overline{C}_k \\ \cdots\cdots\cdots \\ L_A \tilde{C}_k \end{bmatrix} + \begin{bmatrix} \overline{D}_k \\ \cdots\cdots\cdots \\ L_B \tilde{C}_k \end{bmatrix} u = H_{k+1} + J_{k+1}\, u$$

Let $q_{k+1}$ = rank $J_{k+1}$. If $\overline{D}_{k+1}$ is the matrix formed from the first $q_{k+1}$ independent rows of $J_{k+1}$, there then exists a nonsingular matrix $S_{k+1}$, such that

$$S_{k+1} J_{k+1} = \begin{bmatrix} \overline{D}_{k+1} \\ \cdots\cdots\cdots \\ 0 \end{bmatrix}$$

$\Sigma_{k+1}$ is then defined by the equation

$$\dot{x} = A(x) + B(x)\,u$$
$$y_{k+1} = C_{k+1}(x) + D_{k+1}(x)\,u$$

where $y_{k+1} = S_{k+1} M_k y_k$ , $C_{k+1} = S_{k+1} H_{k+1}$ , $D_{k+1} = S_{k+1} J_{k+1}$

It follows from the above that $y_k = N_k y$ where

$$N_k = \Pi_{i=0}^{k-1} S_{i+1} M_i$$

is a sequence of nonsingular matrix differential operators. Furthermore one can define $\bar{N}_k$ and $\tilde{N}_k$ by $\bar{y}_k = \bar{N}_k\, y$ and $\tilde{y}_k = \tilde{N}_k\, y$.

It is clear that the matrices $S_i$ defined above are not unique in general. The following method of constructing the matrices proves to be most convenient and will be utilized in the remainder of the paper.

Let $S^*_{k+1}$ be the (unique) permutation matrix such that $S^*_{k+1} J_{k+1}$ has the following structure. Its first $q_{k+1}$ rows are the first $q_{k+1}$ rows of $J_{k+1}$ with the relative order maintained, and its last $p-q_{k+1}$ rows are the remaining rows of $J_{k+1}$, also with the relative order maintained. It is clear from the form of $J_{k+1}$ that $S^*_{k+1}$ has the form

$$S^*_{k+1} = \begin{bmatrix} I_{q_k} & 0 \\ 0 & \bar{R}_{k+1} \\ 0 & \tilde{R}_{k+1} \end{bmatrix}$$

where $\bar{R}_{k+1}$ has $q_{k+1} - q_k$ rows and $\tilde{R}_{k+1}$ has $p-q_{k+1}$ rows. Furthermore

$$R_{k+1} = \begin{bmatrix} \bar{R}_{k+1} \\ \cdots\cdots \\ \tilde{R}_{k+1} \end{bmatrix}$$

is a permutation matrix and so constant and nonsingular. On the other hand there exists $K^*_{k+1}$ such that

$$\tilde{R}_{k+1} L_B\, \tilde{C}_k = K^*_{k+1}\, \bar{D}_{k+1} \qquad (2.4)$$

So we have that $S_{k+1}$ can be chosen as

$$S_{k+1} = \begin{bmatrix} I_{q_{k+1}} & 0 \\ -K^*_{k+1} & I_{p-q_{k+1}} \end{bmatrix} S^*_{k+1}$$

where $K^*_{k+1} = \tilde{R}_{k+1} L_B\, \tilde{C}_k\, \bar{D}\dagger_{k+1}$; the symbol $\dagger$ denotes the pseudoinverse. Then

$$\bar{D}_{k+1} = \begin{bmatrix} \bar{D}_k \\ \cdots\cdots\cdots \\ \bar{R}_{k+1} L_B\, \tilde{C}_k \end{bmatrix} \qquad \bar{C}_{k+1} = \begin{bmatrix} \bar{C}_k \\ \cdots\cdots\cdots \\ \bar{R}_{k+1} L_A\, \tilde{C}_k \end{bmatrix}$$

and

$$\tilde{C}_{k+1} = \tilde{R}_{k+1} L_A\, \tilde{C}_k - K^*_{k+1}\, \bar{C}_{k+1}$$

This unique explicit representation of the algorithm will be assumed without further comment in the remainder of the paper.

Note that if a is the first integer such that $q_a$ remains constant (it will be proved a little further on (Th.2.1) that a exists), then, because of the nesting properties of the sequences $C_k$ and $D_k$, one can replace $K^*_{k+1}$ by

$$K_{k+1} = \tilde{R}_{k+1} L_B\, \tilde{C}_k\, \bar{D}_a\dagger$$

Thus we have that $\tilde{C}_{k+1} = \tilde{R}_{k+1} L_{\hat{A}}\, \tilde{C}_k$ , where $\hat{A} := A - B\,\bar{D}\dagger_a\, \bar{C}_a$.

3 - Properties of the structure algorithm

We shall first consider the effect of regular static state feedback on the structure algorithm. Let u be a feedback law of the form $u = F(x) + G(x)v$ where $G(x)$ is regular.

Lemma 3.1 The operators $N_i$ and the maps $\tilde{C}_i$ are invariant under regular feedbacks and

$$\overline{C}_i \xrightarrow{F,G} \overline{C}_i + \overline{D}_i F \qquad \overline{D}_i \xrightarrow{F,G} \overline{D}_i G \qquad \text{for } i = 1, 2, \ldots$$

The proof of this lemma is quite similar to that one given in [13 ] , in the linear case , and need not to be repeated here.

Let us define

$$I_k(x) = (\tilde{C}_0{}^t, \tilde{C}_1{}^t, \ldots, \tilde{C}_k{}^t)^t \qquad k \geq 0$$

where $\tilde{C}_0 := C$. These maps will play a key role in what follows.

Lemma 3.2 : If rank $I_k$ = rank $I_{k+1}$, then $\mathcal{L}_k := dI_k$ is $(\hat{A}, BK)$ invariant, where $K := \operatorname{Ker} \overline{D}_a$. Moreover rank $I_k$ = rank $I_{k+l}$ , $\forall\, k \geq 0$

Proof : Assume that rank $I_k$ = rank $I_{k+1}$, then rank $\mathcal{L}_k$ = rank $\mathcal{L}_{k+1}$. So for all $X \in \operatorname{Ker} \mathcal{L}_k$

$$L_X \tilde{C}_{i+1} = 0 \qquad i \in \underline{k}$$

Note that $\overline{R}_{i+1} L_{\hat{A}} \tilde{C}_i$ is an entry of $\overline{C}_{i+1} - \overline{D}_{i+1} \overline{D}_a{}^\dagger \overline{C}_a$

but

$$\overline{C}_{i+1} - \overline{D}_{i+1} \overline{D}_a{}^\dagger \overline{C}_a = 0$$

and so

$$\overline{R}_{i+1} L_{\hat{A}} \tilde{C}_i = 0 \qquad i \in \underline{k}$$

It follows that

$$\begin{bmatrix} 0 \\ \\ \tilde{C}_{i+1} \end{bmatrix} = R_{i+1} L_{\hat{A}} \tilde{C}_i \qquad i \in \underline{k}$$

and

$$0 = L_X \begin{bmatrix} 0 \\ \tilde{C}_{i+1} \end{bmatrix} = R_{i+1} L_X L_{\hat{A}} \tilde{C}_i \qquad i \in \underline{k}$$

or equivalently, since $R_{i+1}$ is non singular

$$L_X L_{\hat{A}} \tilde{C}_i = 0 \qquad i \in \underline{k}$$

for all $X \in \operatorname{Ker} \mathcal{L}_k$. Recall that

$$L_X L_{\hat{A}} \tilde{C}_i = L_{\hat{A}} L_X \tilde{C}_i - L_{ad_{\hat{A}} X} \tilde{C}_i$$

It follows that

$$L_{ad_{\hat{A}} X} \tilde{C}_i = 0 \qquad i \in \underline{k}$$

which points out that

$$[\hat{A}, \operatorname{Ker} \mathcal{L}_k] \subset \operatorname{Ker} \mathcal{L}_k$$

Now, from (2.4) and replacing $K^*_{i+1}$ by $K_{i+1}$, one obtains

$$\tilde{R}_{i+1} L_B \tilde{C}_i . K = K_{i+1} \overline{D}_a K = 0 \qquad i \in \underline{k}$$

with $K := \operatorname{Ker} \overline{D}_a$

On the other hand $\overline{R}_{i+1} L_B \tilde{C}_i$ is a submatrix of $\overline{D}_a$. Thus

$$\bar{R}_{i+1} L_B \tilde{C}_i . K = 0 \qquad i \in \underline{k}$$

It follows that

$$R_{i+1} L_B \tilde{C}_i . K = 0 \qquad i \in \underline{k} \quad (3.1)$$

but also

$$L_B \tilde{C}_i K = 0$$

since $R_{i+1}$ is nonsingular. Consequently

$$BK \in Ker \mathcal{L}_k$$

But $Ker \mathcal{L}_k$ is involutive, which implies that

$$[BK, Ker \mathcal{L}_k] \subset Ker \mathcal{L}_k$$

we have proved the first assertion of the lemma.

For the same reasons, already given, we have

$$\begin{bmatrix} 0 \\ \\ L_X \tilde{C}_{k+2} \end{bmatrix} = R_{k+2} L_X L_{\hat{A}} \tilde{C}_{k+1} \qquad \forall X \in Ker \mathcal{L}_{k+1}$$

we have

$$L_X L_{\hat{A}} \tilde{C}_{k+1} = L_{\hat{A}} L_X \tilde{C}_{k+1} - L_{ad_{\hat{A}} X} \tilde{C}_{k+1}$$

But the right hand side is 0 since $X \in Ker \mathcal{L}_{k+1}$ and $[A, X] \in Ker \mathcal{L}_{k+1}$. It follows that $L_X \tilde{C}_{k+1} = 0$, because of the nonsingularity of $R_{k+2}$. This ends the proof of the lemma.□

Theorem 3.1 : Let a be the first integer such that $q_a$ remains constant ; let β the first integer such that rank $l_\beta$ = rank $l_{\beta+1}$ and denote $\gamma$ = rank $l_0$

Then : $$a \le \beta \le n - \gamma$$

Proof : For the upper bound we have that

$$\text{rank } \mathcal{L}_0 = \gamma$$

$$\text{rank} \mathcal{L}_i - \text{rank} \mathcal{L}_{i-1} \ge 1 \qquad 1 \le i \le \beta$$

So

$$\text{rank} \mathcal{L}_\beta \ge \beta + \gamma$$

But it is clear that rank $\mathcal{L}_\beta \le n$ , which implies that $\beta \le n - \gamma$.

For the lower bound note that

$$\text{rank} \begin{bmatrix} \bar{D}_a \\ \ldots\ldots\ldots \\ L_B \tilde{C}_a \end{bmatrix} = \text{rank } \bar{D}_a$$

Consequently $L_B \tilde{C}_a$ is row linearly dependent of the rows of $\bar{D}_a$. In other words there exist $P_i$ such that

$$L_B \tilde{C}_a = \Sigma_{i=0}{}^{a-1} P_i \bar{R}_i L_B \tilde{C}_i$$
$$= \Sigma_{i=0}{}^{a-1} Q_i L_B \tilde{C}_i$$

but the latter does not imply that

$$d\tilde{C}_a = \Sigma_{i=0}{}^{a-1} Q_i d \tilde{C}_i$$

It follows that $\beta \ge a$.□

Lemma 3.3 Ker $(\mathcal{L}_\beta.B) = \text{Ker}\,\overline{D}_a$

Proof : i) Ker $\overline{D}_a \subset$ Ker $(\mathcal{L}_\beta.B)$. The result comes from (3.1) directly

ii) Ker $(\mathcal{L}_\beta.B) \subset$ Ker $\overline{D}_a$. Let assume that $X \in$ Ker $L_B\widetilde{C}_0$. Then $X \in$ Ker $\overline{D}_1$. Now assume that $X \in \text{Ker} \begin{bmatrix} \overline{D}_k \\ \cdots\cdots \\ L_B\widetilde{C}_k \end{bmatrix}$. Obviously $\overline{D}_{k+1}\, X = \begin{bmatrix} \overline{D}_k \\ \cdots\cdots\cdots\cdots \\ \overline{R}_{k+1}\, L_B\widetilde{C}_k \end{bmatrix} . X = 0$, and so $X \in$ Ker $\overline{D}_{k+1}$. It then follows that $X \in$ Ker $\overline{D}_a$.□

Let us now define $V^+$ and $R^+$ as

$V^+ := \text{Ker}\mathcal{L}_\beta$

$R^+ := \overline{\langle ad_A^j\, BK, \quad j \geq 0 \rangle}$

Lemma 3.4 $R^+$ is the supremal (A, BK) invariant controllability distribution contained in $V^+$. $V^+$ and $R^+$ are invariant under regular static state feedback.

Proof : The first assertion is obvious from lemma 3.3. The second one is a direct consequence of lemma 3.1.□

An important question arises immediately : do there exist respective links between $V^*$ et $V^+$ on the one hand and $R^*$ and $R^+$ on the other hand; $V^*$ and $R^*$ denote the respective maximal regular controlled invariant and controllability distributions contained in ker dC [9]. The answer is

Theorem 3.2 : For any system like (2.1), (2.2) we have $V^* \subset V^+$ and $R^* \subset R^+$

Proof : It is enough to perform the proof for the first inclusion. Let (F,G) be a friend of $V^*$. It is well known (see for instance [9]) that there exist local coordinate transformations which put the closed loop system (C,A+BF,BG) into the form.

$$\Sigma \begin{cases} \dot{x}_1 = A_1(x_1, x_2) + B_1(x_1, x_2)\, v \\ \dot{x}_2 = A_2(x_2) + B_2(x_2)\, v \\ y = C(x_2) \end{cases}$$

$$V^* = \text{span}\,\{\frac{\partial}{\partial x_1}\}$$

Now perform the structure algorithm for this new system. It is clear that $V^* \subset \text{Ker}\mathcal{L}_\beta(\Sigma)$. But by lemma 3.1, we have that $\text{Ker}\mathcal{L}_\beta(\Sigma) = \text{Ker}\mathcal{L}_\beta(\Sigma_0)$, since (F, G) is regular naturally. This ends the proof.□

Before concluding this section, let us establish the following result

Lemma 3.5 : For any system like (2.1) - (2.2), if there exists F such that $\overline{C}_a + \overline{D}_a F = 0$, then

$$[A+BF, V^+] \subset V^+ \qquad \text{and} \qquad [A+BF, R^+] \subset R^+$$

Proof : Start from

$$[\hat{A}, V^+] \subset V^+$$

and note that

$$[B(I_m - \bar{D}_a \dagger \bar{D}_a)F, V^+] \subset V^+$$

for any F, since $I_m - \bar{D}_a \dagger \bar{D}_a$ is the orthogonal projection onto Ker $\bar{D}_a = K$. So

$$[\hat{A} + B(I_m - \bar{D}_a \dagger \bar{D}_a)F, V^+] \subset V^+$$

and

$$[\hat{A} + BF - B\,\bar{D}_a \dagger (\bar{C}_a + \bar{D}_a F), V^+] \subset V^+$$

Finally, using "$\bar{C}_a + \bar{D}_a F = 0$", one obtains

$$[A + BF, V^+] \subset V^+$$

The same proof can be performed for $R^+$. □

Lemma 3.6 : For any F such that $\bar{C}_a + \bar{D}_a F = 0$, we have :

$$V^+ = \cap_{j=0}^{\infty} \text{Ker } d\, L^j_{A+BF}\, C$$

Proof : i) "$V^+ \subset \cap_{j=0}^{\infty} \text{Ker } d\, L^j_{A+BF}\, C$".

It is enough to show that, for any $X \in V^+$, we have

$$L_X L^j_{A+BF}\, C = 0 \qquad (3.2)$$

The property is obviously true for j = 0. Assume that it is also true for some j. Write :

$$L_X L^{j+1}_{A+BF}\, C = L_{A+BF} L_X L^j_{A+BF}\, C - L_{[A+BF,\, X]} L^j_{A+BF}\, C$$

Note, that by lemma 3.5, $[A+BF, V^+] \subset V^+$. Then it is clear, from (3.2), that each term of the right hand side is zero, and so :

$$L_X L^{j+1}_{A+BF}\, C = 0$$

It follows that (3.2) is true for all $j \geq 0$, which establishes the result.

ii) "$\cap_{j=0}^{\infty} \text{Ker } d\, L^j_{A+BF}\, C \subset V^+$".

Since $Im - \bar{D}_a \dagger \bar{D}_a$ is the orthogonal projection on to Ker $\bar{D}_a$, we have for any F.

$$L_{B(Im - \bar{D}_a \dagger \bar{D}_a)F}\, \tilde{C}_k = 0 \qquad (3.3)$$

Using "$\bar{C}_a + \bar{D}_a F = 0$", (3.3) leads to

$$\tilde{C}_{k+1} = \tilde{R}_{k+1} L_{A+BF}\, \tilde{C}_k \qquad \forall k \geq 0 \qquad (3.4)$$

We want to show that, for any $X \in \cap_{j=0}^{\infty} \text{Ker } d\, L^j_{A+BF}\, C$,

$$L_X \tilde{C}_k = 0 \qquad \forall k \geq 0 \qquad (3.5)$$

We still proceed by induction. The property is obviously true for k = 0. Assume that it is true for some k.

Write, by (3.4),

$$L_X \tilde{C}_{k+1} = \tilde{R}_{k+1} L_X L_{A+BF}\, \tilde{C}_k$$
$$= \tilde{R}_{k+1} (L_{A+BF} L_X \tilde{C}_k - L_{[A+BF,\, X]} \tilde{C}_k)$$

It is clear that $\cap_{j=0}^{\infty} \text{Ker } d\, L^j_{A+BF}\, C$ is (A+BF) - invariant. Consequently, the right hand side is zero and thus $L_X \tilde{C}_{k+1} = 0$ , which shows that (3.5) is true. The proof is now complete.□

Theorem 3.2, Lemma 3.5 and Lemma 3.6, will be crucial in the next sections , the purpose of which is to solve the (dynamic) block decoupling problem.

## 4) The Static State Feedback block decoupling problem

The general static state feedback block decoupling problem is defined as follows.

"For a given k-partition of the output $\{C_i\}_K$ find necessary and sufficient conditions under which there exist static state feedbacks $u = F(x) + \Sigma_{i\in \underline{k}} G_i(x)v_i$ such that $v_i$ controls $y_i$ without affecting $y_j$, $j\neq i$, $i \in \underline{k}$".

This problem remains unsolved in its most general setting. Up to now, the only available result deals with the special situation when the static state feedback is regular which means that $G = [G_1|\ldots|G_k]$ has rank m in other words, "regular" means that the number of independent inputs is invariant under the feedback action.

Theorem 4.1 [10] : Let $\Sigma_0$ and a k-partition of the output be given. Then the regular static state feedback decoupling problem is solvable if and only if

$$B = \Sigma_{i\in \underline{k}} B \cap R_i^* \qquad (4.1)$$

$R_i^*$ is the maximal regular controllabilty distibution contained in $\cap$ ker dCj , $j\neq i$.

The framework of the conditions (4.1) is purely geometric and not easy to use for practical calculations. The condition (4.1) can be turned into algebraic conditions based upon the structure algorithm, well suited for practical calculations on the one hand, but also for tackling and solving the dynamic block decoupling on the other hand, as it will be shown a little further on.

Consider a k-partition of the output of $\Sigma_0$ into k nonempty subsets of components.

$$y = \begin{vmatrix} y_1 \\ \ldots \\ y_k \end{vmatrix}$$

and let $p_i$ denote the size of the subvector $y_i$ ($0 < p_i$, $\Sigma_{i\in \underline{k}} p_i = p$). This partition induces a corresponding partition of C.

$$C = \begin{vmatrix} C_1 \\ \ldots.. \\ C_k \end{vmatrix}$$

where $C_i$ has $p_i$ rows

Performing the structure algorithm on the subsystems ($C_i$, A, B), one obtains matrices $\overline{C}_{a_i}$ and $\overline{D}_{a_i}$ , the definitions of which are clear from the section 2, as well as those ones of $q_{a_i}$ , $i \in \underline{k}$.

Define

$$\hat{C} = \begin{bmatrix} \overline{C}_{a_1} \\ \ldots\ldots \\ \overline{C}_{a_k} \end{bmatrix} \qquad \text{and } \hat{D} = \begin{bmatrix} \overline{D}_{a_1} \\ \ldots\ldots \\ \overline{D}_{a_k} \end{bmatrix}$$

We write the set of the non essential rows of $\hat{D}$ relatively to the given k-partition as ness $\hat{D}$ , ness $\hat{D}$ : = (basis matrix of rad $\{\text{Im } D_{a_i}{}^t\}_k)^t$ , where "rad" means "radical" in the sense of [17].

An interesting connection between the geometric and algebraic frameworks is provided by the following lemma

Lemma 4.1 : Let $\Sigma_0$ and a k-partition of the output be given. Then

$$B \ker \text{ness } \hat{D} = \Sigma_{i\in \underline{k}} B \cap \cap_{j\neq i} V_j^+$$

Proof : We have

$$\text{rad } \{\text{Im } \overline{D}_{a_i}{}^t\}_k = \cap_{i\in \underline{k}} (\Sigma_{j\neq i} \text{ Im } \overline{D}_{a_j}{}^t)$$

Taking the orthogonal complement of each side, one obtains

$$\text{Ker ness } \hat{D} = \Sigma_{i\in \underline{k}} \cap_{j\neq i} \text{Ker } \overline{D}_{a_j}$$

So

$$B.\text{Ker ness } \hat{D} = \Sigma_{i\in \underline{k}} B \cap \cap_{j\neq i} V^+ . \square$$

Define $K_i$ as a maximal rank solution of $\overline{D}_{a_j} = 0$, $j\in \underline{k}$ $j\neq i$. Then

Theorem 4.2 : Let $\Sigma_0$ and a k-partition of the output be given. Then the regular static state feedback decoupling problem is solvable if and only if

i) $\hat{D}$ is a full row rank matrix (4.2)

ii) $sp\{BK_i\} + V_i^+$ is involutive for all $i\in \underline{k}$ (4.3)

Moreover if. (4.2) and (4.3) hold $(-\hat{D}\dagger \hat{C}, K)$, with $K := [K_1, ..., K_k]$, is a decoupling control law , where $K_i$ fulfils $[BK_i, V_i^+] \subset V_i^+$, $i\in \underline{k}$.

Proof : 1) Necessity : If the problem is solvable (4.1) holds. Since $R^*_i \subset V^+_j$, $\forall j \neq i$ (see section 3, Theorem 3.2), it is clear that

$$B = \Sigma_{i\in \underline{k}} B \cap \cap_{j\neq 1} V^+_j$$

Using lemma 4.1 and the fact that B is monic, one obtains :

$$\text{Ker ness } \hat{D} = I_m$$

or equivalently

$$\text{ness } \hat{D} = 0$$

which proves the necessity of (4.2)

To establish the necessity of (4.3) , perform the structure algorithm on the decoupled system. One easily obtains that

$$V_i^* = V_i^+$$

Then

$$B \cap R_i^* = BK_i$$

Since the system is decoupled

$$[BK_i, V_i^*] \subset V_i^* \qquad i\in \underline{k}$$

and so

$$[BK_i, V_i^+] \subset V_i^+ \qquad i\in \underline{k}$$

2) Sufficiency

If (4.2) holds, using lemma 3.5, one can assert that

$$[A - B\hat{D}\dagger\hat{C}, V_i^+] \subset V_i^+ \tag{4.4}$$

On the other hand,

$$B = \Sigma_{i\in \underline{k}} sp\{BK_i\} \tag{4.5}$$

Recall that $V_i^+$ is involutive. So

$[BK_j, V_i^+] \subset V_i^+ \qquad i,j \in \underline{k} \qquad j \neq i$ (4.6)

By (4.2) we know that $K_i$ exists such that :

$$[BK_i, V_i^+] \subset V_i^+$$

Then, $V_i^+$ is locally (A,B) -invariant, which is enough to prove the existence of a regular static state feedback that decouples $\Sigma_0$ : from (4.3) - (4.6) it is now obvious that $(-\hat{D}\dagger\hat{C},K)$ is a decoupling control law.□

This last theorem provides an interesting procedure for testing the existence of solutions to the regular decoupling problem. This procedure is mainly based upon algebraic calculations. When decoupling is achievable, a particular feedback F can be computed without solving a partial differential equation system.

Now suppose that (4.2) and/or (4.3) fail. One can search dynamic solutions. For this purpose, an auxiliary system is joined to $\Sigma_0$. It is defined as $\Sigma_a$

$$\Sigma_a \begin{cases} \dot{x}_a = D(x, x_a) + E(x, x_a)v \\ u = F(x, x_a) + G(x, x_a)v \end{cases}$$

where $x_a$ belongs to an auxiliary $n_a$-dimensional analytic manifold P and $v \in \mathbb{R}^m$.

We shall say that decoupling is achievable with dynamic compensation if it is possible to find E, D, F, G and P such that the composite system

$$\Sigma \times \Sigma_a \begin{cases} \begin{bmatrix} \dot{x} \\ \cdots \\ \dot{x}_a \end{bmatrix} = \begin{bmatrix} A(x) + B(x)\,F(x.x_a) \\ \cdots\cdots \\ D(x, x_a) \end{bmatrix} + \begin{bmatrix} B(x)\,G(x) \\ \cdots\cdots \\ E(x, x_a) \end{bmatrix} v \\ \\ y = C(x, x_a) = C(x) \end{cases}$$

defined on the analytic manifold M x P can be decoupled with static state feedback.

In the next section we shall tackle the dynamic block decoupling problem under the assumption $q_a = \Sigma_{i \in \underline{k}}\, q_{a_i}$

<u>5) The dynamic Block Decoupling Problem (preliminary results)</u>

In this section I shall proceed as in [2] for the so called Morgan's Problem

<u>Lemma 5.1</u> Let $\Sigma_0$ and a k-partition of the output be given. If $q_a = \Sigma_{i \in k}\, q_{a_i}$ then
$$B = \Sigma_{i \in \underline{k}}\, B \cap \cap_{j \neq i} V_j^+ \Leftrightarrow V^+ = \cap_{i \in \underline{k}} V_i^+$$

<u>Proof :</u> i )"$V^+ = \cap_{j \in \underline{k}} V_i^+ \Rightarrow B = \Sigma_{i \in \underline{k}}\, B \cap \cap_{j \neq 1} V_j^+$".

The proof is similar to that one of lemma 3.1 of [1], turning p into $q_a$, and $p_i$ into $q_{a_i}$, $i \in \underline{k}$. Remark also that $q_a = \dim B - \dim B \cap V^+$, and $q_{a_i} = \dim B - \dim B \cap V^+_i$, $i \in \underline{k}$.

ii) "$B = \Sigma_{i \in \underline{k}}\, B \cap \cap_{j \neq i} V_j^+ \Rightarrow V^+ = \cap_{i \in \underline{k}} V_i^+$".

Since $B = \Sigma_{i \in \underline{k}}\, B \cap \cap_{j \neq i} V_j^+$ , we know, by lemma 4.1, that $\hat{D}$ is full row rank. Then writing $F = -\hat{D}\dagger\hat{C}$, we have

$[A+BF, V_i^+] \subset V_i^+ \qquad i \in \underline{k}$

Consequently

$$[A+BF, \cap_{i\in \underline{k}} V_i^+] \subset \cap_{i\in \underline{k}} V_i^+$$

Using (A+BF) - invariance it is not difficult to show that

$$\cap_{i\in \underline{k}} V_i^+ \subset \cap_{j=0} \operatorname{Ker} d\, L^j_{A+BF} C \qquad (5.1)$$

(see for instance i) of the proof of lemma 3.6).

On the other hand, from "$q_a = \Sigma_{i\in \underline{k}} q_{a_i}$" and "$\hat{D}$ is full row rank", we have from the structure algorithm, performed for (C, A, B), that $\overline{C}_a + \overline{D}_a F = 0$. Using lemma 3.6, the latter yields from (5.1) $\cap_{i\in \underline{k}} V_i^+ \subset V^+$. As usually, $V^+ \subset \cap_{i\in \underline{k}} \subset V_i^+$ the result follows.□

Now consider the following Dynamic Algorithm (D.A.), which generalizes that one of [2] available when k = p.

Step 1 : From the given ($\{C_i\}_k$, A, B) calculate the $\hat{D}$ matrix. If $\hat{D}$ is a full row rank matrix, Stop !

Step 2 : If $\hat{D}$ is not a full row rank matrix, calculate $G_1$ as a basis matrix of Ker ness $\hat{D}$, and complete $G_1$ with $G_2$ such that $G = [G_1 \mid G_2]$ is a regular matrix. For instance, one can choose the column vectors of $G_2$ as elements of Ker ess $\hat{D}$, where ess $\hat{D}$ denotes the set of essential rows of $\hat{D}$ (ess $\hat{D}$ : = $\hat{D}$/ness $\hat{D}$).

Step 3 : Now consider $\Sigma_0$, and write $u = G_1 v_1 + G_2 v_2$

$$\dot{x} = A + BG_1.v_1 + BG_2 v_2$$

assume that dim $v_2 = q$, and put integrators in series with the q corresponding inputs of $v_2$. One obtains, rewriting $v_1$ as $w_1$.

$$\Sigma_1 \left\{ \begin{array}{l} \begin{bmatrix} \dot{x} \\ \dot{v}_2 \end{bmatrix} = \underbrace{\begin{bmatrix} A + BG_2 v_2 \\ \hline 0 \end{bmatrix}}_{A_e} + \underbrace{\begin{bmatrix} BG_1 & 0 \\ \hline 0 & I_q \end{bmatrix}}_{B_e} \begin{bmatrix} w_1 \\ w_2 \end{bmatrix} \\ \\ y = C(x) = C_e(x) \end{array} \right.$$

Step 4 : Go to step 1 and resume the procedure with $\Sigma_1$ instead of $\Sigma_0$.

Lemma 5.2 : Let $\Sigma_0$ and a k-partition of the output be given. Then, if $q_a = \Sigma_{i\in \underline{k}} q_{a_i}$, "D.A." "converges" after a finite number of "loops", say N, towards an extended system $\Sigma_N$ for which the corresponding $\hat{D}(\Sigma_N)$ matrix is full row rank.

The proof of this lemma being rather long cannot be given here ; it can be found in [18]. It amounts to show that

$$\dim \bigcap_{i\in \underline{k}} V_i^+(\Sigma_{l+1}) < \dim \bigcap_{i\in \underline{k}} V_i^+(\Sigma_l) \qquad (5.2)$$

where l denotes the l-th step of "D.A.". Indeed , if (5.2) holds for increasing l 's, it is clear that , after a finite number of loops , say N , we shall obtain

$$\bigcap_{i\in \underline{k}} V_i^+(\Sigma_N) = V^+(\Sigma_N)$$

By lemma 5.1 , the latter equality is equivalent to say that $\hat{D}(\Sigma_N)$ is a full row rank matrix.

*An illustrative example :*

$$A(x) = \begin{bmatrix} 0 \\ x_4 \\ 0 \\ 0 \end{bmatrix} \quad B(x) = \begin{bmatrix} x_2 & 0 & 0 \\ 0 & 1 & 0 \\ x_3 & 1 & 0 \\ 0 & 0 & 1 \end{bmatrix} \quad C_1(x) = \begin{bmatrix} x_1 \\ x_2 \end{bmatrix} \quad C_2(x) = \begin{bmatrix} x_2 - x_3 \end{bmatrix}$$

Using the structure algorithm one obtains :

$$a_1 = 1 \quad \overline{C}_{a_1} = \begin{bmatrix} 0 \\ x_4 \end{bmatrix} \quad \overline{D}_{a_1} = \begin{bmatrix} x_2 & 0 & 0 \\ 0 & 1 & 0 \end{bmatrix} \quad q_{a_1} = 2$$

$$a_2 = 1 \quad \overline{C}_{a_2} = x_4 \quad \overline{D}_{a_2} = (-x_3\ 0\ \ 0) \quad q_{a_2} = 1$$

One easily verifies that $q_a = 3$ and so $q_a = q_{a1} + q_{a2}$

$$V_1^+ = \operatorname{Ker} d\begin{bmatrix} x_1 \\ x_2 \end{bmatrix} = \begin{bmatrix} 0 & 0 \\ 0 & 0 \\ 1 & 0 \\ 0 & 1 \end{bmatrix} \qquad V_2^+ = \operatorname{Ker} d(x_2 - x_3) = \begin{bmatrix} 1 & 0 & 0 \\ 0 & 1 & 0 \\ 0 & 1 & 0 \\ 0 & 0 & 1 \end{bmatrix}$$

Assume that $x_2 \neq 0$ $x_3 \neq 0$, then ness $\hat{D} = (1\ 0\ 0)$. It follows that one can choose, for D.A.,

$$G_1 = \begin{bmatrix} 0 & 0 \\ 1 & 0 \\ 0 & 1 \end{bmatrix} \text{ and } G_2 = \begin{bmatrix} 1 \\ 0 \\ 0 \end{bmatrix}$$

So we have, putting $x_5 = u_1$,

$$A_e = \begin{bmatrix} x_2 x_5 \\ x_4 \\ x_3 x_5 \\ 0 \\ \hline 0 \end{bmatrix} \qquad B_e = \begin{bmatrix} 0 & 0 & | & 0 \\ 1 & 0 & | & 0 \\ 1 & 0 & | & 0 \\ 0 & 1 & | & 0 \\ \hline 0 & 0 & | & 1 \end{bmatrix}$$

Using again the structure algorithm one obtains,

$$a_1 = 2 \quad \overline{C}_{a_1} = \begin{bmatrix} x_4 x_5 \\ x_4 \end{bmatrix} \quad \overline{D}_{a_1} = \begin{bmatrix} x_2 x_5 & 0 \\ 0 & 1 & 0 \end{bmatrix}$$

$$a_2 = 2 \quad \overline{C}_{a_2} = -x_3 x_5^2 \quad \overline{D}_{a_2} = (-x_3 - x_5\ 1)$$

Now $\hat{D}$ is full row rank

$$V_1^+ = \mathrm{Ker}\, d\begin{bmatrix} x_1 \\ x_2 \\ x_2 x_5 \end{bmatrix} = \begin{bmatrix} 0 & 0 \\ 0 & 0 \\ 0 & 1 \\ \hline 0 & 0 \end{bmatrix} \qquad V_2^+ = \mathrm{Ker}\, d\begin{bmatrix} x_2 - x_3 \\ x_4 - x_3 x_5 \end{bmatrix} = \begin{bmatrix} 1 & 0 & 0 \\ 0 & 1 & 0 \\ 0 & 1 & 0 \\ 0 & 0 & x_3 \\ \hline 0 & -\frac{x_5}{x_3} & 1 \end{bmatrix}$$

For this example it can be verified that decoupling is achievable ($V_1{}^+ = V_1{}^*$, $V_2{}^+ = V_2{}^*$) and even Morgan's Problem is solvable.

In this section , we have proposed an algorithmic dynamic procedure (D.A.) that leads to an extended system $\Sigma_N$ for which the first condition (4.2) of Theorem 4.2 is fulfilled. However the latter condition is not enough for decoupling. It may happen that the second one (4.3) fails for $\Sigma_N$. In such a situation , more auxiliary dynamics are necessary. We shall give the way to procceed in a more complete version of this work.

REFERENCES

[1] DESCUSSE J.
Sur la structure a l'infini des systèmes linéaires découplables: le cas des systèmes inversibles à droite , Outils et Modèles Mathématiques pour l'Automatique , l'Analyse de Systèmes et le Traitement du Signal , Vol. 3 , 489-499 , Editions du CNRS , Paris , 1983.

[2] DESCUSSE J. , MOOG C. H.
Decoupling with dynamic compensation for strong invertible affine nonlinear systems , Intern. Journ. of Contr. , Vol. 42 , 6 , 1387-1398 , 1985.

[3] DESCUSSE J , MOOG C H
Dynamic decoupling for right invertible nonlinear systems , Systems and control letters , Vol. 8 , 345-349 , 1987.

[4] FALB P. , WOLOVICH W. A.
Decoupling in the design and synthesis of multivariable control systems , IEEE Trans. on Automat. Contr. , 12 , 651-659, 1967.

[5] FLIESS M.
A new approach to the noninteracting control problem in nonlinear system theory , Proc. 23rd Allerton Conf. Commun. Cont. Comput. , Monticello IL , 192-197 , USA , 1986.

[6] FREUND E.
The structure of decoupled nonlinear systems , Intern. Journ. of Contr. , Vol. 3 , 3 , 443-450 , 1975.

[7] GRIZZLE J. W. , DI BENEDETTO M. D., MOOG C. H.

Computing the differential output rank of a nonlinear system , Proc. of the 26th IEEE conf. on Dec. and Contr. , 142-145 , Los Angeles , USA , 1987.

[8] HIRSCHORN R. M.
Invertibility of multivariable nonlinear systems , IEEE Trans. on Automat. Contr. , 24 , 855-865 , 1979.

[9] ISIDORI A.
Nonlinear control systems: an introduction , Lecture Notes in Control and Information Scince , Vol. 72 , Springer-Verlag , Berlin , 1985.

[10] ISIDORI A. , KRENER A. J. , GORI GIORGI C. , MONACO S.
Nonlinear decoupling via feedback: a differential geometric approach , IEEE Trans. on Automat. Contr. , 26 , 331-345 , 1981.

[11] NIJMEIJER H.
Noninteracting control for nonlinear systems , Proc. of the 22th IEEE conf. on Dec and Contr. , 131-133 , San Antonio, USA , 1983.

[12] NIJMEIJER H. , RESPONDEK W.
Dynamic input-output decoupling of nonlinear contol systems , Memo 575 , Technical University Twente , 1986.

[13] SILVERMAN L. , PAYNE H. J.
Input-output structure of linear systems with application to the decoupling problem , SIAM Journ. on Contr. and Optim. , Vol. 9 , 2 , 199-233 , 1971.

[14] SINGH S. N.
Decoupling of invertible nonlinear systems with state feedback and precompensation , IEEE Trans. on Automat. Contr. , 25 , 1237-1239 , 1980.

[15] SINGH S. N.
Generalized decoupled-control synthesis for invertible nonlinear systems , IEE Proc. , Part D , 128 , 157-160 , 1981.

[16] SINHA P. K.
State feedback decoupling of nonlinear systems , IEEE Trans. on Automat. Contr. , 22 , 487-489 , 1977.

[17] WONHAM W. M.
Linear multivariable control: a geometric approach , Lecture Notes in Applications of Mathematics , 2nd Edit. Vol 10 , Springer Verlag , Berlin , 1979.

[18] DESCUSSE J.
Towards the solution of the nonlinear block decoupling problem. Preprints de la Conférence Internatinale "Automatique Non-Linéaire" , Année Non-Linéaire du CNRS 13-17 Juin 1988 , Nantes , France.

# LINEARIZATION OF DISCRETE AND DISCRETIZED NONLINEAR SYSTEMS

Hong-Gi Lee
Department of Electrical and Computer Engineering
Louisiana State University
Baton Rouge, Louisiana 70803-5901

Aristotle Arapostathis and Steven I. Marcus
Department of Electrical and Computer Engineering
University of Texas at Austin
Austin, Texas 78712-1084

## I. INTRODUCTION

The problem of linearization of a nonlinear system is that of finding a (nonlinear) state coordinate change (and feedback) such that the resulting closed-loop system behaves as a linear system under the new coordinates. Once such a linearizing transformation and feedback are obtained, we can then use linear system theory in order to control the original system. Thus, linearization when applicable is an extremely powerful technique for the development of efficient control laws for nonlinear systems (i.e., see [8]).

The linearization problem for continuous time systems has been studied extensively (see, e.g., the references in [5]). However, the complexity of the control laws makes almost mandatory the use of computers to perform the necessary on-line calculation. The effect of the discretization on the linearization process has not been fully researched. For example, suppose that a continuous time system (e.g., a robot manipulator) is linearizable by state coordinate change and feedback. Then if a desired feedback is applied, the resulting discretized closed-loop system is no longer a linear one under the new coordinates. This is because the control input should be a constant between the sampling times.

Necessary and sufficient conditions for linearizability of the discrete-time system by state coordinate change (and feedback) can be found in [1,3,5,6]; some of the results of [5] and [6] will be presented in the next section. Some work on the effects of sampling on linearizability has been reported in [1,2,4,10]. Much of the research in this area has been stimulated by the work of J. Grizzle. In particular, he has conjectured the following [1,2].

**Conjecture** [2]: Let $\Sigma$: $\dot{x} = f(x,u)$ be a single-input analytic control system on $\mathbb{R}^n$, $n \geq 2$, such that $f(0,0) = 0$, and let $\Sigma_h$: $x_{k+1} = f_h(x_k,u_k)$ be its sampled data representation for a sampling interval h. Then $\Sigma_h$ is locally feedback linearizable for an open set of sampling times (i.e., it is *sampled feedback linearizable* [4]) if and only if $\Sigma$ is state-equivalent to a controllable linear system.

This conjecture is incorrect if we consider a general nonlinear system of the form $\dot{x} = f(x,u)$. (Suppose that $\dot{x} = F(x) + G(x)u$ is state-equivalent to a controllable linear system. Let $f(x,u) = F(x) + G(x)(u + u^3)$.) Thus we consider an affine control system of the form $\dot{x} = F(x) + G(x)u$ and furthermore, we assume that F(x) and G(x) are complete analytic vector fields.

In this paper, it is shown that the discretized system is linearizable by state coordinate change for an open set of sampling times if and only if the continuous time system is linearizable by state coordinate change. Also, we show that J. Grizzle's conjecture is true for affine control systems when n = 2 and suggest a method of proof when $n \geq 3$. This work is closely related to that of B. Jacubczyk and E. Sontag [4]; they show that sampled feedback linearizability implies that the continuous time system is also linearizable by state coordinate change and feedback, and they have derived necessary conditions for sampled feedback linearizability that are similar to ours.

## II. LINEARIZATION OF DISCRETE-TIME SYSTEMS

Consider the single-input discrete time nonlinear system

$$x(t+1) = f(x(t),u(t)) \tag{2.1}$$

where $x \in \mathbb{R}^n$, $u \in \mathbb{R}$, and f is a smooth function with $f(0,0) = 0$.

**Definition 2.1**: Two discrete time systems are said to be <u>state equivalent</u> if there exists a smooth state coordinate change around $0 \in \mathbb{R}^n$ which transforms one into the other.

**Definition 2.2**: A discrete time nonlinear system of the form (2.1) is said to be <u>linearizable by state coordinate change</u> if it is state equivalent to a reachable linear system.

**Definition 2.3**: A discrete time system is said to be <u>linearizable by state coordinate change and feedback</u> if there exists a smooth nonlinear feedback $u = \gamma(x,v)$ such that the closed-loop system is linearizable by state coordinate change.

Since the problem of linearization is essentially local in nature, the results presented here will be primarily in terms of local coordinates. The following maps naturally describe the dynamics of (2.1) relative to input sequences. Let $\hat{f}^1(x, u) = f(x, u)$, and $\hat{f}^{i+1}(x, u) = f(\hat{f}^i(x, u))$ for $i \geq 1$. This map is the "pulse response" which represents the effect of input u at t = 0 on the state at t = i. Also, we define

$$\Psi_x(u_1, u_2, \dots, u_n) = f(f(\dots(f(x, u_n), \dots), u_2), u_1) \text{ and}$$

$$F_x(u_1, u_2, \dots, u_{n+1}) = f(f(\dots(f(x, u_{n+1}), \dots), u_2), u_1).$$

Clearly then $\Psi_x: \mathbb{R}^n \to \mathbb{R}^n$, while $F_x: \mathbb{R}^{n+1} \to \mathbb{R}^n$ and $\hat{f}^i: \mathbb{R}^{n+1} \to \mathbb{R}^n$. Also, we let $\underline{u} = (u_1,\dots,u_n)$.

The major results of [5] which are relevant to the problem of sampled feedback linearization are summarized below. We consider local linearization about the equilibrium point (x,u) = (0,0).

**Theorem 2.1** [5]: The discrete time system (2.1) is (locally) linearizable by state coordinate change if and only if

(i) $(\Psi_0)_*\big|_{\underline{u}=0}$ is an isomorphism;

(ii) $\left[\frac{\partial}{\partial u_i}, \ker (F_0)_*\right] \subset \ker (F_0)_*$ for $1 \leq i \leq n+1$.

Furthermore, $T = (\Psi_0)^{-1}$ is a linearizing coordinate change.

**Proof**: (Necessity) Let V be an open neighborhood around 0 and T: $V \to \mathbb{R}^n$ be a linearizing coordinate change. It follows that $\tilde{f}(y, u) = T(f(T^{-1}(y), u))$, $y \in \mathbb{R}^n$ is a linear map of the form

$\tilde{f}(y, u) = Ay + bu$, with (A,b) being a reachable pair. Hence, $f(x,u) = T^{-1}(\tilde{f}(T(x), u))$, $x \in V$, and since $T(0) = 0$, we obtain $\Psi_0(u_1, \ldots, u_N) = T^{-1}(A^{n-1}bu_n + \ldots + bu_1)$. Condition (i) follows, since the pair (A,b) is reachable. On the other hand, $F_0(u_1, \ldots, u_{n+1}) = T^{-1}(A^n bu_{n+1} + \ldots + bu_1)$. Hence $(F_0)_*(\frac{\partial}{\partial u_i}) = (T^{-1})_* A^{i-1}b$ is a well defined vector field, and by Lemma 2 of [5], condition (ii) follows.

(Sufficiency) Consider $\tilde{f}(y, u) = \Psi_0^{-1}(f(\Psi_0(y), u))$, which by (i) is a well defined map on a neighborhood of the origin in $\mathbb{R}^n \times \mathbb{R}$. Let $Y_i = (F_0)_*(\frac{\partial}{\partial u_i})$, i=1,2,...,n+1. Note then that $[Y_i, Y_j] = (E_0)_*[\frac{\partial}{\partial u_i}, \frac{\partial}{\partial u_j}] = 0$, for $1 \le i,j \le n+1$. Therefore, $\{Y_1, Y_2, \ldots, Y_{n+1}\}$ is a family of n+1 commuting vector fields defined on a neighborhood of the origin in $\mathbb{R}^n$. Furthermore, since $(F_0)_*(\frac{\partial}{\partial u_i}) = (\Psi_0)_*(\frac{\partial}{\partial u_i})$, i=1,2,...,n, and $(\Psi_0)_*$ is an isomorphism, it follows that $\{Y_1, Y_2, \ldots, Y_n\}$ are linearly independent. Hence, there exist constants $a_i \in \mathbb{R}$, such that $Y_{n+1} = \sum_{i=1}^{n} a_i Y_i$. Observing that $f(\Psi_0(y), u) = F_0(u, y)$, $y \in \mathbb{R}^n$, it follows that

$$\frac{\partial}{\partial u}\tilde{f}(y, u) = \frac{\partial}{\partial u}\Psi_0^{-1}(F_0(u, y)) = (\Psi_0)_*^{-1}(F_0)_*(\frac{\partial}{\partial u}) = (\Psi_0^{-1})_*(\Psi_0)_*(\frac{\partial}{\partial u}) = \frac{\partial}{\partial u}$$

Similarly, $\frac{\partial}{\partial y_i}\tilde{f}(y, u) = (\Psi_0)_*^{-1}(F_0)_*(\frac{\partial}{\partial y_i}) = \frac{\partial}{\partial y_i}$, $i=1,\ldots,n-1$, and $\frac{\partial}{\partial y_n}\tilde{f}(y,u) = (\Psi_0)_*^{-1}(E_0)_*(\frac{\partial}{\partial y_n}) = a_1\frac{\partial}{\partial u} + \sum_{i=2}^{n} a_i \frac{\partial}{\partial y_{i-1}}$. Hence with $(y_n, \ldots, y_1, u)$ an ordered basis for $\mathbb{R}^{n+1}$ we obtain the following representation of $\tilde{f}$:

$$\tilde{f} = \begin{bmatrix} a_n & 1 & 0 & \ldots & 0 & 0 \\ a_{n-1} & 0 & 1 & \ldots & 0 & 0 \\ \vdots & \vdots & \vdots & \vdots & \vdots & \vdots \\ a_2 & 0 & 0 & \ldots & 1 & 0 \\ a_1 & 0 & 0 & \ldots & 0 & 1 \end{bmatrix} \begin{bmatrix} y_n \\ \vdots \\ y_1 \\ u \end{bmatrix}$$

(Q.E.D.)

Notice that condition (i) of Theorem 2.1 is equivalent, in local coordinates, to the statement that $\left\{(\frac{\partial f}{\partial u})_{(0,0)}, (\frac{\partial f}{\partial x})_{(0,0)}(\frac{\partial f}{\partial u})_{(0,0)}, \ldots, (\frac{\partial f}{\partial x})^{n-1}_{(0,0)}(\frac{\partial f}{\partial u})_{(0,0)}\right\}$ are linearly independent.

In order to state a condition equivalent to Theorem 2.1(ii), we need the following definition.

**Definition 2.4**: Let $h: M \to Q$ be a smooth function, from the smooth manifold M to the smooth manifold Q, and suppose that $h_*$, the tangent map from $M_p$ to $Q_{h(p)}$, is surjective. Let X be a smooth vector field on M. Then $h_*(X)$ is said to be a well-defined smooth vector field on Q if $h_*(X_p) = h_*(X_q)$ whenever $h(p) = h(q)$; in this case, the vector fields X and $h_*(X)$ are said to be h-related.

**Theorem 2.2** [5]: Suppose that $(\frac{\partial f}{\partial x})_{(0,0)}$ is nonsingular. Then $\left[\frac{\partial}{\partial u_i}, \ker (F_0)_*\right] \subset \ker (F_0)_*$, i=1,…,n+1 if and only if:

a) $\hat{f}^i_*(\frac{\partial}{\partial u})$ is well defined for i=1,…,n+1;

b) $\left[\hat{f}^i_*(\frac{\partial}{\partial u}), \hat{f}^j_*(\frac{\partial}{\partial u})\right] = 0$ for $1 \le i,j \le n+1$.

**Proof:** Necessity follows easily. For sufficiency, let $Y_i \equiv \hat{f}^i_*(\frac{\partial}{\partial u})$, i=1,…,n+1 and $\Phi^{Y_i}_u(x)$ denote the integral curve of $Y_i$ through $x \in \mathbb{R}^n$. Clearly then $\Phi^{Y_1}_u(0) = f(0,u)$, $\Phi^{Y_1}_{u_1} \circ \Phi^{Y_2}_{u_2}(0) = f(f(0,u_2),u_1)$ and hence $F_0(u_1, \dots, u_{n+1}) = \Phi^{Y_1}_{u_1} \circ \Phi^{Y_2}_{u_2} \circ \dots \circ \Phi^{Y_{n+1}}_{u_{n+1}}(0)$. Since $[Y_i, Y_j] = 0$, it follows that $\Phi^{Y_j}_{u_j} \circ \Phi^{Y_i}_{u_i}(0) = \Phi^{Y_i}_{u_i} \circ \Phi^{Y_j}_{u_j}(0)$, and hence we obtain $(F_0)_*(\frac{\partial}{\partial u_i}) = Y_i$. (Q.E.D.)

Necessary and sufficient conditions for local linearizability by state coordinate change and feedback are given in the following theorem.

**Theorem 2.3** [6]: The system (2.1) is locally linearizable by state coordinate change and feedback if and only if

(i) $(\Psi_0)_*\big|_{u=0}$ is an isomorphism;

(ii) $\Delta_i \equiv f_*(\Delta_0) + \hat{f}^2_*(\Delta_0) + \dots + \hat{f}^i_*(\Delta_0)$ are (locally) well-defined i-dimensional involutive distributions for $1 \le i \le n-1$, where $\Delta_0 \equiv \text{span}\,\{\frac{\partial}{\partial u}\}$.

**Theorem 2.4** [5]: Under condition (i) of Theorem 2.3, condition (ii) of Theorem 2.3 holds if and only if

$$\left[\frac{\partial}{\partial u_i}, \ker\,(\Phi_0)_*\right] \subset \ker\,(\Phi_0)_* + \text{span}\,\{\frac{\partial}{\partial u_1}, \frac{\partial}{\partial u_2}, \dots, \frac{\partial}{\partial u_i}\} \text{ for } 1 \le i \le n-1.$$

**Remarks:** (i) Obviously, the conditions of Theorems 2.1 and 2.2 imply those of Theorems 2.3 and 2.4.
(ii) In addition to the results described above, conditions for approximate linearization are presented in [6], while conditions for linearization of discrete-time systems with outputs are presented in [5].

**Example 2.1** [5]: Consider

$$\Sigma: \begin{bmatrix} x_1(t+1) \\ x_2(t+1) \end{bmatrix} = \begin{bmatrix} x_2(t) - u(t)^2 \\ u(t) \end{bmatrix} = f(x(t), u(t)).$$

Then

$$\Psi_x(u_1, u_2) = f(f(x, u_2), u_1) = \begin{bmatrix} u_2 - u_1^2 \\ u_1 \end{bmatrix}$$

$$\Psi_0(u_1, u_2) = \begin{bmatrix} u_2 - u_1^2 \\ u_1 \end{bmatrix}$$

$$F_x(u_1, u_2, u_3) = f(f(f(x, u_3), u_2), u_1) = \begin{bmatrix} u_2 - u_1^2 \\ u_1 \end{bmatrix}$$

$$F_0(u_1, u_2, u_3) = \begin{bmatrix} u_2 - u_1^2 \\ u_1 \end{bmatrix}$$

Thus $\ker\,(F_0)_* = \text{span}\,\{\frac{\partial}{\partial u_3}\}$.

Since $(\Psi_0)_*\big|_{u=0}$ is nonsingular, condition (i) of Theorem 2.1 is satisfied. Also, since

$\left[\frac{\partial}{\partial u_i}, \ker (F_0)_*\right] \subset \ker (F_0)_*$ for $1 \le i \le 3$, condition (ii) of Theorem 2.1 is satisfied. Hence $\Sigma$ is locally linearizable by state coordinate change. Furthermore, the desired state coordinate transformation can be obtained by solving the following equations:

$$\begin{bmatrix} x_1 \\ x_2 \end{bmatrix} = \Psi_0(\eta_1, \eta_2) = \begin{bmatrix} \eta_2 - \eta_1^2 \\ \eta_1 \end{bmatrix}$$

Thus, $\eta_1 = x_2$ and $\eta_2 = x_1 + x_2^2$. That is,

$$z = T(x) = \begin{bmatrix} \eta_1 \\ \eta_2 \end{bmatrix} = \begin{bmatrix} x_2 \\ x_1 + x_2^2 \end{bmatrix}.$$

It is easy to see that

$$z(t+1) = \begin{bmatrix} 0 & 1 \\ 1 & 0 \end{bmatrix} z(t) + \begin{bmatrix} 1 \\ 0 \end{bmatrix} u(t).$$

**Example 2.2** [5]: Consider

$$\Sigma: \begin{bmatrix} x_1(t+1) \\ x_2(t+1) \end{bmatrix} = \begin{bmatrix} x_2(t) \\ (1 + x_1(t))u(t) \end{bmatrix} = f(x(t), u(t))$$

Then

$$\Psi_x(u_1, u_2) = f(f(x, u_2), u_1) = \begin{bmatrix} (1 + x_1)u_2 \\ (1 + x_2)u_1 \end{bmatrix}$$

$$\Psi_0(u_1, u_2) = \begin{bmatrix} u_2 \\ u_1 \end{bmatrix}$$

$$F_x(u_1, u_2, u_3) = f(f(f(x, u_3), u_2), u_1) = \begin{bmatrix} (1 + x_2)u_2 \\ (1 + (1 + x_1)u_3)u_1 \end{bmatrix}$$

$$F_0(u_1, u_2, u_3) = \begin{bmatrix} u_2 \\ (1 + u_3)u_1 \end{bmatrix}$$

Thus $\ker (F_0)_* = \text{span}\,\{u_1 \frac{\partial}{\partial u_1} - (1 + u_3)\frac{\partial}{\partial u_3}\}$.

Since $(\Psi_0)_*\big|_{u=0}$ is nonsingular, condition (i) of Theorem 2.1 is satisfied. Note that $\left[\frac{\partial}{\partial u_i}, u_1 \frac{\partial}{\partial u_1} - (1 + u_3)\frac{\partial}{\partial u_3}\right] = \frac{\partial}{\partial u_1} \notin \ker (F_0)_*$. Therefore, $\Sigma$ is not locally linearizable by state coordinate change. However, since $\left[\frac{\partial}{\partial u_1}, \ker (F_0)_*\right] \subset \ker (F_0)_* + \text{span}\,\{\frac{\partial}{\partial u_1}\}$, $\Sigma$ is locally linearizable by state coordinate change and feedback by Theorems 2.3 and 2.4.

## III. LINEARIZATION OF DISCRETIZED SYSTEMS

Consider a nonlinear continuous time control system of the form

$$\dot{x}(t) = F(x(t)) + G(x(t))\, u(t) \tag{3.1}$$

where $x \in \mathbb{R}^n$, $u \in \mathbb{R}$, and F(x), G(x) are analytic vector fields with F(0) = 0. Then the discretized system of (3.1) with sampling interval $h > 0$ is as follows:

$$x(t+h) = f_h(x(t),u(t)) \tag{3.2}$$

where $f_h(x,u) = \Phi_h^{F+uG}(x)$, with $\Phi^X$ denoting the flow of X. Define $V_h(u) = \frac{\partial}{\partial\varepsilon}\Big|_{\varepsilon=0}(\Phi_h^{F+(u+\varepsilon)G}\Phi_{-h}^{F+uG})$. Since $f_h(\cdot,u)$ is a diffeomorphism for each h,u, $V_h(u)$ is a well-defined vector field on $\mathbb{R}^N$ for each h,u (vector fields similar to $V_h(u)$ are also used in [4],[11],[12]). The motivation for the introduction of the vector fields $V_h(u)$ is the following: $(f_h)_*(\frac{\partial}{\partial u})$ is a well-defined vector field if and only if $V_h(u_1) = V_h(u_2)$ for all $u_1,u_2 \in \mathbb{R}$, and this is just condition (a) of Theorem 2.2 for i=1. Now (see [9]), $V_h(u) = \sum_{i=0}^{\infty}\frac{(-1)^i h^{i+1}}{(i+1)!}\,ad^i_{F+uG}G = \sum_{j=0}^{\infty}u^j\beta_j(h)$. The vector fields $\beta_j(h)$ take the form $\beta_j(h) = \sum_{i=j+1}^{\infty}(-1)^{i+1}h^i\beta^i_j$, $j \geq 0$ where $\beta^i_0 = \frac{1}{i!}ad_F^{i-1}G$, $i \geq 1$ and $\beta^i_j = \frac{1}{i}(ad_F\beta^{i-1}_j + ad_G\beta^{i-1}_{j-1})$, $j \geq 1$, $i \geq j+2$, with $\beta^{j+1}_j = 0$ for $j \geq 1$.

It is evident that if the system (3.1) is linearizable by state coordinate change only, then the discretized system (3.2) is linearizable by the same state coordinate change. The following theorem establishes a converse to this fact.

**Theorem 3.1**: Suppose that there exists $\delta > 0$ such that, for every $h \in (0,\delta)$, (3.2) is locally linearizable by state coordinate change. Then (3.1) is locally linearizable by state coordinate change.

**Proof**: Suppose that (3.2) is locally linearizable by state coordinate change for every $h \in (0,\delta)$. Then, since $f_h(\cdot,u)$ is a diffeomorphism,

$$X_h^k = \frac{\partial}{\partial u}(\Phi^F_{(k-1)h}\Phi_h^{F+uG}\Phi^F_{-kh})\Big|_{u=0}$$

is a well defined vector field for all $h \in (0,\delta)$, k=1,2,...; in addition, condition (i) of Theorem 2.1 implies that the family $\{X^k_h, 1 \leq k \leq n\}$ consists of linearly independent vector fields. On the other hand, $X^k_h = \sum_{i=0}^{\infty}c_{ik}h^{i+k-1}ad^i_FG$ for suitable constants $c_{ik}$. Therefore, $ad^i_FG$, $0 \leq i \leq n-1$ are linearly independent. Note also that condition (a) of Theorem 2.2 implies that $(\Phi_h^{F+uG})_*(\frac{\partial}{\partial u})$ is a well-defined vector field, and hence $V_h(u_1) = V_h(u_2)$ for all $h \in (0,\delta)$ and $u_1,u_2 \in \mathbb{R}$, which implies that $\beta^i_1 = 0$ for $i \geq 3$. Since $\beta^i_1 = \frac{1}{i!}ad_G(ad_F^{i-2}G)$ for $i \geq 3$, $ad_G(ad^j_FG) = 0$ for $j \geq 1$. Thus (3.1) is locally linearizable by state coordinate change [7]. (Q.E.D.)

However, if the system (3.1) is not locally linearizable by state coordinate change but is locally linearizable by state coordinate change and feedback, then the discretized system (3.2) is no longer guaranteed to be locally linearizable by state coordinate change and feedback. This is because the control u(t) should be a constant between the sampling times. When n = 1, (3.2) is locally linearizable by state coordinate change and feedback if $G(0) \neq 0$. Thus we assume $n \geq 2$ and investigate the effect of sampling on the linearizability by state coordinate change and feedback.

**Lemma 3.1**: Suppose that there exists $\delta > 0$ such that, for every $h \in (0,\delta)$, (3.2) is locally linearizable by state coordinate change and feedback (i.e., (3.1) is *sampled feedback linearizable*). Then $\{\beta_p(h), p \geq 0\}$

are parallel for every $h \in (0,\delta)$.

**Proof:** Suppose that (3.2) is locally linearizable by state coordinate change and feedback for every $h \in (0,\delta)$. Then condition (ii) of Theorem 2.3 implies that $(\Phi_h^{F+uG})_* \,(sp\{\frac{\partial}{\partial u}\})$ is a well-defined distribution. That is, $V_h(u_1)$ and $V_h(u_2)$ should be parallel for every $h \in (0,\delta)$, and $u_1,u_2 \in \mathbb{R}$, which implies that $\{\beta_p(h), p \geq 0\}$ should be parallel for every $h \in (0,\delta)$. (Q.E.D.)

**Lemma 3.2:** Suppose that (3.1) is sampled feedback linearizable, and assume that $ad_F G$ is a complete vector field. Then $ad_G ad_F^i G = 0$ for $1 \leq i \leq 2$.

**Proof:** By Lemma 3.1, $\beta_0$, $\beta_1$, and $\beta_2$ should be parallel. Thus $ad_G ad_F G = \alpha_1 G$ for some analytic scalar function $\alpha_1$. Note that

$$\beta_1^3 = \frac{1}{3!}\, ad_G ad_F G = \frac{1}{3!}(\alpha_1 G)$$

$$\beta_1^4 = \frac{1}{4}(ad_F \beta_1^3 + ad_G \beta_0^3) = \frac{1}{4!}(2L_F(\alpha_1)G + 2\alpha_1 ad_F G)$$

$$\beta_2^4 = \frac{1}{4}\, ad_G \beta_1^3 = \frac{1}{4!} L_G(\alpha_1) G$$

$$\beta_2^5 = \frac{1}{5}(ad_F \beta_2^4 + ad_G \beta_1^4)$$

$$= \frac{1}{5!}\{(L_F L_G(\alpha_1) + 2L_G L_F(\alpha_1) + 2\alpha_1^2)G + 3L_G(\alpha_1) ad_F G\}$$

Now consider

$$0 = \det[\beta_0 \beta_2 \; ad_F^2 G\, ad_F^3 G \ldots ad_F^{n-1} G]$$

$$= -h^5 \det[\beta_0^1 \; \beta_2^4 \; ad_F^2 G \ldots ad_F^{n-1} G]$$

$$+ h^6 \sum_{i=1}^{2} \det[\beta_0^i \; \beta_2^{6-i} \; ad_F^2 G \ldots ad_F^{n-1} G] + 0(h^7)$$

for every $h \in (0,\delta)$. Thus

$$0 = \sum_{i=1}^{2} \det[\beta_0^i \; \beta_2^{6-i} \; ad_F^2 G \ldots ad_F^{n-1} G]$$

$$= \frac{1}{5!\,2} L_G(\alpha_1) \det[G\, ad_F G\, ad_F^2 G \ldots ad_F^{n-1} G]$$

Since $\det[G\, ad_F G \ldots ad_F^{n-1} G] \neq 0$, $L_G(\alpha_1) = 0$. Thus $\beta_2^4 = 0$ and $\beta_2^5 = \frac{1}{5!}(2L_G L_F(\alpha_1) + 2\alpha_1^2)G$. By using the Jacobi identity, it can be easily shown that

$$ad_G ad_G ad_F^3 G = ad_G ad_F ad_G ad_F^2 G - ad_G ad_{ad_F G} ad_F^2 G$$

$$= ad_G ad_F ad_G ad_F^2 G - ad_{ad_F G} ad_G ad_F^2 G + ad_{ad_F^2 G} ad_G ad_F G$$

Thus $ad_G ad_G ad_F^3 G = c_1 G + 3L_G L_F(\alpha_1) ad_F G$ for some scalar function $c_1$. Now, note that

$$\beta_1^5 = \frac{1}{5}(ad_F \beta_1^4 + ad_G \beta_0^4) = \frac{1}{5!}(2L_F^2(\alpha_1)G + 4L_F(\alpha_1) ad_F G + 2\alpha_1 ad_F^2 G + ad_G ad_F^3 G)$$

Thus $\beta_2^6 = \frac{1}{6}(ad_F \beta_2^5 + ad_G \beta_1^5) = c_2 G + \frac{1}{6!}(9L_G L_F(\alpha_1) + 4\alpha_1^2) ad_F G$ for some scalar function $c_2$. Now consider

$$0 = \det[\beta_0 \; \beta_2 \; ad_F^2 G \ldots ad_F^{n-1} G]$$

$$= h^6 \det[\beta_0^1 \; \beta_2^5 \; ad_F^2 G \ldots ad_F^{n-1} G] - h^7 \sum_{i=1}^{2} \det[\beta_0^i \; \beta_2^{7-i} \; ad_F^2 G \ldots ad_F^{n-1} G] + 0(h^8)$$

Thus

$$0=\sum_{i=1}^{2}\det[\beta_0^i\ \beta_2^{7-i}\ ad_F^2G\ldots ad_F^{n-1}G]$$

$$=\frac{1}{6!}(3L_GL_F(\alpha_1)-2\alpha_1^2)\det[G\ ad_FG\ldots ad_F^{n-1}G]$$

$$\therefore L_GL_F(\alpha_1)=\frac{2}{3}\alpha_1^2$$

Since $L_{ad_FG}(\alpha_1)=L_FL_G(\alpha_1)-L_GL_F(\alpha_1)$ and $L_G(\alpha_1)=0$, $L_{ad_FG}(\alpha_1)=-\frac{2}{3}\alpha_1^2$. By assumption, $ad_FG$ is a complete vector field. Therefore, $\alpha_1(x)=0$, which implies $ad_Gad_FG=0$. By the Jacobi identity,

$$ad_G\,ad_F^2G=ad_Fad_Gad_FG-ad_{ad_FG}ad_FG=ad_Fad_Gad_FG=0 \qquad \text{(Q.E.D.)}$$

**Remark:** By using the Jacobi identity, it can be easily shown that $ad_G\,ad_F^iG=0$ for $1\le i\le 2$ implies that every bracket with the same number or more G's than F's must vanish identically.

In order to show that sampled feedback linearizability implies feedback linearizability of (3.1), we must prove that $ad_Gad_F^jG=0$ for $j\ge 1$; Lemma 3.2 shows this for j=1,2. We would like to prove this for all $j\ge 1$ by induction; so we assume that it holds for $j\ge k$, and derive the consequences.

**Lemma 3.3:** Suppose that (3.1) is sampled feedback linearizable. Let $k\ge 2$ and assume $ad_G\,ad_F^iG=0$ for $i\le k$. Then (i) $\beta_1^i=0$ for $i\le k+2$ and $\beta_1^{k+3}=\frac{1}{(k+3)!}ad_Gad_F^{k+1}G$ ; (ii) $ad_Gad_F^{k+1}G=\alpha_{k+1}G$ for some analytic function $\alpha_{k+1}$; (iii) $L_G(\alpha_{k+1})=0$.

**Proof:** (i) and (ii) are obvious. By using the Jacobi identity, it is easy to show that $ad_Gad_F^{k+2}G$ = $\frac{k+2}{2}ad_Fad_Gad_F^{k+1}$. Note that

$$\beta_1^{k+3}=\frac{1}{(k+1)!}\alpha_{k+1}G$$

$$\beta_1^{k+4}=\frac{1}{(k+4)!}\{\frac{k+4}{2}L_F(\alpha_{k+1})G+\frac{k+4}{2}\alpha_{k+1}ad_FG\}$$

$$\beta_2^{k+i}=0 \text{ for } i\le 3$$

$$\beta_2^{k+4}=\frac{1}{k+4}ad_G\beta_1^{k+3}=\frac{1}{(k+4)!}L_G(\alpha_{k+1})G$$

$$\beta_2^{k+5}=\frac{1}{k+5}\{ad_F\beta_2^{k+4}+ad_G\beta_1^{k+4}\}$$

$$=\frac{1}{(k+5)!}\{(L_FL_G(\alpha_{k+1})+\frac{k+4}{2}L_GL_F(\alpha_{k+1}))G+\frac{k+6}{2}L_G(\alpha_{k+1})ad_FG\}$$

Now consider

$$0=\det[\beta_0\ \beta_2\ ad_F^2G\ldots ad_F^{n-1}G]$$

$$=h^6\det[\beta_0^1\ \beta_2^5\ ad_F^2G\ldots ad_F^{n-1}G]-h^7\sum_{i=1}^{2}\det[\beta_0^i\ \beta_2^{7-i}\ ad_F^2G\ldots ad_F^{n-1}G]+0(h^8)$$

Thus

$$0=\sum_{i=1}^{2}\det[\beta_0^i\ \beta_2^{k+6-i}\ ad_F^2G\ldots ad_F^{n-1}G]$$

$$=\frac{1}{(k+5)!\,2}L_G(\alpha_{k+1})\det[G\ ad_FG\ldots ad_F^{n-1}G]$$

Hence $L_G(\alpha_{k+1}) = 0.$ (Q.E.D.)

**Remark:** Under the assumption of Lemma 3.3, if we can show that $\alpha_{k+1} = 0$, then (3.1) is locally linearizable by state coordinate change. Now we show that $\alpha_{k+1} = 0$ when n=2 and k=2.

**Theorem 3.2:** Let n=2, and assume that $ad_F G$ is a complete vector field. The system (3.1) is sampled feedback linearizable if and only if it is locally linearizable by state coordinate change only.

**Proof:** (Sufficiency) Obvious by Theorem 3.1.

(Necessity) By Lemma 3.2, $ad_G ad_F^i G=0$ for $1 \le i \le 2$. By Lemma 3.3, $ad_G ad_F^3 G = \alpha_3 G$ and $L_G(\alpha_3) = 0$. Therefore $\beta^i_2 = 0$ for $i \le 6$ and $\beta^7_2 = \frac{1}{7!}\{3L_G L_F(\alpha_3)G\}$. Let $\beta^q_0 = \sum_{i=1}^{2} \varepsilon^q_i ad_F^{i-1}G$ and $\beta^q_1 = \sum_{i=1}^{2} \lambda^q_i ad_F^{i-1}G$. By considering $0=\sum_{i=1}^{3} \det[\beta^i_0\ \beta^{8-i}_1]$ we obtain $\beta^7_1 = \frac{1}{7!}\{c_3 G+(\frac{21}{2}L_F(\alpha_3)+42\alpha_3\varepsilon^3_2)ad_F G\}$ for some scalar function $c_3$. Thus $\beta^8_2 = \frac{1}{8!}\{c_4(x)G+\frac{27}{2}L_G L_F(\alpha_3)\, ad_F G\}$ for some scalar function $c_4(x)$, because $L_G(\alpha_3)=0$ and $L_G(\varepsilon^3_2)=0$. By considering

$$0=\sum_{i=1}^{2} \det[\beta^i_0\ \beta^{q-i}_2] = \frac{3}{8!\,2!} L_G L_F(\alpha_3) \det[G\ ad_F G],\quad L_G L_F(\alpha_3) = 0.$$

Since $L_G(\alpha_3) = 0$, $L_{ad_F G}(\alpha_3) = 0$. Therefore $\alpha_3$ is a constant. Using the Jacobi identity, it can be easily shown that

$$\beta^8_2 = \frac{1}{8!}\{\sum_{i=0}^{2} ad_G\, ad_F^i\, ad_G\, ad_F^{5-i}G\}=0$$

$$\beta^9_2 = \frac{1}{9!}\{\sum_{i=0}^{3} ad_G\, ad_F^i\, ad_G\, ad_F^{6-i}G\} = \frac{1}{9!}\{18\alpha_3^2 G\}$$

Consider $0=\sum_{i=1}^{5} \det[\beta^i_0\ \beta^{10-i}_1] = \sum_{\lambda=1}^{5} \{\varepsilon^i_1\lambda^{10-i}_2 - \varepsilon^i_2\lambda^{10-i}_1\} \det[G\ ad_F G]$.

Note that $\varepsilon^1_1 = 1$, $\varepsilon^1_2 = 0$, $\varepsilon^2_1 = 0$, and $\varepsilon^2_2 = \frac{1}{2}$. Thus

$$\begin{aligned}0&=\sum_{i=1}^{5} \{\varepsilon^i_1\lambda^{10-i}_2 - \varepsilon^i_2\lambda^{10-i}_1\}\\ &=\lambda^9_2 - \frac{1}{2}\lambda^8_1 + \sum_{i=3}^{5} \varepsilon^i_1\lambda^{10-i}_2 - \sum_{i=3}^{5} \varepsilon^i_2\lambda^{10-i}_1\end{aligned}$$

$$\therefore \lambda^9_2 = \frac{1}{2}\lambda^8_1 + \sum_{i=3}^{5} \varepsilon^i_2\lambda^{10-i}_1 - \sum_{i=3}^{5} \varepsilon^i_1\lambda^{10-i}_2$$

$$\begin{aligned}L_G(\lambda^9_2) &= \frac{1}{2}L_G(\lambda^8_1) + \sum_{i=3}^{5} L_G(\varepsilon^i_2)\lambda^{10-i}_1 + \sum_{i=3}^{5} \varepsilon^i_2 L_G(\lambda^{10-i}_1)\\ &\quad - \sum_{i=3}^{5} L_G(\varepsilon^i_1)\lambda^{10-i}_2 - \sum_{i=3}^{5} \varepsilon^i_1 L_G(\lambda^{10-i}_2)\end{aligned}$$

Since $\beta^q_2 = 0$ for $q \le 8$, $L_G(\lambda^{10-i}_1) = L_G(\lambda^{10-i}_2) = 0$ for $3 \le i \le 5$. Also, since $ad_G ad_F^2 G = 0$,

$ad_G ad_F^3 G = \alpha_3 G$, and $ad_G ad_F^4 G = 2\alpha_3 ad_F G$, $L_G(\varepsilon_1^3) = L_G(\varepsilon_2^3) = 0$, $L_G(\varepsilon_1^4) = \frac{\alpha_3}{4!}$, $L_G(\varepsilon_2^4) = 0$, $L_G(\varepsilon_1^5) = 0$, and $L_G(\varepsilon_2^5) = \frac{2\alpha_3}{5!}$. Also note that $\lambda_1^5 = \frac{2\alpha_3}{5!}$ and $\lambda_2^6 = \frac{\alpha_3}{5!\,2}$. Thus $L_G(\lambda_2^9) = \frac{1}{2} L_G(\lambda_1^8) - \frac{\alpha_3^2}{5!\,5!\,2}$. Since $\beta_2^9 = \frac{1}{9} ad_G \beta_1^8 = \frac{18}{9!}\alpha_3^2 G$, $L_G(\lambda_1^8) = \frac{18}{8!}\alpha_3^2$. Therefore,

$$\begin{aligned}
\beta_2^{10} &= \frac{1}{10}(ad_F \beta_2^9 + ad_G \beta_1^9) \\
&= \frac{1}{10}\{c_5 G + (\frac{18}{9!}\alpha_3^2 + \frac{18}{8!\,2}\alpha_3^2 - \frac{1}{5!\,5!\,2}\alpha_3^2)\, ad_F G\} \\
&= \frac{1}{10} c_5 G + \frac{432}{10!\,5}\alpha_3^2\, ad_F G
\end{aligned}$$

$$0 = \sum_{i=1}^{2} \det[\beta_0^i \;\; \beta_2^{11-i}] = \frac{-18}{10!\,5}\alpha_3^2 \det[G \; ad_F G].$$

Therefore $\alpha_3 = 0$, i.e. $ad_G ad_F^3 G = 0$. Hence, (3.1) is locally linearizable by state coordinate change [7].

(Q.E.D.)

**Remark:** It is conjectured that when $n \geq 3$ (and with the assumption of completeness of $ad_F^j G$, $j = 1, \ldots, n-1$), Theorem 3.2 is true and the same method of proof works. However, the calculations are very complicated and the details are not yet worked out.

## ACKNOWLEDGEMENT

This research was supported in part by the Air Force Office of Scientific Research under Grant AFOSR-86-0029, in part by the National Science Foundation under Grant ECS-8617860, and in part by the DoD Joint Services Electronics Program through the Air Force Office of Scientific Research (AFSC) Contract F49620-86-C-0045.

## REFERENCES

1. J.W. Grizzle, "Feedback linearization of discrete-time systems," in *Lecture Notes in Control and Information Sciences,* Vol. 83, Springer-Verlag, 1986, 273-281.
2. J.W. Grizzle and P.V. Kokotovic, "Feedback linearization of sampled-data systems," preprint, April 20, 1987.
3. B. Jakubczyk, "Feedback linearization of discrete-time systems," *Syst. & Contr. Lett.,* Vol. 9, 1987, 411-416.
4. B. Jakubczyk and E.D. Sontag, "The effect of sampling on feedback linearization," preprint, May 15, 1987.

5. H.G. Lee, A. Arapostathis and S.I. Marcus, "On the linearization of discrete time systems," *Int'l. J. Contr.*, 45, 1987, 1803-1822.

6. H.G. Lee and S.I. Marcus, "Approximate and local linearization of nonlinear discrete-time systems," *Int'l. J. Contr.*, 44, 1986, 1103-1124.

7. H.J. Sussmann, "Lie brackets, real analyticity, and geometric control," in *Differential Geometric Control Theory*, R.W. Brockett, et.al. (ed.), Boston: Birkhauser, 1983, 1-116.

8. T.J. Tarn, A.K. Bejczy, A. Isidori and Y. Chen, "Nonlinear feedback in robot arm control," *Proc. 23rd Conf. on Decision and Control*, Las Vegas, NV, Dec. 1984, 736-751.

9. V.S. Varadarajan, *Lie Groups, Lie Algebras and Their Representations*, Englewood Cliffs, NJ: Prentice-Hall, 1974.

10. H. G. Lee, A. Arapostathis and S. I. Marcus, "Remarks on Discretization and Linear Equivalence of Continuous Time Nonlinear Systems," *Proc. 26th IEEE Conference on Decision and Control*, Los Angeles, CA, December 9-11, 1987.

11. E. D. Sontag, "Orbit Theorems and Sampling," in *Algebraic and Geometric Methods in Nonlinear Control Theory*, M. Fliess and M. Hazewinkel (Eds.), Reidel, Dordrecht, 1986, pp. 441-486.

12. B. Jakubczyk and E. D. Sontag, "Controllability of Nonlinear Discrete Time Systems: A Lie-Algebraic Approach," submitted for publication.

# A CONDITION FOR THE SOLVABILITY OF THE NONLINEAR MODEL MATCHING PROBLEM

M.D. Di Benedetto

Istituto Universitario Navale
Via Acton, 38. 80133 Napoli, Italia
*and*
Dipartimento di Informatica e Sistemistica
Università di Roma "La Sapienza"
Via Eudossiana, 18. 00184 Roma, Italia

**Abstract.** *The problem considered in this paper consists in finding a dynamic state-feedback controller for a given plant such that the output of the compensated system coincides with the output of a prespecified model. A necessary and sufficient condition is found as an application of the zero-dynamics algorithm. An example illustrates this result and the differences between this approach and the geometric one, followed in a previous work.*

## 1. Introduction

The model matching problem consists in designing a compensating control for a given process such that the resulting input-output behaviour matches that of a prespecified model. For linear systems, this problem has been solved in different forms by different authors (see e.g. [11, 12]). For nonlinear systems, many contributions are currently available [2, 3, 7, 13]. A solution to the problem of compensating a nonlinear system in order to match a linear model is proposed in [7] and extends a result by Moore and Silverman [10] for linear systems. In [13] the matching problem is set as a disturbance decoupling problem. In [2], the problem is considered from a differential algebraic point of view and no aspect of causality is taken into account. A general formulation is given in [3] where a geometric condition for the existence of a solution is proposed in terms of controlled invariant distributions related to the process and the model.

In this paper, the problem of matching a linear model is investigated by using a modified version of Singh's inversion algorithm [14], the so-called "zero-dynamics algorithm" [1, 9], and a simple condition for the solvability of this problem is deduced. The application of this result to an example drawn from [3] shows how a compensator that solves the model matching problem can be constructed by means of the proposed procedure, even when the condition given in [3] is not satisfied.

## 2. Preliminaries and problem statement

Consider a nonlinear plant P described by differential equations of the form

$$\dot{x} = f(x) + g(x)\,u$$

$$y = h(x) \tag{1}$$

where the $n\times 1$ vector x is the state (which belongs to some open subset X of $\mathbb{R}^n$), the $m\times 1$ vector u is the control input, and the vector y is the output, which is assumed to be a $p\times 1$ vector. We assume that f and the m columns $g_1, g_2, \ldots, g_m$ of the matrix g are real smooth vector fields and that h is a real smooth function.

The given linear model can be described by differential equations of the form

$$\dot{z} = Az + Bv$$

$$y_M = Cz \tag{2}$$

where the state z belongs to a submanifold Z of $\mathbb{R}^{n_M}$, the $m_M\times 1$ vector v is the control, the $p\times 1$ vector $y_M$ is the output, and A, B, C are matrices of appropriate dimensions. Let $w_M(t)$ denote the impulse response of the model

$$w_M(t) = Ce^{At}B.$$

The output of the model, initialized at $z^0$, takes the form

$$y_M(t) = Ce^{At}z^0 + \int_0^t w_M(t-\tau)\,v(\tau)\,d\tau$$

and will also be denoted by $y_M(z^0, t)$ to evidentiate the dependence on the initial state $z^0$.

The compensator Q used to control P is assumed to be a dynamic state-feedback compensator, with inputs x and v and output u, of the following general form

$$\dot{\xi} = a(\xi, x, v)$$

$$u = b(\xi, x, v) \tag{3}$$

The state $\xi$ belongs to a submanifold $\Xi$ of $\mathbb{R}^\nu$, a is assumed to be a real smooth vector field and b a real smooth function. The system P controlled by Q is denoted by PoQ, and its output by $y^{PoQ}(t)$, or by $y^{PoQ}(x^0, \xi^0, t)$ to make the dependence on the initial states explicit.

The objective of the matching problem is to find a controller Q such that the driven response of the compensated plant PoQ coincides with the driven response of the model. For the sake of clarity, let us formally state the problem.

**Definition 2.1**. Given a plant P = (f, g, h), a model M = (A, B, C) and a point $(x^0, z^0)$ belonging to $X \times Z$, the *model matching problem* consists in finding an integer $\nu$, a compensator Q of the form (3) and an initial state $\xi^0$ for Q such that

(i) $y^{PoQ}(x^0, \xi^0, t) - y_M(z^0, t)$ is independent of v, for $0 \leq t \leq T$,

where T is a real positive number. If, instead of (i), the following stronger property is required,

(ii) $y^{PoQ}(x^0, \xi^0, t) - y_M(z^0, t) = 0$ for $0 \leq t \leq T$,

then the problem is referred to as the *strong model matching problem* .∎

In the given formulation of the model matching problem, the "zero-input" terms of the model and the compensated plant are not required to be the same (see also [3]). In the stronger version of the model matching problem, the exact coincidence of the responses is required, at least for a well chosen initial state of the compensator.

In [3], the model matching problem was extensively investigated using differential geometric tools [8]. A condition was given, which extended to nonlinear systems previous work by Morse [12], concerning the linear case. In this paper, the problem is considered from a different pont of view. It is shown that a condition for the solvability of the strong model matching problem can be derived as an application of the "zero-dynamics algorithm" [1, 9] that is recalled here for the sake of completeness.

**Zero-dynamics algorithm** [1, 9]
*Step 0.* Define $X_0 = h^{-1}(0)$.
*Step k.* Assume $X_{k-1}$ is a smooth submanifold and define $X_k$ as the set of points x belonging to $X_{k-1}$ such that

$$f(x) \in T_x X_{k-1} + \text{span}\,\{g(x)\}$$

If $X_k = X_{k-1}$ then stop.∎

The zero-dynamics algorithm may also be performed locally around a point $x_e \in X$ as follows [1, 9]. Assume $x_e$ is such that $h(x_e) = 0$.
*Step 0.* Define $X_0 = h^{-1}(0)$.
*Step k.* Assume there exists a neighborhood $U_{k-1}$ of $x_e$ such that $X_{k-1} \cap U_{k-1}$ is a smooth manifold and define $X_k$ as the set of points x belonging to $X_{k-1} \cap U_{k-1}$ such that

$$f(x) \in T_x (X_{k-1} \cap U_{k-1}) + \text{span}\,\{g(x)\}.$$

Assume $x_e$ belongs to $X_k$.

The above procedure converges in a finite number of steps, denoted $k^*$, and $k^* \le n$. A point $x_e$ is said to be *regular* for the zero-dynamics algorithm if, locally around $x_e$, $X_k$ is a smooth manifold, for all k, i.e. if the algorithm can be continued at each step [1]. An open subset $U \subset X$ is said to be *regular* for the zero-dynamics algorithm if $U \subset U_{k^*}$.

## 3. Main result

Let us first introduce an extended system, associated with the plant and the model, as follows:

$$\dot{x}^E = f^E(x^E) + \hat{g}(x^E)\,u + \hat{B}\,v$$
$$y^E = h^E(x^E)$$

with state $x^E = (x, z)$, inputs u and v, and

$$f^E(x^E) = \begin{bmatrix} f(x) \\ Az \end{bmatrix}, \quad \hat{g}(x^E) = \begin{bmatrix} g(x) \\ 0 \end{bmatrix}, \quad \hat{B} = \begin{bmatrix} 0 \\ B \end{bmatrix}$$

$$h^E(x^E) = h(x) - Cz$$

Since the objective of the matching problem is to find a control law u leading the output of the extended system to zero for all inputs v, it is quite natural to apply the zero-dynamics algorithm to the extended system as follows.

**Procedure 3.1.**
*Step 0.* Define $M_0$ as the set of points $x^E = (x, z)$ such that

$$h^E(x^E) = 0$$

*Step k.* Assume $M_{k-1}$ is a smooth submanifold and define $M_k$ as the set of points $x^E$ belonging to $M_{k-1}$ such that

$$f^E(x^E) + \hat{B}\,v \in T_{x^E} M_{k-1} + \text{span}\,\{\hat{g}(x^E)\} \qquad \text{for all } v$$

If $M_k = M_{k-1}$ then stop.∎

As it is for the zero-dynamics algorithm, the above procedure can be performed locally around a point $x'^E$ of X x Z. Assume $x'^E$ is such that $h^E(x'^E) = 0$.
*Step 0.* Define $M_0 = \{x^E: h^E(x^E) = 0\}$.
*Step k.* Assume there exists a neighborhood $U_{k-1}$ of $x'^E$ such that $M_{k-1} \cap U_{k-1}$ is a smooth manifold and define $M_k$ as the set of points $x^E$ belonging to $M_{k-1} \cap U_{k-1}$ such that

$$f^E(x^E) + \hat{B}\,v \in T_{x^E}(M_{k-1} \cap U_{k-1}) + \text{span}\,\{\hat{g}(x^E)\} \qquad \text{for all } v$$

Assume $x'^E$ belongs to $M_k$.

The above procedure converges in a finite number of steps, denoted $\kappa^*$ ($\kappa^* \le n + n_M$) to a smooth manifold. Let $M^* = M_{\kappa^*}$ denote the globally defined manifold of Procedure 3.1, and $L^* = M_{\kappa^*} \cap U_{\kappa^*}$ the locally defined one. A point $x^E$ is said to be regular for Procedure 3.1 if, locally around $x^E$, $M_k$ is a smooth manifold for all k. An open subset $U \subset X \times Z$ is said to be regular for the Procedure 3.1 if $U \subset U_{\kappa^*}$. We are now ready to state the following result.

**Theorem 3.2**. The strong nonlinear model matching problem is solvable if $(x^0, z^0)$ belongs to $L^*$. Conversely, if the strong nonlinear model matching problem is solvable in a regular open set of $X \times Z$, then $(x^0, z^0)$ belongs to $L^*$.

*Proof.* (If) By construction, there exists a smooth feedback

$$u = u^*(x^E, v)$$

with the property that the vector

$$f^*(x^E, v) = f^E(x^E) + \hat{B}\, v + \hat{g}\,(x^E)\, u^*(x^E, v)$$

is tangent to $L^*$ for all v. Then, since on the submanifold $L^*$ $y^E(t) = 0$, the output of the system

$$\begin{aligned} \dot{x} &= f(x) + g(x)\, u^*(x, z, v) \\ \dot{z} &= Az + Bv \\ y^E &= h(x) - Cz \end{aligned} \tag{4}$$

whenever the initial state $(x^0, z^0)$ of (4) belongs to $L^*$, is zero, for all v and for $0 \le t \le T$ (i.e. as long as $(x(t), z(t))$ belongs to $L^*$). This means that, if the plant (1) is controlled by a dynamic compensator

$$\begin{aligned} \dot{\xi} &= A\xi + Bv \\ u &= u^*(x, \xi, v) \end{aligned} \tag{5}$$

with $\xi \in Z$ and $\xi^0 = z^0$, the output h(x) equals the output $y_M(t)$ of the model, i.e.

$$\begin{aligned} y^{PoQ}(t) &= y_M(t) \\ &= Cz(t) = Ce^{At}z^0 + \int_0^t Ce^{A(t-\tau)}B\, v(\tau)\, d\tau \end{aligned} \tag{6}$$

and (ii) is satisfied.

(Only if) If the strong model matching problem is solvable for a given initial state $(x^0, z^0)$, there exists a smooth input function $\hat{u}(t)$ such that the output of the extended system when $u(t) = \hat{u}(t)$, i.e. of the system

$$\begin{aligned} \dot{x} &= f(x) + g(x)\, \hat{u}(t) \\ \dot{z} &= Az + Bv \\ y^E(t) &= h(x) - Cz \end{aligned}$$

is zero for all inputs v. By assumption, the initial state $(x^0, z^0)$ belongs to a regular open set, thus $(x^0, z^0)$ belongs to $U_{\kappa^*}$. Since $L^* = M_{\kappa^*} \cap U_{\kappa^*}$ is the largest manifold (in $U_{\kappa^*}$) on which a dynamics yielding $y^E(t)$ to zero can be defined, then, necessarily $(x^0, z^0)$ must belong to $L^*$. ∎

For the Procedure 3.1 to be continued at each step, it is assumed that all the manifolds $M_k$ are smooth. If this is the case, then the manifold $L^*$ in Theorem 3.2 can be replaced by the globally defined manifold $M^*$. The following holds.

**Corollary 3.3.** Assume Procedure 3.1 can be continued at each step. The strong nonlinear model matching problem is solvable if and only if $(x^0, z^0)$ belongs to $M^*$.

Note that, under the hypotheses of Corollary 3.3, $M^*$ is an invariant manifold for (4). Hence, in that case, the strong model matching problem is solvable with $T = \infty$.

**Remark 3.4.** It is worth noting that Theorem 3.2 and Corollary 3.3 hold also in the case the given model is described by nonlinear differential equations of the form

$$\dot{z} = f_M(z) + g_M(z)\, v$$

$$y_M = h_M(z)$$

Indeed, the proof in this more general case is identical to the one illustrated previously. ∎

As far as the model matching problem is concerned, we can state

**Corollary 3.5.** The model matching problem is solvable if $x^0$ is such that there exists $\xi^0 \in Z$ with the property that $(x^0, \xi^0) \in L^*$.

*Proof.* It is seen from the proof of Theorem 3.2, and in particular from (6), that, if the compensator (5) is initialized at a point $\xi^0$ such that $(x^0, \xi^0)$ belongs to $L^*$, one obtains

$$y^{PoQ}(t) = C\xi(t) = Ce^{At}\xi^0 + \int_0^t Ce^{A(t-\tau)}B\, v(\tau)\, d\tau$$

$$= Ce^{At}\xi^0 + \int_0^t w_M(t-\tau)\, v(\tau)\, d\tau$$

i.e.

$$y^{PoQ}(t) = y_M(t) + Ce^{At}(\xi^0 - z^0)$$

Thus, in that case,

$$y^{PoQ}(x^0, \xi^0, t) - y_M(z^0, t) = Ce^{At}(\xi^0 - z^0) \qquad (7)$$

and (i) is satisfied. ∎

The relation (7) indicates that, if the given model is asymptotically stable, the difference between the responses of the model and the controlled system asymptotically tends to zero.

Hence, under the hypothesis that the model is asymptotically stable, the compensator proposed in the (if) part of the proof of Theorem 3.2 not only solves the model matching problem but makes $y^{PoQ}$ *asymptotically* track the output of the model.

**Remark 3.6**. Procedure 3.1 can be written in local coordinates, as done in [1] for the zero-dynamics algorithm, as follows. Let $M_0$ be the set of points $x^E$ such that

$$h^E(x^E) = 0$$

and let $x'^E$ be a point of $M_0$. Set

$$H_1(x^E) = h^E(x^E)$$

Suppose the smooth mapping $h^E$ has constant rank for all $x^E \in M_0$, in a neighborhood $U_0$ of $x'^E$. Then $M_0 \cap U_0$ is a smooth embedded submanifold. At step $k \geq 1$, $M_k$ is the set of points $x^E$ in $M_{k-1} \cap U_{k-1}$ such that the equation

$$<dH_k(x^E), f^E(x^E) + \hat{g}(x^E)\, u + \hat{B}\, v> = 0$$

can be solved for u, for all v. To identify the points of $M_k$, choose a matrix $R_k(x^E)$ whose rows form a basis of the space of solutions $\gamma(x^E)$ of

$$\gamma(x^E)<dH_k(x^E), \hat{g}(x^E)> = 0$$

at each $x^E \in M_{k-1} \cap U_{k-1}$. Set

$$\phi_k(x^E, v) = R_k(x^E)\, [\, L_{f^E} H_k(x^E) + <dH_k(x^E), \hat{B}> v\, ]$$

Then,

$$M_k = \{x^E \in U_{k-1} \cap M_{k-1} : \phi_k(x^E, v) = 0 \quad \text{for all } v\}$$

For v = 0, one has

$$R_k(x^E)\, L_{f^E} H_k(x^E) = 0 \qquad \text{for } x^E \in M_k$$

Therefore,

$$R_k(x^E) < dH_k(x^E), \hat{B} > = 0 \qquad \text{for } x^E \in M_k$$

Then,

$$M_k = M_{k-1} \cap \lambda_k^{-1}(0) \cap U_{k-1}$$

where we have set

$$\lambda_k(x^E) = \begin{bmatrix} R_k(x^E)\, L_{f^E} H_k(x^E) \\ R_k(x^E) < dH_k(x^E), \hat{B} > \end{bmatrix}$$

and

$$\lambda_k^{-1}(0) = \{\, x^E : \ R_k(x^E)\, L_{f^E} H_k(x^E) = 0 \ \& \ R_k(x^E) <dH_k(x^E), \hat{B}> = 0$$

If the matrix $<dH_k(x^E), \hat{g}(x^E)>$ has constant rank for all $x^E \in M_{k-1} \cap U_{k-1}$, the mapping $\lambda_k(x^E)$ is a smooth mapping. Assume $x'^E \in M_k$. If the mapping $\lambda_k$ has constant rank for all $x^E \in M_k$, in a neighborhood $U_k$ of $x'^E$, then $M_k \cap U_k$ is a smooth embedded submanifold.

Set

$$H_{k+1}(x^E) = \begin{bmatrix} H_k(x^E) \\ \lambda_k(x^E) \end{bmatrix}$$

At some step $\kappa^* \leq n + n_M$, $M_{\kappa^*+1} \cap U_{\kappa^*+1} = M_{\kappa^*} \cap U_{\kappa^*}$ and we set $L^* = M_{\kappa^*} \cap U_{\kappa^*}$ and $H^* = H_{\kappa^*+1}$.

This local coordinate version of Procedure 3.1 shows that, to construct the manifold $M_{\kappa^*}$, the condition

$$R_k(x^E) < dH_k(x^E), \hat{B} > = 0 \qquad \text{for } x^E \in M_k$$

has to be satisfied at each step k. This condition can be related to a property of the ranks of the matrices $H_k$, and in particular to the "algebraic structures at infinity" [10] of the model and the process. This point will be dealt with in a forthcoming paper.

The compensator Q which solves the model matching problem, constructed in the proof of Theorem 3.2, is described by differential equations which are linear in v, as it is for the compensator searched for in [3]. Indeed, the control u* is the solution of the equation

$$<dH^*, f^E(x^E) + \hat{g}(x^E)\, u + \hat{B}\, v> = 0$$

It is not difficult to see then that u* is linear in v, i.e. takes the form

$$u^*(x, \xi, v) = c(x, \xi) + d(x, \xi)\, v . \blacksquare$$

In the following section, the above results will be illustrated by an example.

## 4. Example

Consider the example proposed in [3] where the given plant P = (f, g, h) has state $x \in X = \mathbb{R}^4$ and

$$f(x) = \begin{bmatrix} 0 \\ x_4 \\ 0 \\ 0 \end{bmatrix}, \qquad g(x) = \begin{bmatrix} x_3 & 0 \\ 0 & 1 \\ 1 & 1 \\ 0 & 1 \end{bmatrix}$$

$$h(x) = \begin{bmatrix} x_2 - x_3 \\ x_1 \end{bmatrix}$$

Let M = (A, B, C) be a linear model, with state $z \in Z = \mathbb{R}^4$ and

$$A = \begin{bmatrix} 0 & 1 & 0 & 0 \\ 0 & 0 & 0 & 0 \\ 0 & 0 & 0 & 1 \\ 0 & 0 & 0 & 0 \end{bmatrix}, \quad B = \begin{bmatrix} 0 & 0 \\ 1 & 0 \\ 0 & 0 \\ 0 & 1 \end{bmatrix}$$

$$C = \begin{bmatrix} 1 & 0 & 0 & 0 \\ 0 & 0 & 1 & 0 \end{bmatrix}.$$

Note that the model to be matched is decoupled and can be characterized by the transfer matrix $W_M(s)$:

$$W_M(s) = \begin{bmatrix} \frac{1}{s^2} & 0 \\ 0 & \frac{1}{s^2} \end{bmatrix}$$

For that system, the geometric condition given in [3] for the solvability of the model matching problem is not satisfied. On the contrary, the application of Procedure 3.1 leads to a compensator which is a solution to the model matching problem. That compensator coincides with the one derived in [3] in a heuristic way.

$M_0$ is the set of points (x, z) such that

$$x_2 - x_3 - z_1 = 0$$

$$x_1 - z_3 = 0$$

At the first step $M_1$ is the set of all $x^E$ in $M_0$ such that the equation

$$<dh^E(x^E), f^E(x^E) + \hat{g}(x^E)\, u + \hat{B}v> = 0$$

i.e.

$$\begin{bmatrix} x_4 - z_2 \\ -z_4 \end{bmatrix} + \begin{bmatrix} -1 & 0 \\ x_3 & 0 \end{bmatrix} u = 0$$

is solvable in u, for all v. To identify the points of $M_1$, choose a matrix $R_1(x)$ whose rows form a basis of the subspace orthogonal to

$$<dh^E(x^E), \hat{g}(x^E)> = <dh(x), g(x)>$$

$$= \begin{bmatrix} -1 & 0 \\ x_3 & 0 \end{bmatrix}$$

We can choose

$$R_1(x) = \begin{bmatrix} x_3 & 1 \end{bmatrix}$$

Set

$$\lambda_1(x, z) = \begin{bmatrix} x_3 & 1 \end{bmatrix} \begin{bmatrix} x_4 - z_2 \\ - z_4 \end{bmatrix} = x_3 x_4 - x_3 z_2 - z_4$$

$\lambda_1$ is a smooth map and has constant rank in $M_0$. Then,

$$M_1 = M_0 \cap \lambda_1^{-1}(0)$$

is a smooth submanifold of XxZ. Set

$$H_2(x^E) = \begin{bmatrix} h^E(x^E) \\ \lambda_1(x^E) \end{bmatrix}$$

At step 2, $M_2$ is the set of all $x^E$ in $M_1$ such that the equation

$$<dH_2(x^E), f^E(x^E) + \hat{g}(x^E) u + \hat{B} v> = 0$$

is solvable in u, for all v. We obtain

$$\begin{bmatrix} x_4 - z_2 \\ - z_4 \\ 0 \end{bmatrix} + \begin{bmatrix} -1 & 0 \\ x_3 & 0 \\ x_4 - z_2 & x_4 - z_2 + x_3 \end{bmatrix} u + \begin{bmatrix} 0 & 0 \\ 0 & 0 \\ - x_3 & -1 \end{bmatrix} v = 0 \tag{8}$$

The above equation is solvable for a smooth u for all $x^E$ belonging to $M_1$ such that $x_4 + x_3 - z_2 \neq 0$. Hence

$$M_2 = \{ (x, z): \begin{bmatrix} x_2 - x_3 - z_1 \\ x_1 - z_3 \\ x_3 x_4 - x_3 z_2 - z_4 \end{bmatrix} = 0 \quad \& \quad x_4 + x_3 - z_2 \neq 0 \}$$

For any point in $M_2$, there exists a neighborhood U such that $U \cap M_2$ is a smooth manifold. Moreover, since the matrix

$$L_{g^E} H_2(x^E)$$

is full rank for all $x^E$ belonging to $M_2$, $\kappa^* = 2$ and we can set $L^* = U \cap M_2$. The compensator which permits to track the output of the model M can be derived by solving for u the last equation, which can also be written

$$\begin{bmatrix} x_4 - z_2 \\ -x_3 v_1 - v_2 \end{bmatrix} + \begin{bmatrix} -1 & 0 \\ x_4 - z_2 & x_4 - z_2 + x_3 \end{bmatrix} \begin{bmatrix} u_1 \\ u_2 \end{bmatrix} = 0$$

Therefore, the model matching problem is solvable by the dynamic-state feedback

$$\begin{aligned} &\dot{z} = Az + Bv \\ &u_1 = x_4 - z_2 \\ &u_2 = -\frac{(x_4 - z_2)^2}{x_4 + x_3 - z_2} + \frac{x_3}{x_4 + x_3 - z_2} v_1 + \frac{1}{x_4 + x_3 - z_2} v_2 \end{aligned} \tag{9}$$

whenever the initial state $(x^0, z^0)$ belongs to $L^*$, i.e.

$$\begin{aligned} &z_1(0) = x_2(0) - x_3(0) \\ &z_3(0) = x_1(0) \\ &z_4(0) = x_3(0)x_4(0) - x_3(0)z_2(0) \end{aligned} \tag{10}$$

and $x_4(0) + x_3(0) - z_2(0) \neq 0$. An easy calculation shows that, whenever $x_4(0) + x_3(0) - z_2(0) = 0$, the model matching is not solvable. Since only the variable $z_2$ appears in the feedback law u in (9), the dimension of the compensator can be reduced by taking into account only the equation describing $z_2$. By setting $\xi = z_2$, one obtains

$$\begin{aligned} &\dot{\xi} = v_1 \\ &u_1 = x_4 - \xi \\ &u_2 = -\frac{(x_4 - \xi)^2}{x_4 + x_3 - \xi} + \frac{x_3}{x_4 + x_3 - \xi} v_1 + \frac{1}{x_4 + x_3 - \xi} v_2 \end{aligned} \tag{11}$$

The compensator (11) solves the model matching problem for any x(0), provided $z_2(0) \neq x_4(0) + x_3(0)$.

The compensator (11) coincides with the one derived in [3]. Indeed, it is rather simple to compute the Volterra kernels of P composed with (11) and verify that its driven response coincides with the one of the model, as indicated here as an application of Theorem 3.2.

If the strong model matching problem is considered, then the initial states of the model and the process have to satisfy (10) for the whole response of the process controlled by (11) to coincide with that of the given model. Let us show that the equality of the zero-output terms leads to (10).

Let us consider the plant compensated with (11), having state $x' = (x, \xi)$, which can be described by the vector fields

$$f'(x') = \begin{bmatrix} x_3(x_4 - \xi) \\ x_4 - \dfrac{(x_4-\xi)^2}{x_4+x_3-\xi} \\ \dfrac{x_3(x_4-\xi)}{x_3+x_4-\xi} \\ \dfrac{-(x_4-\xi)^2}{x_3+x_4-\xi} \\ 0 \end{bmatrix}$$

$$g'(x') = \frac{1}{x_3+x_4-\xi}\begin{bmatrix} 0 & 0 \\ x_3 & 1 \\ x_3 & 1 \\ x_3 & 1 \\ 0 & x_3+x_4-\xi \end{bmatrix}$$

and by the output function

$$h'(x') = \begin{bmatrix} x_2-x_3 \\ x_1 \end{bmatrix}$$

The zero-input response of the model , denoted $y_{ML}(t)$, takes the form:

$$y_{ML}(t) = Ce^{At}z^0 = \begin{bmatrix} 1 & 0 & 0 & 0 \\ 0 & 0 & 1 & 0 \end{bmatrix}\begin{bmatrix} 1 & t & 0 & 0 \\ 0 & 1 & 0 & 0 \\ 0 & 0 & 1 & t \\ 0 & 0 & 0 & 1 \end{bmatrix} z^0$$

$$= \begin{bmatrix} z_1(0) + z_2(0)\,t \\ z_3(0) + z_4(0)\,t \end{bmatrix}$$

The zero-input response $w(t, x'^0)$ of the compensated plant, initialized at $x'^0$, takes the form:

$$w(t, x'^0) = \sum_{j=0}^{\infty} L^j_{f'}\, h'(x'^0)\frac{t^j}{j!}$$

Let us compute this response. We have

$$L_f h' = \begin{bmatrix} \xi \\ x_3(x_4 - \xi) \end{bmatrix}, \qquad L_f^2 h' = \begin{bmatrix} 0 \\ 0 \end{bmatrix}$$

Therefore,

$$w(t, x'^0) = \begin{bmatrix} x_2(0) - x_3(0) \\ x_1(0) \end{bmatrix} + \begin{bmatrix} \xi(0) \\ x_3(0)\,[x_4(0) - \xi(0)] \end{bmatrix} t$$

From the last equation and the expression of $y_{ML}(t)$, the relation (10) immediately follows.

## 5. Concluding remarks

In this paper, the model matching problem has been investigated by using a modification of the zero-dynamics algorithm. A necessary and sufficient condition has been proposed, based on the existence of an appropriate manifold of the state space, to which the initial states of the process and the model have to belong if the exact coincidence of the responses is required. The geometric condition given in [3] was related to an equality of the geometric structures at infinity of the process and the extended system. A similar result can be established for the condition proposed in this paper by considering the algebraic structures at infinity of the process and the extended system. This will be illustrated in a forthcoming paper.

**Acknowledgments.** I wish to stress the importance of many ideas, discussions and suggestions offered, during the preparation of this work, by Professor A. Isidori, to whom go my deepest thanks.

## References

[1] Byrnes, C.I., and Isidori, A. "Local stabilization of minimum-phase systems", *Syst. & Contr. Lett.*, vol.10, 9-17.

[2] Conte, G., Moog, C.H., and Perdon, A.M., 1987. "The model matching problem using a differential algebraic approach". Submitted to *Syst. & Contr. Lett.*.

[3] Di Benedetto, M.D., and Isidori, A. ,1986. "The matching of nonlinear models via dynamic state-feedback", *SIAM J. Control & Optimiz.*, 24, 1063 - 1075.

[4] Di Benedetto, M.D., and Slotine, J.-J.E., 1987. "Robust trajectory control for multi-input nonlinear systems", MIT Report NSL 871002.

[5] Fliess, M., 1986. "Nonlinear control theory and differential algebra", Proc. I.I.A.S.A. Conf. Modelling Adaptive Control, Sopron, Hungary.

[6] Hirschorn, R.M., 1979. "Invertibility of multivariable nonlinear control systems", *IEEE Trans. on Automat. Contr.,* AC-24, 855 - 865.

[7] Isidori, A., 1985. "The matching of a prescribed linear input-output behavior in a nonlinear system", *IEEE Trans. on Automat. Contr.*, AC-30, 258 - 265.

[8] Isidori, A., 1985. "Nonlinear control systems: an introduction", Lect. Notes in Control and Info. Scie., Springer-Verlag, vol.72, 1 - 297.

[9] Isidori, A., and Moog, C.H., 1986. "On the nonlinear equivalent of the notion of transmission zeros", Proc. I.I.A.S.A. Conf. Modelling Adaptive Control, Sopron, Hungary.

[10] Moog, C.H., 1988. "Nonlinear decoupling and structure at infinity", *Math. Contr. Sign. & Syst.,* 1, 257 - 268.

[11] Moore, B.C., and Silverman, L.M., 1972. "Model matching by state feedback and dynamic compensation", *IEEE Trans. on Automat. Contr.*, AC-17, 491 - 497.

[12] Morse, A.S., 1973. "Structure and design of linear model following systems", *IEEE Trans. Autom. Contr.*, AC-18, 346 - 354.

[13] Okutani, T., and Furuta, K., 1984; "Model matching of nonlinear systems". Preprints IFAC 9th World Congress, Budapest, vol.IX, pp.168-172.

[14] Singh, S.N., 1981. "A modified algorithm for invertibility in nonlinear systems", *IEEE Trans. on Automat. Contr.*, AC-26, 595 - 598.

# OBSERVERS

# SYNTHESIS OF NON LINEAR OBSERVERS:

# A HARMONIC ANALYSIS APPROACH

**F.CELLE** Laboratoire d'automatique de Grenoble (UA CNRS 228 )
BP 46 38402 SAINT-MARTIN D'HERES FRANCE

**JP GAUTHIER** Université de Lyon I Claude Bernard, Laboratoire D'Automatique
43 Bld du 11 novembre 1918 , 69622 VILLEURBANNE

**G.SALLET** Laboratoire de méthodes mathématiques d'analyse des systèmes (UA CNRS 399)
Université de METZ 57045 METZ cedex

## INTRODUCTION

A system whose task is state estimation is called an observer. The observer is driven by the inputs and the outputs of the initial system.The problem of approximating the state of a linear system is solved by the celebrated Luenberger observer. The Luenberger observer has the property that the dynamics of the error estimation is independant of the inputs.For non linear systems this is no more true.An input which has the property to distinguish any pair of distinct points is called a universal input. For a great class of systems, including analytical systems, universal inputs do exist and are generic.Even if the system is observable, the non universal inputs are unable to distinguish some states. This difficulty has to be overcome for observers. In our opinion, these bad inputs constitute the singularity of the problem and have to be taken into account.
Recently some authors have proposed observers for nonlinear systems. These observers fall in two classes:

The system is nonlinear but can be linearized by input-output injection and nonlinear change of coordinates. An observer can be built which works independently from the inputs. Actually if a nonlinear system can be linearized by a procedure preserving observability properties, it is not difficult to compute a "Luenberger like" observer. The difficulty is in the linearization but not in the construction procedure of the observer. For these kind of systems there are not bad inputs.[KI,KR,LM]

The system is bilinear but has no bad inputs, or bad inputs exists but the states that are not distinguishable are asymptotically stable to the origin.
For both classes the problem posed by bad inputs is squezed.[BI,BZ,F,HF,W]

We propose effectively computable observers for a class of systems for which bad inputs generically do exist: Killing systems. We define Killing systems as systems such that the vector fields are complete and generate a finite dimensional Lie algebra An equivalent requirement is : the group of the system is a Lie group. Bilinear systems, right invariant systems on Lie groups are example of Killing systems. These systems constitute a class for which the real problem of bad inputs exists, but computational aspects are simplified.

# I ASYMPOTIC OBSERVERS FOR BILINEAR SYSTEMS.

## I.1 universal initialized observer for bilinear systems.[GK]

The observer is another system driven by the inputs and the outputs of the initial system. The output of the observer is an estimation of the state of the observed system . If the convergence of the estimation to the state depends on the initial state of the observer we call such an observer an <u>initialized observer.</u> If the convergence is obtained for a special class V of inputs ,the observer is called a <u>V-observer.</u>

We consider observable bilinear systems of the following kind:

$$\dot{x}(t) = \left(A + \sum_{i=1}^{m} u_i B_i\right) x(t)$$

$$y = Cx \quad x(o) = x_0 \qquad x \in \mathbb{R}^n - \{0\} \quad u \in \Omega \subset \mathbb{R}^m$$

Where A, $B_i$ , C are matrices of the appropriate dimension.

There is no loss of generality to consider only homogeneous bilinear systems.

If we denote by S(n, $\mathbb{R}$ ) the vector space of symmetric matrices on $\mathbb{R}^n$ by $|||M|||$ the norm of a matrix M subordinated to the euclidian norm we have the following result:

The following system on $GL^+(n, \mathbb{R}) \times S(n, \mathbb{R}) \times \mathbb{R}^{n+1}$ :

$$\begin{aligned}
&\dot{X}(t) = \left(A + \sum u_i(t) B_i\right) X(t) \\
&\dot{Y}(t) = {}^tX\,{}^tC\,C\,X \\
(1) \qquad &\dot{Z}(t) = {}^tX\,{}^tC\,y \\
&\dot{\lambda}(t) = -\alpha\,(1 + |||X|||)\,\lambda \\
&\hat{x} = X\,(Y + \lambda I)^{-1} Z
\end{aligned}$$

is an asymptotic observer converging for any universal input, when $Z(0) = Y(0).X(0)^{-1}.x_0$.
Of course, because $x_0$ is supposed to be unknown we can choose $Y(0) = 0$, then $Z(0) = 0$.

If we define

$$i(u,T) = \min_{\|x\|=1} \int_0^T \| C.\Phi_u(t).x \| \, dt$$

An input is universal on [0,T] iff $i(u,T) > 0$ . For the preceding observer we can precise the convergence by

$$\| \hat{x}(t) - x(t) \| \leq K \|x_0\| \; e^{-\alpha t} \frac{1}{i(u,T)}$$

## I.2 Universal non initialized observers do not exist [CGKS]

We give an easy example proving that observers for bilinear systems working for any initial point of the state of the observer an any universal input cannot exist. The point is that an input can be universal on an interval , and non universal for the remaining time.

We consider the system:

$$\dot{x} = u\,Ax + vBx$$

$$x = (x_1 , x_2) \quad (u,v) \in \mathbf{R}^2$$

$$y = x_1$$

where $A = \begin{pmatrix} 0 & ,-1 \\ 1 & , \ 1 \end{pmatrix}$ and B is any matrix.

If we consider the input defined by $u(t) = 1$ if $0 \le t \le \pi/2$ , $u(t) = 0$ otherwise, $v \equiv 0$
this input is universal, and it is not difficult to prove that if a non-initialized universal observer would exist ,by choosing conveniently two initial states for the observer, the observer is driven to converget oward the distinct points (0,1) and (0,1+ε) for ε sufficiently small,which is a contradiction.

## I.3 Persistent inputs and observers for skew-symmetric systems.

Persistent inputs , loosely speaking are inputs that are universal from time to time in the future. To be more precise, let K(u,T) be the set of translation of the control $u \in L^\infty_{\mathbb{R}}$ restricted to [ 0,T], ie $K(u,T) = \{ u_\theta | [0,T] \mid \theta \in \mathbb{R} \}$ where $u_\theta(t) = u(\theta+t)$. This set is relatively compact for the weak - $*$ topology on $L^\infty_{[0,T]}$. We say that an input u of $L^\infty_{\mathbb{R}}$ is an universal input if there is a universal input $u^*$ which is an accumulation point of K(u,T) for the the weak - $*$ topology on $L^\infty_{[0,T]}$. In other words there is a sequence $\theta_n$ of real ,$\theta_n \to \infty$ , s.t $\operatorname{Lim} u_{\theta_n} = u^*$ and $i(u,T)>0$.

**Theorem** : Persistent inputs are generic

For bilinear systems the set of persistent inputs is an open dense subset of the admissible control sets $C^\infty_{\mathbb{R}+}$ and $L^\infty_{\mathbb{R}+}$ with their respective topology.

### I.4 Non initialized observer for persistent inputs of skew-symmetric systems [GK]

A bilinear system is said skew-symmetric iff the linear vector fields of the systems are skew-symmetric for some scalar product on $\mathbb{R}^n$.

**Theorem:**
for a bilinear system (1) with skew-symmetric matrices, the system

$$\begin{cases} \dot{\xi} = (A + \sum_{i=1}^{m} u_i B_i)\xi - r\,{}^tC(C\xi - y) \\ \hat{x} = \xi \quad \xi \in \mathbb{R}^n \quad r > 0 \end{cases}$$

is a persistent non initialized observer.

### I.5 Non existence of persistent observers

Is the class of persistent inputs the right class to be considered in observers problems? The answer is no ! The Kalman 's filter (in a determinist view point) for time dependant system provides a non-initialized observer for bilinear systems. One can construct persistent inputs for which the observer does not work. It is sufficient to consider an input having an inobservable (for that input !) unstable subspace. In fact the question is a ill-posed problem, the state going to the $\infty$, the question of the convergence of the observer has no real meaning.

## II. OBSERVERS AND IMMERSION

We recall the following facts:

An analytic system $\Sigma$, is constituted by the following data: an analytic manifold M, a family $\mathcal{F}$ of analytic vectors field, an output vector space E, an analytic observation map h from M into E, a set U of value for the input, a set of admissible control $\mathcal{U}$.

The observation space $O(\Sigma)$ of a system is the smallest vector space of $C^{\omega}(M,E)$ containing h and stable under the action of the vector fields of $\mathcal{F}$.

An analytic system is observable iff $o(\Sigma)$ separates the points of M.

**Definition: (immersion) [FK]**

A system $\Sigma = (M, \mathcal{F}, U, \mathcal{U}, h, E)$ is said to be immersed into a system $\Sigma'$ if

1/ $\Sigma$ and $\Sigma'$ have the same control space U, admissible control set $\mathcal{U}$, output space E.

2/ there is an analytic map $\tau$ from M into M' such that for any control u we have

$$y_{\Sigma}(x,u,t) = y_{\Sigma'}(\tau(x),u,t)$$

where $y_{\Sigma}(x,u,t)$ denote the output of $\Sigma$ for the initial state x, the control u , at time t. This equality must be true for any t for which the left side is defined.

3/ $h(x) \neq h(y)$ implies $h'(x) \neq h'(y)$.

**definition: weak homomorphism of systems** [S]

Given two systems $\Sigma_i = (M_i, \mathcal{F}_i, U, \mathcal{U}, h_i, E)$ a weak homorphism of $\Sigma_1$ into $\Sigma_2$ is a mapping F from $M_1$ into $M_2$ such that from any u, $x \in M_1$, $t > 0$ for which the solution of the first system $\pi_{\Sigma 1}(x,u,t)$ is defined in $M_1$ we have

$$F( \pi_{\Sigma 1}(x,u,t) ) = \pi_{\Sigma 2}(F(x),u,t) \text{ and}$$
$$h_2 \circ F = h_1.$$

Weak homomorphism of systems is stronger than immersion.

There is a natural action of the group ( for Killing systems , pseudo-group in general) G of a system on $C^{\omega}(M,E)$ with the usual left action :

$$\text{if} \quad g \in G, \ h \in C^{\omega}(M,E): \quad (g.h)(x) = h(g^{-1}.x)$$

We denote by $\langle G.h \rangle_{LS}$ the vector space generated by the orbit of h under G. In general $\langle G.h \rangle_{LS}$ is different from $O(\Sigma)$ but we have the

**Theorem:**

the following conditions are equivalent for an analytic system $\Sigma$

i) $O(\Sigma)$ is finite dimensional then $O(\Sigma) = \langle G.h \rangle_{LS}$

ii) G.h generates a finite dimensional vector space then $O(\Sigma) = \langle G.h \rangle_{LS}$ .

iii) $\Sigma$ is weakly homomorphic to a state affine observable system, and the homomorphism is unique.

In view to built observers for systems, it is certainly desirable to immersed systems into skew-symmetric bilinear systems (affine skew-symmetric ). The answer is given by:

**Theorem:** [CGK]

A control affine Killing system, observable and orbit minimal can be immersed into a bilinear observable system iff

i) $O(\Sigma)$ is finite dimensional

ii) a lift of h on G $\mathbf{h}_x$ is almost periodic.

( we define a lift $\mathbf{h}_x$ of h on G by $\mathbf{h}_x(g) = h(g.x)$, $\mathbf{h}_x$ is almost periodic on G for any x iff it is almost periodic for some x. )

We have obstruction at the level of the dynamic , represented by the group of the system, for immersion:

**Theorem:** [CGKS]

A necessary condition for an observable Killing system to be immersed into a bilinear system is the group is the semi-direct product of a reductive group by a normal simply connected solvable group.

The corresponding obstruction for immersion into skew-symmetric systems is the direct product of a compact group by a vector space.

**Approximate observers for compact Killing systems.**

We consider $\Sigma$ a minimal Killing system, whose group is a compact Lie group. Let h be the observation map of $\Sigma$, and any lift $\mathbf{h}_x$ of h on G. By the celebrated Peter-Weyl theorem, $\mathbf{h}_x$ can be approximated by an almost periodic function $h_n$. For n sufficiently large the system $\Sigma_n=(M,\mathcal{F},U,h_n, E)$ is observable and by the previous theorem can be immersed into a skew-symmetric observable bilinear system. An observer can be built for $\Sigma_n$, we call this observer an approximate observer of $\Sigma$.

## III. Infinite dimensional unitary immersion of Killing systems.

Good observers can be built for Killing systems that can be immersed into skew-symmetric observable bilinear system. That is for systems for which $\langle G.h\rangle_{LS}$ is finite dimensional and can be provided with an inner product. The idea is that if $\langle G.h\rangle_{LS}$ is anymore finite dimensional but is a Hilbert space then the same result is obtained.

**Definition: Unitary immersion**

A unitary immersion of a minimal Killing system $\Sigma = (M,\mathcal{F},U,\mathcal{U},h,E)$ is the following data: (we denote as usual by $\Phi_u(t)$ the solution in G for the input u)

1) a unitary representation $\pi$ of the Lie group G of the system into a Hilbert space $\mathcal{H}(\pi)$.

2) An analytic vector $\psi$ of $\mathcal{H}^{\omega}(\pi)$.

3) a continuous map $\tau : M \to \mathcal{H}(\pi)$ such that for any $x \in M$, any $u \in \mathcal{U}$:

$$\langle \pi(\Phi_u(t) . \tau(x) , \psi\rangle = h(\Phi_u(t) .x)$$

moreover if $\{ \pi(g).\psi \mid g \in G \}^{\perp} = \{0\}$ the immersion is said observable.

Now let us associate to a unitarily immersed Killing system a differentiable system a differentiable system on $\mathcal{H}(\pi)$. A s usual we deal with control affine Killing systems :

$$(2)\quad \begin{cases} \dot{x}(t) = (X_0(x(t)) + \sum_{i=1}^{m} u_i(t)\, X^i(x(t)) \\ x(0) = x_0 \\ y = h(x);\ u \in \Omega \subset \mathbb{R}^m \ ;\ x \in M \end{cases}$$

and the right invariant system associated on G:

$$(3)\quad \begin{cases} \dot{g}(t) = ( X^0 + \sum_{i=1}^{m} u_i(t)\, X^i )\, g(t) \\ g(o) = e \\ y = h_{x_0}(g) = h(g.x_0) \quad g \in G \quad X^j \in \underline{g} = \mathrm{Lie}(G) \end{cases}$$

To be perfectly strict the $X^i$ in (3) would have to be written $X^{i*}$, the Killing vector field associated to $X^i \in \mathfrak{g}$.

For any element X in the Lie algebra $\mathfrak{g}$ of G and $\xi$ in $\mathcal{H}^{\omega}(\pi)$, the analytic vector of the representation, the derivative at t=0 of $\pi(\exp tX).\xi$, is denoted by $\pi(X).\xi$. Then $\pi(X)$ is a linear transformation of $\mathcal{H}^{\omega}$, and defines an essentially skew-adjoint operator on $\mathcal{H}(\omega)$.

To $\Sigma$ unitarily immersed we associate the $\Sigma_\pi$ equation on $\mathcal{H}(\pi)$

$$\begin{cases} \dot{\xi}(t) = (\pi(x_0) + \sum_{i=1}^{m} u_i(t)\, \pi(X^i))\, \xi(t) \\ \xi(0) = \xi_0 \end{cases}$$

$$y = <\xi, \psi> \quad \xi \in \mathcal{H}(\pi)$$

and the **observer equation** on $\mathcal{H}(\pi)$ :

$$(4) \quad \begin{cases} \dot{\zeta}(t) = (\pi(X_0) + \sum_{i=1}^{m} u_i(t)\, \pi(X^i))\, \zeta(t) \; - r\,(<\zeta,\psi> - y)\,\psi \\ \zeta(0) = \zeta_0 \;\; ; \;\; \zeta \in \mathcal{H}(\pi) \;\; ; \;\; r > 0 \end{cases}$$

To the observer equation is associated the equivalent **error equation** $\Sigma_\varepsilon$**:**

$$\Sigma_\varepsilon \begin{cases} \dot{\varepsilon}(t) = (\pi(X_0) + \sum_{i=1}^{m} u_i(t)\, \pi(X^i))\, \varepsilon(t) \; - r <\varepsilon,\psi> \psi \\ \varepsilon(0) = \varepsilon_0 \qquad ; \qquad \varepsilon \in \mathcal{H}(\pi) \end{cases}$$

**Definition:** regularly persistent inputs

An input is said to be $\Sigma$ regularly persistent if there is a time interval $T > 0$, a real sequence $\theta_n$, $\theta_n \to \infty$, with $\theta_{n+1} - \theta_n$ bounded, such that the translated inputs $u_{[\theta n]}$ converge *-weakly to a universal input u* on [0,T] for the system $\Sigma$.

In other words,information is required to appear with some regularity when the time is passing by.

**Theorem (main result)** [CGKS]

For a $\Sigma_\pi$ regularly persistent input, the solution of $\Sigma_\varepsilon$ the error equation converges weakly to zero in $\mathcal{H}(\pi)$ .

We do not know how to prove in general that universal inputs exist. However in all the example encoutered we can exhibit many.

**Approximate observers for Killing systems:**

Let $\Sigma$ be a minimal Killing system and K an arbitrary subset of the group G. There is a sequence of functions $h_n$ converging uniformly on K to **h** such that $\Sigma_n=(M,\mathcal{F},U,h_n, E)$ has a unitary stable K-observable immersion.

This notion is exactly the generalization of the notion of immersion of Fliess-Kupka. This result is simply based on the density of the finite linear combinations of positive type functions, for the uniform convergence on compact sets. Since the problem of observation near the infini is ill-posed, and moreover of no practical interest, this is not a lack of generality.

The prototype of the situation, which is described before, is: Let $\Sigma$ a minimal Killing system $(M,\mathcal{F},\mathcal{U},h,E)$ whose associated Lie group is G (minimal means observable, with G transitive on M).Then choosing any point x of M, M can be identified with the homogeneous space G/K, where K is the isotropy subgroup of G, of the point x.Let $\pi$ be the canonical mapping from G onto G/K, and $\tilde{h} = h \circ \pi$ the lift of h on G. If $\tilde{h}$ is a positive type function on G, the space generated by the translated of $\tilde{h}$, $\langle G.\tilde{h}\rangle_{LS}$ ,gives rise by quotient and completion to a Hilbert space $\mathcal{H}$ whose inner product is invariant by G. We obtained then easily an unitary immersion of the Killing system.

we consider the system:

$$\begin{cases} \begin{pmatrix} \dot{x}_1 \\ \dot{x}_2 \end{pmatrix} = \begin{pmatrix} 0 & 1 \\ -1 & 0 \end{pmatrix} \begin{pmatrix} x_1 \\ x_2 \end{pmatrix} + \begin{pmatrix} 0 \\ 1 \end{pmatrix} u \\ x = (x_1, x_2) \in \mathbb{R}^2 \quad ; \quad y = x_1^2 + x_2^2 \end{cases}$$

without changing observability we can consider $h(x) = J_0(\|x\|)$, where $J_0$ is the O $^{th}$ Bessel function.In the physical problem of synthesis of an observer it is obviously allowed to make some change in the output function ,if the observability is preserved.

This system is evidently lifted on the Lie group of motion group of the plane ,ie the semi-direct product of $\mathbb{R}^2$ by the group of rotation SO(2).
The space $\langle G.h\rangle_{ls}$ is $L^2(S^1)$, and the action of G is an irreducible Unitary representation .
( because $J_0$ is a positive type function)
The immersion is observable , as can be seen easily.

The immersion $\Sigma_\pi$ is given by the system:

$$\begin{cases} \dot{\xi}_t = -\dfrac{d}{d\theta}\xi_t + u\, i \sin\theta\, \xi_t \\ y = \langle \xi_t , 1\rangle = \dfrac{1}{2\pi}\displaystyle\int_0^{2\pi} \xi_t(\theta)\, d\theta \end{cases}$$

The following system is a non initialized observer for regularly inputs,

$$\begin{cases} \zeta_t = \frac{d}{d\theta}\zeta_t + 2\,u\,(\,l\cos\theta\,\zeta_t \;-2\sin\theta\,\frac{d}{d\theta}\,\zeta_t\,) - r\,(\,\frac{1}{2\pi}\int_0^{2\pi}\zeta_t\,(\theta)\,d\theta\;-y\,) \\ \zeta_0 \in \mathcal{H} \qquad r > 0 \end{cases}$$

We can prove that there are numerous regularly persistent inputs.
It is straighforward to prove that for a constant input u on $[\,0,2\pi]$ , u is universal on this interval if $J_n(u) \neq 0$ for any integern, where $J_n$ is the $n^{th}$ Bessel function. By analiticity u is universal on any interval of lenght $T > 0$, hence u is regularly persistent. Many other regularly persistent inputs can be found in the basis of constant control...

Numerical computation have been conducted for these equation and are satisfactorily good.[CG]

**References:**

[BZ] D.BESTLE- M.ZEIST
"Canonical form design for nonlinear observers with linearizable error dynamics"
Internat.J. Control, 1981,23, pp 419-431

[CG] F.CELLE-J.P. GAUTHIER
"Theory od dynamic observers for a class of nonlinear systems"
MTNS, 1987 Phoenix

[CGK] F.CELLE- J.P.GAUTHIER- K.KAZAKOS
Orthogonal representations of nonlinear systems and input-output maps"
Systems Control Lett 7, 1986, pp 365-372

[CGKS] F.CELLE-J.P.GAUTHIER-K.KAZAKOS-G.SALLET
"Synthesis of nonlinear observers: a Harmonic analysis approach"
to appear in MCSS.

[F] FLIESS
"Quelques remarques sur les observateurs nonlinéaires"
colloque GRETSI Nice 1987

[FK] M.FLIESS-I.KUPKA
"A finiteness criterion for nonlinear input-output differential systems"
SIAM J. Control Optim. 21 , 1983 , pp721-728.

[Fu] Y.FUNAHASHI
"stableState estimator for bilinear systems"
Internat. J. Control , 1979 ,29, pp181-188

[GG] JP.GAUTHIER-JP.GUERIN
"Unitary immersion of nonlinear systems"
Math.Syst. Theory 19, 1986 , pp 135-153

[GI] O.GRASSELLI- A.ISIDORI
"An existence theorem for observers of bilinear systems"
IEEE Trans. Automat. Control AC 26, 1981, pp 1299-1301

[HF] S.HARA- K.FURUTA
"Minimal order state observers for bilinear systems"
Internat. J. Control , 24 ,1976, pp 705-718

[KI] A.J.KRENER- A.ISIDORI
"Linearization by output injection and nonlinear observers"
Systems Control Lett. 3 , 1983 pp 47-52

[KR] A.J.KRENER-W.RESPONDEK
"Nonlinear Observers with linear error dynamics"
SIAM J. Control Optim. 23 , 1985 , pp197-216

[L] D.G. LUENBERGER
"Observers for multivariable systems"
IEEE Trans. Automat. Control, AC 11, 1966, pp190-197

[LM] J.LEVINE-R.MARINO
"Nonlinear systems immersion, observers and finite dimensional filters"
Systems Control Lett. 7, 1986 , pp133-142

[S] H.J. SUSSMANN
"Existence and uniqueness of minimal realizations of nonlinear systems"
Math. Systems Theory 10, 1977 , pp 263-284 .

[W] D.WILLIAMSON
"Observability of bilinear systems, with applications to biological control"
Automatica 13 ,1977 , pp243-254

## REGULARLY PERSISTENT OBSERVERS FOR BILINEAR SYSTEMS

G. BORNARD, N. COUENNE, F. CELLE

Laboratoire d'Automatique de Grenoble, LA CNRS 228
BP 46 - 38 402 Saint-Martin d'Hères - FRANCE

**ABSTRACT** : This paper deals with the problem of synthesis of observers for bilinear systems working for the class of regularly persistent inputs. Moreover we show how this result is related to the Kalman observer for time-varying linear systems.

**KEY WORDS** : bilinear systems, observer, linear time dependent systems.

### I - INTRODUCTION

This paper deals with the problem of synthesis of observers for bilinear systems of the form :

$$(B) \qquad \dot{x}(t) = (A_0 + \sum_{i=1}^{m} u_i(t) B_i) x(t) \quad , \qquad y(t) = C x(t)$$

where $x(t) \in R^n - \{0\}$, $u(t) \in \Omega \subset R^m$, $y(t) \in R^p$ and the matrices being of the appropriate dimension.

That is, we search for a differentiable system (0) whose inputs are u(t) and y(t) :

$$(0) \qquad \dot{z}(t) = f(z, u, y) \quad , \quad \hat{x}(t) = \varphi(z(t)) \, , \quad z(t_0) = z_0$$

the output $\hat{x}(t)$ being an approximation to the state x(t) for any initial state of (B) and (0) in the sense : $\lim_{t \to +\infty} || \hat{x}(t) - x(t) || = 0$ for any $\hat{x}(t_0), x(t_0)$.

(0) is called an observer of (B).

Let us recall that the observability of a nonlinear system ($\Sigma$) does not imply that every input distinguishes points of ($\Sigma$). An input having the property of distinguishing any couple of distinguishable points is called a universal input. The study of universal inputs has been initiated by E. SONTAG [20] for the discrete case. Moreover universal inputs do exist and are generic in the analytic case [22].

A crucial point for the problem of synthesis of observers is the fact that the class of non-universal inputs is in general non empty. This class of bad inputs constitues the real "obstruction" to the existence of observers. In most of the works about observers of nonlinear systems the problem of the bad inputs is not taken into account [23], [11], [5], [2], [16], [17], [19]. Their consideration for observer's synthesis appears in [8], [4], [7].

Let us precise different kinds of observers :

- the non initialized observer is an observer which works for any initial state of the observer. For example, (0), as described above, is a non initialized observer of (B).
- in the other case we will say an initialized observer (an observer which works for some initial state of the observer). Practically the "good" observer is the non initialized one.
- Finally if the observer works for some subclass U of the inputs, the observer is called U-observer. For example if (0) works only for universal inputs, we will say (0) is a universal observer

In the first section we review some results obtained in [8] on initialized observers of bilinear systems and in [4] on the non existence of non initialized universal observers. These results motivated our research on non initialized observers of bilinear systems for some class of inputs, the regularly persistent inputs. In [4] two classes of inputs are studied, the persistent and regularly persistent inputs. Notice that the notion of regularly persistent input corresponds to the notion of "persistent excitation" in adaptive control. These two classes will be defined in section II.

The first result is given in section III : a non initialized observer (of bilinear systems) working for the class of regularly persistent inputs is presented. The estimation error goes asymptotically to zero with an arbitrarily chosen speed, as soon as the inputs are uniformly bounded, the bound being known.

In section IV we connect our result with the KALMAN observer [13] for time varying linear systems. Namely we restate in the case of bilinear system, the condition of "uniform complete observability" given by KALMAN : any bounded regularly persistent input applied to the observable bilinear system (B) generates a time varying system which is completely uniformly observable in the KALMAN sense. Thus the KALMAN observer is a non initialized observer for bilinear systems, when the input is bounded (the bound being unknown) and regularly persistent.

The proofs of the theorems are given in [3].

## II - UNIVERSAL OBSERVER FOR BILINEAR SYSTEMS

To the bilinear system (B), one can associate the system on $GL^+(n, R)$ :

(F) $\quad \dot{\phi}_u(t, t_0) = (A_0 + \sum_{i=1}^{m} u_i(t) B_i) \quad \phi_u(t, t_0) \quad , \quad$ with $\phi_u(t_0, t_0) = Id$

where $\phi_u(t, t_0) \in GL^+(n, R)$ is the fundamental matrix.

Let U be a subset of $L^\infty[t_0, t_1]$ $t_1 > t_0$, the space of bounded measurable functions defined on $[t_0, t_1]$.

Let $u \in U$. The unobservability subspace of (B) related to u is the space :

$$U_u = \bigcap_{t \in [t_0, t_1]} \ker C \phi_u(t, t_0)$$

It is clear that u is universal in $[t_0, t_1]$ iff $U_u = \{0\}$.

### II - 1 Universal initialized observer for bilinear systems [8]

J.P. GAUTHIER and D. KAZAKOS [8] present a universal initialized observer for bilinear systems.

- If u(t) is not universal, the convergence of the error $\| \hat{x}(t) - x(t) \|$ is not guaranteed. In the other case, the speed of convergence is arbitrarily chosen.

- Behind the proof of their theorem is hidden the following fact : At each time t, the initial state $z_0$ of the observer (O) is estimated by minimizing with respect to $z_0$ a least square criterion of the form :

(C1) $\quad J(z_0) = \int_{t_0}^{t} \| y(\tau) - C \phi_u(\tau, t_0) z_0 \|^2 d\tau$

with the regularization factor $\lambda$ . Id.

## II - 2 Universal non initialized observers do not exist

Consider the system :

$$(E)\qquad \dot{x} = \begin{pmatrix} 0 & 1 \\ -1 & 0 \end{pmatrix} xu \qquad u \in R$$

$$y = x_1 \qquad\qquad x \in R^2 - \{0\}$$

The system is observable, the only non-universal input is the constant input $u(t) = 0$.

Let us apply to the system the following input :

$u(t) = 1 \quad t < 0 \quad ; \quad u(t) = 0 \quad t \geq 0$

Suppose there exists a universal non initialized observer (O).

Let us denote $x_0^\sigma = \begin{pmatrix} 0 \\ \sigma \end{pmatrix}$ for $\sigma \in R$.

$x^\sigma(t) = x(t, 0, x_0^\sigma, u)$ solution of (E) , $y^\sigma(t) = x_1^\sigma(t)$,

$\hat{x}^\sigma(t) = \hat{x}(t, 0, \hat{x}_0, y^\sigma, u)$ solution of (O) for some $\hat{x}_0 \in R^2$.

Clearly one has :

$$y^{\sigma_1}(t) = y^{\sigma_2}(t) \qquad \forall t \geq 0 \quad \sigma_1, \sigma_2 \in R$$

then $\hat{x}^{\sigma_1}(t) = \hat{x}^{\sigma_2}(t) \qquad \forall t \geq 0 \quad \sigma_1, \sigma_2 \in R$

Consider now $\sigma_1, \sigma_2 \in R, \sigma_1 - \sigma_2 = \varepsilon \neq 0$ there exits $t_0 < 0$ such that $\hat{x}^{\sigma_1}(t)$ and $\hat{x}^{\sigma_2}(t)$ exist for every $t \geq t_0$.

Consider the following initialization :

$x(t_0) = x^{\sigma_1}(t_0)$ for (E)

$\hat{x}(t_0) = \hat{x}^{\sigma_1}(t_0)$ for (O)

(O) being a universal non initialized observer and u being universal over $[t_0, +\infty[$ , one has :

$$\| \hat{x}^{\sigma_1}(t) - x^{\sigma_1}(t) \| \to 0 \quad \text{when } t \to +\infty .$$

But with the initialization :

$$x(t_0) = x^{\sigma_2}(t_0) \quad \text{for (E)}$$

$$\hat{x}(t_0) = \hat{x}^{\sigma_2}(t_0) \quad \text{for (O)}$$

one has $\| \hat{x}^{\sigma_2}(t) - x^{\sigma_2}(t) \| \to \varepsilon$ when $t \to +\infty$,

since $\forall\, t \geq 0 \quad \hat{x}^{\sigma_1}(t) = \hat{x}^{\sigma_2}(t)$ and

$$x^{\sigma_1}(t) = x_0^{\sigma_1}, \quad x^{\sigma_2}(t) = x_0^{\sigma_2} \quad \text{and} \quad \| x_0^{\sigma_2} - x_0^{\sigma_1} \| = \varepsilon.$$

Then the situation is the following one :

- u is universal over $[t_0, +\infty[$
- the observer does not converge for the initial condition $x(t_0) = x^{\sigma_2}(t_0)$, $\hat{x}(t_0) = \hat{x}^{\sigma_2}(t_0)$

This is a contradiction with the existence of such observer.

Let us only notice that the input $u(t)$ over $[t_0, +\infty[$ is universal but the input $u(t)$ over $[0, +\infty[$ is not universal.

## III - PERSISTENT INPUT, REGULARLY PERSISTENT INPUT

Given a bilinear system (B), we compute for some input $u(t) \in L^\infty [t_0, t_1], t_1 > t_0$ :

- the matrix : $W_u[t_0, t_1] = \int_{t_0}^{t_1} \phi_u^T(\tau, t_0)\, C^T C\, \phi_u(\tau, t_0)\, d\tau$

where $\phi_u(\tau, t_0)$ is the fundamental matrix of (B) for the control u (.).

- the index : $i_u[t_0, t_1] = \min\limits_{\|x\|=1} x^T W_u[t_0, t_1]\, x$

$i_u[t_0, t_1]$ is called the universality index of u on $[t_0, t_1]$.

Clearly the input u is universal on $[t_0, t_1]$ iff $i_u[t_0, t_1] > 0$.

Let us notice that if u is universal on $[t_0, t_1]$ and if $u \in L^\infty [t_0, +\infty[$ then it is universal on $[t_0, +\infty[$.

Let $u_\delta(t)$ (for $\delta > 0$) denote $u(t+\delta)$.

Definition 1 : $u(t) \in L^\infty [t_0, +\infty[$ is a persistent input for (B) if there exists $T > 0$ such that the following limit : $\lim_{\delta \to +\infty} i_{u_\delta} [t_0, t_0 + T]$

either is strictly positive or does not exist.

Définition 2 : $u(t) \in L^\infty [t_0, +\infty[$ is regularly persistent for (B) if there exist $T > 0$ and $\alpha > 0$ such that $i_{u_\delta} [t_0, t_0 + T] > \alpha \quad \forall \delta \geq 0$

Remarks :

- This definition is valid for bilinear system, only.
- The universal input of example II-2 is not persistent since as soon as $\delta \geq 0$, $u_\delta(t)$ is not universal.

(The authors thank Professor Hassan HAMMOURI for his judicious remark on definition 2).

## IV - NON INITIALIZED OBSERVER FOR BILINEAR SYSTEMS

Let $U_\beta$ be the subset of $L^\infty [t_0, +\infty[$ whose elements are bounded by $\beta$ and $S^*(n,R)$ be the space of positive definite symmetric matrices of dimension n.

For any $u \in U_\beta{}^m$, (B) generates a time varying linear system $\dot{x}(t) = A_u(t)\, x(t)$ with :

$$A_u(t) = A_0 + \sum_{i=1}^{m} u_i(t)\, B_i \quad , \quad t \geq 0$$

Each element of the family of matrices $\{A_u(t)\}_{u \in U_\beta{}^m}$ is measurable and uniformly bounded.

The problem of the observer is to estimate the initial state of (B). We propose to do it by minimizing at each time t with respect to $z_0$ (the initial condition of the observer (O)) the index performance (C2) :

$$\text{(C2)} \quad J(z_0) = e^{-\theta t}\left[e^{\theta t_0}\|z_0 - \hat{x}_0\|^2_{S_0} + \int_{t_0}^{t} e^{\theta \tau}\|y(\tau) - C\,\phi_u(\tau, t_0)\,z_0\|^2_R \, d\tau\right]$$

where $S_0 \in S^*(n, R)$, $R \in S^*(p, R)$, $\theta > 0$, $\|x\|_{S_0} = x^T S_0 x$.

From (C2) it is clear that $\hat{x}(t) = \phi_u(t, t_0)\, z_0(t)$ and $\hat{x}_0$ corresponds to the inital state of the observer (O) : $z_0$.

The index performance (C2) differs from (C1) in :

- the exponential forgetting factor $e^{-\theta t}$
- the term $\|z_0 - \hat{x}_0\|_{S_0}$ [instead of the regularization term $\lambda$.Id in (C1)].

Then we state :

<u>Theorem</u> : Assume that the inputs u of (B) belong to $U_\beta{}^m$ ($\beta > 0$ being given) and are regularly persistent, then $\exists\, \theta_0$ such that $\forall\, \theta \geq \theta_0$ , the system (O) defined on $R^n \times S^*(n,R)$ by :

$$\text{(O)} \quad \begin{cases} \dot{\hat{x}}(t) = A_u(t)\,\hat{x}(t) + P(t)\,C^T R\,[\,y(t) - C\,\hat{x}(t)\,] , & \hat{x}(t_0) = \hat{x}_0 \\ \dot{P}(t) = \theta\, P(t) + P(t)\,A_u{}^T(t) + A_u(t)\,P(t) - P(t)\,C^T R\, C\, P(t) , & P(t_0) = P_0 \end{cases}$$

is a non initialized observer for the system (B).

Moreover the estimation error $\|\varepsilon(t)\|$ tends exponentially to 0 with an arbitrarily chosen speed that is :

$$\forall \lambda > 0,\ \exists\, \theta \ / \ \|\varepsilon(t)\| = \|x(t) - \hat{x}(t)\| \leq \alpha\, e^{-\lambda(t - t_0)}\,\|\varepsilon(t_0)\| \quad \text{with } \alpha > 0 \text{ for } t \geq t_0.$$

<u>Remarks</u> :

- $\theta_0$ is function of $\beta$.
- The value of the norm of the matrix P(t) represents an index of quality for the input. Thus if the input contains a small amount of information, the norm of the matrix increases.

## V - KALMAN OBSERVER AND BILINEAR SYSTEMS

Let us recall some facts about Kalman observer. Let (L) be the time-dependent linear system :

$$(L)\quad \begin{cases} \dot{x}(t) = A(t)\,x(t) \\ y(t) = C(t)\,x(t) \end{cases}$$

where $A(t)$ is regulated [13], $x(t) \in R^n$.

The state vector $x(t)$ may be estimated, via an estimate $\hat{x}(t)$, available at time t. The equations describing the calculation of $\hat{x}(t)$ are :

$$\dot{\hat{x}}(t) = A(t)\,\hat{x}(t) - P(t)\,C^T(t)\,R^{-1}(t)\,[\,C(t)\,\hat{x}(t) - y(t)\,] \quad , \quad \hat{x}(t_0) = \hat{x}_0$$

$$\dot{P}(t) = P(t)\,A^T(t) + A(t)\,P(t) - P(t)\,C^T(t)\,R^{-1}(t)\,C(t)\,P(t) + Q \quad , \quad P(t_0) = P_0$$

where $R > 0$, $Q \geq 0$.

Remark : These equations are the Kalman-Bucy filter equation without stochastic terms.

Sufficient conditions for this filter to be exponentially stable are :

1) The entries of $A(t)$, $C(t)$, $Q(t)$, $R(t)$ and $R^{-1}(t)$ are bounded for $t \geq t_0$.
2) With $D(t)$ any matrix such that $D(t)\,D^T(t) = Q(t)$, the pair $[A(t), D(t)]$ is uniformly completely controllable.
3) The pair $[A(t), C(t)]$ is uniformly completely observable.

Definition :

The pair $[A(t), C(t)]$ is said to be uniformly completely observable if there exist positive constants $T$, $\alpha_0$, $\alpha_1$, $\beta_0$ and $\beta_1$ such that :

(a) $\quad \alpha_0\, Id \leq \Gamma(t, t+T) \leq \alpha_1\, Id \quad$ for all t

(b) $\quad \beta_0\, Id \leq \phi^T(t, t+T)\,\Gamma(t, t+T)\,\phi(t, t+T) \leq \beta_1\, Id \quad$ for all t

$$\text{where } \Gamma(t, t+T) = \int_t^{t+T} \phi^T(\tau, t+T)\,C^T(\tau)\,C(\tau)\,\phi(\tau, t+T)\,d\tau$$

and Id is the identity matrix.

The term uniform complete controllability can most readily be defined as the dual of the uniform complete observability.

For any $u \in U_\beta{}^m$, (B) generates a time-varying linear system $\dot{x} = A_u(t)\, x(t)$. (see Section IV). We recall that u (.) is bounded.

Theorem : The pair $[A_u(t), C]$ is uniformly completely observable iff u (t) is a regularly persistent input.

Corollary : The Kalman observer is asymptotically stable for the class of bilinear system iff the inputs are regularly persistent.

Remarks :

* For the Kalman observer, the knowledge of the bound of the inputs is not necessary.
* The differences between the two observers are :
  - In the Kalman's case the equations of the observer are obtained by minimizing a performance index on an infinite-dimensional functional space ; in our case, on a finite dimensional space $R^n$.
  - In the Kalman's case, the constant term Q in the Ricatti equation must be different from 0.

In order to illustrate the corollary, we propose a bilinear system to which a persistent input is applied. We show that the Kalman observer diverges.

Let consider :

$$(B)\quad \begin{aligned} \dot{x} &= \left[\begin{pmatrix} 0 & 0 \\ 0 & 1 \end{pmatrix} + u \begin{pmatrix} 0 & 1 \\ 0 & 0 \end{pmatrix}\right] x = A_u(t)\, x \\ y &= [\,1 \quad 0\,]\, x = C x \end{aligned}$$

Clearly (B) is observable and any constant input $u \neq 0$ is universal.

Let us choose for the Kalman observer $R = 1$ $\quad Q = \begin{pmatrix} 1 & 0 \\ 0 & 1 \end{pmatrix}$.

From the corollary it is already known that the Kalman observer converges iff the applied input u (t) is regurarly persistent, that is for the input u (t) $\quad \|\varepsilon(t)\| = \|x(t) - \hat{x}(t)\| \underset{t \to +\infty}{\longrightarrow} 0$.

First look at the behavior of the observer for u = 0.

Let $P(t) = \begin{pmatrix} P_{11}(t) & P_{12}(t) \\ P_{12}(t) & P_{22}(t) \end{pmatrix}$ with $P(t_0) > 0$.

One obtains :

$$\dot{p}_{11} = 1 - p_{11}^2$$

$$\dot{p}_{12} = (1 - p_{11})\, p_{12}$$

$$\dot{p}_{22} = 2\, p_{22} - p_{12}^2 + 1$$

This system is forward complete and $\dot{\varepsilon}(t) = \begin{pmatrix} -p_{11} & 0 \\ -p_{12} & 1 \end{pmatrix} \varepsilon(t)$.

Hence one has $\| \varepsilon(t) \| \underset{t \to +\infty}{\longrightarrow} +\infty$.

We propose a persistent input such that $\| \varepsilon(t) \| \underset{t \to +\infty}{\longrightarrow} +\infty$.

Let $\Delta t > 0$ and a sequence $\tau = (t_0, t_1, \dots, t_i, \dots)$ with $\begin{cases} t_i \in R^+ \\ t_i + \Delta t \le t_{i+1} \end{cases}$ for any $i \in N$.

Let $u_\tau(t)$ be the persistent input for (B) associated to t defined by $u_\tau(t) \begin{cases} 1 & \text{if } t_i \le t \le t_i + \Delta t \quad i \in N \\ 0 & \text{otherwise} \end{cases}$

For such an input, the observer is forward complete.

Let us choose the sequence $\tau$ in the following way : for any $t_0 \ge 0$ and $t_{i+1} > 0$ such that

$$\| \varepsilon(t_{i+1}) \| = 2 \sup \| \varepsilon(t) \| \quad t_i \le t \le t_i + \Delta t$$

($t_{i+1}$ exists since $\| \varepsilon(t) \| \underset{t \to +\infty}{\longrightarrow} +\infty$ when $u = 0$ and $t_{i+1} \ge t_i + \Delta t$).

Clearly, the observer does not converge (since $\| \varepsilon(t_{i+1}) \| \ge 2 \| \varepsilon(t_i) \| \quad \forall i \in N$) for the persitent input $u_\tau$. Moreover this property is open with respect to the initial conditions.

**REFERENCES**

[ 1] B.D. ANDERSON, J. MOORE: Linear optimal control, Prentice Hall Network series, 1971.

[ 2] D. BESTLE - M. ZEITZ: Canonical form design for nonlinear observers with linearizable error dynamics, Internat. J. Control, 198, 23, p.419-431.

[ 3] G. BORNARD, N. COUENNE, F. CELLE: Proceedings of the International Conference "Automatique Non linéaire", Nantes, June 1988, France.

[ 4] F. CELLE, J.P. GAUTHIER, D. KASAKOS, G. SALLET: Synthesis of nonlinear observers : a harmonic analysis approach, submitted to MCSS.

[ 5] Y. FUNAHASHI: Stable state estimator for bilinear systems, Internat. J. Control, 1979, 29, p.181-188.

[ 6] J.P. GAUTHIER, G. BORNARD: Observability for any u(t) of a class of bilinear systems, IEEE Trans. Automat. Control, 1981, AC 26, p.922-926.

[ 7] J.P.GAUTHIER, F. CELLE: Theory of dynamic observers for a class of nonlinear systems, MTNS, Janvier 1987, Phoenix.

[ 8] J.P. GAUTHIER, D. KAZAKOS: Observabilité et observateurs de systèmes non linéaires, RAIRO APII 21, 1987.

[ 9] O. GRASSELI, A. ISIDORI: Deterministic reconstruction and reachability of bilinear control processes, Roc. JACC, San Francisco, June 22-25, 1977.

[10] O. GRASSELI, A. ISIDORI: An existence theorem for observers of bilinear systems, IEEE Trans. Automat. Control, AC 26, 1981, p.1299-1301.

[11] S. HARA and K. FURUTA: Minimal order state observers for bilinear systems, Internat. J. Control 24, 1976, p.705-718.

[12] R. KALMAN: Contributions to the theory of optimal control, Proceedings of the Conference on Differential Equations, Mexico City, Mexico,1959 ; Bol. Soc. Mat. Mex. 1961.

[13] R. KALMAN, R. BUCY: New results in linear filtering and prediction theory, J. of Basic Engineering 82 D (1960) 35-45.

[14] R. KALMAN, P. FALB, M. ARBIB: Topics in mathematical systems theory, Mac Graw Hill, 1969.

[15] S.R. KOU, D.L. ELLIOT, T.J. TARN: Exponential observers for nonlinear dynamics systems, Inform. and Control 29, 1975, p.204-216.

[16] A.J. KRENER, A. ISIDORI: Linearization by output injection and nonlinear dynamics systems, Systems Control Lett. 3, 1983, p.47-52.

[17] A.J. KRENER, W. RESPONDEK: Nonlinear observers with linear error dynamics, SIAM J. Control Optim. 23, 1985, p.197-216.

[18] M. KWAKERNAAK: Linear optimal control theory, New York, J. Wiley cop. 1972.

[19] J. LEVINE, R. MARINO: Nonlinear system immersion, observers and finite dimensional filters, Systems and Control Letters, 7, 1986, p.137-142.

[20] E. SONTAG: On the observability of polynomial systems, SIAM J. on Control and Optimization 17, 1979.

[21] E. SONTAG: Nonlinear regulation : the piecewise linear approach, IEEE Trans. on Aut. Control, vol. AC 26, n° 2, April 1981.

[22] H. J. SUSSMANN : Single input observability of continuous time systems, Math. Systems Theory 12, 1979, p.371-393.

[23] D. WILLIAMSON : Observability of bilinear systems, with applications to biological control, Automatica 13, 1977, p.243-254.

# II - ALGEBRAIC SYSTEM THEORY

# A SHORT INTRODUCTION TO DIFFERENTIAL GALOIS THEORY

**RAMIS J.P.**
**Université de Strasbourg 1**
**Strasbourg, France**

*...le trait essentiel du visage* était *son* ambiguïté.
*John David Morley, Pictures from the water trade.*

*"Julien Torma n'existe pas".*
*Henri Thomas, Une saison* volée.

In the first part of these notes we will give a brief description of the "*classical*" differential Galois theory (for more details see [Pi], [Ve], [Kap], [Ko1], [Be1], [Sin1]). One problem with the classical theory is the difficulty of explicit calculations : from the birth of our subject (late 19th century) until to very recent work ([Kat4], [KP], [B.B.H], [B.H], [Ra5], [Ra8], [DM]) the only explicit computations we know are for Airy equation [Kap], and Bessel equations [Ko2] (and in fact Airy equation can be reduced to a special case of Bessel equation [AS]...), but for evident situations. In the second part of our paper we will give a new description of the differential Galois theory (when the "field of constants" is the complex field **C**) in relation with recent progress on the problem of *classification* of analytic differential equations in the complex domain up to analytic transformations ([Si], [Ma1], [Ma2], [BJL1], [BJL2], [J], [BV1], [BV2], [Ra7], [Ra2], for the linear case, and [M.R.1], [M.R2], [MR3], [E3] for the non linear case), and with a new theory of *asymptotics* ([Ra1], [Ra2], [Ra4], [Ra7], [RS1], [MR1], [E1], [E2] [E3], [E4], [EMMR1], [EMMR2], [MR4]). Using this description it is in particular possible to get a method of computation for the case of Meijer $G$-functions, that is for more or less all the cases of *special functions* solutions of linear differential equations [Er] (*cf.* [DM], [Ra8], [BH]).

## I. Differential Galois theory : the classical theory.

### *1. Algebraic Galois theory.*

Algebraic Galois theory is the model for differential Galois theory. It is a whole subject and we will limit ourselves to basic definitions and one example. We will also give a "geometric interpretation" of Galois groups a little sophisticated for the algebraic case, but very useful for the generalisations. For more details it is certainly well worth to read the original paper of Galois [Gal1] and a classical book (as [L]).

*Example.*

Let $P(x) = x^3 + bx + c = 0$, with $b, c \in \mathbf{Q}$. We suppose the polynomial $P$ is irreducible on $\mathbf{Q}$ ($P$ has no rational root). Galois theory is related to the study of "rational invariants" that is of rational functions of the roots $\alpha_1, \alpha_2, \alpha_3$ of $P$ (in a splitting field) which are in fact in the "rationality field" $\mathbf{Q}$.

One example of rational function of the roots is the polynomial

$$\delta = (\alpha_1 - \alpha_2)(\alpha_2 - \alpha_3)(\alpha_3 - \alpha_1),$$

that is the discriminant. The polynomial $\Delta = \delta^2$ is clearly invariant by the group $\mathfrak{S}_3$ of permutations of the roots, so $\Delta \in \mathbb{Q}$; and for $\sigma \in \mathfrak{S}_3$, $\sigma(\delta) = \pm\delta$. If $\Delta = -4b^3 - 27c^2$ has a square root $\sqrt{\Delta}$ in $\mathbb{Q}$, then $\delta = \pm\sqrt{\Delta} \in \mathbb{Q}$ and $\delta$ is a rational invariants in this case the Galois group $G$ is not trivial : $G = \mathfrak{A}_3$; if $\Delta$ has no square root in $\mathbb{Q}$ then $G = \mathfrak{S}_3$.

DÉFINITION. — *Let $P \in \mathbb{Q}[X]$ and $K$ a splitting field for $P$. The Galois group of $P$ is the group of $\mathbb{Q}$-automorphisms of field of $K$ (that is the group of automorphisms leaving fixed the elements of $\mathbb{Q} \subset K$).*

In the case of our example :

$$G = \mathfrak{S}_3 \text{ if } b = c = 1, \text{ and } G = \mathfrak{A}_3 \text{ if } b = -3,\ c = 1.$$

Let $\alpha$ be a root of $P(x) = 0$ and $s \in G$ (degree of $P = n$) : $P(s(\alpha)) = s(P(\alpha)) = 0$; $s(\alpha)$ is a root and we get an injection of groups $G \hookrightarrow \mathfrak{S}_n$.

We will consider the roots of $P$ as "*living on the Galois group* $G$", that is as functions of $\sigma \in G$, or functions of a "*hidden variable*" $\sigma \in G$, using the definition :

$$\alpha_i(\sigma) = \alpha_{\sigma(i)}, \text{ for } i \in [1, \ldots, n], \sigma \in G \subset \mathfrak{S}_n.$$

Clearly rational functions of the roots can be interpreted as functions on $G$, *rational invariants* corresponding to *constant* functions.

From the action of $G$ as permutations of the roots we get an action of $G$ on the functions, corresponding to the translation on the left on $G$ :

$$g_2(\alpha(g_1)) = g_2\alpha(g_1) = \alpha(g_2 g_1).$$

*2. Differential Galois Theory.*

We will limit ourselves to basic definitions and examples, and to the statement of the most important results for the linear case (Picard-Vessiot extensions). We will explain how to deal with non linear equations in another paper [MR4,], with new methods involving a lot of Analysis.

DÉFINITION. — *A differential ring $(A, \partial)$ is a ring $A$ with a derivation $\partial$ : that is a map $\partial : A \to A$, satisfying*

$$\partial(x + y) = \partial x + \partial y \text{ and}$$
$$\partial(xy) = (\partial x)y + x(\partial y), \text{ for every } x, y \in A.$$

The sub-ring of constants of $A$ is

$$C = \{x \in A \,/\, \partial x = 0\}.$$

In the following we will suppose $A$ commutative and $C = \mathbb{C}$. Then the field of constants has characteristic zero and is algebraically closed.

When $A = K$ is a field, $(K, \partial)$ is a differential field. A morphism of differential rings $f : (A_1, \partial_1) \to (A_2, \partial_2)$ is a map $f : A_1 \to A_2$ commuting with the derivations. When $A_1 \subset A_2$ and $\partial_2 | A_1 = \partial_1$ we will say that $(A_2, \partial_2)$ is a differential extension of $(A_1, \partial_1)$.

Classical differential Galois theory uses *differential fields* by analogy with the algebraic case. In fact it is better to work with *differential algebras*. An important example is the linear case : if $f$ is a solution of a first order linear differential equation, then $\frac{1}{f}$ is also a solution of a first order linear differential equation (that is the adjoint equation), this is no longer true in general for a second order linear differential equation : $f(z) = \sin z$ is a solution of the linear differential equation

$$y'' + y = 0,$$

but $\frac{1}{\sin z}$ satisfies no linear differential equation ([HS], [Sin2]). Then if $(f_1, \ldots, f_n)$ is a fondamental system of solutions of a linear differential equation $D$, with coefficients in the differential field $K$ ($f_1, \ldots, f_n \in L$ differential extension of $K$), it is better to work with the $C$-algebra $A_D = K\langle W(f_1, \ldots, f_n)\rangle [f_1, \ldots, f_n]$ than (like in the classical way) with the differential field $K\langle f_1, \ldots, f_n\rangle$ ($K\langle g\rangle$ denote the differential field generated by $K$ and $g$, and $W(f_1, \ldots, f_n)$ is the *Wronskian* of $(f_1, \ldots, f_n)$); then each $f \in A_D$ is a solution of a linear differential equation.

DÉFINITION. — *Let $K$ be a differential field. We will say that a differential extension $M$ of $K$ is a Picard-Vessiot extension if*

*(i) $M = K\langle u_1, \ldots, u_n\rangle$ where $u_1, \ldots, u_n$ are $n$ solutions of a differential linear equation $Dy = 0$ (in a differential extension of $K$) independant over constants (i.e. $W(u_1, \ldots, u_n) \neq 0$).*

*(ii) $M$ admits the same field of constants as $K$.*

THEOREM ([Ko1]). — *If $K$ has characteristic 0 and has an algebraically closed field of constants there exists a Picard-Vessiot extension associated to each linear differential equation, unique up to differential isomorphisms.*

*Examples.*

1. Extension "*by an integral*".

Let be $u' = a$, $a \in K$ (such that there is no $b$ in $K$ such that $b' = a$); $(1, u)$ is a fondamental solution of the second order linear differential equation

$$Dy = y'' - (a'/a)y' = 0$$

and $K\langle u\rangle$ is a Picard-Vessiot extension.

2. Extension "*by the exponential of an integral*".

$$Dy = y' - ay \ , \ u \in K.$$

If $K\langle u\rangle$ has the same field of constants than $K$, it is a Picard-Vessiot extension.

DÉFINITION. — *Let $K$ be a differential field and $L$ a Picard-Vessiot extension associated to a linear differential equation $Dy = 0$. The Galois differential group of $D$ is the group of $K$-automorphisms of differential fields of $L$.*

We will denote $\mathrm{Gal}_K(D) = \mathrm{Aut}_K L$.
For the above examples :

1. We get (*cf.* [Kap])

$$\mathrm{Gal}_K(D) \approx C \text{ (additive group of constants) ;}$$

$L = K\langle u\rangle$ and

$$\begin{array}{l}\sigma_c \in \mathrm{Gal}_K(D) \\ \sigma_c : u \mapsto u + c \quad (c \in C)\end{array}.$$

If $K = \mathbf{C}(x)$, $D = x(\frac{d}{dx})^2 + \frac{d}{dx}$, we get $L = \mathbf{C}(x)\langle\log x\rangle$ and $\mathrm{Gal}_K(D) \approx \mathbf{C}$.

2. We get (*cf.* [Kap]).
$\mathrm{Gal}_K(D)) \approx G \subset C^*$, $G$ subgroup of the multiplicative group of constants. For $C = \mathbf{C}$, we get $G = \{\mathrm{id}\}$, $G \approx \mathbf{Z}/q\mathbf{Z}$, or $G = \mathbf{C}^*$. The following examples are simple but very important :

a) $K = \mathbf{C}(x)$, $D_\alpha = x\frac{d}{dx} - \alpha(\alpha \in \mathbf{C})$; the general solution is $y = Cx^\alpha$;
- if $\alpha \notin \mathbf{Q} : \mathrm{Gal}_K(D) \approx \mathbf{C}^*$
- if $\alpha \in \mathbf{Q} - \mathbf{Z} : \mathrm{Gal}_K(D) \approx \mathbf{Z}/q\mathbf{Z}$ $(q \in \mathbf{N}^*)$
- if $\alpha \in \mathbf{Z} : \mathrm{Gal}_K(D) = \{\mathrm{id}\}$.

b) $K = \mathbf{C}(x)$, $D = x^2\frac{d}{dx} + 1$; the general solution is $y = Ce^{\frac{1}{x}}$; $\mathrm{Gal}_K(D) \approx \mathbf{C}^*$.

Let $D$ be a linear differential operator of order $n$ and $(u_1, \dots, u_n)$ a fundamental system of solutions of $D$ (in a differential extension $M$) : Let $\sigma \in \mathrm{Gal}_K(D)$ and $u$ a solution of $D$ ($u$ is in the $C$-vector space generated by $u_1, \dots, u_n$). Then $D(\sigma u) = \sigma(D_u) = 0$, and $\sigma u$ is again a solution of $D$. The map $\sigma$ : Solutions $\to$ Solutions is clearly $C$-linear and we get a map

$$\mathrm{Gal}_K(D) \to GL(\mathit{solutions}) \approx GL(n; C).$$

This map is injective and $\mathrm{Gal}_K(D)$ can be considered as a subgroup of the *linear group* $GL(n; C)$, just like in ordinary Galois Theory $\mathrm{Gal}(P)$ can be considered as a subgroup of the *permutation groups* $\mathfrak{S}_n$.
The following result ([*Kap*], [*Kol*], [*Be*2]) is fundamental :

THEOREM. — *Let $D$ be a linear differential equation of order $n$. Then $\mathrm{Gal}_K(D)$ is an algebraic subgroup of $GL(n; C)$.*

That is $\mathrm{Gal}_K(D)$ is defined by (a finite number of) algebraic equations (in $n^2$ variables) in $GL(n; C)$.

Algebraic equations satisfied by the elements of $\mathrm{Gal}_K(D)$ correspond to "differential invariants" (similar to the "rational invariants" of the ordinary Galois Theory). In some cases it is easy to get such an invariant using the Wronskian :

Let $Dy = y'' + ay' + b = 0$; $a, b \in K$.

Let $(y_1, y_2)$ be a fundamental system of solutions of $D$ and $W = \begin{vmatrix} y_1 & y_2 \\ y_1' & y_2' \end{vmatrix}$.

We get $W' + aW = 0$.

If $u = 0$, $W \in C \subset K$ and $\mathrm{Gal}_K(D) \subset SL(2; C)$. This result remains true if $a = d'$ with $d \in K$ : for

$$\sigma \in \mathrm{Gal}_K(D), \sigma(W) = \mathrm{D\acute{e}t}(c_{ij})W = W \Rightarrow \mathrm{D\acute{e}t}(c_{ij}) = 1$$
$$(\sigma \text{ with matrix } (c_{ij})).$$

Just like in ordinary Galois theory we have also a "*Galois correspondance*" :

THEOREM. — *Let $M$ be a Picard-Vessiot extension of $K$ associated to the linear differential equation $D$. Then there is a natural bijective correspondance between intermediate differential fields*

$$K \subset L \subset M$$

*and algebraic subgroups*

$$H \subset \mathrm{Gal}_K(D).$$

*Moreover normal extensions $K \subset L$ ($K \subset L \subset M$) correspond to normal subgroups $H \subset \mathrm{Gal}_K(D)$ and $G/H = \mathrm{Gal}_K(L)$.*

Among important applications of Galois differential theory we just mention the following result (similar to solution of the problem of "*integration by quadratures*" which motivated the work of Galois) :

THEOREM. — *Let $M$ be a Picard-Vessiot extension of $K$ associated to the linear differential equation $D$. Then the following statements are equivalent :*

*(i) It is possible to get $M$ from $K$ by a succession of "elementary extensions":*

*a) Algebraic extensions, or*

*b) Extension "by an integral"*

*c) Extension "by an exponential of an integral".*

*Such an extension is called a "generalised Liouville extension".*

*(ii) The connected component (for the Zariski topology*) of the identity in $\mathrm{Gal}_K(D)$ is solvable.*

## II. Differential Galois Theory : new methods.

Analytic differential equations have been divided in *two classes* by nineteen century mathematicians (Fuchs, Thomé, Frobenius ...) :

* In this topology closed sets are algebraic subsets.

a) *Regular singular* (Fuchs) equations

b) *Irregular* equations.

For the first class a quite complete theory was known at the end of the nineteen century (for a modern exposition and generalisations see [De], [GL], [Kat1], [Kat3]). But for the second class a long time was necessary to get a satisfying theory : *cf.* the remarks of [Gar], p. 100. Today such a theory is awailable in works of [J], [S], [Mal2], [Mal3], [Mal4], [BV1], [BV2], [Ra2], [Ra4], [Ra5], [Ra7], [MR1], [BJL1], [BJL2],... Unfortunately there exists no synthetic exposition ([Mal5] and [RS] are in preparation).

Our purpose now is to explain the principles of the *analytic classification* of linear differential equations (for the first class it comes from works of Riemann, Fuchs, Schlessinger, ..., and for the second one it needs very recent tools), that is to give a method to describe all the *analytic invariants* of an equation. We will see afterwards how to derive the Galois differential group (that is *a particular set* of analytic invariants) from the knowledge of the analytic classification (that is from *all* the analytic invariants), which seems, from an abstract view-point reasonable ... There last results have been proved as far as we know by Schlessinger [Schl] for regular singular equations and us for general equations [Ra3], [Ra5].

*1. Regular singular equations.*

Let $Dy = a_n y^{(n)} + \cdots + a_0 y = 0$ be a germ at $0 \in \mathbf{C}$ of linear differential equation, with $a_0, \ldots, a_n \in \mathbf{C}\{x\}$ (or $\mathbf{C}\{x\}[x^{-1}]$).

THEOREM. — *For $D$ the following conditions are equivalent :*
*(i) $D$ can be written (up to multiplication by a meromorphic germ $\not\equiv 0$)*

$$\left(x\frac{d}{dx}\right)^n + b_{n-1}\left(x\frac{d}{dx}\right)^{2-1} + \cdots + b_0.$$

*with $b_0, \ldots, b_{n-1} \in \mathbf{C}\{x\}$.*

*(ii) Every solution $y$ of $Dy = 0$ admits a "meromorphic growth" near the origin, that is, for $\alpha < \beta$, there exists $C, \mu > 0$ (depanding only of $y$ and $\alpha, \beta$) such that :*

$$|x|^{\mu}|y(x)| < C \text{ for } \alpha < \operatorname{Arg} x < \beta.$$

If these conditions are satisfied $D$ is *regular-singular* (or *Fuchsian*).

A fundamental example of regular-singular equation is the famous differential equation of Euler and Gauss, the *hypergeometric* equation (*cf.* [R], [Gou], [Poo], [Lu]) :

$$D_{a,b,c}\, y = x(1-x)y'' + [c - (a+b+1)x]y' - aby = 0 \begin{pmatrix} a, b, c \in \mathbf{C}; \\ "x \in \mathbf{P}^1(\mathbf{C})" \end{pmatrix}.$$

This equation admits a convergent solution in a disk centered at the origin with radius one, the sum of the hypergeometric pover serie :

$$F(x) = F(a,b;c;x) = 1 + \frac{ab}{1!c}x + \frac{a(a+1)b(b+1)}{2!c(c+1)}x^2 + \cdots (F(0) = 1).$$

Let $Dy = 0$ be now a linear differential equation on the Riemann sphere $\mathsf{P}^1(\mathsf{C}) = X$ (or more generally on a connected Riemann surface $X$). Let $a \in X$ be a regular point for $D$; by Cauchy's Theorem $D$ admits a fundamental system of solutions $\{y_1, \ldots, y_n\}$ holomorphic on a small disk centered at $a$. If $b \in y$ is another regular point for $D$ and if $\gamma$ is a continuous path on $X$, $\gamma : [0,1] \to X$, with $\gamma(0) = a$, $\gamma(1) = b$ and $\gamma(t)$ regular for $D$ for $t \in [0,1]$, each solution $y$ admits an *analytic continuation* along $\gamma$ and gives $y_\gamma$ holomorphic on a small disk centered at $b$ and solution of $D$. If we deform continuously the path $\gamma$ leaving fixed $\gamma(0) = a$ and $\gamma(1) = b$ (homotopy with origin and extremity fixed) $y_\gamma$ remains unchanged. We can in particular do that for a *loop* $\gamma$ (i.e. $\gamma(0) = \gamma(1) = a$); then $\gamma$ induces a *linear permutation* of the solutions in a neighborhood of $a$, the *monodromy transformation* along the loop $\gamma$ (in fact "along the homotopy class" of $\gamma$) :

$$y\mathcal{M}_\gamma = y_\gamma.$$

We have, using the classical composition of loops :

$$y\mathcal{M}_{\gamma_1}\mathcal{M}_{\gamma_2} = y\mathcal{M}_{\gamma_1\gamma_2}.$$

If we denote $\mathrm{Sol}_a(D)$ the C-vector space of germs of solutions of $D$ at $n$, we have

$$\mathcal{M}_\gamma \in GL(\mathrm{Sol}_a(D)),$$

and

$$\rho_a : \text{Homotopy class of } \mapsto \mathcal{M}_\gamma$$
$$\rho_a : \pi_1(X - S; a) \to GL(\mathrm{Sol}_a(D))$$

(where $S \subset X$ is the discrete set of singular points for $D$, and $\pi_1(X - S; a)$ is the *fundamental group* of $X - S$ with base point $a$).

We have obtained a *linear representation* of the fundamental group, the monodromy representation (with base point $a$). If we change the base point $a$ in $b$, each homotopy class of continuous path $\gamma$ from $a$ to $b$ gives isomorphisms

$$\begin{array}{ccc} \pi_1(X - S; a) & \xrightarrow{\rho_a} & GL(\mathrm{Sol}_a) \\ \downarrow & & \downarrow \\ \pi_1(X - S; b) & \xrightarrow{\rho_b} & GL(\mathrm{Sol}_b) \end{array}$$

so we will speak improperly of "the" monodromy representation

$$\rho : \pi_1(X - S) \to GL(\mathrm{Sol}).$$

If we express $\mathcal{M}_\gamma$ through a fundamental system $\{y_1, \dots, y_n\}$ we get a "monodromy matrix" $M_\gamma$, and a linear representation :

$$\rho : \pi_1(X - S; a) \to GL(n; \mathbf{C}).$$

There is a local version of this situation : $X = D$ is a germ of disk centered at 0, $S = \{0\}$, $X - S = D^* = D - \{0\}$ : punctured disk. Then $\pi_1(X - S; a) \approx \mathbf{Z}$ (the class of a simple loop turning around 0 in the positive sense corresponding to 1) and we get a representation

$$\rho : \mathbf{Z} \to GL(n; \mathbf{C}).$$

*Ex 1.*

Let $X = \mathbf{P}^1(\mathbf{C})$, $D = D_\alpha = x\frac{d}{dx} - \alpha$ ($\alpha \in \mathbf{C}$). Then $S = \{0, \infty\}$, $X - S = \mathbf{C}^*$, $\pi_1(\mathbf{C}^*) \approx \mathbf{Z}$ and $\rho(1) = \exp(2i\pi\alpha) \in \mathbf{C}^* = GL(1; \mathbf{C})$.

*Ex 2.*

Let $X = \mathbf{P}^1(\mathbf{C})$, $D = x^2\frac{d}{dx} + 1$. Then $S = \{\infty\}$, $X - S = \mathbf{C}$, $\pi_1(\mathbf{C}^*) = \{e\}$ and $\rho(e) = 1 \in GL(1; \mathbf{C}) = \mathbf{C}^*$. The monodromy is trivial.

*Ex 3.*

Let $X = \mathbf{P}^1(\mathbf{C})$, $D = x\left(\frac{d}{dx}\right)^2 + \frac{d}{dx}$. Then $S = \{0, \infty\}$, $X - S = \mathbf{C}^*$, $\pi_1(\mathbf{C}^*) \approx \mathbf{Z}$ and $\rho(1) = \begin{pmatrix} 1 & 2i\pi \\ 0 & 1 \end{pmatrix}$ in the basis $\{1, \log x\}$.

*Ex 4.*

Hypergeometric equation : $D_{a,b,c} = x(1-x)\left(\frac{d}{dx}\right)^2 + [c - (a+b+1)x]\frac{d}{dx} - ab$ ; $X = \mathbf{P}^1(\mathbf{C})$, $S = \{0, 1, \infty\}$. The computation of the monodromy of the hypergeometric equation is due to Riemann [Ri] (*cf.* [Gou], [Poo]). The "exponents of monodromy" (exp (exponent) = eigenvalue of the monodromy around a singularity) are given by the table :

| 0 | 1 | $\infty$ |
|---|---|---|
| 0 | $a$ | 0 |
| $1-c$ | $b$ | $c-a-b$ |

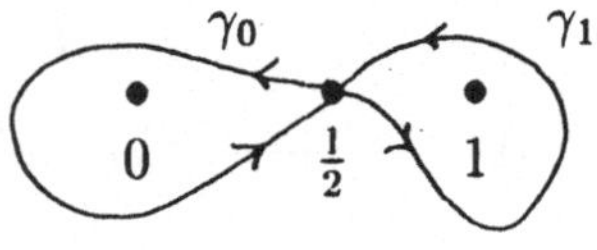

We get a representation

$$\rho : \pi_1(\mathbf{P}^1(\mathbf{C}) - \{0, 1, \infty\}; \frac{1}{2}) \to GL(2; \mathbf{C})$$
$$\wr\wr$$
$$\mathbf{Z} * \mathbf{Z}.$$

For a well chosen basis we get :

$$\rho(\gamma_0) = \begin{pmatrix} 1 & 0 \\ 0 & \omega_0 \end{pmatrix} \text{ and } \rho(\gamma_1) = \begin{pmatrix} 1 + (\omega_1 - 1)C_1 & (\omega_1 - 1)C_2 \\ (\omega_1 - 1)C_1 & 1 + (\omega_1 - 1)C_2 \end{pmatrix};$$

with $\omega_0 = e^{-2i\pi c}, \omega_1 = e^{2i\pi(c-a-b)}$, and

$$C_1 = \frac{-\sin(\pi a)\sin(\pi b)}{\sin(\pi c)\sin[\pi(c-a-b)]}, \quad \frac{\sin\pi(c-a)\sin\pi(c-b)}{\sin(\pi c)\sin[\pi(c-a-b)]}.$$

It is also possible to compute the monodromy in a basis consisting of "Kummer solutions". The formulas are more complicated, using $\Gamma$ function (*cf.* [Poo]).

By a well known elementary method it is possible to replace a linear differential equation of order $n$ by a linear system of order one :

$$Dy = y^{(n)} + \cdots + a_0 y = 0 ;$$

$$Y = \begin{pmatrix} y \\ y' \\ \vdots \\ y^{(n-1)} \end{pmatrix} \text{ and } \Delta Y = Y' - AY = 0,$$

with

$$A = \begin{pmatrix} 0 & 1 & 0 & \dots & 0 \\ 0 & 0 & 1 & \dots & 0 \\ \dots & \dots & \dots & \dots & \\ -a_0 & \dots & \dots & \dots & -a_{n-1}. \end{pmatrix}$$

Conversely it is possible, using a cyclic vector, to replace a differential system $\Delta = \frac{d}{dx} - A$, $A \in \mathrm{End}(n; K)$ ($K$ differential field) by an equation (*cf.* [De], [Ra7], [Kat2]).

A system $\Delta = \frac{d}{dx} - A$ being given it is possible to transform it by an analytic invertible "change of unknown vector" or "*analytic gauge transformation*" : $Z = PY, P \in GL(n; K)$. We get $\frac{dZ}{dx} = P\frac{dY}{dx} + \frac{dP}{dx}Y$, and $\Delta$ becomes $\Delta^P = \frac{d}{dx} - A^P$, with $A^P = PAP^{-1} - \frac{dP}{dx}P^{-1}$.

If $(Y_1, \dots, Y_n)$ is a fundamental matrix for $\Delta$ and $M(\gamma)$ a monodromy matrix :

$$(Y_1, \dots, Y_n) \mapsto (Y_1, \dots, Y_n)M(\gamma),$$
$$(Z_1, \dots, Z_n) \mapsto P(Y_1, \dots, Y_n)M(\gamma)$$
$$(Z_1, \dots, Z_n)M(\gamma),$$

and the monodromy is clearly invariant by *gauge transformation.* It is *an analytic invariant.*

An equation, or system, *modulo gauge transformation* can be interpreted as *an analytic* connection (flat with singularities). If $a \in X - S$ is fixed, we get a "*fibre functor*" :

Connection $\nabla \mapsto$ Sol at a (**C**-vector space) (*cf.* [De M]).

For *regular singular* equations we get the "*Riemann-Hilbert correspondance* ".

| Regular singular equations modulo gauge group $GL(n; K)$ (or regular – singular connections) | $\leftrightarrow$ | Representations of monodromy (*i.e.* linear representations of the fundamental group $\pi_1(X - S; a)$). |
|---|---|---|

For these equations this correspondance is a *bijection* (and more : an *equivalence* of *Tannakian categories*). So monodromy gives in this case the analytic classification of equations.

This result is no longer true for general equations :

Ex. $D_1 = \frac{d}{dx}$, $D_2 = x^2\frac{d}{dx} + 1$ admit the same monodromy (the identity).

Anyway the explicit computation of the representation of monodromy, well known for classical equations, remains an open problem, even for algebraic equations on $\mathbf{P}^1(\mathbf{C})$...

The Galois differential group is a particular set of "analytic invariants"; for Fuchsian equations it can be derived from the analytic classification by the following result [Schl] :

THEOREM. — *Let be D a regular singular equation with coefficients in K (local or global field of meromorphic functions). Then* $\mathrm{Gal}_K(D)$ *is the Zariski closure of the group generated by the image of the representation of monodromy in* $GL(n; \mathbf{C})$.

Using $D = x^2\frac{d}{\partial x} + 1$ it is easy to verify that this result can be false if we remove the hypothesis "regular singular".

2. *Asymptotic expansions and resommation; savage* $\pi_1$.

The key to extend the preceding results to general equations with a similar "geometric flavour" is a new theory of asymptotic expansions due to J. Ecalle and J.P. Ramis (following ideas of Euler, Borel, Stieltjes, Watson, Nevanlinna, Carleman, Perron, Turrittin ...).

Because lack of space I will be very sketchy and refer to [E1], [E2], [E3], [E4], [Ra1], [Ra2], [Ra4], [Ra6], [MR1], [RS], [EMMR1], [EMMR2], and for a detailed review to [MR4]. The classical theory (Poincaré's theory) of asymptotic expansions presents a well-known flaw :

The correspondance : asymptotic expansions $\leftrightarrow$ actual fonctions (i.e. "formal functions") is not bijective (infinitely flat "errors") and not "Galois". A more precise theory is "Gevrey asymptotics expansions theory"(then there are only "exponentially flat errors"). But in fact it is possible to get a "perfect" theory, using "polarised resummations" and taking into account if necessary "exponential corrections" (or "worse" for non linear complicated cases) :

Working with asymptotic expansions we get natural correspondance.

Polarised summation.

"Transasymptotic expansions" $\leftrightarrow$ Actual functions.

For "generic situations" the polarised summation is based upon Laplace-Borel summation (up to some ramification). For "non generic cases" or "multilevel cases" (the phenomenum of multiplicity of "levels" was first noticed in [Ra2], [Ra7]) we need the "acceleration theory" of J. Ecalle [E4].

Example (with three levels : $k_3 < k_2 < k_1$) :

Analytic continuation

$$f = \rho_{k_1}^{-1} \quad L * \quad \mathrm{P}_{k_1/k_2} * \mathrm{P}_{k_2/k_3} * \quad B \quad \rho_{k_3} \quad \hat{f}$$

| $\rho_{k_1}^{-1}$ | $L*$ | $\mathrm{P}_{k_1/k_2} * \mathrm{P}_{k_2/k_3} *$ | $B$ | $\rho_{k_3}$ |
|---|---|---|---|---|
| $\uparrow$ | $\uparrow$ | $\nwarrow \ \nearrow$ | $\uparrow$ | $\uparrow$ |
| Ramification $k_1$ | Laplace | Accelerations | Borel | Ramification $k_3$ |

"Polarisation" refers to a "direction of resummation" $\alpha$ starting from the origin and a sign $+$ or $-$ (intuitively $\alpha^+$ and $\alpha^-$ are directions "infinitely close to" $\alpha$; $\alpha^+$ is "after" and $\alpha^-$ "before", turning in the positive sense). We have "natural summation processes" $S_{\alpha+}$ and $S_{\alpha-}$, each one is "Galois" in any reasonable sense (for large classes of linear or non linear functional equations). Then $(S_{\alpha+})^{-1} \circ S_{\alpha-}$ and $S_{\alpha+} \circ (S_{\alpha-})^{-1}$ are Galois automorphisms : "formal Stokes monodromy" and "Stokes monodromy" ("pointed" at $\alpha_-$). Intuitively (and intuition can be transformed in a formal mathematical description) our Stokes monodromy correspond to a "loop" around "singularities infinitely close to the origin".

For various $q_1, \ldots, q_n \in \frac{1}{x}\mathbb{C}[\frac{1}{x}]$ (or more generally in $\frac{1}{t}\mathbb{C}[\frac{1}{t}]$ with $t^\nu = x$, $\nu \in \mathbb{N}^*$) we get an "exponential torus"

$$T = \mathrm{Gal}_K \langle e^{q_1}, \ldots, e^{q_n} \rangle \approx (\mathbb{C}^*)^\mu \qquad (\mu \in \mathbb{N}; \mu \leq n) ;$$

$K = \mathbb{C}\{x\}[x^{-1}]$). These tori are abelian groups and we will use their abelian Lie algebras Lie $T$.

Using resummation theory and the projective limit of all the exponential tori we built a sort of "local fundamental group" taking into account not only regular singularities but all singularities. The exponential torus (i.e. the limit of) is a sort of Cartan-subgroup of this "*savage* $\pi_1$" (for a formal description see [Ra6]). Stokes monodromy admits an *infinitesimal generator*, by a "*Fourier analysis*" of this generator, (which is a *Galois derivation*) associated to the *adjoint action* of exponential tori we get "*pointed alien derivations*" in the sense of J.Ecalle [E1]. So we build the Lie algebra of the savage $\pi_1$ by a direct product of Lie $T$ and a free infinite dimensional Lie algebra, the *resurgence algebra.* To get the savage $\pi_1$ we "add" by a direct product a formal monodromy to exp (Lie savage $\pi_1$).

Finally it is possible to get a natural generalisation of the Riemann-Hilbert correspondance (for local or global cases) :

| | | |
|---|---|---|
| Analytic linear differential equations modulo gauge group $GL(n;K)$ (or connexions). | $\leftrightarrow$ | Linear repesentations of savage $\pi_1$ |

This correspondance is bijective (and an *equivalence of Tannakian categories*); it gives an analytic classification of analytic differential equations. It is possible to *compute explicitely* the representations for large classes of equations (Meijer equations in particular [DM]).

There is also a generalisation of Schlessinger's theorem for the computation of Galois differential group ([Ra3], [Ra5]) :

THEOREM. — *Let be $D$ a linear differential equation with coefficients in $K$ (local or global field of meromorphic functions). Then* $\mathrm{Gal}_K(D)$ *is the Zariski closure of the image of the group generated by the representation of the savage $\pi_1$ corresponding to $D$ in $GL(n;\mathbf{C})$.*

COROLLARY. — *In local or global situations the following statements are equivalent:*

*(i)* $\mathrm{Gal}_K(D)$ *is Zariski-connected.*

*(ii) The Zariski closure of the group generated by the image of the representation of (ordinary) monodromy is Zariski connected.*

*(iii) (Only in the local case). The Zariski closure of the group generated by the image of the representation of "formal monodromy" is Zariski connected.*

For global situations $(i) \Leftrightarrow (ii)$ is proved by O. Gabber using another method [Kat4].

Now it is possible to interpret solutions of differential equations as functions* not only of the usual variable but also of "*hidden variables*". Considering first the Fuchsian case : for $\gamma \in \pi_1(X - S; a)$ we denote $f(x)\gamma = f(x, \gamma)$; when $\gamma$ varies in $\pi_1(X - S; a)$ we get the different "*branches*" of the solution $f$; now $f$ is a function of $f$ and $\gamma$ and the action of $\pi_1(X - S; a)$ on this function is given by $f(x, \gamma_1)\gamma_2 = f(x, \gamma_1\gamma_2)$, that is by the right translation on the group $\pi_1(X - S; a)$. Example : $\log(z, n) = \log z + 2i\pi n, n \in \mathbf{Z}$. Every "reasonable" analytic relation *remains true* by analytic continuation along a continuous path, so if $f(x)$ satisfies an analytic functional equation ("reasonable"), $f(x, \gamma)$ satisfies

* More formally thase functions are living on a *principal bundle* with basis $X$ and structure group savage $\pi_1$.

the same equation : if $g$ is "uniform" or "rational", $g(x,\gamma) = g(x)$. If $H$ is the group generated by the monodromy representation associated to $D$ (a Fuchsian linear differential equation), for $Df = 0$, we have $Df(x,\gamma) = 0$ for $\gamma \in H$ and $g(x,\gamma)$ makes sense for $\gamma \in H$ and $g$ obtained from $f$ by "any construction of differential algebra". Then $f$ (and more generally $g$) can be "extended by continuity"(in Zariski's topology sense) to the closure $\overline{H}$ of $H$ in $GL(n; \mathbb{C})$, but by Schlessinger's theorem : $\overline{H} = \mathrm{Gal}_K(D) = G$, and $f, g$ are "living" on the Galois differential group $G$ of $D$.

It is possible to perform the same constructions in the general case, replacing $\pi_1$ by the savage $\pi_1$. The variable $\gamma \in \pi_1$ is discrete but the variable $\gamma \in$ savage $\pi_1$ is "partly continuous" : Lie (savage $\pi_1$) is *non trivial* (it is infinite dimensional ...), and it is possible to develop an *infinitesimal calculus* with the "*hidden variables*"$\gamma$. This was done (and discovered) by J.Ecalle: he has introduced "*alien calculus*", "*alien derivations*"; his "hidden variables" correspond to the dual of the envelopping algebra of Lie (savage $\pi_1$). The "geometric ideas" developped above can be extended, with a lot of technical difficulties but quite little changes, to *non linear situations. cf.* [MR1], [MR2], [MR3], [E4],.... For a description of this extension (sketched in Nantes lecture) see [MR4].

As a conclusion I would ask the reader to read the "*Letter to Auguste Chevallier*" written by E. Galois [Gal2] before his duel and in particular the paragraph at the end of the letter about the "*Théorie de l'ambiguïté*" which was considered a long time as a puzzle. My feeling is that the "*Théorie de l'ambiguïté*" is what I have tried to describe above (my conviction is reinforced by the text of E. Picard in [Gal1] and of J. Dieudonné in [Gal2], and still more by the comments at the end of the Thesis of J. Drach [Dr]).

## REFERENCES

[AS] M. ABRAMOWITZ, I. STEGUN, *Handbook of Mathematical Functions*, National Bureau of Standards U.S.A (1964).

[BJL1] W. BALSER, W. JURKAT, D.A. LUTZ, *A general theory of invariants for meromorphic differential equations*, Part I Funkcialaj Ekvacioj 22 (1979), 197–221.

[BJl2] W. BALSER,W. JURKAT, D.A. LUTZ, *A general theory of invariants for meromorphic differential equations*; Part II, Proper invariants, Funkcialaj Ekvacioj 22 (1979), 257–283.

[BV1] D.G. BABBIT, V.S. VARADARAJAN, *Local moduli for meromorphic differential equations*, Bull. Amer. Math. Soc. 12 (New Series) (1985), 95-98.

[BV2] D.G. BABBIT, V.S. VARADARAJAN, *Local moduli for meromorphic differential equations.I. The Stokes sheaf and its cohomology*, UCLA Preprint, 1985.

[BV3] D.G. BABBIT, V.S. VARADARAJAN, *Local isoformal deformation theory for meromorphic differential equations near an irregular singularity*, to appear in the Proceedings of the NATO Advanced Study Institute on Deformation Theory II, Ciocco, Italy, 1986.

[BV4] D.G. BABBIT, V.S. VARADARAJAN, *Structure and moduli for meromorphic differential equations near an irregular singularity*, preprint (1988).

[Bel] D. BERTRAND, *Travaux récents sur les points singuliers des équations différentielles lineaires,* Sém. Bourbaki 1978/ 1979, Exp. 525–542, Lecture Notes in Mathematics, N° 770, Springer-Verlag, 1980.

[BE2] D. BERTRAND, cours de troisième cycle.

[BBH] F. BEUKERS, W.D. BROWNAWELL, G. HECKMAN, *Siegel Normality*, to appear in Annals of Math.

[BH] F. BEUKERS, G. HECKMAN, *Monodromy for the Hypergeometric Function* ${}_nF_{n-1}$, University of Utrecht Preprint N° 483, 1987.

[De] P. DELIGNE, *Equations Différentielles à points singuliers Réguliers*, Lecture Notes in Mathematics, N° 163, Springer-Verlag, 1970.

[De M] P. DELIGNE, J.S. MILNE, *Tannakian Categories* in Lecture Notes in Mathematics, n° 900, Springer-Verlag, 1980.

[DM] A. DUVAL and C. MITSCHI, *Matrices de Stokes et groupe de Galois des équations hypergéométriques confluentes généralisées*, to appear in the Pacific Journal of Mathematics.

[Dr] D. DRACH, *Essai sur une Théorie Générale de l'Intégration,* Annales École Norm. Sup. (1898), 243–384.

[E1] J. ECALLE, *Les Fonctions Résurgentes.* t.I, Publications Mathématiques d'Orsay (1981).

[E2] J. ECALLE, *Les Fonctions Résurgentes.* t. II, Publications Mathématiques d'Orsay(1981)

[E3] J. ECALLE, *Les Fonctions Résurgentes.* t. III, Publications Mathématiques d'Orsay (1985).

[E4] J. ECALLE, *L'Accélération des fonctions résurgentes,* manuscrit 1987.

[EMMR1] J. ECALLE, J. MARTINET, R. MOUSSU, J.P. RAMIS : *Non accumulation des cycles limites (I)*, C.R.A.S. Acad. Sc. Paris, 304 (1987), 375–377.

[EMMR2] J. ECALLE, J. MARTINET, R. MOUSSU, J.P.RAMIS : *Non accumulation des cycles limites (II)*, C.R.A.S. Acad. Sc. Paris, 304 (1987), 431–434.

[E] A. ERDÉLYI, *Higher transcendental functions*, Vol. I. Bateman Manuscript Project, McGraw-Hill, New York 1953.

[Gal 1] E. GALOIS, *Oeuvres Mathématiques d'Evariste Galois, avec une introduction d'Emile Picard,* Gauthiers-Villars, Paris (1897).

[Gal 2] E. GALOIS, *Ecrits et mémoires mathématiques d'Evariste Galois, avec une préface de Jean Dieudonné*, Gauthiers-Villars, Paris (1962).

[Gar] R. GARNIER, *Sur les singularités irrégulières des équations différentielles linéaires*, J. Math. Pures et Appl., 8e série, 2 (1919), 99–198.

[GL] R. GÉRARD, A.H.M. LEVELT, *Invariants mesurant l'irégularité en un point singulier des systèmes d'équations différentielles linéaires*, Ann. Inst. Fourier, t. XXIII, 1(1973).

[Gou] E. GOURSAT, *Leçons sur les séries hypergéométriques I, Propriétés générales de l'équation d'Euler et de Gauss*, Hermann, Paris 1936

[HS] W.A. HARRIS, J.R., Y. SIBUYA, *The reciprocals of solutions of Linear Ordinary Differential Equations*, Advances in Math., 58 (1985), 119–132.

[J] W. JURKAT, *Meromorphe Differentialgleichungen*, Lecture Notes in Mathematics, N° 637, Springer-Verlag (1978).

[Kap] I. KAPLANSKY, *An Introduction to Differential Algebra*, Hermann, Paris 1957.

[Kat1] N.M. KATZ, *Nilpotent connections and the monodromy theorem : application of a result of Turrittin*, Pub. Math. I.H.E.S., 39 (1970), 207–238.

[Kat2] N.M. KATZ, *A Simple Algorithm for Cyclic Vector*,

[Kat3] N.M. KATZ, *A conjecture in the arithmetic theory of differential equations*, Bull. Soc. Math. Fr. (1982).

[Kat4] N.M.KATZ, *On the calculation of some differential Galois groups*, Inventiones Math., 87 (1987).

[KP] N.M. KATZ, R.PINK, *A note on Pseudo-CM representations and differential galois groups*, Duke Math. J., 54, N° 1 (1987).

[Ki] T. KIMURA, *On Riemann's equations which are solvable by quadratures*, Funkcialaj Ekvacioj, 12 (1969), 269–281.

[Kl] F. KLEIN, *Vorlesungen über die hypergeometrische Funktion*, Springer, Berlin 1933.

[Ko1] E.R. KOLCHIN, *Differential Algebra and Algebraic Groups*, Academic Press, New York, 1973.

[Ko2] E.R. KOLCHIN, *Algebraic groups and algebraic dependance*, American Journal of Math., 90 (1968), 1151–1164.

[L] S. LANG, *Algebra, Addison-Wesley* (1985).

[Lu] Y.L. LUKE, *The special functions and their approximations,* vol. 1, Academic Press (1969).

[Ma] Yu. MANIN, *Moduli fuchsiani*, An. Sc. Norm. Sup. Pisa, 19 (1965), 113–126.

[Mal1] B. MALGRANGE, *Sur les points singuliers des équations différentielles*, l'Enseignement Math. 20, (1974), 147–176.

[Mal2] B. MALGRANGE, *Sur la réduction formelle des équations à singularités irrégulières*, Grenoble preprint, 1979.

[Mal3] B. MALGRANGE, *Remarques sur les équations différentielles à points singuliers irréguliers,* in Equations Différentielles et Systèmes de Pfaff dans le champ complexe, R. Gérard et J.P. Ramis, Eds., Lecture Notes in Mathematics N° 712, Springer-Verlag, 1979.

[Mal4] B. MALGRANGE, *La classification des connexions irrégulières à une variable,* "in Mathématique et Physique" : Sém. Ecole Norm. Sup. 1979–1982, Birkhäuser (1983).

[Mal5] B. MALGRANGE, Livre en préparation.

[MR1] J. MARTINET, J.P. RAMIS, *Problèmes de modules pour des équations différentielles non linéaires du premier ordre*, Publications Math. de l'I.H.E.S., 55 (1982), 64–164.

[MR2] J. MARTINET, J.P. RAMIS, *Classification analytique des équations différentielles non linéaires résonnantes du premier ordre*, Annales Scient. École Normale Supérieure, 16 (1983), 571–62.

[MR3] J. MARTINET, J.P. RAMIS, *Analytic Classification of Resonant Saddles and Foci*, Singularities and Dynamical Systems, North-Holland Math. Studies, 103 (1985), 109–135.

[MR4] J. MARTINET, J.P. RAMIS, *Théorie de Galois différentielle et resommation*, Proceedings of "Cade 1", Academic Press, to appear.

[Pi] PICARD, *Analogies entre la Théorie des équations différentielles linéaires et la Théorie des équations algébriques*, Gauthier-Villars, Paris 1936.

[Poo] E.G.C. POOLE, *Introduction to the theory of linear differential equations*, Dover Publications (1960). (Oxford University Press 1936).

[Ra1] J.P. RAMIS, *Devissage Gevrey*, Astérisque 59-60 (1978), 173–204.

[Ra2] J.P. RAMIS, *Les séries k-sommables et leurs applications*, Analysis, Microlocal Calculus and Relativistic Quantum Theory, Proceedings Les Houches 1979, Springer Lecture Notes in Physics, 126 (1980), 178–199.

[Ra3] J.P. RAMIS, *Phénomène de Stokes et filtration Gevrey sur le groupe de Picard-Vessiot*, C.R. Acad. Sc. Paris, 301 (1985), 165–167.

[Ra4] J.P. RAMIS, *Phénomène de Stokes et resommation*, C.R. Acad. Sc. Paris, 301 (1985), 99–102.

[Ra5] J.P. RAMIS, *Filtration Gevrey sur le groupe de Picard-Vessiot d'une équation différentielle irrégulière*, preprint, Instituto de Matematica Pura e Aplicada (IMPA), Rio de Janeiro, 45 (1985), 38 pages.

[Ra6] J.P. RAMIS, *Irregular connections, savage $\pi_1$, and confluence*, Proceedings of a Conference at Katata, Japan 1987, Taniguchi Fundation (1988).

[Ra7] J.P.RAMIS, *Théorèmes d'indices Gevrey pour les équations différentielles ordinaires*, Memoirs of the American Mathematical Society, 296 (1984), 95 pages.

[Ra8] J.P. RAMIS, *On the calculation of some Galois group by a new method,* in preparation.

[RS1] J.P. RAMIS, Y. SIBUYA, *Hukuhara's domains and fundamental existence and uniqueness theorems for asymptotic solutions of Gevrey type*, to appear in Asymptotics.

[RS2] J.P.RAMIS, Y. SIBUYA, *Asymptotic expansions with Gevrey estimates and cohomological methods.* Book in preparation.

[R] B.RIEMANN, *Gesammelte mathematische Werke.*

[Schl] L. SCHLESINGER, *Handbuch der Theorie der linearen Differentialgleichungen*, Teubner, Leipzig, 1895.

[Sc] H.A. SCHWARZ, *Über diejenigen Fälle in welchen die Gaussische hypergeometrische Reihe eine algebraische Funktion ihres vierten Elementes darstellt*, Crelle J. Bd 75 (1873), 292-335.

[Sib] Y. SIBUYA, *Stokes phenomena*, Bull. Amer. Math. Soc, 83 (1977), 1075-1077.

[Sin1] M.F. SINGER, *An outline of Differential Galois Theory*, in this volume.

[Sin2] M.F. SINGER, *Algebraic relations among solutions of linear differential equations*, Trans. Am. Math. Soc., 295(1986).

[Ve] VESSIOT, *Sur les équations différentielles linéaires*, Thèse, Annales de l'École Normale, 1891.

[W] W. WASOW, *Asymptotic Expansions for Ordinary Differential Equations*, Interscience, New-York, 1965.

# Linear Algebra and Nonlinear Control

Claude H. MOOG*
Laboratoire d'Automatique de Nantes
Equipe de Recherche Associée au C.N.R.S.
1, rue de la Noë, 44072 Nantes Cedex 03, France.

Abstract : This paper first summarizes a recently introduced framework for the analysis of nonlinear systems, and then shows that it can provide insights that are complementary to those that one gets from the more classical differential geometric approach, when applied to some important control problems.

## 1. INTRODUCTION

The goal of this paper is to present some algebraic tools that have been recently introduced for the analysis of nonlinear systems, and to apply them to some well-known control problems. The approach proposed in Section 2 is centered around the study of a finite chain of ordinary vector spaces consisting of differentials of functions constructed from the output of a nonlinear system [3]. It will seem to be inspired by the differential algebraic approach of Fliess [6-9] which will be briefly recalled in Section 3. Whereas the differential algebraic approach has been able to give a fundamentally new undertanding of system theory, the derived "linear algebraic approach" focuses on the usual state space description of a nonlinear system. In neither case, as opposed to the geometric approach, do the ever present singularities of a nonlinear system pose any difficulties ; they exist in the equations of interest, but do not inhibit any of the algebraic operations that are needed. However, in all fairness, it must be said that the algebraic approach does not seem to be able to deal with certain aspects of static state feedbacks [23], that have already been nicely treated by geometric methods [13]. The two approaches are thus complementary.

Section 4 is devoted to the definition of infinite and finite zeros for nonlinear systems in terms of the framework of section 2.

Important motivation for the introduction of these new tools appears in the "algebraic" statements of such well known control problems as Disturbance Decoupling, Noninteracting Control or Model Matching. Certain technical assumptions that one has to impose in the differential geometric approach, and that are due to the existence of singularities, disappear in this algebraic framework (section 5). The obtained solutions are then, in this sense, global.

---

* Work performed while a NATO Visiting Professor at the University of Michigan.

## 2. LINEAR ALGEBRA

Consider a nonlinear control system of the form

$$\Sigma: \begin{array}{l} \dot{x} = f(x) + g(x)\,u \\ y = h(x) \end{array} \tag{1.1}$$

where, for simplicity, $x(t) \in \mathbb{R}^n$, $u(t) \in \mathbb{R}^m$, $y(t) \in \mathbb{R}^p$, and $f(\cdot)$, the columns of $g(\cdot)$ and the rows of $h(\cdot)$ are *meromorphic* functions of x ; that is, they are elements of the fraction field $\mathcal{F}(x)$ of the ring of analytic functions of x. Following [3], one may associate to such a system a chain of (nondifferential) vector spaces, which recovers in a unified way, the inversion algorithm of Singh [29], the generic ranks of Nijmeijer [22], the dynamic decoupling algorithms of Descusse-Moog [2] and Nijmeijer-Respondek [24], and the differential output rank of Fliess [7, 9]. These results can be found in a joint work with Di Benedetto and Grizzle [3].

To proceed, suppose that the input function u(t) to the system (1.1) is N times continuously differentiable. Then by Taylor's Theorem,

$$u(t) = \sum_{k=0}^{N} u^{(k)} \frac{(t-t_0)^k}{k!} + R_N(t-t_0),$$

where $t_0$ is some initial point in time, $u^0 := u(t_0)$, $u^{(i+1)} := \frac{d}{dt}u^{(i)}(t)|_{t=t_0}$, and $R_N$ is a remainder term.

View $x, u, \ldots, u^{(n-1)}$ as indeterminates, and let $\mathcal{K}$ denote the field of meromorphic functions of $(x, u, \ldots, u^{(n-1)})$. Recall that given such a field, say in the indeterminates $(v_1, \ldots, v_j)$, one defines $\frac{\partial}{\partial v_i}$ acting on a meromorphic function $\eta(v) = \frac{p(v)}{q(v)}$, where $p(\cdot)$ and $q(\cdot)$ are analytic, by the usual quotient rule of calculus,

$$\frac{\partial}{\partial v_i} \frac{p(v)}{q(v)} := \left( q(v) \frac{\partial}{\partial v_i} p(v) - p(v) \frac{\partial}{\partial v_i} q(v) \right) / q^2(v) . \tag{1.2}$$

Then one defines the differential of $\eta$ by

$$d\eta(v) := \sum_{i=1}^{j} \frac{\partial \eta(v)}{\partial v_i} dv_i . \tag{1.3}$$

Returning to the system (1.1), one defines in a natural way,

$$\dot{y} = \dot{y}(x,u) = \frac{\partial y}{\partial x}[f(x) + g(x)\,u\,] \tag{1.4a}$$

$$y^{(k+1)} = y^{(k+1)}(x,u,\dots,u^{(k)})$$

$$= \frac{\partial y^{(k)}}{\partial x}[f(x) + g(x)\,u\,] + \sum_{i=0}^{k-1} \frac{\partial y^{(k)}}{\partial u^{(i)}}\, u^{(i+1)} \tag{1.4b}$$

Note that $\dot{y}, \dots, y^{(n)}$ so defined are meromorphic functions of x, u,..., $u^{(n-1)}$.

Let $\mathcal{E}$ denote the vector space (over $\mathcal{K}$ ) spanned by $\{dx, du, \dots, du^{(n-1)}\}$. One defines the subspaces $\mathcal{E}_0 \subset \mathcal{E}_1 \subset \cdots \subset \mathcal{E}_n$ of $\mathcal{E}$ by

$$\mathcal{E}_k := \mathrm{span}\{dx, d\dot{y}, \dots, dy^{(k)}\}\,. \tag{1.5}$$

In [3] it is shown that the above chain gives a linear algebraic framework that clarifies many structural properties of nonlinear systems and leads to a synthesis of many previous works on rank invariants of nonlinear systems [11, 22, 24, 28, 29]. It gives moreover a convenient framework within which to state and solve in a global way such control problems as decoupling and thus seems to be a tool that is well adapted to these problems. This will be further explored in section 5.

## 3. DIFFERENTIAL ALGEBRA

In 1985, Fliess introduced an approach to the analysis of nonlinear systems [6, 9], centered around differential algebra. He was able to define, for the first time ever in a clear and precise way, the fundamental notion of the *rank* of a nonlinear system, generalizing the usual notion of the rank of a transfer matrix of a linear system. Moreover, Fliess was then able to give basic definitions of right-invertibility, left-invertibility, and important new results on dynamic feedback [6, 9].

To aid the reader, a few notions from differential algebra are briefly recalled. Let $\mathbb{K}\langle y\rangle$ be the set of all rational functions of $y_i^{(\ell)}$, $1 \le i \le p$, $\ell \ge 0$, with coefficients in the field $\mathbb{K}$. Here, $\mathbb{K}$ is the set of all rational functions in t (over $\mathbb{R}$), and the $y_i^{(\ell)}$ are defined as in (1.4). A finite set of elements $\zeta_1,\dots,\zeta_k$ of $\mathbb{K}\langle y\rangle$ is *differentially algebraically dependent* if there exists a nonzero differential polynomial **P**, with coefficients in $\mathbb{K}$, such that $\mathbf{P}(\zeta_1,\dots,\zeta_k) = 0$ ; that is, a nonzero polynomial in $\zeta_1,\dots,\zeta_k$ and a finite number of their time derivatives. The *differential output rank*, denoted $d^o(\Sigma)$, is the maximum number of differentially algebraically independent elements of $\mathbb{K}\langle y\rangle$. This number is well-defined [18].

In [3] it is shown that the integer $\rho^* = \dim \mathcal{E}_n - \dim \mathcal{E}_{n-1}$ is actually the differential output rank, for those systems that meet both the requirements of Section 1 and those of the differential algebraic approach. More precisely, suppose that the system (1.1) satisfies Assumptions : (A1) f, g and h are meromorphic; (A2) f, g and h are differentially algebraic (i.e., elementary transcendental) functions of their arguments [7, 26, 27] ; (A3) $\mathbb{K}\langle y\rangle$ is a differential field. It is apparently unknown whether (A1) implies (A3).

The result relating the $d^o(\Sigma)$ to the chain $\mathcal{E}_0 \subset \mathcal{E}_1 \subset \cdots \subset \mathcal{E}_n$ is as follows.

**Theorem 3.1 [3]**

Suppose that the system (1.1) satisfies Assumptions A1-A3. Then the differential output rank is given by,

$$d^o(\Sigma) = \dim \mathcal{E}_n - \dim \mathcal{E}_{n-1}. \tag{3.1}$$

Within this framework, the following definitions of left- and right-invertibility are natural.

**Definition 3.2 [7]**

The system (1.1) is said to be left-invertible if its differential output rank $d^o(\Sigma)$ is equal to the number of scalar input components.

It is said to be right-invertible if its differential output rank $d^o(\Sigma)$ is equal to the number of scalar output components.

These definitions can be related to previous ones known in the litterature - see [21] and [3] for left- and right-invertibility respectively.

For latter purposes, let us introduce here the notation $\mathbb{K}\{v_1, \ldots, v_k\}$ for the ordinary field consisting of all rationnal functions of $v_1, \ldots, v_k$.

## 4. INFINITE AND FINITE ZEROS

The notion of infinite zeros or structure at infinity has proven to be useful, in the class of linear systems, for tackling such problems as noninteracting control and model matching [1, 19]. The algebraic approach of section 2 allows one to give an abstract definition of the structure at infinity [20]. In section 6, it will be used to give solvability conditions for certain control problems.

**Definition 4.1 [20]**

The number $\sigma_k$ of zeros at infinity (counted without multiplicity) of order less than or equal to k, for $k \geq 1$, is defined to be

$$\sigma_k = \dim \mathcal{E}_k - \dim \mathcal{E}_{k-1}.$$

Remark : The number of zeros at infinity of order k is then $\sigma_k - \sigma_{k-1}$, and the total number $N_\infty$ of zeros at infinity (counted with multiplicity) is

$$N_\infty = \sum_{k\geq 1} k\,(\sigma_k - \sigma_{k-1})$$

Definition 4.1 is the abstract version of the one formerly given in [20] based upon Singh's Inversion Algorithm. It is important to note that the integers $\sigma_k$ can be easily calculated by using the latter algorithm or, equivalently [3, 10], by computing the sequence of Jacobian matrices defined in [22].

The structure at infinity can also be defined in the field theory framework ; $\mathbb{K}\{x, \dot{y},..., y^{(k)}\}$ being a field extension of $\mathbb{K}\{x, \dot{y},..., y^{(k-1)}\}$ one can note that $\sigma_k$ is the (nondifferential) transcendence degree of $\mathbb{K}\{x, \dot{y},..., y^{(k)}\}/\mathbb{K}\{x, \dot{y},..., y^{(k-1)}\}$ for $k\geq 1$.

Let us now consider the related topic of finite zeros dynamics. In the sequel we will call the order of these dynamics the number $N_z$ of finite zeros. In [16] it has been shown that at least three different notions of finite zeros dynamics can be derived in the nonlinear setting. Here we restrict our interest to the definition that generalizes the fact that for a SISO linear system, the finite zeros are nothing else than the poles of the inverse system . This definition was related in [16, §4] to the order of a reduced inverse system and deduced from the Singh's Inversion Algorithm. Note that it seems not yet clear that the dynamics of different reduced inverse systems are equivalent, diffeomorphic or not. However, its order is well defined. Using the chain $\mathcal{E}_0 \subset \mathcal{E}_1 \subset \cdots \subset \mathcal{E}_n$ one can give the following abstract result, which follows immediately from [16, 20] :

**Corollary 4.2**
The number $N_z$ of finite zeros is

$$N_z = n - N_\infty$$

In linear system theory, the number of finite zeros can be given a geometric interpretation since $N_z$ is equal to the dimension of the largest controlled invariant subspace V* contained in the kernel of the output map. In the nonlinear case, when V* denotes the largest controlled invariant distribution contained in ker dh, this is no longer true ; in general,

$$\dim V^*(\Sigma) \neq n - N_\infty(\Sigma) \qquad (4.1)$$

with any definition of the structure at infinity [20, 23].

However it is possible to derive a geometric interpretation of these finite zeros, using the notion of regular dynamic feedback used in [21]. In order to focus on the most important aspect here, let us consider the case of right-invertible systems (Definition 3.2).

**Theorem 4.3**

Assume that $\Sigma$ (1.1) is right-invertible and let $\Sigma_e$ be the extended system obtained as the limit of the dynamic extension algorithm of [1]. Then,

$$\dim V^*(\Sigma_e) = n - N_\infty(\Sigma) \qquad (4.2)$$

The proof of Theorem 4.3 is based on the fact that $\Sigma_e$ can be decoupled via regular static state feedback and in that case, $\dim V^*(\Sigma_e) = n(\Sigma_e) - N_\infty(\Sigma_e)$ [16]. Furthermore, the dynamic extension algorithm as proposed in [2] changes the structure at infinity so that

$$N_\infty(\Sigma_e) - N_\infty(\Sigma) = n(\Sigma_e) - n. \qquad (4.3)$$

The right-hand side of (4.3) denotes the total number of integrators added to the input channels in the dynamic extension procedure.

## 5. DISTURBANCE DECOUPLING, NONINTERACTING CONTROL AND MODEL MATCHING

In this section we are interested in control problems that are related to Disturbance Decoupling. The latter will be stated here but no solution is given. The aim is to introduce *global* problem statements and solutions for Noninteracting Control and Model Matching ; this is achieved in the context of the linear algebraic framework presented in Section 2.

### 5.1 Disturbance Decoupling (Problem Statement)

Consider the following system

$$\begin{aligned} \dot{x} &= f(x) + g(x)\, u + p(x)\, w \\ y &= h(x) \end{aligned} \qquad (5.1)$$

where $w(t) \in \mathbb{R}^q$ is a disturbance. The Disturbance Decoupling Problem [12] can be stated as follows : find, if possible, a regular static state feedback $u = \alpha(x) + \beta(x)\, v$ such that for any $k \geq 1$,

$$d(y^{(k)}) \in \text{span}\ \{ dx, dv, \ldots, d(v^{(k-1)}) \} \qquad (5.2)$$

The condition (5.2) means that the disturbance w does not affect the output y. This problem is stated in a global fashion in the sense that, considering the field of meromorphic functions, the singularities do not enter the picture.

### 5.2 Noninteracting Control

Consider the static Morgan's Problem for a square system (p=m) ; i.e. the noninteracting control problem in the case where the outputs to be decoupled are scalar. The latter is stated as follows :
Find, if possible, a regular static state feedback $u = \alpha(x) + \beta(x)\, v$ such that for any $k \geq 1$,

$$d(y_i^{(k)}) \in \text{span}\ \{ dx, dv_i, \ldots, d(v_i^{(k-1)})\},\ \text{for } i = 1,\ldots,m \qquad (5.3)$$

For each single output subsystem, it is possible to define the chain of subspaces $\mathcal{E}_0^i \subset \mathcal{E}_1^i \subset \cdots \subset \mathcal{E}_n^i$ and the integers $\sigma_k^i$ for $k \geq 1$ from the structure at infinity. Note that from the scalar output assumption, $\{\sigma_k^i\}_{k=1}^n$ contains the same information as the usual characteristic number associated to the output $y_i$ ; it is merely a convenient notation that is consistent with the linear algebraic setting of Section 2. A necessary and sufficient condition for the solvability of the static Morgan's Problem is given by the equality of the local and global structures at infinity . More precisely, one obtains the following theorem.

**Theorem 5.1 [20]**

There exists a regular static state feedback solution to the static Morgan's Problem if and only if

$$\sigma_k = \sum_{i=1}^{p} \sigma_k^i \qquad \text{for } k \geq 1.$$

When there does not exist such a static state feedback, it may be interesting to look for a "regular" dynamic state feedback. One must first specify what is meant by regular. For linear systems, a dynamic compensator is said to be regular if its transfer function matrix is invertible. For nonlinear systems, regularity of a compensator is also related to its invertibility. More precisely,

**Definition 5.2**

The dynamic compensator described by

$$\dot{\xi} = M(x, \xi) + N(x, \xi)\, v \qquad (5.4)$$

$$u = \alpha(x, \xi) + \beta(x, \xi)\, v \qquad (5.5)$$

is said to be regular if its rank equals m (= dim u = dim v).

The compensator is just viewed as a usual system with state $(x, \xi)$, input v and output u. The Dynamic Morgan's Problem is then stated as follows.

Find, if possible, a regular dynamic state feedback (5.4 - 5.5) such that for any $k \geq 1$,

$$d(y_i^{(k)}) \in \text{span}\ \{ dx, d\xi, dv_i, \ldots, d(v_i^{(k-1)})\},\ \text{for } i = 1,\ldots,m \qquad (5.6)$$

Then one establishes the following result.

**Theorem 5.3 [2, 6]**

There exists a regular dynamic state feedback solution to Morgan's Problem if and only if the rank $\sigma_n$ $(=\rho^*)$ of system (1.1) equals the number of scalar outputs (i.e. iff (1.1) is right-invertible).

## 5.3 Right Model Matching (RMM)

The reader is referred to [2] for a differential geometric problem statement and solution to the Model Matching Problem. Here we redefine this problem in an algebraic way.

Consider the model T described by

$$\dot{x} = f_1(x) + g_1(x)\, u \tag{5.7}$$

$$y_T = h_T(x) \tag{5.8}$$

and the system G given by

$$\dot{z} = f_2(z) + g_2(z)\, v \tag{5.9}$$

$$y_G = h_G(z) \tag{5.10}$$

We restrict our attention to the simplified case where T and G are square systems and G is invertible (its rank is equal to m = dim u = dim v). The general case will be published elsewhere. Then the Right Model Matching Problem is stated as :

Find, if possible, a regular dynamic compensator C

$$\dot{\xi} = M(z, \xi) + N(z, \xi)\, u \tag{5.11}$$

$$v = \gamma(z, \xi, u, \dot{u}, \ldots) \tag{5.12}$$

and an initial condition $\xi(0)$ such that,

$$d[(y_T - y_{GH})^{(k)}] \in \text{span}_{\mathcal{K}(x, z, \xi)} \{ dx, dz, d\xi \} \text{ for every } k \geq 0. \tag{5.13}$$

where $y_{GH}$ denotes the output of the cascade GH and $\mathcal{K}(x, z, \xi)$ denotes the field of meromorphic functions of $(x, z, \xi)$.

Remark

In the case where the model is equal to the idendity system, RMM reduces to Right-Invertiblity.

Let the composite system (T,G) be given by the dynamics (5.7) and (5.9) and the output $(y_T - y_G)$ by

$$y_T - y_G = h_T(x) - h_G(z) \tag{5.14}$$

then one obtains :

**Theorem 5.4**

Under the assumption that G is invertible, there always exists a dynamic compensator (5.11- 5.12) solving the RMM Problem.

*Proof:*

Call $\chi$ the state (x, z) of (T, G) and write (T, G) as

$$\dot{\chi} = F(\chi) + G(\chi) \binom{u}{v}$$

$$\mathcal{Y} = y_T - y_G$$

Apply the so-called Singh's Inversion Algorithm to (T, G), but considering u, $\dot{u}$,... as parameters. Then one gets

$$\mathcal{Y}_1^{(i_1)} = F_1(\chi, u, \dot{u}, \ldots, v)$$

$$\vdots$$

$$\mathcal{Y}_m^{(i_m)} = F_m(\chi, u, \dot{u}, \ldots ; \mathcal{Y}_1^{(.)}, \ldots, \mathcal{Y}_{m-1}^{(i_{m-1})}, v)$$

where $F_1, \ldots, F_m$ are affine in v and fom the invertibility assumption on G, the matrix $\frac{\partial(F_1, \ldots, F_m)}{\partial v}$ has full rank.

Solve for v in the equation

$$\begin{pmatrix} \mathcal{Y}_1^{(i_1)} \\ \vdots \\ \mathcal{Y}_m^{(i_m)} \end{pmatrix} = C$$

where C is a vector of constants $C_k$, $1 \leq k \leq m$, such that $\det \frac{\partial(F_1, \ldots, F_m)}{\partial v} \neq 0$ for $\mathcal{Y}_k^{(i_k)} = C_k$.

Then v is obtained as a meromorphic function of $(x, z, u, \dot{u}, \ldots, \mathcal{Y}_k^{(.)}, \ldots)$ :

$$v = \gamma(x, z, u, \dot{u}, \ldots, \mathcal{Y}_k^{(.)}, \ldots).$$

Now, the dynamic compensator

$$\dot{\xi}_0 = f_1(x) + g_1(x)\, u$$

$$\dot{\xi}_k = \begin{pmatrix} 0 & 1 & \cdot & \cdot & \cdot \\ 0 & 0 & 1 & \cdot & \cdot \\ \cdot & \cdot & \cdot & \cdot & \cdot \\ \cdot & \cdot & \cdot & \cdot & 1 \\ 0 & \cdot & \cdot & \cdot & 0 \end{pmatrix}_{(i_k+1)x(i_k+1)} . \xi_k \qquad , 1 \leq k \leq m$$

$$v = \gamma(\xi_0, z, u, \dot{u}, \ldots, \xi_1, \ldots, \xi_m)$$

solves the RMM Problem.

## 5.4 Left Model Matching (LMM)

In order to recall from [5] a differential algebraic solution to the Left-Model Matching, let us adopt here an input-output point of view. Consider a Model T given by a set of polynomial input-output differential equations $P_T(y, \dot{y}, \ldots, u, \dot{u}, \ldots)$ and the system H is supposed to be given by a set of polynomial input-output differential equations $P_H(w, \dot{w}, \dots, u, \dot{u}, \dots)$. The LMM Problem consists in finding, if possible, a post compensator G in the form of a set of polynomial input-output differential equations $P_G(y, \dot{y}, \dots, w, \dot{w}, \dots)$ such that the cascade (G, H) has the same input-output behavior as the Model T. Then one obtains,

**Theorem 5.5 [5]**

Under the general assumptions (A1 - A3), the LMM Problem is solvable if and only if

$$d^o(H, T) = d^o(H).$$

*Proof:*

Consider the tower of differential fields $\mathbf{K} \subset \mathbf{K}\langle w\rangle \subset \mathbf{K}\langle w,y\rangle$, let tr. $d^o$ denote the differential transcendence degree of a differential field extension, then

$$\text{tr. } d^o\, \mathbf{K}\langle w,y\rangle/\mathbf{K} = \text{tr. } d^o\, \mathbf{K}\langle w,y\rangle/\mathbf{K}\langle w\rangle + \text{tr. } d^o\, \mathbf{K}\langle w\rangle/\mathbf{K}$$

The condition of Theorem 5.5 implies

$$\text{tr. } d^o\, \mathbf{K}\langle w,y\rangle/\mathbf{K}\langle w\rangle = 0$$

which yields the result.

## 6. CONCLUSION

This paper, has underlined the fact that a point of view which is more algebraic than geometric can help in the analysis of nonlinear systems. With the *differential algebraic approach*, one works with fields of nonlinear (rationnal or meromorphic) functions whereas in the related *"linear algebraic"* framework, the analysis is performed on the differentials of these nonlinear functions ; these define linear vector spaces over a *field of scalars* which is the *field of meromorphic functions*. A major consequence is that the usual singularities do not cause trouble neither in the differential nor in the "linear" algebraic approach of nonlinear systems analysis. The study of these singularities reduces to the study of the domain of the involved scalars (the meromorphic functions). In this sense, the proposed algebraic setting yields *global* solutions to control problems.

Moreover, these recently introduced tools can be thought as good candidates for building a *unified* framework for Systems Analysis.

## REFERENCES

[1] J. DESCUSSE, J.F. LAFAY and M. MALABRE, on the Structure at Infinity of Linear Block-Decouplable Systems : the General Case, *I.E.E.E. Trans. Aut. Contr.*, 28, 1983, pp.1115-1118.

[2] J. DESCUSSE and C.H. MOOG, Dynamic decoupling for right-invertible nonlinear systems, *Syst. Contr. Lett..*, 8, 3, 1987, pp. 345-349.

[3] M.D. DI BENEDETTO, J.W. GRIZZLE and C.H. MOOG, A Unified Notion of Rank for a Nonlinear System, *27th CDC*, Austin, Dec. 1988.

[4] M.D. DI BENEDETTO and A. ISIDORI, The Matching of Nonlinear Models via Dynamic State Feedback, *S.I.A.M. J. Contr. Opt.*, 24, 5, 1986, pp. 1063-1075.

[5] G. CONTE, C.H. MOOG and A.M. PERDON, The Model Matching Problem Using a Differential Algebraic Approach, *Rapport Interne du Laboratoire d'Automatique de Nantes*, 1986, and *1st Int. Indus. Applied Math.*, Paris, 1987.

[6] M. FLIESS, A new approach to the noninteracting control problem in nonlinear systems theory, *Proc. 23rd Allerton Conf.*, Monticello, IL, 1985, pp. 123-129.

[7] M. FLIESS, A note on the invertibility on nonlinear input-output differential systems, *Syst. Contr. Lett..*, 8, 2, 1986, pp. 147-151.

[8] M. FLIESS, A new approach to the structure at infinity of nonlinear systems, *Syst. Contr. Lett.*, 7, 5, 1986, pp. 419-421.

[9] M. FLIESS, Some remarks on nonlinear invertibility and dynamic state-feedback, in Theory and Applications of Nonlinear Control Systems, MTNS 85, Stockholm, C.I. Byrnes and A. Lindquist, eds, North-Holland, Amsterdam, 1986, pp. 115-121.

[10] J.W. GRIZZLE, M.D. DI BENEDETTO and C.H. MOOG, Computing the differential output rank of a nonlinear system, *Proc. 26th CDC*, LA, Dec. 1987, pp. 142-145.

[11] R.M. HIRSCHORN, Invertibility of multivariable nonlinear control systems, *I.E.E.E.Trans. Auto. Contr.*, 24, 1979, pp. 855-865.

[12] A. ISIDORI, Sur la Théorie Structurelle et le Problème de la Réjection des Perturbations dans les Systèmes Non Linéaires, in Outils et Modèles Mathématiques pour l'Automatique, l'Analyse des Systèmes et le Traitement du Signal, vol. 1, I.D.Landau, Ed., Editions du C.N.R.S., Paris, 1981, pp. 245-294.

[13] A. ISIDORI, Nonlinear Control Systems : An Introduction, Lect. Notes Contr. Inf. Sci., vol. 72, Springer Verlag, Berlin, 1985.

[14] A. ISIDORI, Control of nonlinear systems via dynamic state-feedback, in Algebraic and Geometric Methods in Nonlinear Control Theory, *Proc. Conf. Paris*, 1985, M. Fliess and M. Hazewinlel, eds, Reidel, Dordrecht, 1986.

[15] A. ISIDORI, C.H. MOOG and A. DE LUCA, A Sufficient Condition for Full Linearization via Dynamic State Feedback, *Proc. 25th CDC*, Athens, 1986, pp. 203-208.

[16] A. ISIDORI and C.H. MOOG, On the equivalent of the notion of transmission zeros, in Modelling and Adaptive Control, *Proc. IIASA Conf.*, Sopron, 1986, C.I. Byrnes and A. Kurszanski, eds, Lect. Notes Contr. Inf. Sci., Vol. 105, to appear.

[17] J. JOHNSON, Kähler differentials and differential algebra, Annals of Math., 89, 1969, pp. 92-98.

[18] E.R. KOLCHIN, Differential Algebra and Algebraic Groups, Academic Press, New York, 1973.

[19] M.MALABRE, Structure à l'Infini des Triplets Invariants - Application à la Poursuite Parfaite de Modèle, *Proc. INRIA Conf.*, Versailles, Lect. Notes Contr. Inf. Sci., vol. 44, Springer Verlag, Berlin, 1982, pp. 43-53.

[20] C.H. MOOG, Nonlinear decoupling and structure at infinity, *Math. Contr. Sign. Syst.*, 1, 1988, pp. 257-268.

[21] C.H.MOOG, Note on the Left-Invertibility of Nonlinear Systems, *Proc. MTNS 87*, to appear.

[22] H. NIJMEIJER, Right-invertibility for a class of nonlinear control systems: a geometric approach, *Syst. Contr. Lett.*, 7, 1986, pp. 125-132.

[23] H. NIJMEIJER and J.M. SCHUMACHER, Zeros at infinity for affine nonlinear control systems, *I.E.E.E.Trans. Auto. Contr.*, 30, 1985, pp. 566-573.

[24] H. NIJMEIJER and W. RESPONDEK, Decoupling via dynamic compensation for nonlinear control systems, *Proc. 25th CDC,* Athens, 1986, pp. 192-197.

[25] J.F. POMMARET, Géométrie différentielle algébrique et théorie du contrôle, *C.R.A.S., Série I*, 302, 15, Paris, 1986, pp. 547-550.

[26] J.F. POMMARET, Differential Galois Theory, Gordon and Breach, New York, 1983.

[27] L.A. RUBEL and M.F. SINGER, A differentially algebraic elimination theorem with application to analog computability in the calculus of variations, *Proc. Amer. Math. Soc.*, 94, 4, 1985, pp. 653-658.

[28] L.M. SILVERMAN, Inversion of multivariable linear systems, *I.E.E.E. Trans. Auto. Contr.*, 14, 1969, pp. 270-276.

[29] S. N. SINGH, A modified algorithm for invertibility in nonlinear systems, *I.E.E.E. Trans. Auto. Contr.*, 26, 1981, pp. 595-598.

[30] S.H. WANG, Design of precompensator for decoupling problem, *Electronics Letters*, 6, pp. 739-741.

# SOME REMARKS ON NONLINEAR INPUT-OUTPUT SYSTEMS WITH DELAYS

---

Michel FLIESS

Laboratoire des Signaux et Systèmes
CNRS-ESE
Plateau du Moulon
91192 Gif-sur-Yvette Cedex
France

**Abstract** We outline a general theory of nonlinear input-output differential systems with delays by employing difference-differential algebra.

## INTRODUCTION

After the introduction by Kalman, Rouchaleau and Wyman [15, 16] of the study of linear systems over commutative rings, Kamen [11] had the most brilliant idea of applying this new theory to linear systems with delays. This gave birth to an abundant literature (see, e.g., Sontag [17], Byrnes [4], Datta and Hautus [7], Brewer, Bunce and Van Vleck [3]), which until now could not be extended to nonlinear systems. The aim of this communication is to outline a new algebraic approach for nonlinear delay systems. We will be employing difference-differential algebra, which is a mixture of difference and differential algebras. The differential part will take into account the derivatives, while the difference part will reflect the delays.

We recall that differential algebra was introduced in control theory for solving the long-standing problem of inverting multivariable nonlinear systems and has since permitted a complete new understanding of continuous-time systems [9]. Thanks to difference algebra a quite analogous theory for discrete-time systems has been proposed [8, 10], which in some sense solves once and for all the intriguing problem of the relationship between discrete and continuous times. Difference-differential algebra will permit us to treat differential systems with delays along the same lines. Questions like the characterization of input and output, invertibility, cascade connection, feedback decoupling and linearization are investigated.

**Acknowledgements** The author would like to thank Professor R.M. Cohn for some helpful comments.

## I. DIFFERENCE-DIFFERENTIAL ALGEBRA

I.1. We recall that differential, difference and difference-differential algebras were introduced by the American mathematician J.F. Ritt and some of his pupils between the two World Wars. Their aim was to give for differential, difference and difference-differential equations a quite analogous background to the one which is provided by commutative algebra for algebraic equations. Contrarily to differential and difference algebras, where books by Ritt [14], Kaplansky [12], Cohn [5] and Kolchin [13] and numerous papers offer a great deal of information, there are only a few published works on the mixed case (see, e.g., Białynicki-Birula [1], Cohn [6]). It will therefore be necessary in the final full version of this paper to provide detailed proofs of basic facts, which will not be given here.

I.2. Let R be a commutative ring with 1, $\Delta = \{\partial_0, \partial_1, ..., \partial_n\}$ be derivations of R into R and $T = \{\tau_1, ..., \tau_m\}$ be monomorphisms of R into R, such that all pairs of members of $\Delta \cup T$ commute. Then R is a *difference-differential ring*, or a $\Delta \cup T$-*ring*. When T (resp. $\Delta$) is empty, R is a differential (resp. difference) ring. A *difference-differential field* is a difference-differential ring which is a field.

I.3. A *difference-differential morphism*, or a $\Delta \cup T$-*morphism* $\phi : R_1 \to R_2$ between two difference-differential rings is a morphism which commutes with the elements of $\Delta \cup T$.

I.4. Let k be a difference-differential field and $w = \{w_i \mid i \in I\}$ be a set of difference-differential quantities. Then $k\{w\}$ (resp. $k\langle w\rangle$) is the difference-differential ring (resp. field) generated by k and the components of w.

I.5. Let K/k be a difference-differential field extension. Denote by $\Theta$ the set of products (including the identity) of elements in $\Delta \cup T$. A set $\xi = \{\xi_i \mid i \in I\}$ of elements in K is said to be $\Delta \cup T$-k-*algebraically independent* if, and only if, the set $\{\theta \xi_i \mid i \in I, \theta \in \Theta\}$ is k-algebraically independent. A maximal $\Delta \cup T$-k-independent set is called a *difference-differential transcendence basis*, or a $\Delta \cup T$-*transcendence basis* of K/k. Any two such bases have the same cardinality, which is called the *difference-differential transcendence degree*, or the $\Delta \cup T$-*transcendence degree*, of K/k. It is denoted by $\Delta \cup T$ tr d° K/k.

I.6. If $\Delta \cup T$ tr d° K/k = 0, the extension K/k is said to be $\Delta \cup T$-*algebraic*. Any element of K is $\Delta \cup T$-*algebraic* over k. This is equivalent to saying it satisfies an algebraic difference-differential equation with coefficients in k.

If $\Delta \cup T$ tr d° K/k $\geq 1$, the extension K/k is said to be $\Delta \cup T$-*transcendental*. Then there exists at least one element $a \in K$ which is $\Delta \cup T$-*transcendental* over k, i.e., which is not $\Delta \cup T$-algebraic over k.

I.7. Let $K \subset L \subset M$ be three difference-differential fields. A classic relation on transcendence degree can be generalized to this situation:

$$\Delta \cup T \text{ tr d° } M/k = \Delta \cup T \text{ tr d° } M/L + \Delta \cup T \text{ tr d° } L/K.$$

I.8. Suppose now that $\Delta$ contains only one derivative, i.e., that we are considering ordinary differential structures and not partial ones. Suppose also that the difference-differential field extension K/k is finitely generated, i.e., K = k<w> where w is a finite set. The next two conditions are then equivalent:

(i) K/k is $\Delta \cup T$-algebraic.

(ii) With respect to the difference structure T, K/k has a finite degree of transcendence[1].

---

[1] Or *transformal* transcendence degree, as in [5].

## II. DIFFERENTIAL SYSTEMS WITH DELAYS

II.1. Let k be a given difference-differential base field. A *systemic field* K is a finitely generated difference-differential extension K/k. In this setting, the operators of Δ (resp. T) play the role of derivatives (resp. delays).

II.2. An *input-output differential system with delays* is a $\Delta \cup$ T-algebraic field extension k<u, y>/k<u>, where

- $u = (u_1, ..., u_m)$ is the *input*,
- $y = (y_1, ..., y_p)$ is the *output*.

When u is a $\Delta \cup$ T-transcendence basis of k<u, y>/k, u is said to be *independent*.

II.3. **Remarks.**

(i) When T (resp. Δ) is empty, we recover the field-theoretic definition of continuous-time (resp. discrete-time) systems [8, 9, 10].

(ii) As with difference algebra applied to discrete-time systems [8, 10], non-causal systems may occur in our field theoretic description, as shown by $y(t) = u(t+1)$ ($m = p = 1$). Although such systems may appear in some realistic case studies, this is an important topic, which will be dealt with in section V.5.

II.4. Suppose we have a finite set $w = (w_1, ..., w_s)$ of quantities which are related by differential equations with delays. Following Willems [18], we will choose among the components of w a possible input and a possible output. Set m = $\Delta \cup$ T tr d° k<w>/k. Then w can be divided in a non-unique way into two disjoint subsets

-$u = (u_1, ..., u_m)$ which is a difference-differential transcendence basis of k<w>/k,

- the remaining components $y = (y_1, ..., y_p)$, $m + p = s$, are $\Delta \cup$ T-algebraic over k<u>.

It is clear that u and y should be termed input and output respectively.

II.5. Consider a general cascade connection of differential systems with delays

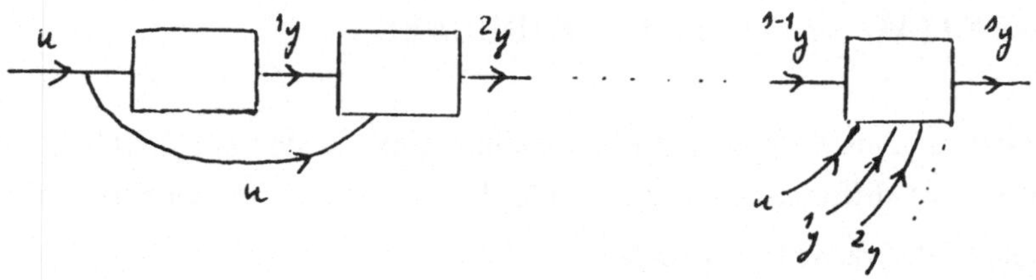

It corresponds to a tower of $\Delta \cup T$-algebraic extensions:

$$k\langle u\rangle \subset k\langle u, {}^1y\rangle \subset k\langle u, {}^1y, {}^2y\rangle \subset \ldots \subset k\langle u, {}^1y, \ldots, {}^sy\rangle.$$

The converse problem of decomposing a given system in "simple" components is then obviously related to a Galois theory of difference-differential fields [1].

## III. INVERTIBILITY

III.1. Not much seems to be known on the invertibility of differential systems with delays. The solution we propose here just mimics what had been done in the continuous-time and discrete-time cases [8, 9, 10].

III.2. The $\Delta \cup T$ *output rank* $\rho$ of an I/O system $k\langle u, y\rangle/k\langle u\rangle$ is the $\Delta \cup T$-transcendence degree of $k\langle y\rangle/k$. The following property is easy:

**Proposition.** $\rho \leq \inf(\mu, p)$, where $\mu = \Delta \cup T$ tr d° $k\langle u\rangle/k$.

III.3. The I/O system $k\langle u, y\rangle/k\langle u\rangle$ is said to be

-left-invertible if, and only if, $\rho = \mu$,
- right-invertible if, and only if, $\rho = p$.

When the input is independent, the left-invertibility condition reads $\rho = m$. With a square system, i.e., $\mu = p$, left and right invertibility are equivalent.

III.4. **Theorem.** An I/O system $k\langle u, y\rangle/k\langle u\rangle$ is left-invertible if, and only if, the extension $k\langle u, y\rangle/k\langle y\rangle$ is $\Delta \cup T$-algebraic.

**Proof.** We apply the equality of section I.7 to two towers of extensions. For $k \subset k\langle u\rangle \subset k\langle u, y\rangle$, we obtain

$$\Delta \cup T \text{ tr } d^\circ\, k\langle u, y\rangle/k = \Delta \cup T \text{ tr } d^\circ\, k\langle u, y\rangle/k\langle u\rangle + \Delta \cup T \text{ tr } d^\circ\, k\langle u\rangle/k.$$

Since $k\langle u, y\rangle/k\langle u\rangle$ is $\Delta \cup T$-algebraic, $\Delta \cup T \text{ tr } d^\circ\, k\langle u, y\rangle/k\langle u\rangle = 0$. Therefore

$$\Delta \cup T \text{ tr } d^\circ\, k\langle u, y\rangle/k = \Delta \cup T \text{ tr } d^\circ\, k\langle u\rangle/k = \mu.$$

To the tower $k \subset k\langle y\rangle \subset k\langle u, y\rangle$ corresponds

$$\Delta \cup T \text{ tr } d^\circ\, k\langle u, y\rangle/k = \Delta \cup T \text{ tr } d^\circ\, k\langle u, y\rangle/k\langle y\rangle + \Delta \cup T \text{ tr } d^\circ\, k\langle y\rangle/k.$$

Since $\Delta \cup T \text{ tr } d^\circ\, k\langle u, y\rangle/k = \mu$, the condition $\rho = \mu$ is equivalent to $\Delta \cup T \text{ tr } d^\circ\, k\langle u, y\rangle/k\langle y\rangle = 0$. □

In a more down-to-earth, but less precise, language, the preceding theorem reads:

**Property.** An I/O system is left-invertible if, and only if, the components of the input can be recovered from the components of the output by a finite set of differential equations with delays.

III.5. Right invertibility of the I/O system $k\langle u, y\rangle/k\langle u\rangle$ is equivalent to saying that the components of y are $\Delta \cup T$-k-algebraically independent, i.e., are not related by any delay-differential equation with coefficients in k.

III.6. **Example.** Suppose we have only one output channel, i.e., $p = 1$. There are then two possibilities for the $\Delta \cup T$-output rank $\rho$:

- $\rho = 0$. The extension $k\langle y\rangle/k$ is $\Delta \cup T$-algebraic and the system is not right-invertible.
- $\rho = 1$. The system is right-invertible. If, moreover, $\mu = 1$, the system is left- and right-invertible.

## IV. FEEDBACK DECOUPLING

IV.1. Let K be a difference-differential field containing k. A $\Delta \cup T$-k-*specialization* $\phi$ : $k<u, y> \rightarrow K$ is a $\Delta \cup T$-morphism $k\{u, y\} \rightarrow K$ leaving the elements of k fixed.

IV.2. A *feedback loop* is a $\Delta \cup T$-k-specialization $\phi : k<u, y> \rightarrow K$. Let m', $0 \leq m' \leq \mu = \Delta \cup T$ tr d° $k<u>/k$, be the difference-differential transcendence degree of the quotient field of $\phi(k\{u, y\})$ over k. Two cases should be examined:

- $m' = 0$. The components $\phi u_1, \ldots, \phi u_m, \phi y_1, \ldots, \phi y_p$ are $\Delta \cup T$-algebraic over k.
- $m' \geq 1$. Any finite set $v = (v_1, \ldots, v_s)$ of difference-differential quantities, such that $\phi u_1, \ldots, \phi u_m, \phi y_1, \ldots, \phi y_p$ are $\Delta \cup T$-algebraic over $k<v>$, is called a *new input*, which is *independent* if, and only if, $s = m'$. If m' is maximal, i.e., equal to $\mu$, the loop is said to be *square*, or *regular*. The loop is said to be *non-degenerated* if, and only if, the difference-differential transcendence degree of $k<\phi y>/k$ is equal to the $\Delta \cup T$-output rank $\rho$.

IV.3. **Remark.** As in II.3, non-causal feedback loops may occur with the preceding definition. This disturbing fact may be avoided by replacing v by $\tau_1^{\nu 1} \ldots \tau_m^{\nu m} v = (\tau_1^{\nu 1} \ldots \tau_m^{\nu m} v_1, \ldots, \tau_1^{\nu 1} \ldots \tau_m^{\nu m} v_s)$ for sufficiently large integers $\nu_1, \ldots, \nu_m$ (see [5]).

IV.4. Suppose now that the input u is independent. The I/O system $k<u, y>/k<u>$ is said to be *decoupled*, or *non-interacting*, if, and only if, after a possible renumbering of the components of u and y, $y_i$ is only affected by $u_i$, $i = 1, \ldots, p$, $p \leq m$. This property should be expressed in a more precise mathematical language.

- The extensions $k<u_1, y_1>/k, \ldots, k<u_\mu, y_p>/k, k<u_{p+1}, \ldots, u_m>/k$ are k-algebraically disjoint [2].
- The components of y are $\Delta \cup T$-transcendental over k.

IV.5. The next result is copied from the continuous-time and discrete-time cases:

**Theorem.** A decoupled I/O system is right-invertible. Any right-invertible I/O system can be decoupled by a non-degenerated feedback loop, which can be chosen to be square.

**Proof.** (i) Take a decoupled system. Set $\bar{u} = (u_1, \ldots, u_p)$. Since

$$\begin{aligned}\Delta \cup T \operatorname{tr} d^{\circ}\, k<y>/k &= \Delta \cup T \operatorname{tr} d^{\circ}\, k<\bar{u}, y>/k - \Delta \cup T \operatorname{tr} d^{\circ}\, k<\bar{u}, y>/k<y> \\ &= p - 0,\end{aligned}$$

the system is right-invertible.

(ii) For the converse, take a right-invertible system. For the specialization, take the identity. Let $v_i$ be an element of $k<y_i>$, $i = 1, \ldots, p$, which is $\Delta \cup T$-transcendental over k. It is easy to complete $v_1, \ldots, v_p$ in a $\Delta \cup T$-transcendental basis of $k<u, y>/k$. This new control gives the feedback decoupling.

## V. REALIZATION

V.1. In this chapter, only ordinary differential structures will be considered. Denote by $\frac{d}{dt}$ = "$\cdot$ " the derivative.

V.2. Take first a usual state-space description

$$(\Sigma)\begin{cases} \dot{x}_i = \alpha_i\,, \; i = 1, \ldots, \,, \\ y_j = \beta_j\,, \; j = 1, \ldots, p \end{cases}$$

where $\alpha_i$ and $\beta_j$ are polynomial functions, with coefficients in k, of the components of x and u and of a finite number of their delays. As in [8, 9, 10], it is quite straightforward to prove the following.

**Property.** $(\Sigma)$ defines an I/O system in the sense of II.2. This means that the extension $k<u, y>/k<u>$ is $\Delta \cup T$-algebraic.

V.3. **Remark.** It is possible to extend the preceding result to more general situations, i.e., to the cases where $\alpha_i$ and $\beta_j$ are rational functions or contain elementary transcendental functions.

V.4. Take now a general I/O system k<u, y>/k<u>. From I.8, we know that the T-transcendence degree of k<u, y>/k<u> is finite. Choose a T-transcendence basis $\xi = (\xi_1, \ldots, \xi_d)$. This means that $\xi_1, \ldots, \xi_d, y_1, \ldots, y_p$ are T-algebraic over k<u, ξ>:

$$
(S) \begin{cases} F_i(\text{f.n.d. } \xi_i, \text{ f.n.d. } \xi, \text{ f.n.d.} u) = 0, \; i = 1, \ldots, d, \\ H_j(\text{f.n.d. } y_j, \text{ f.n.d. } \xi, \text{ f.n.d.} u) = 0, \; j = 1, \ldots, p. \end{cases}
$$

where $F_i$, $H_j$ are polynomial functions, with coefficients in k, of a finite number of delays (f.n.d.) of their arguments. ξ will be called a *state*, although it is clear that in general (S) cannot be written in the form of (Σ). Note that (S) does not need to be causal.

V.5. Let K be an arbitrary difference-differential field. Its *inversive closure* [5], with respect to T, is the difference-differential field $K^* \supset K$, which contains all the *advances* $\{\tau_1^{-\nu_1} \ldots \tau_m^{-\nu_m} a \mid a \in K; \nu_1, \ldots, \nu_m \in \mathbb{N}\}$ of the elements of K. It is possible to show that the T-transcendence degree of k<u, y>*/k<u>* is smaller than or equal to the one of k<u, y>/k<u>.

V.6. In the same way as for discrete-time systems [8, 10], it is possible to show that causality is equivalent to the equality of the T-transcendence degrees of k<u, y>*/k<u>* and k<u, y>/k<u>. Moreover, the dimension of any state is greater than or equal to this degree.

## References


[1] Białynicki-Birula, A., On Galois theory of fields with operators, Amer. J. Math., 84, 1962, 89-109.

[2] N. Bourbaki, Algèbre, chap.4 to 7, Masson, Paris, 1981.

[3] Brewer, J.W., Bunce, J.W., and Van Vleck, F.S., Linear Systems over Commutative Rings, Marcel Dekker, New York, 1986.

[4] Byrnes, C.I., On the control of certain deterministic infinite-dimensional systems by algebro-geometric techniques, Amer. J. Math., 100, 1978, 1333-1381.

[5] Cohn, R.M., Difference Algebra, Interscience, New York, 1965.

[6] Cohn, R.M., A difference-differential basis theorem, Canad. J. Math., 22, 1970, 1224-1237.

[7] Datta, K.B., and Hautus, M.L.J., Decoupling of multivariable control systems over unique factorization domains, SIAM J. Control Optimiz., 22, 1984, 28-39.

[8] Fliess, M., Esquisses pour une théorie des systèmes non linéaires en temps discret, Proc. Conf. Linear Nonlinear Math. Control Theory, Torino, July 1986, Rend. Semin. Mat. Univ. Politec. Torino, Fasc. spec. 1987, pp.55-67.

[9] Fliess, M., Automatique et corps différentiels, Forum Math., 1, 1989.

[10] Fliess, M., Automatique en temps discret et algèbre aux différences, to appear.

[11] Kamen, E.W., On an algebraic theory of systems defined by convolution operators, Math. Systems Theory, 9, 1975, 57-74.

[12] Kaplansky, I., An Introduction to Differential Algebra, Hermann, Paris, 1957.

[13] Kolchin, E.R., Differential Algebra and Algebraic Groups, Academic Press, New York, 1973.

[14] Ritt, J.F., Differential Algebra, Amer. Math. Soc., New York, 1950.

[15] Rouchaleau, Y., and Wyman, B., Linear dynamical systems over integral domains, J. Comput. Systems Sci., 9, 1974, 129-142.

[16] Rouchaleau, Y., Wyman, B., and Kalman, R.E., Algebraic structure of linear dynamical systems III: Realization theory over a commutative ring, Proc. Nat. Acad. Sci. USA, 69, 1972, 3404-3406.

[17] Sontag, E.D., Linear systems over commutative rings: A survey, Ricerche Automatica, 7, 1976, 1-33.

[18] Willems, J.C., From time series to linear systems, I: Finite dimensional linear time invariant systems, Automatica, 22, 1986, 561-580.

# NONLINEAR STATE SPACE AND INPUT OUTPUT DESCRIPTIONS USING DIFFERENTIAL POLYNOMIALS.

S. T. Glad
Department of Electrical Engineering
Linköping University
S-581 83 Linköping, Sweden

## 1. Introduction.

For a linear system with one input and one output one can choose between a time domain description based on the state space, i.e. a system of first order differential equations, or one based on a single higher order differential equation that directly relates input and output. There exists well developed procedures for passing from one description to the other. For nonlinear systems there has recently been a great interest in developing similar procedures. In [1] the existence of an input-output relation is proved, in [2], [3] and [4] the relations between the state space and input-output relations is discussed, using differential geometric concepts. The paper [5] treats realization of an input-output relation. The aim of this paper is to discuss the use differential algebraic procedures to do the computations explicitly. Another differential algebraic approach is described at this conference in [6].

## 2. Some differential algebraic notions.

The basic references for differential algebra are [7] and [8]. Some of the notions that are needed here are presented briefly. The basic objects will be *differential polynomials*, i. e. polynomials in several variables that also contain derivatives of the variables.

### 2.1 Ranking.

The derivatives of variables are ranked according to a system that satisfies

$$u \leq \theta u, \quad u \leq v \Rightarrow \theta u \leq \theta v \tag{1}$$

where u and v are variables or derivatives of variables and $\theta$ is an arbitrary derivation. Examples of rankings of the two variables $u$ and $y$ are

$$u < \dot{u} < \ddot{u} < \cdots < y < \dot{y} < \ddot{y} < \cdots$$

$$u < y < \dot{u} < \dot{y} < \ddot{u} < \ddot{y} < \cdots$$

The *leader* of a polynomial is the highest ranking derivative of that polynomial (it might be a derivative of order zero). For a polynomial $A$ the leader is often denoted $u_A$.

The *initial* of the polynomial $A$ is the coefficient of the highest power of $u_A$. The *separant* of $A$ is $\partial A/\partial u_A$. They are often denoted $I_A$ and $S_A$ respectively.

If $u_A$ is lower than $u_B$ or if $u_A = u_B$ and $deg_{u_A} A < deg_{u_A} B$, then the polynomial $A$ is said to be of lower rank than $B$.

### 2.2 Autoreduced sets, differential ideals.

Let the leader of the polynomial $A$ be $u_A$. The polynomial $F$ is *partially reduced* with respect to $A$ if there is no proper derivative of $u_A$ in $F$. If $F$ is partially reduced with respect to $A$ and the degree of $u_A$ in $F$ is less than the degree of $u_A$ in $A$, then $F$ is said to be *reduced* with respect to $A$. A set of polynomials that are all reduced with respect to each other, is called an *autoreduced set*.

Two autoreduced sets, $\mathbf{A} = A_1, \ldots, A_r$ and $\mathbf{B} = B_1, \ldots, B_s$ are ranked according to the following principle. If there is an integer $k$, $0 \le k \le \min(s, r)$ such that

$$\text{rank}A_j = \text{rank}B_j, \quad j = 0, \ldots, k-1, \quad \text{rank}A_k < \text{rank}B_k$$

then $\mathbf{A}$ is said to be of lower rank than $\mathbf{B}$. If $r > s$ and

$$\text{rank}A_j = \text{rank}B_j, \quad j = 0, \ldots, s$$

then $\mathbf{A}$ is also said to be lower. The lowest autoreduced set that can be formed among a given set of differential polynomials, is called the *characteristic set*.

A set of differential polynomials that forms an ideal and is also closed under differentiation, is called a *differential ideal*. A differential ideal $\Sigma$ is called *prime* if, whenever a product of two differential polynomials, $AB$ is a member of $\Sigma$, then either $A$ or $B$ belongs to $\Sigma$.

### 2.3 Remainder.

If $\mathbf{A}$ is an autoreduced set and $F$ a polynomial then there exist numbers $i_A$, $j_A$ and a polynomial $F_0$ such that

$F_0$ is reduced with respect to $\mathbf{A}$,

the rank of $F_0$ is lower than or equal to that of $F$

$$\prod_{A \in \mathbf{A}} I_A^{i_A} S_A^{j_A} F - F_0$$

is a linear combination of polynomials $\theta A$ with $A \in \mathbf{A}$ and $\theta u_A$ lower than or equal to $u_F$.

The polynomial $F_0$ is called the *remainder* of $F$ with respect to $\mathbf{A}$.

## 3. Transformation from state space to input output polynomial.

Consider a state space description of a system.

$$\begin{aligned} \dot{x}_1 &= f_1(x_1,\dots,x_n,u) \\ &\vdots \\ \dot{x}_n &= f_n(x_1,\dots,x_n,u) \\ y &= h(x_1,\dots,x_n) \end{aligned} \tag{2}$$

where the right hand side consists of polynomials. Alternatively consider the set of differential polynomials in $x$, $u$ and $y$

$$A_i = \dot{x}_i - f_i, \quad i = 1,\dots,n, \quad y - h \tag{3}$$

Let $\Sigma$ be the differential ideal generated by the equations (3). Then $\Sigma$ is readily seen to be prime. Consider the ranking

$$u < \dot{u} < \ddot{u} < \cdots < x_1 < x_2 < \cdots < x_n < \dot{x}_1 < \dot{x}_2 < \cdots < \dot{x}_n < \cdots < y < \dot{y} < \ddot{y} < \cdots \tag{4}$$

The following fact is an immediate consequence of the definitions.

**Proposition 1** *The polynomials (3) form a characteristic set of $\Sigma$ with respect to the ranking (4).*

In many cases one would like to have an equation that directly relates input and output. For the system (2) it is possible to compute such a relation, which is a differential polynomial.

**Proposition 2** *$\Sigma$ contains a differential polynomial in $y$ and $u$ only whose highest derivative of $y$ is at most of order $n$.*

**Proof.** Define $\rho_0 = h$ and let $\rho_1$ be the polynomial formed by differentiating $h$ with respect to time and substituting the polynomials $f_j$ for the derivatives $\dot{x}_j$. It is clear that $\dot{y} - \rho_1$ is in $\Sigma$. Continuing derivation and substitution one forms the polynomials

$$y^{(j)} - \rho_j, \quad j = 0,\dots,n$$

belonging to $\Sigma$, where the polynomials $\rho_j$ contain the $x_j$ but no derivatives of them. Now consider the polynomials

$$y^{j_0}\dot{y}^{j_1}\cdots(y^{(n)})^{j_n} - \rho_0^{j_0}\cdots\rho_n^{j_n}, \quad \sum j_k \le \nu \tag{5}$$

that also belong to $\Sigma$. Then consider the expressions

$$x_1^{j_1}\cdots x_n^{j_n} \tag{6}$$

Assume that the highest degree of such an expression in the $\rho_i$ is $m$. Then the highest degreee of such an expression in (5) is $\nu m$. The number of expressions (6) with degree less than or equal to $\nu m$ is

$$\binom{\nu m + n}{n} \tag{7}$$

The number of polynomials in (5) is

$$\binom{\nu + n + 1}{n + 1} \tag{8}$$

If $\nu$ is large enough, then the latter expression is the greatest. It follows that it is possible to form a linear combination of the polynomials in (5) ( with coefficients depending on $u$ and its derivatives ) such that all expressions of the form (6) disappear. What remains is a nonzero differential polynomial in $y$ and $u$.

■

An immediate consequence is

**Proposition 3** *A characteristic set of* $\Sigma$ *with respect to the ordering*

$$u < \dot{u} < \ddot{u} < \cdots < y < \dot{y} < \ddot{y} < \cdots < x_n < \dot{x}_n < \cdots < x_1 < \dot{x}_1 < \cdots \tag{9}$$

*has a differential polynomial*

$$p(y, u) \tag{10}$$

*whose order in* $y$ *is at most* $n$*, as its lowest element.*

### 3.1 Computational aspects.

Ritt, [7], describes an algorithm for computing the characteristic set of a given set of polynomials. Applied to the present problem it would consist of the following steps.

1. Form the set **A** consisting of the polynomials (3) and adopt the ranking (4).
2. Let $\Sigma$ be the lowest autoreduced set of **A**.
3. If the remainders of all polynomials in **A** with respect to $\Sigma$ are zero, then go to 5.
4. Include the nonzero remainders in **A** and return to step 2.
5. Check if $\Sigma$ is the characteristic set of a prime differential ideal. If not use the procedure described in [7].

### 3.2 Relation between solutions.

It follows from [7] that every solution of the system defined by the characteristic set $\Sigma$, for which none of the separants of $\Sigma$ vanish, is also a solution of the original system. To investigate the possible solutions for which separants vanish, the separants can be added to the differential polynomials and new characteristic sets computed.

### 3.3 An example.

Consider the system whose state space description is given by the polynomials

$$\dot{x}_1 - x_2^2 - x_2, \quad \dot{x}_2 - u, \quad y - x_1$$

With respect to the ranking (4) the characteristic set is

$$\ddot{y}^2 - 4u^2\dot{y} - u^2, \quad \ddot{y} - 2ux_2 - u, \quad y - x_1$$

Note that the characteristic set gives not only the input output relation, but also polynomials that express the state variables in terms of the input and output.

## 4. Realization of an input output relation.

Now consider the realization of an input output relation where the highest derivatives of $u$ and $y$ appear linearly.

$$y^{(n)} - p\left(y, \dot{y}, \ldots, y^{(n-1)}, u, \dot{u}, \ldots, u^{(m-1)}\right) u^{(m)} - q\left(y, \dot{y}, \ldots, y^{(n-1)}, u, \dot{u}, \ldots, u^{(m-1)}\right) = 0 \tag{11}$$

The functions $p$ and $q$ are assumed to be continuously differentiable. Define the $m$-vector

$$w^T = \left(u, \dot{u}, \ddot{u}, \ldots, u^{(m-1)}\right)^T \tag{12}$$

It is now possible to form a state space description

$$\begin{aligned} \dot{z} &= f(z,w) + g(z,w)u^{(m)} \\ \dot{w} &= Aw + bu^{(m)} \end{aligned} \tag{13}$$

where $z$ is an $n$-vector and

$$A = \begin{pmatrix} 0 & 1 & 0 & \ldots & 0 \\ 0 & 0 & 1 & \ldots & 0 \\ \vdots & \vdots & & \ddots & \vdots \\ 0 & 0 & 0 & \ldots & 1 \\ 0 & 0 & 0 & \ldots & 0 \end{pmatrix}, \qquad b = \begin{pmatrix} 0 \\ 0 \\ \vdots \\ 0 \\ 1 \end{pmatrix} \tag{14}$$

where $u^{(m)}$ is regarded as the input. The most straightforward way would be to use

$$z_1 = y, \; z_2 = \dot{y}, \ldots, z_n = y^{(n-1)} \tag{15}$$

giving the state space equations

$$\begin{aligned} \dot{z}_1 &= z_2 \\ \dot{z}_2 &= z_3 \\ &\vdots \\ \dot{z}_n &= q(z) + p(z)\, u^{(m)} \\ \dot{w}_1 &= w_2 \\ &\vdots \\ \dot{w}_m &= u^{(m)} \end{aligned} \tag{16}$$

Note that this is a description in terms of differential polynomials if (11) is a differential polynomial. The disadvantage of this state space realization for some purposes is that it uses the $m$:th derivative of $u$ explicitly. If for instance the state space description were used as the base for a physical realization of the system, then it would be necessary to differentiate the physical input signal. For linear systems it is well known that it is possible to find state space descriptions that do not involve the explicit differentiation of the input ( if $n > m$ ). This can be achieved by using controllabel or observable canonical form.

For the description (13) one could try to improve the situation by a change of variables

$$x = T(z, w), \tag{17}$$

where the continuously differentiable transformation $T$ from $R^n \times R^m$ to $R^n$ can be inverted to give $z$

$$z = \tilde{T}(x, w) \tag{18}$$

If the transformation satisfies

$$T_z g + T_w b = 0 \tag{19}$$

i.e. if each component $T_i$ satisfies the partial differential equation

$$\sum_{k=1}^{n} g_i \frac{\partial \phi}{\partial z_k} + \frac{\partial \phi}{\partial w_m} = 0 \tag{20}$$

then the differential equations in the new variables become

$$\begin{aligned} \dot{x} &= F(x, w) \\ \dot{\bar{w}} &= \bar{A}\bar{w} + \bar{b}u^{(m-1)} \end{aligned} \tag{21}$$

Here $\bar{w}$ and $\bar{b}$ contain the first $m-1$ elements of $w$ and $b$ respectively, while $\bar{A}$ contains the first $m-1$ rows and columns of $A$. The function $F$ is given by

$$F(x, w) = (T_z f)(\tilde{T}(x, w), w) + T_w(\tilde{T}(x, w), w)Aw \tag{22}$$

Equation (21) can be regarded as a state space description where $u^{(m-1)}$ is the input. There is thus one differentiation less of the input.

**Proposition 4** *Let $(z_0, w_0)$ be a given point. Then there exists a neighborhood of that point where a transformation satisfying (19) exists.*

**Proof.** One can regard (20) as a partial differential equation in $(z, w_m)$ with the other components of $w$ as parameters. It follows from the theory of first order partial differential equations, see e. g. [9], that there exist, in a neighborhood of $(z_0, w_0)$, $n$ independent integrals, i.e. functions

$$\phi_1(z, w_m), \ldots, \phi_n(z, w_m)$$

whose Jacobian with respect to $z$ is nonsingular, that satisfy (20). Taking $T_i = \phi_i$, the implicit function theorem will gurantee the existence of a continuously differentiable $\tilde{T}$, in a neighborhood.

If $F$ is linear in $w_m$, then the same procedure can be repeated to give a new state space representation with $u^{(m-2)}$ as the input. If this representation is linear in $u^{(m-2)}$, the procedure can be repeated once more and so on. It might thus be possible to arrive at a state space description with $u$ as the input, with no derivative of $u$ appearing explicitly.

### 4.1 Computational procedure.

One way of computing the integrals of (20) referred to in the proof of proposition 4 is to solve the equation of the characteristics

$$\frac{d\xi}{d\tau} = g(\xi, w_1, \ldots, w_{m-1}, \tau) \tag{23}$$

If the solution of (23) passing through $\xi = z$, $\tau = w_m$ is denoted

$$\phi(\tau, z, w)$$

then its components will satisfy (20).

### 4.2 Two examples.

Consider the input output relation

$$\ddot{y} = \dot{y}\dot{u} + u \tag{24}$$

With the state variables $z_1 = y$, $z_2 = \dot{y}$, $w_1 = u$, the state equations are

$$\begin{aligned} \dot{z}_1 &= z_2 \\ \dot{z}_2 &= w_1 + z_2\dot{u} \\ \dot{w}_1 &= \dot{u} \\ y &= z_1 \end{aligned} \tag{25}$$

with $\dot{u}$ regarded as input. Solving (23) gives the integrals $z_1$ and $z_2e^{-w_1}$. With the new state variables

$$x_1 = z_1, \quad x_2 = z_2e^{-w_1}$$

the state equations become

$$\begin{aligned} \dot{x}_1 &= x_2e^{u} \\ \dot{x}_2 &= ue^{-u} \\ y &= x_1 \end{aligned} \tag{26}$$

i.e. a state space description without any explicit dependence on $\dot{u}$. Note that this is a non polynomial representation, despite the fact that (24) is polynomial.

Now consider the input output relation

$$y^{(3)} = y\ddot{u} + u \tag{27}$$

For the standard description (16) one has

$$p = z_1, \quad q = w_1$$

A solution of (23) and change of state variables gives the description

$$\begin{aligned} \dot{z}_1 &= z_2 \\ \dot{z}_2 &= z_3 + z_1\dot{u} \\ \dot{z}_3 &= w_1 - z_2\dot{u} \\ \dot{w}_1 &= \dot{u} \\ y &= z_1 \end{aligned} \tag{28}$$

Solving (23) for this case gives the coordinate change

$$x_1 = z_1, \quad x_2 = z_2 - z_1 w_1, \quad x_3 = z_3 + z_2 w_1 - z_1 w_1^2/2$$

with the new state space description

$$\begin{aligned} \dot{x}_1 &= x_2 + x_1 u \\ \dot{x}_2 &= x_3 - 2x_2 u - \tfrac{3}{2}x_1 u^2 \\ \dot{x}_3 &= u + x_3 u - \tfrac{3}{2}x_2 u^2 - x_1 u^3 \\ y &= x_1 \end{aligned} \tag{29}$$

## References.


[1] G. Conte, C. H. Moog, A. Perdon, Un théorème sur la représentation entrée - sortie d'un système non linéaire, C. R. Acad. Sci. Paris, to appear.

[2] A. J. van der Schaft, On realization of nonlinear systems described by higher-order differential equations, Math. Systems Theory, 19, 1987, 239-275.

[3] A. J. van der Schaft, Transformation of nonlinear systems under external equivalence, this conference.

[4] A. J. van der Schaft, Representing a nonlinear state space system as a set of higher-order differential equations in the inputs and outputs. Memorandum No. 698, University of Twente, 1988.

[5] P. E. Crouch, F. Lamnabhi - Lagarrigue, State space realizations of nonlinear systems defined by input - output differential equations. 8th Internat. Conf. Analysis and Optimization of Systems, Antibes, June 1988.

[6] S. Diob, A state elimination procedure for nonlinear systems, this conference.

[7] J. F. Ritt, Differential algebra, American Mathematical Society, Providence, RI, 1950.

[8] E.R.Kolchin, Differential Algebra and Algebraic Groups, Academic Press, New York, 1973.

[9] E. Kamke, Differentialgleichungen reeller Funktionen, Chelsea Publishing Company, 1947.

# A STATE ELIMINATION PROCEDURE FOR NONLINEAR SYSTEMS

**Sette DIOP**(†)

ABSTRACT: We solve a long-standing problem on the elimination of the state in the state-space representation of a nonlinear system. For this, we employ techniques from differential algebra.

## 0. INTRODUCTION

The following question is encountered in control theory literature and particularly in WILLEMS' work, see [1983,1986]. Let

(S) $\dot{x}_i = F_i(u, \dot{u},\ldots,x)$ , $y_j = H_j(u, \dot{u}, \ldots, x)$

be a state-space representation of a system. Is there an input-output representation

(I/O) $q_1(u,\dot{u}, \ldots, y,\dot{y}, \ldots) = 0,\ q_2(u,\dot{u}, \ldots, y,\dot{y}, \ldots) = 0, \ldots$

which is *equivalent to* ( S ) ?

WILLEMS [1983] has solved this problem for linear, time-invariant systems; see also the book by BLOMBERG & YLINEN [1983] where the linear case is treated by means of polynomial systems theory. In a nonlinear context, there are contributions by CONTE, MOOG & PERDON [1988], CROUCH & LAMNABHI-LAGARRIGUE [1988], GLAD [1988] and van der SCHAFT [1988].

It is easy, in general, to derive, from (S), equations of type (I/O) whose solutions contain (in an obvious way) those of (S). FLIESS [1989] has shown that this is always possible for a large class of equations (S).

But without some caution, the equations (I/O) thus obtained possess many more solutions than (S). The above question is then meaningful, since it asks how to derive (I/O) such that (S) and (I/O) have the *same* external behavior, provided this is possible.

The aim of this communication is to show that, using differential algebra, there is an elementary but very general answer to the question. A constructive algorithm is given for deriving (I/O) from (S).

This note establishes a link between the control theory problem stated above and the algebraic elimination theory which is a key concept in algebraic geometry (see e.g JACOBSON [1974], TARSKI [1951] and SEIDENBERG [1956a,1956b]). It shows that it is always possible to construct finitely many systems of equations and inequations

($R_j$) $q_{j_1}(u, \ldots, y, \ldots) = 0,\ q_{j_2}(u, \ldots, y, \ldots) = 0, \ldots;\ q_j(u, \ldots, y, \ldots) \neq 0,$

(†)Laboratoire des Signaux & Systèmes
CNRS ESE
Plateau du Moulon
91192 GIF-SUR-YVETTE CEDEX
FRANCE (mailing address)

Universite d'Orléans, UFR Sciences
Laboratoire d'Enseignement Electronique-Automatique
BP 6759
45067 ORLEANS CEDEX
FRANCE

whose disjunction(*) is equivalent to (S). This is a particular case of a general result found by SEIDENBERG in differential algebra.

The following two facts are then clarified: firstly, in general, it is necessary to add an inequation in u, y to (I/O), and, secondly, it is not always possible to derive from (S) a global representation of type (I/O) equivalent to (S). Indeed, the inequation in ($R_j$) is there to define exactly a kind of differential algebraic open set in which the associated equations are the input-output representation equivalent to (S).

The very basic notions of differential algebra which are needed will be recalled below. Only sketches of proof are given.

It must be recalled that differential algebra was introduced to control theory about three years ago by FLIESS [1989].

***Acknowledgements:***

*The author would like to express his gratitude to Professor FLIESS whose assistance was quite indispensable when preparing this Note.*

## I. SOME DIFFERENTIAL ALGEBRA:

The following basic notions of differential algebra should be recalled for the convenience of the reader.

Let A be a ring. A derivation on A is a map of A into itself, denoted by " $\dot{}$ " and such that the following three axioms are satisfied

$$\forall\ a \in A, \quad \dot{a} \in A;$$

$$\forall\ a, b \in A, (a + b)^{\cdot} = \dot{a} + \dot{b};$$

$$\forall\ a, b \in A, (ab)^{\cdot} = \dot{a}b + a\dot{b}.$$

A *differential ring* is a commutative ring with 1 equipped with a derivation. Similarly, a *differential field* is a field equipped with a derivation.

An element a is called a *constant* if

$$\dot{a} = 0.$$

Usual rings such as **Z**, the ring of integers, and fields such as $\mathbb{R}$, the field of real numbers, may be converted into differential ones by associating to them the *trivial derivation* defined by

$$\forall\ a, \ \dot{a} = 0.$$

An example of a field which is not of constants is given by any field of meromorphic functions with the usual differentiation of functions as derivation.

A differential field *extension* K of a differential field k is simply a differential field which contains k as subfield.

Similarlarly, a *differential algebra* is an algebra equipped with a derivation which commutes with the three operations of this algebra.

Given a differential ring A, and an integer n > 0 there exists a differential algebra over A denoted by

$$A\{X_1,\ldots,X_n\}$$

which is the *differential polynomial algebra in the n differential indeterminates* $X_1,\ldots,X_n$ and with coefficents in A. The usual polynomial algebra, $A[X_1,\ldots,X_n]$, is a subalgebra of $A\{X_1,\ldots,X_n\}$; the latter being much bigger. When denoting as

(*) see definition on p. 4

$$X_i = X_i^{(0)},\ \dot{X}_i = X_i^{(1)} \text{ and } (X_i^{(j)})^{\cdot} = X_i^{(j+1)}$$

the respective derivatives of the differential indeterminates, a typical element of $A\{X_1,...,X_n\}$ is then for example

$$P = aX_1X_2^{(2)^2} + bX_3^4X_3^{(1)}$$

The derivative of P is then

$$\dot{P} = \dot{a}X_1X_2^{(2)^2}+\dot{b}X_3^4X_3^{(1)}+aX_1^{(1)}X_2^{(2)^2}+2aX_1X_2^{(2)}X_2^{(3)}+4bX_3^3X_3^{(1)^2}+bX_3^4X_3^{(2)}$$

(in the particular case where the *ground ring*, i.e. the differential ring A, only consists of constants, only the derivatives of the indeterminates are considered when differentiating polynomials).

Let P be a differential polynomial in the single indeterminate X. Suppose P depends on X i.e. P is not in A.

What is called the *order* of P is the greatest integer, say r, such that P depends effectively on $X^{(r)}$.

The *initial* of P is simply the coefficient of $X^{(r)}$ in P, provided that r is the order of P.

The *separant* of P is the formal partial derivative of P with respect to $X^{(r)}$, r being the order of P.

Details on these notions can be found in the early work of RITT [1950].

Initial and separant need not of course be simply elements of the ground ring, but an important fact is that they are of lower order or degree than P.

Now let

$$\omega(P) = (r,d)$$

where r and d are the order and degree of P.

## II. ALGEBRAIC ELIMINATION PROCEDURE

Following the work of KALMAN, control theorists used to manipulate state representation such as

$$(S) \qquad \begin{cases} \dot{x}_i = F_i(u, \dot{u}, \ldots, x) \\ y_j = H_j(u, \dot{u}, \ldots, x) \end{cases}$$

where the $F_i$'s and $H_j$'s are polynomials with coefficients, here, in $\mathbb{R}$ in the variables $u_1, \ldots, u_m, \dot{u}_1, \ldots, \dot{u}_m, \ldots$ and $x_1, \ldots, x_n$; formally, that is :

$$F_i, H_j \in \mathbb{R}\{U\}[X]$$

where $U = (U_1, \ldots, U_m)$, $X = (X_1, \ldots, X_n)$ and $Y = (Y_1, \ldots, Y_p)$ are indeterminates.

As announced earlier, the task is to derive from (S) an *equivalent* representation where the variables x are eliminated.

The notion of equivalence thus invoked is not so simple to define. One of the great problems of algebra is exactly to classify algebraic equations within specified equivalence classes.

A relatively good definition is obtained when it is stated that the *reduced*(*) $\mathbb{R}$-algebras respectively associated with the two systems of equations are isomorphic.

If $P_1, \ldots, P_s$ are the differential polynomials which define the equations, then the differential ideal

(*) An algebra A is said to be *reduced* if whenever $a \in A$ and $a^n = 0$ then $a = 0$.

associated with the equations is the *perfect*(**) differential ideal generated by the $P_i$'s; and the differential reduced $\mathbb{R}$-algebra corresponding to the equations is the differential quotient ring of $\mathbb{R}\{X_1, ..., X_n\}$ by the preceding ideal.

**Remark:** The main inadequacy of the latter definition of equivalence of systems of equations is that it ignores some infinitesimal phenomena in systems. GROTHENDIECK (see GROTHENDIECK & DIEUDONNE [1971]) has noticed this and proposed a more general theory using sophisticated concepts which would be difficult to insert in this note.

Thus, a less general but more intuitive notion of equivalence will be used in what follows. It is implied by the notion of "resultant systems" defined below.

A *solution of equations (S) in a differential extension field K of* $\mathbb{R}$ is an element (u, y, x) of an affine space over K which satisfies relations (S). It must be emphasized that the differential extension field K must be specified when speaking of solutions of (S)! The problem is not , here, to find solutions of equations (which is handled in algebra by means of adjunction or quotient ring theory) but to classify equations within a suitable criterion of equivalence. In that sense the preceding notion of solution is clear enough and efficient.

The *set of solutions* of a system of equations, commonly called the *variety of solutions* then depends on the differential extension field of $\mathbb{R}$ actually considered.

## 1. Notion of resultant systems

*A finite family of systems of equations of type*

$$(R_j) \qquad \begin{cases} f_{j_1}(u,y) = 0 \\ f_{j_2}(u,y) = 0 \\ \quad\vdots \\ g_j(u,y) \neq 0 \end{cases}$$

*where the* $f_{j_l}$*'s (in a finite number) and* $g_j$*'s are in* $R\{U,Y\}$*, is said to be a resultant system for (S) with respect to the x's if and only if, for any differential extension field K of* $\mathbb{R}$*, any* $u, y \in K$*, a necessary and sufficient condition for (S) to possess a solution (in x) in a differential extension field of K, is that u, y be solutions in K of at least one of the* $(R_j)$.

(S) is said to be *equivalent to the disjunction of the* $(R_j)$*'s* if the latter are a family of resultant systems for (S) with respect to the x's.

According to this definition the following equations

$$\text{(a)} \qquad \begin{cases} \dot{x} = u \\ y = x \end{cases}$$

$$\text{(b)} \qquad \dot{y} = u$$

are equivalent. Nevertheless they do not have the same set of solutions since they do not invoke the same number of indeterminates.

In the algebraic context specified above, an answer to the question is the following theorem.

## 2. Theorem

(**) An ideal I of a ring R is said to be *perfect* if whenever $a \in R$ and $a^n \in I$ then $a \in I$.

*For any system defined by the following equations*

$$(S)\qquad \begin{cases} \dot{x}_i = F_i(u, \dot{u}, \ldots, x) & 1 \le i \le n \\ y_j = H_j(u, \dot{u}, \ldots, x) & 1 \le j \le p \end{cases}$$

*where $F_i$, $H_j \in R\{U\}[X]$, and $U = (U_1, \ldots, U_m)$, $X = (X_1, \ldots, X_n)$ and $Y = (Y_1, \ldots, Y_p)$ are the indeterminates, a finite family of resultant systems of (S) with respect to the x's can be computed within a finite number of steps depending only on the $F_i$'s and $H_j$'s.*

This theorem can be compared to the one usually encountered in elementary courses in algebra where there is a question of eliminating a variable between two polynomials in this variable. The word "resultant" comes from this early result (EULER, BEZOUT, ...) on algebraic elimination theory and applies to a polynomial which is a determinant obtained very simply from the coefficients.

The above theorem was obtained for the first time by TARSKI [1951] in (non differential) algebra (see JACOBSON [1974] p.307 and the references on p. 324). The sketch of proof given here is inspired by SEIDENBERG [1956a, 1956b] who obtained the result in the general context of differential algebra. Its merit is that it is elementary.

## 3. Proof

It is clear that it is enough to be able to eliminate one variable, since, if such a procedure is available it will be applied to each variable to be eliminated.

So, in what follows, X is assumed to represent a single indeterminate.

Assume the following notations for the sake of simplicity:

$$P_j(U,Y,X) = H_j(U,X) - Y_j \qquad j = 1, 2, \ldots, p;$$
$$P_{p+1}(U,Y,X) = X^{(1)} - F_1(U,X).$$

The equations (S) then become

$$(S)\qquad P_j(u,y,x) = 0, \qquad j = 1, 2, \ldots, p+1.$$

Change notation again in the following way:

$$(S)\qquad \begin{cases} \tilde{P}_1(u,y) = 0 \\ \tilde{P}_2(u,y) = 0 \\ \cdot \\ \cdot \\ \cdot \\ \tilde{P}_t(u,y) = 0 \\ P_1(u,y,x) = 0 \\ P_2(u,y,x) = 0 \\ \cdot \\ \cdot \\ \cdot \\ P_s(u,y,x) = 0 \\ Q(u,y,x) \neq 0 \end{cases}$$

In the latter notation the $\tilde{P}$'s are assumed not to involve the variable x to be eliminated; thus they do not play an important role. The P's do involve the variable x to be eliminated. It is also supposed that $P_1$ has the *least* ω (denoted by (r,d)) among the P's. The unexpected Q polynomial is initialised with 1 ( such that the corresponding inequation is the always true one: $1 \neq 0$) and updated by the algorithm . Now let

$$\pi(S) = (r,d,s),$$

and make induction on $\pi(S)$.

The stopping rule is clearly the case where $s = 0$. Starting with an arbitrary system of equations of type (S) the decreasing induction on $\pi$ involves essentially simple arguments of polynomial division. This will be reported with full details in a paper to appear. See also WU [1985] for a similar point of view.

## III. EXAMPLES:

### 1. Linear systems

For constant linear systems, the procedure is particularly easy to use. But it already gives results which, *a priori*, might not be obvious.

Consider the following example:

$$(S) \qquad \begin{cases} \dot{x}_1 = x_1+x_2+u_1 \\ \dot{x}_2 = x_1-x_2-u_2 \\ y = x_1+2x_2 \end{cases}$$

We first eliminate $x_2$ then we eliminate $x_1$.

According to notations adopted earlier (S) may be written as

$$(S) \qquad \begin{cases} P_1(U,Y,X) = 2X_2+X_1-Y \\ P_2(U,Y,X) = X_2-X_1^{(1)}+U \\ P_3(U,Y,X) = X_2^{(1)}+X_2-X_1+\dot{U}_2 \\ Q(U,Y,X) = 1 \end{cases}$$

$\pi = (0,1,3)$ $\qquad$ $I=S=2$ (Initial and separant of $P_1$).

By dividing $2P_2$ by $P_1$ the remainder is $-2X_1^{(1)}+X_1+2U_1+Y$. By dividing $2P_3$ by $\dot{P}_1$ then the remainder by $P_1$ the final remainder is $-X_1^{(1)}-3X_1-2U_2+Y^{(1)}+Y$. The system of equations (S) is then equivalent to the following one

$$(S') \qquad \begin{cases} P_1(U,Y,X) = X_1^{(1)}+3X_1-2U_2-Y-Y^{(1)} \\ P_2(U,Y,X) = 2X_1^{(1)}-X_1-2U_1-Y \\ Q(U,Y,X) = 2 \end{cases}$$

now: $\pi = (1,1,2)$ $\qquad$ $I=S=1$.

By dividing $P_2$ by $P_1$ we obtain the following equivalent system of equations

$$(S') \qquad \begin{cases} P_1(U,Y,X) = 7X_1+2U_1-4U_2-Y-2Y^{(1)} \\ P_2(U,Y,X) = X_1^{(1)}+3X_1-2U_2-Y-Y^{(1)} \\ Q(U,Y,X) = 1 \end{cases}$$

now: $\pi = (0,1,2)$ $\qquad$ $I=S=7$.

By dividing $7P_2$ by $\dot{P}_1$ and then the remainder by $P_1$ we finally get the family of the resultant system of (S) with respect to the x's:

$$(R) \qquad F = 3U_1 +U_1^{(1)}+U_2-2U_2^{(1)}+2Y-Y^{(2)},$$

or, which is the same thing:

$$(I/O) \qquad \ddot{y}-2y+\dot{u}_1-2\dot{u}_2+3u_1+u_2 = 0.$$

The very simple example we have just detailed throws light on how the procedure works.

## 2. Nonlinear systems

Let a system be represented by the following equations

$$(S)\qquad \begin{cases} \dot{x} = ux^2+u^2x \\ y = x^2 \end{cases}$$

Application of the procedure gives successively

$$(S')\qquad y = 0$$

$$(S'')\qquad \begin{cases} x^2 - y = 0 \\ 2uyx - \dot{y} + 2u^2y = 0 \\ y \neq 0 \end{cases}$$

Finally, a family of resultant systems for (S) is given by the following three systems of equations:

$$(R_1)\qquad y = 0$$

$$(R_2)\qquad \begin{cases} u = 0 \\ \dot{y} = 0 \\ y \neq 0 \end{cases}$$

$$(R_3)\qquad \begin{cases} (\dot{y} - 2u^2y)^2 - 4u^2y^3 = 0 \\ uy \neq 0 \end{cases}$$

Remark: the particular solutions (($R_1$) and ($R_2$)) are clearly exhibited by the procedure. Except for these, (S) is equivalent to the equation in ($R_3$) in the region defined by the inequation in ($R_3$).

This example shows the possible complexity of calculation even for simple cases.

# IV.CONCLUSION

It may be stated that for systems defined by algebraic state equations, the question of deriving an equivalent input-output representation is completely solved. There are still many things to say about the volume of calculations generated by the algorithm which is only possible with a computer whenever the number of variables to be eliminated is too big. This is no great defect since the calculations may be reduced for more particular equations.

WILLEMS [1986] often considers implicit systems with external variables not necessarily the state. Of course our procedure works again for such equations.

The procedure given above also works for state equations (S) where the $\dot{x}$'s are multiplied by nonconstant coefficients, that is, for implicit differential equations. This is an important feature.

Is it possible to give an analytic version in which polynomials are supplemented by formal power series? This is one open question to which this note leads. If this is the case, it would be possible to treat equations with elementary transcendental functions such as sinus, exponential, etc.

**REFERENCES**


Hans BLOMBERG & Raimo YLINEN

1983: Algebraic Theory for Multivariable Linear Systems

*Academic Press, London, 1983.*

Giuseppe CONTE, Claude H. MOOG & Annamaria PERDON

1988: Un théorème sur la représentation entrée - sortie d'un système non linéaire,

*C.R. Acad. Sci. Paris, 307, Série I (1988), 363-366.*

P.E. CROUCH & Françoise LAMNABHI-LAGARRIGUE

1988: State space realizations of nonlinear systems defined by input-output differential equations,

*Proc. 8th Internat. Conf. Analysis Optimiz. Systems, Antibes, 1988, A. Bensousan and J.L. Lions eds,*

*Lect. Notes Control Inform. Sci., 111 (1988), 138-149, Springer-Verlag, Berlin.*

Michel FLIESS

1989: Automatique et corps differentiels,

*Forum Math.,1, 1989.*

S.T. GLAD

1988: Nonlinear state space and input output descriptions using differential polynomials,

*this volume.*

A. GROTHENDIECK & J.A. DIEUDONNE

1971: Eléments de Géométrie Algébrique, I,

*Springer - Verlag, Berlin, Heidelberg, New York, 1971.*

Nathan JACOBSON

1974: Basic Algebra, I,

*W. H. Freeman & Co, San Francisco, 1974.*

Joseph Fels RITT

1950: Differential Algebra,

*Amer. Math. Soc., New York, 1950.*

A.J. van der SCHAFT

1988: Transformations of nonlinear systems under external equivalence,

*this volume.*

Abraham SEIDENBERG

1956a: Some remarks on Hilbert's Nullstellensatz,

*Arch. Math., Vol.7, 1956, 235-240.*

1956b: An elimination theory for differential algebra,

*Univ. California Publications in Math., (N. S.), 3, n°2, 1956, 31-65.*

Alfred TARSKI

1951: A decision method for elementary algebra and geometry (prepared for publication with the assistance of J.C.C. McKinsey),
*University of California Press, Berkeley and Los Angeles 1951,*
also in *Collected Papers, Vol. 3: 1945-1957, 297-367, S.R. Givant & R.N. McKenzie, eds. Birkhäuser Verlag, Basel,1986.*

Jan C. WILLEMS

1983: Input - output and state space representation of finite dimensional linear time invariant systems,
*Lin. Alg. and its Appl., 50 (1983), 581-608.*

1986: From time series to linear system, Part I: Finite dimensional linear time invariant systems,
*Automatica, Vol. 22 (1986), 561-580.*

Wen-Tsun WU

1985: A constructive theory of differential algebraic geometry based on works of J.F. RITT with particular applications to mechanical theorem-proving of differential geometries,
*Proc. Conf. Differential Geometry and Differential Equations, SHANGHAI, 1985, Lect. Notes in Math., 1255(1987), 173-189, Springer-Verlag, Berlin Heidelberg.*

# III - OPTIMAL CONTROL

# A NEW REGULARITY THEOREM FOR BANG-BANG TRAJECTORIES

*H. J. Sussmann**

Mathematics Department

Rutgers University

New Brunswick, N.J. 08903

U.S.A.

*sussmann@pisces.bitnet*     *sussmann@math.rutgers.edu*

**Abstract**

For systems of the form $\Sigma: \ \dot{x} = f(x) + ug(x), \ \ x \in M, \ \ u \in U$, where $M$ is a real analytic manifold, $f$ and $g$ are real analytic vector fields on $M$, and $U$ is the interval $[-1,1]$, we show that, if $p$ and $q$ are points of $M$ that can be joined by a bang-bang trajectory, then they can be joined by a bang-bang trajectory whose set of switching points has measure zero.

## §1. Introduction.

One of the most basic questions in optimal control theory is that of the *regularity* of trajectories. Specifically, if $\gamma$ is an optimal trajectory, and $\gamma$ corresponds to some control $\eta$, one wants to know whether it is possible to conclude that the control $\eta$ has some extra regularity property, e.g. whether $\eta$ is piecewise continuous, or piecewise smooth. In order to take into account degenerate cases, e.g. when every trajectory is optimal, we rephrase the question as that of knowing whether, if there is an optimal trajectory $\gamma$, then there is one that has some extra

* Partially supported by NSF Grant No. DMS83-01678-01.

regularity properties. (In the particular case when the optimal trajectory is unique, an answer to this new question clearly answers the original question as well.) Moreover, we may forget about optimality altogether, and simply ask whether, when two points $p$ and $q$ can be joined by means of a trajectory of a control system, one can then conclude that they can be joined by a trajectory with some regularity property. (If we have an answer $\mathcal{A}$ to this question, then we can use $\mathcal{A}$ to get an answer to the optimality question as well, by applying $\mathcal{A}$ to the control system obtained by "adding the cost as an extra variable.")

In previous papers ([1], [2], [3]) we have considered systems of the form

$$\Sigma : \qquad \dot{x} = f(x) + ug(x), \quad x \in M, \quad u \in U, \tag{1.1}$$

where $M$ is a real analytic manifold, $f$ and $g$ are real analytic vector fields on $M$, and $U$ is the interval $[-1,1]$. We have shown that, for every such system, if $p$, $q$ are points of $M$ such that there is a trajectory $\gamma$ from $p$ to $q$, then there is a trajectory from $p$ to $q$ that corresponds to a control $\eta$ that has the following regularity property:

**(R)** *$\eta$ is real analytic on an open dense subset of the interval $I(\eta)$ on which it is defined.*

(We have proved —but not yet published— a similar result for more general systems

$$\Sigma : \qquad \dot{x} = f(x,u), \quad x \in M, \quad u \in U, \tag{1.2}$$

where $M$ is a real analytic manifold, $f(x,u)$ is real analytic on $M \times U$, and $U$ is a compact subanalytic subset of some real analytic manifold $N$. The conclusion may fail if $U$ is not compact.)

We also showed (cf., e.g., [2]) that a similar conclusion is not true for $C^\infty$ systems, i.e. systems $\Sigma$ such that $M$, $f$ and $g$ are only required to be $C^\infty$. (In fact, given *any* measurable function $\eta : [0,T] \to [-1,1]$, we show in [2] how to construct a $C^\infty$ system 1.1, with $M = \mathbb{R}^3$, and two points $p$, $q$ in $M$, such that $\eta$ steers $p$ to $q$, and no other control does.) Thus the results of [1] and [2] show that there is a fundamental difference between the $C^\infty$ case and the real-analytic one, in that for $C^\infty$ systems no general regularity conclusion is possible, whereas for real analytic systems *some regularity follows from the mere assumption of real analyticity*, without any extra special hypotheses on $f$ and $g$.

However, the regularity property **(R)** is quite weak. It is an open problem whether one can prove a similar conclusion with "open dense" replaced by "open dense with a complement of measure zero," or perhaps "open dense with a countable complement."

In this note we prove a stronger form of the conclusion, for the particular case of systems of the form 1.1, and for *bang-bang trajectories*. Specifically, we show that, if $p$ and $q$ are points of $M$ that can be joined by a bang-bang trajectory, then they can be joined by a bang-bang trajectory whose set of switching points has measure zero.

## §2. The main theorem.

Consider a control system 1.1, where $M$ is a real analytic manifold, $f$ and $g$ are real analytic vector fields on $M$, and $U$ is the interval $[-1,1]$.

As usual, a *control* is a measurable $U$-valued function $\eta$, defined on an interval $[a,b] \subseteq \mathbb{R}$. (The interval $[a,b]$ is then known as the *domain* of $\eta$, and denoted by $I(\eta)$.) A control $\eta$ is *bang-bang* if $|\eta(t)| = 1$ for almost every $t \in I(\eta)$. A *switching time* of a bang-bang control $\eta$ is a $t \in I(\eta)$ with the property that for every $\varepsilon > 0$ the sets

$$E^+(\eta,t,\varepsilon) = \{s : s \in I(\eta)\,,\ |s-t| < \varepsilon\,,\ \eta(t) = +1\} \tag{2.1}$$

$$E^-(\eta,t,\varepsilon) = \{s : s \in I(\eta)\,,\ |s-t| < \varepsilon\,,\ \eta(t) = -1\} \tag{2.2}$$

have strictly positive Lebesgue measure. We use $SW(\eta)$ to denote the set of switching times of $\eta$. It is clear that $SW(\eta)$ is a closed set.

A *trajectory* of a control $\eta$ is an absolutely continuous curve $\gamma : I(\eta) \to M$ such that

$$\dot\gamma(t) = f(\gamma(t)) + \eta(t)g(\gamma(t)) \text{ for a.e. } t \in I(\eta). \tag{2.3}$$

If $I(\eta) = [a,b]$, and $p = \gamma(a)$, $q = \gamma(b)$, then we call $p$, $q$ the *initial point* and the *terminal point* of $\gamma$, and we say that $\gamma$ (or $\eta$) *steers $p$ to $q$*, or that *$p$ can be reached from $q$ by means of $\gamma$ (or $\eta$)*. We use $\mathcal{R}_\Sigma(p)$ to denote the set of all points that can be reached from $p$ by means of some trajectory of $\Sigma$. If $\gamma$ is trajectory of a bang-bang control $\eta$, then we call $\gamma$ a *bang-bang trajectory*.

We will prove:

**Theorem 2.1.** *Let $\Sigma$ be a system given by 1.1. Let $p$, $q$ be points of $M$ such that $q$ can be reached from $p$ by means of a bang-bang trajectory. Then $q$ can be reached from $p$ by means of a bang-bang trajectory of a control $\eta$ whose set $SW(\eta)$ of switching times has Lebesgue measure zero.*

*PROOF.* If $\Sigma$ is any system given by 1.1, use $X_-$, $X_+$ to denote the vector fields $f-g$, $f+g$, respectively. It will be convenient to work with systems for which

(1) *the vector fields $X_-$, $X_+$ never vanish.*

If $\Sigma$ does not satisfy Condition (1), let $\Sigma'$ be the system on $M' = M \times \mathbb{R}$ obtained by "adding time as a new variable," i.e. the system whose state variable is $(x, x_0)$, with $x$ obeying Equation 1.1, while $x_0$ satisfies $\dot{x}_0 = 1$. It is clear that, if we prove the desired conclusion for $\Sigma'$, then it will follow for $\Sigma$ as well. Moreover, $\Sigma'$ does satisfy (1). Hence we may assume, without loss of generality, that (1) holds for $\Sigma$.

Assume that $\bar{\gamma} : [0,T] \to M$ is a trajectory of $\Sigma$, corresponding to a bang-bang control $\bar{\eta}$, such that $\bar{\gamma}(0) = p$ and $\bar{\gamma}(T) = q$. Let $\Lambda$ be the Lie algebra of vector fields generated by $f$ and $g$. Since $\Lambda$ is a Lie algebra of real analytic vector fields, it is well known that through every point of $M$ there passes a unique maximal integral manifold (m.i.m.) of $\Lambda$. Moreover, every trajectory of $\Sigma$ is entirely contained in some m.i.m. Let $M'$ be the m.i.m. that contains $p$. Then $f$ and $g$ are tangent to $M'$, and $\Sigma$ has a well defined restriction $\Sigma'$ to $M'$. Moreover, $\bar{\gamma}$ is entirely contained in $M'$. So, if we prove the theorem for $\Sigma'$, it will follow that it holds for $\Sigma$ as well. Hence we may assume without loss of generality that $M' = M$. Equivalently, we may assume that $\Sigma$ has the *accessibility property*, i.e. that

(2) *if $\Lambda(x) = \{X(x) : X \in \Lambda\}$, and $T_xM$ is the tangent space of $M$ at $x$, then $\Lambda(x) = T_xM$ for all $x$.*

Notice that, if $\Sigma$ satisfies (1), then $\Sigma'$ also satisfies (1). Hence we may actually assume that both (1) and (2) hold. *We make these assumptions from now on.*

It is well known that, if $q$ is any point in the interior of $\mathcal{R}_\Sigma(p)$, then $q$ can be reached from $p$ by means of a bang-bang trajectory that corresponds to a control with finitely many switchings. Hence the only case where something has to be proved is when $q$ is on the boundary $\partial\mathcal{R}_\Sigma(p)$ of the reachable set. So from now on we will assume that $q \in \partial\mathcal{R}_\Sigma(p)$.

It then follows that we can apply the Pontryagin Maximum Principle (P.M.P.) to $\bar{\gamma}$ and $\bar{\eta}$. For our purposes, it will be convenient to state the P.M.P. in the language of symplectic geometry and Hamiltonian vector fields, as we now proceed to do.

Let $\hat{M}$ be the cotangent bundle of $M$ with the zero section removed. Then $\hat{M}$ has a canonical symplectic structure, which makes it possible to associate to each smooth real-valued function $H$ on $\hat{M}$ a Hamiltonian vector field $V_H$. It is well known that, if $H$, $K$ are smooth functions on $\hat{M}$, then

$$[V_H, V_K] = V_{\{H,K\}}, \tag{2.4}$$

where "[...,...]" and "{...,...}" denote Lie bracket and Poisson bracket, respectively.

For each smooth vector field $X$ on $M$, let $H_X : \hat{M} \to \mathbb{R}$ be the function given by $H_X(\lambda, x) = \langle \lambda, X(x) \rangle$, for $x \in M$ and $\lambda$ a covector at $x$. Let $\hat{X}$ be the *Hamiltonian lift* of $X$, i.e. the Hamiltonian vector field $V_{H_X}$. It is easy to see that the identity

$$\{H_X, H_Y\} = H_{[X,Y]} \tag{2.5}$$

holds whenever $X$, $Y$ are smooth vector fields on $M$. If we combine 2.5 with 2.4 we get

$$[\hat{X}, \hat{Y}] = \widehat{[X,Y]} \ . \tag{2.6}$$

We can then consider the *lifted system* $\hat{\Sigma}$, given by

$$\hat{\Sigma}: \qquad \dot{\xi} = \hat{f}(\xi) + u\hat{g}(\xi), \quad \xi \in \hat{M}, \quad u \in U, \tag{2.7}$$

where $\xi$ is a variable in $\hat{M}$, i.e. $\xi = (\lambda, x)$, $x \in M$, $\lambda \in T_x^* M$, $\lambda \neq 0$. (We use $T_x^* M$ to denote the cotangent space of $M$ at $x$.) It is clear that, if $\Gamma$ is a trajectory of $\hat{\Sigma}$ for a control $\eta$, and $\pi : \hat{M} \to M$ is the canonical projection, then $\pi \circ \Gamma$ is a trajectory of $\Sigma$ for $\eta$. On the other hand, if $\gamma$ is trajectory of $\Sigma$ for $\eta$, then we can construct a trajectory $\Gamma$ of $\hat{\Sigma}$ for $\eta$ such that $\pi \circ \Gamma = \gamma$ by choosing a $t_0 \in I(\eta)$ and then letting $\Gamma(t_0) = (\lambda, \gamma(t_0))$, where $\lambda$ is an arbitrary point of $T^*_{\gamma(t_0)} M$ such that $\lambda \neq 0$. The trajectory $\Gamma$ is then uniquely determined. Any such $\Gamma$ will be called a *Hamiltonian lift* of $(\gamma, \eta)$.

Call a trajectory $t \to \Gamma(t) = (\lambda(t), x(t))$ of $\hat{\Sigma}$ and corresponding control $\eta$ *null minimizing* if the equality

$$\langle \lambda(t), f(x(t)) + \eta(t) g(x(t)) \rangle = \min \{ \langle \lambda(t), f(x(t)) + u g(x(t)) \rangle : -1 \leq u \leq 1 \} = 0 \tag{2.8}$$

holds for almost all $t$. Equivalently, $(\Gamma, \eta)$ is null-minimizing if, for almost all $t$, the function $u \to \mathcal{H}(\Gamma(t), u)$, $u \in [-1, 1]$, is minimized by $u = \eta(t)$ and the minimum value is equal to zero. Here $\mathcal{H}$ is the *control theory Hamiltonian* of $\Sigma$, i.e. the function $\mathcal{H} : \hat{M} \times U \to \mathbb{R}$ given by

$$\mathcal{H}(\xi, u) = \langle \lambda, f(x) + u g(x) \rangle \ , \ \xi = (\lambda, x) \in \hat{M}. \tag{2.9}$$

Let $H_-(\xi) = \mathcal{H}(\xi, -1)$, $H_+(\xi) = \mathcal{H}(\xi, 1)$. Recall that $X_-$, $X_+$ are the vector fields $f - g$, $f + g$. For $\xi \in \hat{M}$, let $\Xi(\xi)$ denote the convex hull of $X_-(\xi)$ and $X_+(\xi)$. Then $H_-$, $H_+$ are the Hamiltonian functions that correspond to $X_-$ and $X_+$, respectively. A null minimizing trajectory of $\hat{\Sigma}$ is an absolutely continuous curve $\Gamma$ such that, for almost every $t$, the vector $\dot{\Gamma}(t)$ satisfies

$$\dot{\Gamma}(t) = \hat{X}_-(\Gamma(t)) \text{ and } 0 = H_-(\Gamma(t)) < H_+(\Gamma(t)), \tag{2.10}$$

or

$$\dot{\Gamma}(t) = \hat{X}_+(\Gamma(t)) \text{ and } 0 = H_+(\Gamma(t)) < H_-(\Gamma(t)), \tag{2.11}$$

or

$$\dot{\Gamma}(t) \in \Xi(\Gamma(t)) \text{ and } 0 = H_-(\Gamma(t)) = H_+(\Gamma(t)). \tag{2.12}$$

With this terminology, the P.M.P. simply says that

(P.M.P) $(\bar{\gamma},\bar{\eta})$ *has a Hamiltonian lift* $\Gamma$ *which is null minimizing.*

Fix such a $\Gamma$. We now want to make a more detailed analysis of the structure of $\Gamma$. For this purpose, we first have to construct a stratification $\mathcal{S}$ of $\hat{M}$.

Recall that a $C^\omega$ *stratification* of a $C^\omega$ manifold $N$ is a locally finite partition $\mathcal{S}$ of $N$ into connected embedded $C^\omega$ submanifolds of $N$ such that, if $S \in \mathcal{S}$, then the closure Clos $S$ of $S$ is the union of $S$ and a collection of members $T$ of $\mathcal{S}$ such that $\dim T < \dim S$. We let $\mathcal{S}$ be a $C^\omega$ stratification of $\hat{M}$ such that

**S1.** *if* $S \in \mathcal{S}$, *and* $\varphi$ *is any of the functions* $H_-$, $H_+$, *then either* $\varphi$ *vanishes identically on* $S$, *or it never vanishes on* $S$;

**S2.** *if* $S \in \mathcal{S}$, *and* $V$ *is any of the vector fields* $\hat{X}_-$, $\hat{X}_+$, *then either* $V$ *is everywhere tangent to* $S$, *or it is never tangent to* $S$.

(The existence of $\mathcal{S}$ follows from general properties of real analytic stratifications. Details of the construction are given in [1].)

We now pick a stratum $S$ of $\mathcal{S}$, and let $J(S)$ be the set of those $t \in I(\bar{\eta})$ such that $\Gamma(t) \in S$. We let $B(S) = SW(\eta) \cap J(S)$. We will prove that $B(S)$ has Lebesgue measure zero. Since this is true for each $S$, and $\Gamma$ visits finitely many strata $S$ (because $\mathcal{S}$ is locally finite), this will imply our desired conclusion.

Let $K$ be the set of all points $t \in I(\eta)$ such that

a. $\dot{\Gamma}(t)$ *exists and equals* $\hat{f}(\Gamma(t)) + \bar{\eta}(t)\hat{g}(\Gamma(t))$,

b. $|\bar{\eta}(t)| = 1$,

and

c. *one of the three conditions 2.10, 2.11, 2.12 holds.*

We know that $\Gamma$ is absolutely continuous and $|K| = |I(\bar{\eta})|$. (Recall that $\bar{\eta}$ is bang-bang, so Condition b holds almost everywhere. We use "$|\ldots|$" to denote Lebesgue measure.) Hence

$$|K \cap B(S)| = |B(S)|. \tag{2.13}$$

Let $B_d(S)$ be the set of points of density of $K \cap B(S)$. (Recall that a *point of density* of a measurable subset $E$ of $\mathbb{R}$ is a point $t \in E$ with the property that

$$\lim_{h \to 0+} \frac{|E \cap [t-h, t+h]|}{2h} = 1. \tag{2.14}$$

It is well known that almost every point of a measurable set $E$ is a point of density of $E$.) In particular, the set $B_d(S)$ satisfies $|B_d(S)| = |B(S)|$.

Now assume $B_d(S)$ is nonempty. Pick a point $\bar{t} \in B_d(S)$. Let $z = \Gamma(\bar{t})$. Then, in particular, there are times $\tau \neq \bar{t}$, arbitrarily close to $\bar{t}$, such that $\Gamma(\tau) \in S$. This implies that $\dot{\Gamma}(\bar{t})$ is tangent to $S$ at $z$. On the other hand, $\dot{\Gamma}(\bar{t})$ is one of the vectors $\hat{X}_-(z)$, $\hat{X}_+(z)$. Assume that $\dot{\Gamma}(\bar{t}) = \hat{X}_-(z)$. Then $\hat{X}_-$ is tangent to $S$ at $z$, and therefore $\hat{X}_-$ is tangent to $S$ at every point of $S$.

We now show that $H_-(z) = H_+(z) = 0$. Indeed, since one of the three conditions 2.10, 2.11, 2.12 is satisfied at $t = \bar{t}$, it is clear that the inequalities $H_-(z) \geq 0$, $H_+(z) \geq 0$ hold, and at least one of them is an equality. Assume that $H_+(z) > 0$, so that $H_-(z) = 0$. Then 2.11 and 2.12 cannot hold for any $t$ sufficiently close to $\bar{t}$. This means that 2.10 holds for almost all $t \in [\bar{t} - \varepsilon, \bar{t} + \varepsilon]$, for some $\varepsilon > 0$. But then, if $\tilde{\Gamma}$ denotes the restriction of $\Gamma$ to $[\bar{t} - \varepsilon, \bar{t} + \varepsilon]$, it follows that $\tilde{\Gamma}$ is an integral curve of $\hat{X}_-$. But, since $\bar{t} \in SW(\eta)$, there is a set $G \subseteq [\bar{t} - \varepsilon, \bar{t} + \varepsilon]$ such that $|G| > 0$ and $\eta(t) = 1$ for $t \in G$. For almost every $t \in G$ we have $\dot{\Gamma}(t) = \hat{X}_-(\Gamma(t))$ and also $\dot{\Gamma}(t) = \hat{X}_+(\Gamma(t))$. Hence

$$\hat{X}_+(\Gamma(t)) = \hat{X}_-(\Gamma(t)) \tag{2.15}$$

for almost all $t \in G$. Since both sides of 2.15 are analytic functions of $t$ on $[\bar{t} - \varepsilon, \bar{t} + \varepsilon]$ (because $\tilde{\Gamma}$ is analytic), it follows that 2.15 holds in fact for all $t \in [\bar{t} - \varepsilon, \bar{t} + \varepsilon]$. This can only happen if $g(\gamma(t)) = 0$ for all $t \in [\bar{t}-\varepsilon, \bar{t}+\varepsilon]$. But then it follows from the definition of the Hamiltonian that $H_-(\Gamma(t)) = H_+(\Gamma(t))$ for all $t \in [\bar{t}-\varepsilon, \bar{t}+\varepsilon]$. In particular, $H_-(z) = H_+(z)$. But then $H_+(z) = 0$, because $H_-(z) = 0$. This contradicts the hypothesis that $H_+(z) > 0$. The possibility that $H_-(z) > 0$ is excluded in a similar way.

We now show that

**(A)** *$\hat{X}_+(z)$ is not tangent to $S$.*

To see this, assume that $\hat{X}_+(z)$ is tangent to $S$. Then $\hat{X}_+$ is tangent to $S$ at every point. Also, since $H_-(z) = H_+(z) = 0$, the functions $H_-$ and $H_+$ vanish identically on $S$. Let $\mathcal{L}_V$ denote Lie differentiation in the direction of the vector field $V$. Then $\mathcal{L}_V H \equiv 0$ on $S$, if $V$ is $\hat{X}_-$ or $\hat{X}_+$, and $H$ is $H_-$ or $H_+$. Therefore, an easy induction shows that, if $\alpha = (\alpha_0, \ldots, \alpha_m)$ is any sequence of indices such that each $\alpha_i$ is either $+$ or $-$, then

$$\mathcal{L}_{\hat{X}_{\alpha_m}} \cdots \mathcal{L}_{\hat{X}_{\alpha_1}} H_{\alpha_0} \equiv 0 \text{ on } S. \tag{2.16}$$

Equivalently, we have

$$\{H_{\alpha_m}, \{H_{\alpha_{m-1}}, \ldots, \{H_{\alpha_1}, H_{\alpha_0}\} \ldots\}\} \equiv 0 \text{ on } S. \tag{2.17}$$

But this says that

$$H_V \equiv 0 \text{ on } S, \tag{2.18}$$

where

$$V = [X_{\alpha_m}, [X_{\alpha_{m-1}}, \ldots, [X_{\alpha_1}, X_{\alpha_0}] \ldots]]. \tag{2.19}$$

If $z = (\lambda, x)$, then 2.18 and 2.19 imply

$$\langle \lambda, V(x) \rangle = 0. \tag{2.20}$$

As $\alpha$ varies over all possible indices, it is clear that the corresponding vector fields $V$ span the Lie algebra $\Lambda$. Hence 2.20 says that

$$\langle \lambda, v \rangle = 0 \text{ for all } v \in \Lambda(x). \tag{2.21}$$

By the accessibility property, $\Lambda(x) = T_x M$. Hence $\lambda = 0$. But this contradicts the fact that $z = (\lambda, x) \in \hat{M}$, which implies that $\lambda \neq 0$.

This contradiction shows that **(A)** holds, as stated.

We now show

**(B)** *There exists a neighborhood $W$ of $z$ in $\hat{M}$, such that there is a smooth function $\psi : W \to \mathbb{R}$ that satisfies $\psi \equiv 0$ on $S \cap W$, $\mathcal{L}_{\hat{X}_-}\psi \equiv 0$ on $W$, and $\mathcal{L}_{\hat{X}_+}\psi > 0$ on $W$.*

Indeed, since (1) holds, we have in particular $\hat{X}_-(z) \neq 0$ (because $d\pi(\hat{X}_-(z)) = X_-(x) \neq 0$). Also, we know that $\hat{X}_-$ is tangent to $S$. Pick an embedded submanifold $T$ of $\hat{M}$ of codimension 1, such that $z \in T$ and $\hat{X}_-$ is not tangent to $T$ at any point of $T$. Let $\sigma = \dim S$, $\mu = \dim \hat{M}$. Then $T$ and $S$ are transversal. So $T \cap S$ is an embedded submanifold of $T$, of dimension $\sigma - 1$. Pick a cubic chart $\mathbf{x} = (x^2, \ldots, x^\mu)$ of $T$, centered at $z$, with domain $W_0$, such that $S \cap W_0$ is the subset of $W_0$ defined by the equations $x^{\sigma+1} = \ldots = x^\mu = 0$. Let

$$\Phi(x^1, y) = e^{x^1 \hat{X}_-} y, \tag{2.22}$$

where we use exponential notation fot the flow of a vector field, i.e. $s \to e^{sX} r$ is the integral curve of $X$ that goes through $r$ at time 0.

By shrinking $W_0$, if necessary, we can assume that there exists a $\delta > 0$ such that

*(a)* $\mathbf{x}$ *maps* $W_0$ *onto the cube* $]-\delta, \delta[^{\mu-1}$;

*(b)* $\Phi$ *is defined on* $W' = ]-\delta, \delta[\times W_0$, *and maps* $W'$ *diffeomorphically onto an open subset* $\tilde{W}$ *of* $\hat{M}$;

*(c)* $\Phi$ *maps* $]-\delta, \delta[\times(S \cap W_0)$ *diffeomorphically onto an open subset of* $S$,

*(d)* $\hat{X}_-(y) \neq \hat{X}_+(y)$ *for all* $y \in W$.

We define coordinates $\tilde{x}^i$, $i = 1, \ldots, \mu$ on $\tilde{W}$ as follows. If $r \in W$, the $r = \Phi(x^1, r')$ for a unique $r' \in W_0$, $x^1 \in ]-\delta, \delta[$. Then we let $\tilde{x}^1(r) = x^1$, $\tilde{x}^i(r) = x^i(r')$ for $i = 2, \ldots, \mu$.

With respect to these coordinates $\tilde{W}$ is a cube centered at $z$, and the subset $S'$ of $\tilde{W}$ defined by the equations $\tilde{x}^{\sigma+1} = \ldots = \tilde{x}^\mu = 0$ is open in $S$. For $0 < \theta \leq \delta$, let $W(\theta)$ be the subset of $\tilde{W}$ defined by $|\tilde{x}^i| < \theta$, $i = 1, \ldots, \mu$. Since $S'$ is open in $S$, and $S$ is embedded, we have $W(\theta) \cap S = W(\theta) \cap S'$ if $\theta$ is small enough. Let $\tilde{\psi} : W \to \mathbb{R}$ be a linear combination with constant coefficients of the functions $\tilde{x}^{\sigma+1}, \ldots, \tilde{x}^\mu$, and let $\psi_\theta$ be the restriction of $\tilde{\psi}$ to $W(\theta)$ . Then $\tilde{\psi}$ is constant on the integral curves of $\hat{X}_+$, so $\mathcal{L}_{\hat{X}_-}\tilde{\psi} \equiv 0$. Also, $\tilde{\psi} \equiv 0$ on $S'$, and therefore $\psi_\theta \equiv 0$ on $S \cap W(\theta)$, if $\theta$ is small enough. Since $\hat{X}_+(z)$ is not tangent to $S$, we have $\hat{X}_+ = \sum_{i=1}^{\mu} \zeta^i \frac{\partial}{\partial \tilde{x}^i}$, where the $\zeta^i$ are smooth functions and $\zeta^i(z) \neq 0$ for at least one $i \in \{\sigma+1, \ldots, \mu\}$. After relabelling the coordinates, and changing $\tilde{x}^\mu$ into $-\tilde{x}^\mu$ if necessary, we may assume that $\zeta^{\mu}(z) > 0$. Now choose $\psi = \tilde{x}^{\mu}$. Then $\mathcal{L}_{\hat{X}_+}\psi(z) > 0$. Therefore $\mathcal{L}_{\hat{X}_+}\psi_\theta > 0$ thoughout $W(\theta)$, if $\theta$ is small enough. So, if we choose $\theta$ sufficiently small, and let $W = W(\theta)$, $\psi = \psi_\theta$, all our conditons are satisfied.

We now choose $W$, $\psi$ as specified in **(B)**. Pick $\varepsilon > 0$ so small that $\Gamma(t) \in W$ for $|t - \bar{t}| < \varepsilon$. Since $\bar{t}$ is a point of density of $K \cap B(S)$, there exists a point $t_1 \in ]\bar{t}, \bar{t} + \varepsilon[\cap B(S)$. Then $\Gamma(t_1) \in S \cap W$ and so $\psi(\Gamma(t_1)) = \psi(\Gamma(\bar{t})) = 0$. On the other hand, if we let $\rho(t) = \psi(\Gamma(t))$, we have

$$\dot{\rho}(t) = \mathcal{L}_{\dot{\Gamma}(t)}\psi \quad \text{for } \bar{t} \le t \le t_1 \ . \tag{2.23}$$

But $\dot{\Gamma}(t)$ is a convex combination of $\hat{X}_-(\Gamma(t))$ and $\hat{X}_+(\Gamma(t))$. Since $\mathcal{L}_{\hat{X}_-}\psi \equiv 0$ and $\mathcal{L}_{\hat{X}_+}\psi > 0$, we see that $\dot{\rho}(t) \ge 0$ for almost every $t \in [\bar{t}, t_1]$, and equality holds almost everywhere if and only if $\dot{\Gamma}(t) = \hat{X}_-(\Gamma(t))$ for a.e. $t$. Since $\rho(\bar{t}) = \rho(t_1)$, equality must indeed hold almost everywhere, and so $\dot{\Gamma}(t) = \hat{X}_-(\Gamma(t))$ a.e. on $[\bar{t}, t_1]$. So, if we let $\Gamma^*$ be the restriction of $\Gamma$ to $[\bar{t}, t_1]$, we see that $\Gamma^*$ is an integral curve of $\hat{X}_-$, and so in particular $\dot{\Gamma}(t) = \hat{X}_-(\Gamma(t))$ for all $t \in [\bar{t}, t_1]$. Since $\hat{X}_-(y) \ne \hat{X}_+(y)$ for all $y \in W$, we must have $\eta(t) = -1$ for almost all $t \in [\bar{t}, t_1]$. But then no point in the open interval $]\bar{t}, t_1[$ can be a switching point of $\eta$. On the other hand, the fact that $\bar{t}$ is a point of density of $B(S)$ implies that there is a switching point of $\eta$ in $]\bar{t}, t_1[$.

So we have arrived at a contradiction, from the assumption that $\bar{t}$ was a point in $B_d(S)$, and $\dot{\Gamma}(\bar{t}) = \hat{X}_-(z)$. Clearly, a similar argument yields a contradiction if $\dot{\Gamma}(\bar{t}) = \hat{X}_+(z)$. For a point $\bar{t} \in B_d(S)$ one of the two equalities $\dot{\Gamma}(\bar{t}) = \hat{X}_-(z)$, $\dot{\Gamma}(\bar{t}) = \hat{X}_+(z)$ must hold, because $\bar{t} \in K$. Hence the assumption that $B_d(S)$ is nonempty leads to a contradiction. So $B_d(S) = \emptyset$. But then $|B(S)| = 0$. Since this is true for every stratum $S$ of $\mathcal{S}$, we have shown that $|SW(\eta)| = 0$. ▮

## REFERENCES

[1] Sussmann, H.J., *A weak regularity theorem for real analytic optimal control problems*, Revista Matemática Iberoamericana **2**, No. 3 (1986), pp. 307-317.

[2] Sussmann, H.J., *Recent developments in the regularity theory of optimal trajectories*, to appear in the *Proceedings of the conference on Linear and Nonlinear Mathematical Control Theory* held in Torino, Italy, June 1986.

[3] Sussmann, H.J., *Trajectory regularity and real analyticity: some recent results*, in *Proceedings of the 25th IEEE Conference on Decision and Control*, Athens, Greece, Dec. 1986, pp. 592-595.

# ON VOLTERRA APPROXIMATIONS

Gianna Stefani
Dipartimento di Matematica e Applicazioni
Via Mezzocannone 8 - 80134 Napoli - Italia

## 0 - Introduction

The aim of the paper is to review and discuss some results on the Volterra approximation of the time dependent control system

$$(\Sigma)\qquad \begin{cases} \dot{\xi}(t) = F(t,\xi(t),u(t)) \\ \eta(t) = \alpha(\xi(t)) \,. \end{cases}$$

We shall state the results under precise mathematical assumptions in order to see the Volterra approximation as a Taylor approximation of a suitable map between Banach spaces.

Using the methods in [1] , the Volterra kernels will be given in a very simple way as directional derivatives with respect to suitable time-dependent vector fields. Some links with the Hamiltonian of the system will be also given.

Volterra expansions have been applied to obtain high order necessary conditions for a singular extremal trajectory $\hat{\xi}$ to be minimal under the assumption that F and $\hat{\xi}$ are $C^{\infty}$ (see [2] , [3]). Here we want to consider a more general case to obtain, hopefully, a first step to unify the methods from the geometric theory and those from the calculus of variations in the study of optimal control problems.

An example of how to use this formulation of Volterra kernels to prove a sufficient condition for a trajectory to lie on the boundary of the reachable set is given in [4].

## 1 - Notations and preliminary results

We shall give the properties of the solutions of ($\Sigma$) following [5] , where extensive and complete proofs of all the statements are given (see also [6] , [7]) .

Let $E_1,\ldots,E_{s+1}$ be Banach spaces, $\mathcal{U}$ an open subset of $E \equiv E_1 \times \ldots \times E_s$ and $\phi : \mathcal{U} \to E_{s+1}$ a map. $\phi$ is said to be *locally bounded* if for each $x \in \mathcal{U}$ there is a neighbourhood $\mathcal{V}$ of x in $\mathcal{U}$ such that $\phi_{|\mathcal{V}}$ is bounded.

We denote by $D_i\phi(x)$ the Frechet derivative of the map $\phi$ with respect to the $i^{th}$ variable evaluated at the point x, so that $D_i\phi$ is a map from $\mathcal{U}$ to the space $L(E_i,E_{s+1})$ of the linear continuous maps from $E_i$ to $E_{s+1}$.

Let I be an interval in $\mathbb{R}$ and U an open set in $\mathbb{R}^n$.

**Definition 1.1** - A map $F : I \times U \times \mathbb{R}^m \to \mathbb{R}^n$ is said to be quasi-$C^r$ if the following assumptions are satified.

i) For each $t \in I$ the map $(x,w) \mapsto F(t,x,w)$ is $C^r$ .

ii) For each $i=0,1,\dots,r$ the map $\bar{D}^iF$ is locally bounded on $I \times U \times \mathbb{R}^m$ ($\bar{D}^iF$ is the $i^{th}$-derivative with respect to the coupled variables in $U \times \mathbb{R}^m$ and $\bar{D}^0F \equiv F$)

iii) For each $i=0,1,\dots,r$ and each $(x,w) \in U \times \mathbb{R}^m$ the map $t \to \bar{D}^iF(t,x,w)$ is measurable.

Let $L^\infty(I,\mathbb{R}^m)$ denote the Banach space of the measurable essentially bounded maps from $I$ to $\mathbb{R}^m$ with the norm $\|u\|_\infty = \text{ess sup}\{\|u(t)\| : t \in I\}$ . Recall that two maps $u_1$ and $u_2$ are to be considered the same one if they coincide almost everywhere.

In the following we shall use also the $L^p$-norms on a closed interval $[t_0,t_1]$ and we denote them by

$$\|u\|_p = \left(\int_{t_0}^{t_1} \|u(t)\|^p dt\right)^{1/p}.$$

For each $(s,x,u) \in I \times U \times L^\infty(I,\mathbb{R}^m)$ there is a unique solution $\xi_\Sigma(\cdot,s,x,u)$ of $(\Sigma)$ satisfying the initial condition $\xi(s)=x$ and defined on a maximal subinterval $I(s,x,u)$ . $\xi_\Sigma(\cdot,s,x,u)$ is a locally Lipschitz map and we shall look at its restriction to any compact interval $J \subset I(s,x,u)$ as an element of the Banach space $C(J,\mathbb{R}^n)$ of the continuous functions from $J$ to $\mathbb{R}^n$ with the norm:

$$\|\xi\|_C = \max\{\|\xi(t)\| : t \in J\}.$$

Let the domain of the "flow of F" be defined by

$$\mathcal{D}(F) = \left\{(t,s,x,u) \in I \times I \times U \times L^\infty(I,\mathbb{R}^m) : t \in I(s,x,u)\right\}.$$

The flow of F is the map

$$\xi_\Sigma : \mathcal{D}(F) \to U \text{ , given by } (t,s,x,u) \to \xi_\Sigma(t,s,x,u).$$

**Theorem 1.1** - [5] - Let F be quasi-$C^r$ , then

i) $\mathcal{D}(F)$ is open in $I \times I \times U \times L^\infty(I,\mathbb{R}^m)$.

ii) $\xi_\Sigma : \mathcal{D}(F) \to U$ is continuous and moreover it is $C^r$ in the last two variables

iii) If for a given $(t_0,x_0,\hat{u}) \in I \times U \times L^\infty(I,\mathbb{R}^m)$ , $I(t_0,x_0,\hat{u})$ contains a compact interval $J=[t_0,t_1]$ , there is a neighbourhood $\mathcal{U}$ of $(t_0,x_0,\hat{u})$ in $I \times U \times L^\infty(J,\mathbb{R}^m)$ such that $J \times \mathcal{U} \subseteq \mathcal{D}(F)$ and the map $\phi : \mathcal{U} \to C(J,\mathbb{R}^n)$ given by $(s,x,u) \to \xi_\Sigma(\cdot,s,x,u)$ is $C^r$ with respect to the last two variables.

**Remark 1.1** - We shall consider $L^\infty(J,\mathbb{R}^m)$ as the control space. The reason for it is that $u \in L^1(J,\mathbb{R}^m)$ does not imply $t \to F(t,x,u(t))$ integrable on J. Moreover it is known that a "minimal trajectory" need not satisfy the maximum principle if it is not Lipschitz (see [8]).

If we apply the above arguments to the case $m=0$, we get analogous properties for the flow of a quasi-$C^r$ time dependent vector field $f : I \times U \rightarrow \mathbb{R}^n$. Here below we list some other properties of the flow $\gamma : \mathfrak{D}(f) \rightarrow U$ of $f$ which will be used later on. Some of them are well known some others can be derived in a standard way.

Let us suppose that for a given $(t_0,x_0) \in I \times U$ the compact interval $J=[t_0,t_1]$ belongs to $I(t_0,x_0)$.

(P.1) $D_1\gamma(t,s,x) = f(t,\gamma(t,s,x))$.

(P.2) $\gamma(t,s,\cdot)$ is defined in an open set $V(t,s)$ and it is a $C^r$ map

(P.3) $D_3\gamma(\cdot,s,x)$ satisfies the linear differential equation on $L(\mathbb{R}^n,\mathbb{R}^n)$

$$\dot{\chi}(t) = D_2 f(t,\gamma(t,s,x)) \circ \chi(t) \ , \ \chi(s) = id$$

(P.4) $\gamma(s,t,\cdot)$ is defined on $\gamma(t,s,V(t,s))$, it is equal to $[\gamma(t,s,\cdot)]^{-1}$ and

$$D_3\gamma(s,t,x) = [D_3\gamma(t,s,\gamma(s,t,x))]^{-1}.$$

(P.5) $\gamma(t,\cdot,x)$ is defined on an open set, it is locally Lipschitz and it satisfies

$$D_2\gamma(t,s,x) = -D_3\gamma(t,s,x)(f(s,x))$$

(P.6) $D_2D_3\gamma(t,s,x) = D_3D_2\gamma(t,s,x) =$

$$= -D_3^2\gamma(t,s,x)(f(s,x),\cdot) - D_3\gamma(t,s,x) \circ D_2 f(s,x)$$

(P.7) For each $t \in J$, $D_3\gamma(t,s,\gamma(s,t_0,x_0))$ satisfies the linear differential equation

$$\dot{\chi}(s) = -\chi(s) D_2 f(s,\gamma(s,t_0,x_0)) \ , \ \chi(t) = id.$$

(P.8) For each covector $\omega$ at $\gamma(t,t_0,x_0)$, $\omega \circ D_3\gamma(t,s,\gamma(s,t_0,x_0))$ is the solution of the adjoint equation

$$\dot{p}(s) = -p(s) \circ D_2 f(s,\gamma(s,t_0,x_0))$$

with boundary condition $p(t)=\omega$ .

For each quasi-$C^r$ time dependent vector field $g$ , we define a vector field $g^* : J \times V(t_1,t_0) \rightarrow \mathbb{R}^n$ by

$$g^*(s,x) \equiv g^*(s)(x) = D_3\gamma(t_0,s,\gamma(s,t_0,x)(g(s,\gamma(s,t_0,x))).$$

For each $s$ , $g^*(s)$ is the pull-back of $g(s)$ through $\gamma(s,t_0,\cdot)$ :

$$\begin{array}{ccc} & D_3\gamma(t_0,s,\cdot) & \\ T\mathbb{R}^n & \leftarrow & T\mathbb{R}^n \\ g^*(s) \uparrow & & \uparrow g(s) \\ V(s,t_0) & \rightarrow & V(t_0,s) \\ & \gamma(s,t_0,\cdot) & \end{array}$$

In particular from (P.5) it follows that $\gamma(t_0,s,x)$ is solution of the differential equation

$$\dot{\xi}(s) = -f^*(s,\xi(s)) \ , \ \xi(t_0) = x.$$

**Remark 1.2** - If $f$ is time-independent , $f^*(s,x)=f(x)$ for each $s$ and $x$.

By the properties of the pull-back (see for example [9]) the following additional properties hold:

(P.9) If $g(\cdot,x)$ is locally Lipschitz $g^*(\cdot,x)$ is locally Lipschitz too and

$$D_1g^*(s,x)=(ad_f g)^*(s,x), \text{ where } ad_f g(s,x)=[f(s),g(s)](x)+D_1g(s,x).$$

(P.10) If $g$ and $g_1$ are quasi-$C^r$ , then

$$[g^*(s),g_1^*(s)] = [g(s),g_1(s)]^*.$$

(P.11) For each $C^r$ function $\alpha$

$$(g^*(s)\cdot\alpha)(x) \equiv d\alpha(x)(g^*(s,x)) = \Big[g(s)\cdot(\alpha\circ\gamma(t_0,s,\cdot))\Big](\gamma(s,t_0,x)).$$

## 2 - Volterra approximations

Let F be quasi-$C^r$ and let us choose a reference control $\hat{u}$ and let the reference solution $t\to\hat{\xi}(t)\equiv\xi_\Sigma(t,t_0,x_0,\hat{u})$ be defined on the compact interval $J=[t_0,t_1]$.

We want to determine a Taylor approximation of the map $\xi_\Sigma(t,t_0,x_0,\cdot)$ at $\hat{u}$ for each $t\in J$. Such an approximation turns out to be also a Taylor approximation at $\hat{u}$ of the map $\phi(t_0,x_0,\cdot)$ defined in Theorem 1.1 iii).

To do this we need some notations. For simplicity let us consider the scalar input case, i.e. m=1.

Let us define

$$f(t,x) = F(t,x,\hat{u}(t)) \quad , \quad g_i(t,x) = D_3^i F(t,x,\hat{u}(t)) \ , \ i=1,..,s\leq r-1$$

and let us consider now a system, linear with respect to the control, given by :

$$(\Sigma_s) \qquad \dot{\xi}(t) = f(t,\xi(t)) + \sum_{i=1}^{s}\frac{u_i(t)\,g_i(t,\xi(t))}{i!}.$$

**Lemma 2.1** - Let $u\in L^\infty(J,\mathbb{R})$ and $\mathbf{u}=(u,u^2,\ldots,u^s)$. There is a neighbourhood $\mathcal{V}$ of 0 in $L^\infty(J,\mathbb{R})$ such that $\xi_\Sigma(t,t_0,x_0,\hat{u}+u)$ and $\xi_{\Sigma_s}(t,t_0,x_0,\mathbf{u})$ are defined for each $(t,u)\in J\times\mathcal{V}$. Moreover there are L and L' such that

$$\left\|\xi_\Sigma(t,t_0,x_0,\hat{u}+u)-\xi_{\Sigma_s}(t,t_0,x_0,\mathbf{u})\right\|\leq L\,\|u\|_{s+1}^{s+1}\leq L'\,\|u\|_\infty^{s+1}\ , \quad \forall(t,u)\in J\times\mathcal{V}.$$

**Proof** - The existence of $\mathcal{V}$ is a consequence of Theorem 1.1 . The existence of L can be proved in a standard way using the properties of F and the Gronwall's inequality . □

From Lemma 2.1 it follows that the Taylor approximation of order s of $\xi_\Sigma(t,t_0,x_0,\hat{u}+u)$ at 0 can be deduced from the one of $\xi_{\Sigma_m}(t,t_0,x_0,\mathbf{u})$. Therefore we are led to consider the Volterra approximation of a control system , linear with respect to the control.

Again for simplicity we consider a scalar input system defined by:

$$(\Sigma_1) \qquad \dot{\xi}(t) = f(t,\xi(t)) + u(t)g(t,\xi(t)).$$

Let $\gamma$ be the flow of the time-dependent vector field f as in the previous section . Using the properties of $\gamma$ it is easy to see that the map $t\to\gamma(t_0,t,\xi_{\Sigma_1}(t,t_0,x,u))$ is defined on J for (x,u) belonging to a suitable neighbourhood of $(x_0,0)$ in $U\times L^\infty(J,\mathbb{R})$ and it is the solution of the control system

$$(\Sigma^*) \qquad \dot{\xi}(t) = u(t)g^*(t,\xi(t))\ , \quad \xi(t_0)=x.$$

If $\alpha : U \to \mathbb{R}$ is a $C^r$ function $\alpha\circ\xi_{\Sigma^*}(t,t_0,x_0,\cdot)$ is $C^r$ too and we can calculate its Taylor expansion.

Let $y(t,x)$ denote $\xi_{\Sigma^*}(t,t_0,x,u)$. It is not difficult to see that

$$D_1(\alpha \circ y)(t,x) = d\alpha(y(t,x))D_1y(t,x) = u(t)d\alpha(y(t,x))g^*(t,y(t,x)) = \\ = u(t)(g^*(t)\cdot\alpha)(y(t,x)).$$

From the above equality we get

$$\alpha(y(t,x_0)) = \alpha(y(t_0,x_0)) + \int_{t_0}^{t} D_1(\alpha \circ y)(s,x_0)ds = \alpha(x_0) + \int_{t_0}^{t} u(s)(g^*(s)\cdot\alpha)(y(s,x_0))ds.$$

Applying the same arguments to the function $g^*(s)\cdot\alpha$ (here s must be thought as fixed) we get $\alpha(y(t,x_0)) =$

$$= \alpha(x_0) + \int_{t_0}^{t} u(s)(g^*(s)\cdot\alpha)(x_0)ds + \int_{t_0}^{t} u(s)\int_{t_0}^{s} u(\tau)(g^*(\tau)\cdot g^*(s)\cdot\alpha)(y(s,x_0))d\tau ds.$$

Defining $S_k = (s_1,\ldots,s_k)$ , $\upsilon_k(\alpha,S_k) = \left[g^*(s_1)\cdots g^*(s_k)\cdot\alpha\right](x_0)$ , $u(S_k) = u(s_1)\ldots.u(s_k)$ and $dS_k = ds_1\cdots ds_k$ , we get

$$\alpha(\xi_{\Sigma^*}(t,t_0,x_0,u)) = \alpha(x_0) + \sum_{k=1}^{r-1}\int_0^t\int_0^{s_k}\cdots\int_0^{s_2} \upsilon_k(\alpha,S_k)\,u(S_k)\,dS_k + \mathcal{R}_r(t,u)$$

with

$$|\mathcal{R}_r(t,u)| = \left|\int_0^t\int_0^{s_r}\cdots\int_0^{s_2}\left[g^*(s_1)\cdots g^*(s_r)\cdot\alpha\right](y(s_r,x_0))\,u(S_r)\,dS_r\right| \le M\|u\|_1^r \le M'\|u\|_\infty^r$$

for suitable $M, M' \in \mathbb{R}$.

Applying the above approximation to the function $(\alpha \circ \gamma(t,t_0,\cdot))$ we get the Volterra kernels $w_k(t,\alpha,S_k) = \upsilon_k((\alpha \circ \gamma(t,t_0,\cdot)),S_k)$ and

$$\alpha(\xi_{\Sigma_1}(t,t_0,x_0,u)) = \alpha(\hat{\xi}(t)) + \sum_{k=1}^{r-1}\int_0^t\int_0^{s_k}\cdots\int_0^{s_2} w_k(t,\alpha,S_k)\,u(S_k)\,dS_k + \mathfrak{R}_r(t,u)$$

with

$$|\mathfrak{R}_r(t,u)| = \left|\int_0^t\int_0^{s_r}\cdots\int_0^{s_2}\left[g^*(s_1)\cdots g^*(s_r)\cdot(\alpha \circ \gamma(t,t_0,\cdot)\right](y(s_r,x_0))\,u(S_r)\,dS_r\right| \le$$

$$\le M\|u\|_1^r \le M'\|u\|_\infty^r$$

for suitable $M, M' \in \mathbb{R}$.

Applying the same arguments to a multiinput system, we have the derivative of $\alpha(\xi_\Sigma(t_1,t_0,x_0,\cdot))$ at $\hat{u}$ as

$$u \to \int_{t_0}^{t_1}\sum_{i=1}^{m} u_i(s)(g_i^*(s)\cdot(\alpha\cdot\gamma(t_1,t_0,\cdot)))(x_0)\,ds.$$

Choosing as function $\alpha$ the canonical projections $\pi_j: \mathbb{R}^n \to \mathbb{R}$, $j=1,..,n$, we obtain:

$$D_4\xi_\Sigma(t_1,t_0,x_0,\hat{u}): u \mapsto \int_{t_0}^{t_1} \left[D_3\gamma(t_1,t_0,x_0) \circ D_3\gamma(t_0,s,\hat{\xi}(s))\right](D_3F(s,\hat{\xi}(s),\hat{u}(s)))u(s)ds.$$

If we linearize $(\Sigma)$ along $(\hat{\xi},\hat{u})$, we get the linear system

$$(\Sigma_L) \qquad \dot{\xi}(t) = D_2f(t,\hat{\xi}(t))\xi(t) + D_3F(t,\hat{\xi}(t),\hat{u}(t))u(t) \quad , \quad \xi(t_0) = 0.$$

The solutions of such a system are:

$$\xi_{\Sigma_L}(t_1,t_0,0,u)) = \int_{t_0}^{t_1} \left[D_3\gamma(t_1,t_0,x_0) \circ D_3\gamma(t_0,s,\hat{\xi}(s))\right](D_3F(s,\hat{\xi}(s),\hat{u}(s)))u(s)\,ds.$$

That is $D_4\xi_\Sigma(t_1,t_0,x_0,\hat{u})(\cdot) = \xi_{\Sigma_L}(t_1,t_0,0,\cdot)$ . In other words *the derivative of the solution map of* $(\Sigma)$ *is the solution map of the linearized system* $(\Sigma_L)$.

Another consequence of the above arguments is the following Lemma on the linear independence of the constraints.

**Lemma 2.2** - Let $\varphi = (\varphi_1,..,\varphi_p) : \mathbb{R}^n \to \mathbb{R}^p$ be a $C^1$ map. The map $\varphi \circ \xi_\Sigma(t_1,t_0,x_0,\cdot) : L^\infty(J,\mathbb{R}^m) \to \mathbb{R}^p$ has maximum rank at $\hat{u}$ iff there is no vector $(\lambda_1,..,\lambda_p) \in \mathbb{R}^p \setminus \{0\}$ such that

$$\sum_{i=1}^{p} \lambda_i D\varphi_i(\hat{\xi}(t_1))$$

is orthogonal at the space which is reachable at time $t_1$ by means of the trajectories of $\Sigma_L$.

**Corollary 2.2** - If $(\Sigma_L)$ is controllable at time $t_1$, $\varphi \circ \xi_\Sigma(t_1,t_0,x_0,\cdot)$ has maximum rank at $\hat{u}$ iff $\varphi$ has maximum rank at $\hat{\xi}(t_1)$.

## 3 - Volterra approximations and optimal problems

Let us give , as an example , an application of the above results in deriving well known $2^{nd}$ order necessary conditions for $\hat{\xi}$ being minimal .

Let F be quasi $C^3$. Setting

$$D_3F(t,x,\hat{u}(t))(a) = \sum_{i=1}^{m} g_i(t,x)a_i$$

and

$$D_3^2F(t,x,\hat{u}(t))(a,a) = \sum_{i,j=1}^{m} g_{ij}(t,x)a_ia_j$$

arguments analogous to the above ones lead for the $2^{nd}$ approximation of $\alpha(\xi_\Sigma(t_1,t_0,x_0,\hat{u}+u)$ to

$$\alpha(\xi_\Sigma(t_1,t_0,x_0,\hat{u}+u) =$$

$$= \alpha(\hat{\xi}(t_1)) + \int_{t_0}^{t_1} \sum_{i=1}^{m} u_i(s)w_1^i(t_1,\alpha,s)ds + \frac{1}{2}\int_{t_0}^{t_1} \sum_{i,j=1}^{m} u_i(s)u_j(s)\kappa_2^{ij}(t_1,\alpha,s)ds +$$

$$+\int_{t_0}^{t_1}\int_{t_0}^{s}\sum_{i,j=1}^{m} u_i(s)u_j(\tau)w_2^{ij}(t_1,\alpha,s,\tau)\,d\tau ds \;+\; \mathfrak{R}_2(t_1,u)$$

with

$$w_1^i(t_1,\alpha,s) \;=\; (g_i^{*}(s)\cdot(\alpha\circ\gamma(t_1,t_0,\cdot)))(x_0)$$

$$\kappa_2^{ij}(t_1,\alpha,s) \;=\; (g_{ij}^{*}(s)\cdot(\alpha\circ\gamma(t_1,t_0,\cdot)))(x_0)$$

$$w_2^{ij}(t_1,\alpha,s,\tau) \;=\; (g_j^{*}(\tau)\cdot g_i^{*}(s)\cdot(\alpha\circ\gamma(t_1,t_0,\cdot)))(x_0)$$

and

$$|\mathfrak{R}_2(t_1,u)| \;<\; K\|u\|_3^3 \;\le\; K'\|u\|_\infty^3 \;,$$

for some constants K , K'.

For each $C^3$ function $\alpha$ and each time dependent vector field g , we get

$$g^{*}(s)\cdot(\alpha\cdot(\gamma(t_1,t_0,\cdot)))(x_0)=\left[d\alpha(\hat\xi(t_1))\circ D_3\gamma(t_1,t_0,x_0)\circ D_3\gamma(t_0,s,\hat\xi(s))\right](g(s,\hat\xi(s))=$$

$$=\left[d\alpha(\hat\xi(t_1))\circ D_3\gamma(t_1,s,\hat\xi(s)))\right](g(s,\hat\xi(s)) \;=\; \hat p_\alpha(s)(g(s,\hat\xi(s))) \;,$$

where $\hat p_\alpha$ is the solution of the adjoint equation with boundary condition $p(t_1)=d\alpha(\hat\xi(t_1))$.

Therefore we get

$$w_1^i(t_1,\alpha,s) \;=\; \hat p_\alpha(s)(g_i(s,\hat\xi(s)))$$

$$\kappa_2^{ij}(t_1,\alpha,s) \;=\; \hat p_\alpha(s)(g_{ij}(s,\hat\xi(s))).$$

Introducing the Hamiltonian

$$\mathcal{H}(t,\omega,x,w) \;=\; \omega(F(t,x,w)),$$

and setting

$$D_4\hat{\mathcal{H}}(s) \;=\; D_4\mathcal{H}(s,\hat p_\alpha(s),\hat\xi(s),\hat u(s)) \;,\; D_4^2\hat{\mathcal{H}}(s) \;=\; D_4^2\mathcal{H}(s,\hat p_\alpha(s),\hat\xi(s),\hat u(s))$$

we can write the $2^{nd}$ approximation of $\alpha(\xi_\Sigma(t_1,t_0,x_0,\hat u+u)$ as

$$\alpha(\hat\xi(t_1)) + \int_{t_0}^{t_1} D_4\hat{\mathcal{H}}(s)(u(s))ds \;+\; \frac{1}{2}\int_{t_0}^{t_1} D_4^2\hat{\mathcal{H}}(s)(u(s),u(s))ds \;+$$

$$+\int_{t_0}^{t_1}\int_{t_0}^{s} w_2^{ij}(t_1,\alpha,s,\tau)u_i(s)u_j(\tau)d\tau ds.$$

It is clear that

$$\int_{t_0}^{t_1} D_4^2\hat{\mathcal{H}}(s)(u(s),u(s))ds \;=\; O(\|u\|_2^2)$$

and

$$\int_{t_0}^{t_1}\int_{t_0}^{s}\sum_{i,j=1}^{m} w_2^{ij}(t_1,\alpha,s,\tau)u_i(s)u_j(\tau)d\tau ds \;=\; O(\|u\|_1^2).$$

If the control $u_\epsilon$ is equal to $\epsilon$ in a set $J_\epsilon$ of measure $\epsilon$ and 0 on $J\backslash J_\epsilon$ we get

$$\|u_\epsilon\|_2^2 = O(\epsilon^3) \quad , \quad \|u_\epsilon\|_1^2 = O(\epsilon^4),$$

therefore in a standard way we can prove the following

**Theorem 3.1** - Let $\mathcal{U}$ be an open neighbourhood of 0 in $L^\infty(J,\mathbb{R}^m)$. If

$$\alpha(\hat{\xi}(t_1)) = \min\left\{\alpha(\xi_\Sigma(t_1,t_0,x_0,\hat{u}+u)) : u \in \mathcal{U}\right\}$$

then for almost all $s \in J$

$$D_4\hat{\mathcal{H}}(s) = 0$$

and

$$D_4^2\hat{\mathcal{H}}(s) \text{ is a symmetric nonnegative bilinear form.}\square$$

To obtain high order conditions similar to the ones in [3], we need stronger smoothness assumptions on the data.

**Theorem 3.2** - Let $\mathcal{U}$ be an open neighbourhood of 0 in $L^\infty(J,\mathbb{R}^m)$ and

$$\alpha(\hat{\xi}(t_1)) = \min\left\{\alpha(\xi_\Sigma(t_1,t_0,x_0,\hat{u}+u)) : u \in \mathcal{U}\right\}.$$

If

(A.1) $t \to D_3F(t,x,\hat{u}(t))$ is locally Lipschitz for $x$ belonging to a neighbourhood of the reference trajectory

(A.2) $D_4^2\hat{\mathcal{H}}(s) \equiv 0$

then

i) $\hat{p}_\alpha(s)\left([g_i(s),g_j(s)](\hat{\xi}(s))\right) \equiv 0$

ii) $a \to \sum_{i=1}^{m}\hat{p}_\alpha(s)\left([ad_f g_j(s),g_i(s)](\hat{\xi}(s))\right)a_i a_j$ is a nonnegative quadratic form for a.a $s \in J$.

**Proof** - Using the integral identities in [10] we get for the 2[nd] approximation of $\alpha(\xi_\Sigma(t_1,t_0,x_0,\hat{u}+u)$ :

$$\alpha(\hat{\xi}(t_1)) + \int_{t_0}^{t_1}\int_{t_0}^{t_1}\sum_{i,j=1}^{m} u_i(s)u_j(\tau)w_2^{ij}(t_1,\alpha,s,\tau)\,d\tau\,ds \; +$$

$$+ \int_{t_0}^{t_1}\sum_{i,j=1}^{m} u_i(s)\int_{t_0}^{s} u_j(\tau)[g_j^\bullet(\tau),g_i^\bullet(s)]\cdot(\alpha\circ\gamma(t_1,t_0,\cdot))(x_0)\,d\tau\,ds.$$

Let $s_0 \in [t_0,t_1]$ and let $u \in L^\infty([0,1],\mathbb{R}^m)$ be a nonzero control such that

$$\int_0^1 u(s)\,ds = 0. \tag{3.1}$$

If we define

$$u_\epsilon(s) = \begin{cases} \epsilon^3 u((s-s_0)/\epsilon) & s \in [s_0, s_0+\epsilon] \\ 0 & s \in J \setminus [s_0, s_0+\epsilon] \end{cases}$$

we get

$$\alpha(\xi_\Sigma(t_1,t_0,x_0,\hat{u}+u_\epsilon) = \alpha(\hat{\xi}(t_1)) + \epsilon^8 \int_0^1\int_0^1 \sum_{i,j=1}^{m} u_i(s)u_j(\tau)w_2^{ij}(t_1,\alpha,s_0+\epsilon s,s_0+\epsilon\tau)d\tau ds$$

$$+\epsilon^8\int_0^1 \sum_{i,j=1}^{m} u_i(s)\int_0^s u_j(\tau)\left[g_j^\bullet(s_0+\epsilon\tau),g_i^\bullet(s_0+\epsilon s)\right]\cdot(\alpha\circ\gamma(t_1,t_0,\cdot))(x_0)\,d\tau ds + O(\epsilon^{10}).$$

By (A.1) we can integrate by parts. Setting $v_i(t)=\int_0^1 u_i(s)ds$ and taking into account (3.1), we obtain

$$\alpha(\xi_\Sigma(t_1,t_0,x_0,\hat{u}+u_\epsilon) = \alpha(\hat{\xi}(t_1)) +$$

$$+\epsilon^{10}\int_0^1\int_0^1 \sum_{i,j=1}^{m} v_i(s)v_j(\tau)D_3D_4w_2^{ij}(t_1,\alpha,s_0+\epsilon s,s_0+\epsilon\tau)\,d\tau ds +$$

$$+\epsilon^8\int_0^1 \sum_{i,j=1}^{m} u_i(s)v_j(s)\hat{p}_\alpha(s)\left(\left[g_j(s_0+\epsilon s),g_i(s_0+\epsilon s)\right]\right)(\hat{\xi}(s))\,ds +$$

$$-\epsilon^9\int_0^1 \sum_{i,j=1}^{m} u_i(s)\int_0^s v_j(\tau)\left[(ad_f g)_j^\bullet(s_0+\epsilon\tau),g_i^\bullet(s_0+\epsilon s)\right]\cdot(\alpha\circ\gamma(t_1,t_0,\cdot))(x_0)d\tau ds +$$

$$+O(\epsilon^{10}) = \alpha(\hat{\xi}(t_1)) +$$

$$+\epsilon^8\int_0^1 \sum_{i,j=1}^{m} u_i(s)v_j(s)\hat{p}_\alpha(s)\left(\left[g_j(s_0+\epsilon s),g_i(s_0+\epsilon s)\right]\right)(\hat{\xi}(s))\,ds +$$

$$+\epsilon^9\int_0^1 \sum_{i,j=1}^{m} v_i(s)v_j(s)\,\hat{p}_\alpha(s)\left(\left[(ad_f g)_j(s_0+\epsilon s),g_i(s_0+\epsilon s)\right]\right)(\hat{\xi}(s))\,ds + O(\epsilon^{10}).$$

From the above equalities the statement follows in a standard way.□

**Remark 3.1** - The assumption (A.1) is necessary for the existence of $ad_f g_i$ , while the statement i) could be obtained without it. Moreover with stronger regularity assumptions we can obtain other necessary conditions for $\hat{\xi}$ to be minimal in more degenerate cases.

**Remark 3.2** - Some other high order conditions could be given considering higher approximation of $\alpha(\xi_\Sigma(t_1,t_0,x_0,\cdot))$.

**Remark 3.3** - With the same techniques it is possible to get necessary conditions also in partially degenerate cases .

**4 - References**

[1] Lesiak C., Krener J.A. - "The existence and uniqueness of Volterra series for nonlinear systems" *IEEE Transaction on Automatic Control* AC-23 (1978) p. 1090-1095.

[2] Lamnabhi-Lagarrigue F. - "Series de Volterra et commande optimale singuliere" *These d'Etat* Université Paris XI

[3] Lamnabhi-Lagarrigue F., Stefani G. - "Singular optimal problems: on the necessary conditions of optimality" Preprint.

[4] Stefani G. - "A sufficient condition for extremality" in "Analysis and Optimizations of Systems, *Lect. Notes in Control and Information Sciences* 111 - Springer-Verlag, Berlin, Heidelberg 1988.

[5] Grasse K.A. - "Controllability and accessibility in nonlinear control systems" *Thesis of PHD* University of Illinois at Urbana-Champaign, 1979.

[6] Lee E.B.,Markus L. - "Fundations of optimal control theory", John Wiley, New York, 1967.

[7] Dieudonné J. - "Foundations of modern analysis" Academic Press, New York, 1969.

[8] Cesari L. - "Optimization - Theory and applications" *Applications of Math.* Springer-Verlag, New York 1983.

[9] Godbillon C. - "Géométrie différentielle et mécanique analytique" *Collection Méthodes* Herman, Paris 1969.

[10] Crouch P., Lamnablii-Lagarrigue F. - "Algebraic and multiple integral identities" Preprint.

# The Lie Saturate and its Applications to Singular Control Problems

Velimir Jurdjevic
Department of Mathematics
University of Toronto
M5S 1A1

## 0. Introduction

In this paper I will describe some of the fundamental ideas required for the Optimal Synthesis of linear systems with quadratic integral cost for the fixed two point boundary value problem. The main issue is the degeneracy of the Legendre-Clebsch condition and its precise connections to the nature of optimal solutions. This question is certainly very classical and there is a substantial number of studies dealing with its various aspects ([AC], [BJ], [HS], [K], [KSW], [O'M-J] and [Y]). The focus of this study is on the optimality with respect to the fixed end conditions and the fixed time interval, without any explicit assumptions on the positivity of the cost functional. The approach taken here is slightly different from the existing ones, in the sense that it makes systematic use of the Hamiltonian formalism and of the symplectic geometry.

The basic results, whose details will appear in [JK2], are as follows: For any points $a, b$ in $R^n$, and any $T > 0$ there exist unique points $\overline{a}$ and $\overline{b}$ in $R^n$ such that there exist an optimal trajectory of the system which initiates at $\overline{a}$ at $t = 0$ and terminates at $\overline{b}$ at $t = T$. The cost of this trajectory is smaller than the cost of any trajectory which initiates at $a$ and terminates at $b$ in $T$ units of time. Finally, for any $\varepsilon > 0$ there is a trajectory of the system which connects $a$ to $b$ in $T$ units of time and whose cost is within $\varepsilon$ of the cost of the optimal path from $\overline{a}$ to $\overline{b}$. We term such synthesis The Generalized Turnpike Synthesis.

## I. Statement of The Problem

The basic setting consists of a linear autonomous system in $R^n$

$$\frac{dx}{dt} = Ax + \sum_{j=1}^{m} b_j u_j \tag{1}$$

and a quadratic cost functional $c$.

We will use $u$ to denote $(u_1, \ldots, u_m)$. The cost functional $c$ is of the form

$$c(x,u) = \tfrac{1}{2}\langle u, Pu\rangle + \langle u, Qx\rangle + \tfrac{1}{2}\langle x, Rx\rangle$$

in terms of an inner product $\langle\ ,\rangle$ on $R^n$ (respectively $R^m$) where $P, Q$ and $R$ are matrices of appropriate dimensions with $P$ and $R$ symmetric. The main problem is the following:

Given $a, b$ in $R^n$ and $T > 0$ find a control $\overline{u}$ on $[0,T]$ with the corresponding state trajectory $\overline{x}$ of (1) which satisfies $\overline{x}(0) = a$, $\overline{x}(T) = b$ and which minimizes $\int_0^T c(x,u)dt$ among all trajectory pairs $(x,u)$ which satisfy $x(0) = a$ and $x(T) = b$.

$v(a,b,T)$ will denote the infimum of $\int_0^T c(x,u)dt$ over the trajectories of (1) which satisfy the given boundary conditions. The study is based on the following natural assumptions.

(A1) It is assumed that eq(1) is controllable and that

(A2) $v(a,b,T) > -\infty$ for any $a, b$ in $R^n$ and each $T > 0$. We refer to (A1) as the *Well-Posedness Assumption* and to (A2) as the *Existence Assumption.*

## II. Preliminaries

It follows from the Existence Assumption that $v(0,0,T) \geq 0$ which in turn implies that the zero trajectory corresponding to the zero control is optimal for $a = b = 0$. By applying the Maximum Principle to this optimal pair it follows that matrix $P$ is *positive semi-definite.*

Let $H(x,p,u) = -c(x,u) + \langle p, Ax + \sum_{j=1}^{m} b_j u_j\rangle$ be the Hamiltonian of the system. Since $P \geq 0$, it follows that $H$ is concave in $u$ for a fixed $x$ and $p$, and hence its maximum is given by $\frac{\partial H}{\partial u} = 0$. Triples of curves $(x,p,u)$ which satisfy

$$\frac{dx}{dt} = \frac{\partial H}{\partial p}(x,p,u)\ , \quad \frac{dp}{dt} = -\frac{\partial H}{\partial x}(x,p,u) \text{ and } \frac{\partial H}{\partial u}(x,p,u) = 0 \tag{2}$$

will be referred to as the extemal triples. It follows from the Existence Assumption that projections $(x,u)$ of the extremal triples are optimal for $a = x(0)$ and $b = x(T)$. So our problem reduces to a study of the extremals. The equation $\frac{\partial H}{\partial u} = 0$ is given by

$$-Pu - Qx + B'p = 0. \tag{3}$$

We term the problem *regular* if ker $P = 0$. In such a case (3) determines a unique Optimal Feedback Law $u = P^{-1}(Bp - Qx)$, which in turn, determines the system Hamiltonian $H_0(x,p) = H(x,p,u(x,p))$. The *regular synthesis* result is that for any $a, b$ in $R^n$ and $T > 0$ there is a unique extremal triple $\overline{x}, \overline{p}, \overline{u})$ such that $\overline{x}(0) = a$ and $\overline{x}(T) = b$.

The problem is *singular* if ker $P \neq 0$. In such a case equation (3) is not solvable for $u$ in terms of $x$ and $p$. For instance if $u \in$ ker $P$ then (3) reduces to a linear constraint equation $Qx - B'p = 0$.

The maximal linear variety $\Omega$ in $R^n \times (R^n)'$ which contains all the extremal triples $(x,p,u)$ is in general determined by additional linear constraints. The central issue of this paper is the precise connection between $\Omega$ and the linear constraints which define it.

## III. Examples

It will be natural to consider some examples before proceeding with the theory.

**Example 1.** Minimize $\frac{1}{2}\int_0^T x_2^2 dt$ over the trajectories of $\frac{dx_1}{dt} = x_2$, $\frac{dx_2}{dt} = u$ which satisfy $x_1(0) = a_1$, $x_2(0) = a_2$ and $x_1(T) = b_1$, $x_2(T) = b_2$.

The Hamiltonian for this system is $H = -\frac{1}{2}x_2^2 + p_1 x_2 + p_2 u$. $\frac{\partial H}{\partial u} = 0$ means that $p_2 = 0$. In addition

$$\frac{dx_1}{dt} = x_2, \quad \frac{dx_2}{dt} = u, \quad \frac{dp_1}{dt} = -\frac{\partial H}{\partial x_1} = 0, \quad \frac{dp_2}{dt} = -\frac{\partial H}{\partial x_2} = x_2 - p_1. \tag{E}$$

Differentiating $p_2 \equiv 0$ along extremals we get another algebraic relation $x_2 - p_1 = 0$.

Let $\Omega = \{(x,p),\ x_2 - p_1 = 0,\ p_2 = 0\}$. $\Omega$ is a 2-dimensional variety which carries the extremals (optimal paths). The control $u$ which ensures that $\Omega$ is invariant is obtained by differentiating the relation $x_2 - p_1 = 0$ along the flow. We get that $u = 0$ on $\Omega$. The extremals are given by

$$\frac{dx_1}{dt} = x_2, \quad \frac{dx_2}{dt} = 0, \quad \frac{dp_1}{dt} = 0, \quad \frac{dp_2}{dt} = x_2 - p_1.$$

To obtain the entire picture we need note that $e_2$ is a special direction, which we call the *jump direction.* The system can move in this direction quickly with small cost.

Given $a = (a_1, a_2)$, $b = (b_1, b_2)$ and $T > 0$ let $\bar{a} = \left(a_1, \frac{b_1 - a_1}{T}\right)$ and $\bar{b} = \left(b_1, \frac{b_1 - a_1}{T}\right)$.

The trajectory

$x_1(t) = a_1 + \frac{b_1 - a_1}{T} t$

is the "right turnpike" for this problem.

$a - \bar{a}_1$ and $b - \bar{b}$ are in the jump direction; they represent the points of entry (see the diagram).

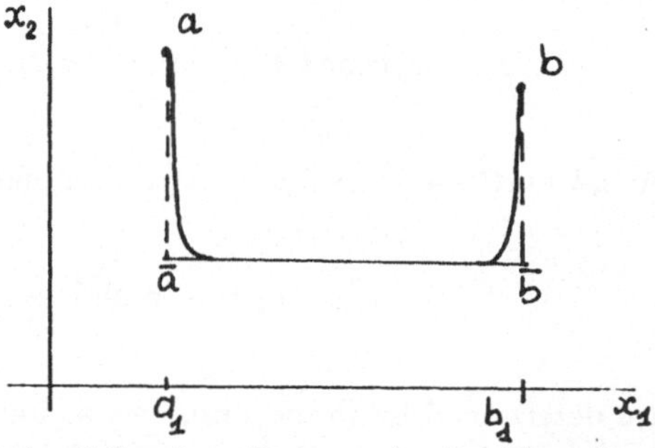


The above picture is the projection of the following diagram in $\Omega$: The diagonal line represents the vertical fiber displaced by the extremal flow $T$ units later. "The right turnpike" is given by the intersection of the displaced fiber with the vertical fiber above $b$.

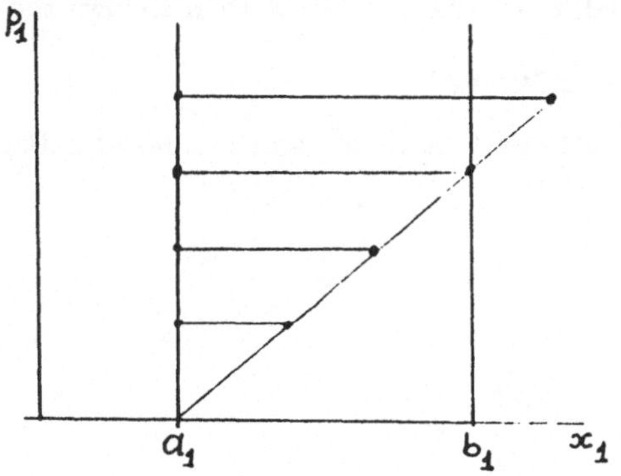


**Example 2.** Minimize $\frac{1}{2} \int_0^T (2x_1^2 + 2x_1 x_3 + x_3^2 + x_4^2) dt$ over the solution curves of

$$\frac{dx_1}{dt} = u_1, \quad \frac{dx_2}{dt} = u_2, \quad \frac{dx_3}{dt} = x_2, \quad \frac{dx_4}{dt} = x_1$$

satisfying the fixed boundary conditions.

The jump directions for this problem coincide with the vector space spanned by $e_1$, $e_2$ and $e_3$. (That is to say, the system can respond quickly in these directions with small cost.)

The optimal variety $\Omega$ which carries the extremals is given by

$$\Omega = \{(x, p) \in \mathbf{R}^8 : \; p_1 = p_2 = p_3 = 0, \; 2x_1 + x_3 - p_4 = 0, \quad x_1 + x_3 = 0, \; x_2 + x_4 = 0\}.$$

The feedback control which makes $\Omega$ invariant is given by $u_1 = \frac{1}{2}(x_4 - x_2)$ and $u_2 = -x_1$.

$\Omega$ is a 2 dimensional symplectic variety - the extremal flow in terms of the coordinates $x_4$ and $p_4$ is:

$$\frac{dx_4}{dt} = p_4, \frac{dp_4}{dt} = x_4 \tag{E}$$

The integral curves of (E) are given by:

$$x_4 = a_4 \cosh t + p_4^0 \sinh t, \quad p_4(t) = p_4 \cosh t + p_4^0 \sinh t.$$

Given $a, b$ in $\mathbf{R}^4$ and $T > 0$ the "right turnpike" is given by the particular $p_4^0$ for which

$$b_4 = a_4 \cosh T + p_4^0 \sinh T.$$

$\overline{a}$ and $\overline{b}$ are determined by $\Omega$ and are given as follows:

$$\overline{a} = (p_4^0, \ -a_4, p_4^0, a_4) \ , \ \overline{b} = (p_4(T), \ -b_4, -p_4(T), \ b_4).$$

The system moves quickly with small cost from $a$ to $\overline{a}$, it then follows the optimal path from $\overline{a}$ to $\overline{b}$ and finally it exists quickly from $\overline{b}$ to $b$.

This matching of different optimal paths on $\Omega$ is shown on the following diagram:

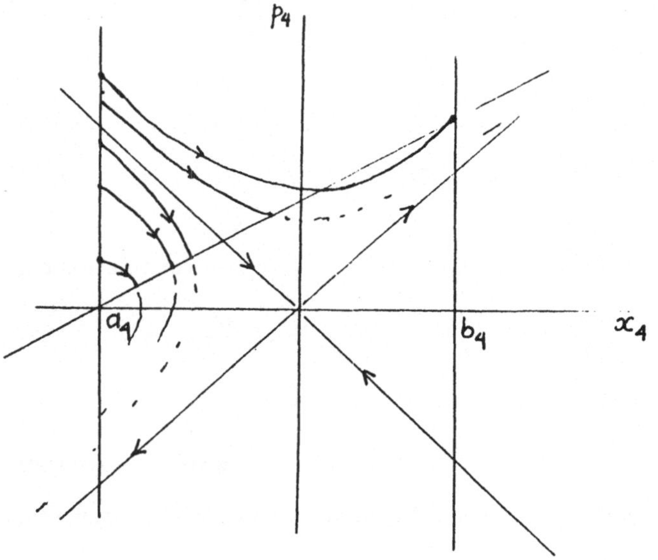

**IV. Extended Systems, their Lie Saturates and the space of Jump Fields.**

It is natural to regard the system (1) along with its cost functional as an extended differential system in $\mathbf{R}^{n+1}$:

$$\frac{dx_0}{dt} = c(x,u)\ , \ \frac{dx}{dt} = Ax + \sum_{j=1}^{m} b_j u_j \tag{4}$$

and will be convenient to express the basic ideas in terms of families of vector fields rather than differential equations. For that reason $F$ will be used to denote the family $\{(c(x,u),\ Ax + \sum b_j u_j) : (u) \in \mathbf{R}^m\}$ defined by (4). Points of $R^{n+1}$ will be denoted by $\tilde{x} = (x^0, x)$. Then $A_F(\tilde{a}, T)$ denotes the set of reachable points in exactly $T$ units of time in $\mathbf{R}^{n+1}$ along the trajectories of $F$. $A_{\mathcal{F}}(\tilde{a}, \leq T)$ will denote $\bigcup_{a \leq t \leq T} A_F(\tilde{a}, t)$.

Lie $(F)$ stands for the Lie algebra spanned by the elements of $F$. We now recall the motion of Lie saturate introduced in our earlier paper [JK1]: $LS(F)$, the Lie saturate of $F$, is the largest family of vector fields contained in Lie $(F)$ with the property that

$$c\ell A_F(\tilde{a}, \leq T) = c\ell A_{LS(F)}(\tilde{a}, \leq T) \quad \text{for all} \quad \tilde{a} \in \mathbf{R}^{n+1} \quad \text{and all} \quad T > 0.$$

The basic properties of the Lie saturate are:

(i) $LS(F)$ is a closed convex subset of all (in this case analytic) vector fields. The space of all vector fields is regarded as a vector space topologized with any topology stronger than the $C^1$ topology.

(ii) if $V$ is any vector space in $LS(F)$, then Lie $(\dot{V}) \subset LS(F)$; and

(iii) if $\mathbf{R} \subset LS(F)$ then

$$exp\ Y_{\#}(X) = \sum_{k=0}^{\infty} \frac{1}{k!} ad^k Y(X) \quad \text{is in} \quad LS(F)$$

provided that the infinite sum is in Lie $(F)$. In particular that is the case if $ad^k Y(X) = 0$ after some finite index $N$.

**Definition 1.** A constant vector field $d$ in $\mathbf{R}^n$ is called a *jump direction* if there exists $v \in (\mathbf{R}^n)^*$ such that $\lambda(\langle v, x\rangle, d)\ \forall x \in \mathbf{R}^n$ is an element of $LS(F)$ for each $\lambda \in \mathbf{R}$.

We will refer to the extended field $(\langle v, x\rangle, d)$ as the *jump field*. It is easy to see that for each $u \in kerP$, $\left(\sum_{j=1}^m u_j q_j(x) \ \sum_{j=1}^m b_j u_j,\right)$ is a jump field. It is also easy to show that the space of jump fields is zero when $kerP = 0$. so the enlargement of $F$ by addition of jump fields exists only for the degenerate problems .

The key properties of the space of jump fields are:

(a) Let $D$ be the space of all jump directions in $\mathbf{R}^n$. There exists $M : D \to D^*$ such that each jump field is of the form $(\langle Md, x\rangle, d)$.

(b) Any two jump fields commute. Since

$$[(\langle Md_1, x\rangle, d_1), \quad (\langle Md_2, x\rangle, d_2)] = (\langle Md_2, d_1\rangle - \langle Md_1, d_2\rangle, 0)$$

it follows that $\langle Md_2, d_1\rangle =< Md_1, d_2\rangle$. Hence $M$ is a self-adjoint operator on $D$

## V. Construction of the Space of Jump Directions

The construction is fundamentally based on the following fact: For any jump field $V$ $expt\ V(\tilde{a})$ belongs to the boundary of $A_{\mathcal{LS}(\mathcal{F})}(\tilde{a}, \leq t)$. Therefore for any $\tilde{a} = (a^0, a)$ in $R^{n+1}$ the integral curves of $V$ satisfy the Maximum Principle for the Lie Saturate of $\mathcal{F}$. More explicitly this means the following: Let $V = (\langle Md, x\rangle, d)$ be any jump field, and let $\tilde{x}(t) = (x^0(t), x(t))$ be an integral curve of $V$. The Hamiltonian $h$ of $V$ is given by $h(x,p) = -\langle Md, x\rangle + \langle p, d\rangle$. Then there exists an adjoint curve $p(t)$ which satisfies $\frac{dp}{dt} = -\frac{\partial h}{\partial x}$ such that

$$-f_0(x(t)) + \langle p(t), \ F(x(t))\rangle \leq h(x(t), \ p(t))$$

for almost each $t > 0$ and each vector field $(f_0, F)$ in the Lie saturate of $\mathcal{F}$. Thus the Hamiltonian lifts of jump fields are contained among the extremals of the Hamiltonians of the Lie saturate.

The space of the jump fields will be constructed in terms of their Hamiltonians. Let $\mathcal{L}_0$ be the space of the Hamiltonians induced by

$$\sum_{j=1}^m (\langle q_j, x\rangle, b_j) u_j$$

with $u \in ker\ P$.

The construction will be based on an increasing sequence $\mathcal{L}_0 \subset \mathcal{L}_1 \subset \cdots$ of vector spaces of linear Hamiltonians which correspond to the Hamiltonian lifts of the jump fields. It follows that $(\langle Md, x\rangle, d)$ is a jump field if and only if $h(x,p) = -\langle Md, x\rangle + \langle p, d\rangle$ is an element of $\bigcup_{k\geq 0}\mathcal{L}_k$. The starting point for the construction is the equation $-Pu - Qx + B'p = 0$.

Let $V$ be any space complementary to ker $P$. Let $u = v + w$ be the corresponding splitting of the control space with $v \in V$ and $w \in \ker P$. $-Pu - Qx + B'p = 0$ splits into $-P_1v - Q_1x + B_1'p = 0$ and $-Q_2x + B_2'p = 0$. The first equation is the projection on $V$ and the second is the projection on ker $P$. Let $v(x,p) = P_1^{-1}(B_1'p - Q_1x)$. This choice of feedback determines a quadratic Hamiltonian $H_0$,

$$H_0(x,p) = -c(x, v(x,p)) + \langle p,\ Ax + B_1v(x,p)\rangle.$$

If $\mathcal{L}$ denotes the space of the Hamiltonian lifts of the space of jump fields then $\mathcal{L}$ is isotropic; that is, $\{h, g\} = 0$ for any $h, g$ in $\mathcal{L}$. Let $\beta$ be the symmetric bilinear form $\beta(h,g) = adgadh(H_0)$. The sequence $\{\mathcal{L}_k\}$ is defined as follows

$$\mathcal{L}_{k+1} = \mathcal{L}_k + \{adH_0(h) : h \in \mathcal{L}_k\ \beta(h,h) = 0\} \qquad k = 0, 1, \cdots$$

It follows that $\mathcal{L} = U\mathcal{L}_k$ and it also follows that the preceding construction does not depend on the choice of the complementary space $V$ which defines $H_0$.

Alternatively $\mathcal{L}$ could be defined as the smallest vector space which contains $\mathcal{L}_0$ and satisfies $adH_0(\mathcal{L} \cap ker\beta) \subset \mathcal{L}$.

The space $\mathcal{L}_0$ is partitioned according to certain indexes $i_1 < i_2 < \cdots < i_q \leq \infty$. They are defined as follows: $i_j$ is the first integer $k$ with the property that

$$ad^j H_0(h) = 0 \qquad j \leq k \text{ and}$$

$$ad\ h\ ad^{2k+1}H_0(h) \neq 0 \quad \text{for some element } h \text{ in } L_0.$$

These indexes are quite analogous to the Kronecker indexes in the theory of linear systems.

It follows that the optimal variety $\Omega$ is symplectic if and only if the maximal index $i_q$ is less than $\infty$. In such a case there is a finite reduction procedure enumerated by the above set of indexes which determines the optimal variety $\Omega$.

## References


[AC] B.D.O Anderson and D.J. Clements, "Singular Optimal Control: The linear-quadratic problem", Lecture Notes in Control and Information Sciences No. 5, Springer-Verlag.

[BJ] D.J. Bell and D.H. Jacobson, *Singular optimal control problems*, Academic Press 1975.

[H] G.H. Hardy, J.E. Littlewood, G. Polya, *Inequalities,* Cambridge Press (1934).

[HS] M.L.I. Hautus and L.M. Silverman, "Systems Structure and singular control", Linear Algebra and Its Applications **50** (1983), 369-402.

[JK1] V. Jurdjevic and I. Kupka, "Polynomial Control Systems", Math. Ann. **272** (1986), 361-368.

[JK2] V. Jurdjevic and I. Kupka, "Linear systems with Singular Quadratic Cost", to appear.

[K] J. Kogan, "Bifurcation of Extremals in optimal control", Lecture Notes in Mathematics, No. 1216, Springer-Verlag 1986.

[Kr] A. Krener, "The High Order Maximum Principle and its applications to singular extremals", SIAM Journal on Control and Optim., Vol. 151, No. 2 (1977), 256-292.

[KSW] A. Kitapcu, L.M. Silverman and J.C. Willems, "Singular Optimal Control: a geometric approach", SIAM Journal on Control and Opt. 24 (1986), 369-402.

[KT] P.V. Kokotović, "Applications of singular perturbations techniques to control problems", SIAM Review, vol. 26 (1984), 501-550.

[O'M-J] R.E. O'Malley, Jr. and A. Jameson, "Singular perturbations and singular arcs", part II, IEEE Trans. Ant. Control **22** (1977), 328-337.

[Y] V.A. Yakubovich, "Optimization and invariance of linear stationary control systems", Automatika and Telemehanika no. 8 (1984), 5-44.

# DEGENERATE LINEAR SYSTEMS WITH QUADRATIC COST UNDER FINITENESS ASSUMPTIONS

I.A.K. Kupka
Department of Mathematics, University of Toronto
Toronto, Ontario, Canada, M5S 1A1

## 0. Introduction

A classical optimal control problem has the following setting: on a linear state space $X$, we are given a linear system $\frac{dx}{dt} = Ax + Bu$, having as control space, the vector space $U$. A quadratic cost function $c : X \times U \to \mathbf{R}$ is also given: $c(x,u) = \frac{1}{2}\langle Pu, u\rangle + \langle Qx, u\rangle + \frac{1}{2}\langle Rx, x\rangle$ where $Q$ is a linear map $X \to U'$, the dual of $U$, $P$ is a symmetric linear mapping $U \to U'$, $R$ a symmetric linear mapping: $X \to X'$, the dual of $X$

The problem of optimizing the cost $\int_0^T c(x(t), u(t))dt$ along any trajectory $(x,u)$ : $[0,T] \to X \times U$ of the system for a fixed time $T$ and a fixed initial point $x(0) = x^0$ has been analysed thoroughly in the literature ([AC], [BJ], HS], [KE] [KSW], [Wi], [Wo], [Y]) under strong positivity assumptions on $c$. The two point boundary value problem where we are fixing the time $T$, the initial and final points $x(0) = x^0$, $x(T) = x^1$, has attracted much less attention. It was studied in ([BJ], [Wi].)

In our study [JK3] in collaboration with V. Jurdjevic, we analysed the fixed time, two point boundary value problem under a different point of view. We assumed that for any time $T > 0$, any couple of points $x^0, x^1$ in $X$, the optimum cost $v(x^0, x^1, T)$ for going from $x^0$ to $x^1$ in $T$ units of time is always a number, that is, it is never $-\infty$ or $+\infty$. Under this, in our opinion, very natural assumption, we were able to obtain a complete solution of the optimal control synthesis problem. One of the main new features here is that there does not always exist an optimal trajectory going from $x^0$ to $x^1$ in $T$ unit of times. One has to allow jumps in certain well defined directions. The space of these directions forms a vector subspace $J$ of $X$, which is reduced to 0 in the classical case when $\partial^2 c/\partial u^2 = P$, is positive definite. Under some mild non-degeneracy condition, given any $T > 0$, any two affine planes $J^0$, $J^1$ parallel to $J$, there exists a unique optimal trajectory going from $J^0$

to $J^1$ in $T$ units of time.

When $P$ is positive definite, we get back the classical result.

## 1. Basic definitions and properties

**Definition 0.** A trajectory $(\hat{x}, \hat{u}) : [0,T] \to X \times U$ of the system $\frac{dx}{dt} = Ax + Bu$ is a pair of curves such that:

(i) $\hat{u}$ is square integrable on $[0,T]$

(ii) $\hat{x}$ is absolutely continuous and for almost every $t$ in $[0,T]$, $\frac{d\hat{x}}{dt}(t) = A\hat{x}(t) + B\hat{u}(t)$

**Definition 1.** For any couple $x^0, x^1$ in $X$ any time $T > 0$, the optimum cost $v(x^0, x^1, T)$ for going from $x^0$ to $x^1$ in $T$ units of time is: $v(x^0, x^1, T) = +\infty$ if there is no trajectory of the system going from $x^0$ to $x^1$ in time $T$, otherwise

$$v(x^0, x^1 T) = \inf\left\{ \int_0^T c\big(x(t), u(t)\big)dt \,\Big|\, (x,u) : [0,T] \to X \times U \text{ trajectory} \right.$$
$$\left. x(0) = x^0, \quad x(T) = x^1 \right\}$$

where we allow $-\infty$ as infinity.

Our basic assumptions will be the following:

(C) The system $\frac{dx}{dt} = Ax + Bu$ is controllable.

This implies that $-\infty \leq v(x^0, x^1, T) < +\infty$ for all triple $(x^0, x^1, T) \in X \times X \times ]0, +\infty$

(E) For all triples $(x^0, x^1, T) \in X \times X \times ]0, +\infty[$. $v(x^0, x^1, T) > -\infty$. Hence $v(x^0, x^1, T) \in \mathbb{R}$, always.

Finally we shall call the triple $(A, B, C)$ formed by the system and its cost $c$, *non degenerate* if the following assumption is satisfied.

(U) For any time $T > 0$, $v(0,0,T) = 0$ and the only optimal trajectory going from 0 back to 0 in time $T$ is the zero trajectory.

The following proposition is quite trivial but important.

**Proposition 0.** Let $(A, B, c)$ be a linear system $(A, B)$ with quadratic cost, $LQC$ system for shorts, satisfying assumptions (C) and (E). Then $P = \frac{\partial^2 c}{\partial u^2}$ is positive semi-definite.

**Remark 0.** This is analogous to a theorem of Tonelli [T] in the calculus of variations.

**(B) Main concepts and their properties** The concept introduced in the next definition is crucial in our study.

**Definition 2.** Let $(A, B, c)$ be an $LQC$ system satisfying (C) and (E). A line $\delta$ in $X$ through 0 is called a jump direction if for any $x^0$, $x^1$ in $X$ such that $x^1 - x^0 \in \delta$, the lower limit $v_\infty(x^0, x^1)$ of $v(x^0, y^1, T)$ as $T$ tends to 0 and $y^1$ to $x^1$ is a number $(< +\infty)$. The next two propositions state the most important properties of jump directions.

**Proposition 1.**

(i) The union of all the jump lines is a vector subspace $J$ of $X$. $J$ contains the kernel of $\frac{\partial^2 c}{\partial u^2}$ and $J$ is reduced to 0 if that kernel if 0.

(ii) There exists a quadratic form $q : X \to \mathbf{R}$ (not unique in general) such that $v_\infty(x^0, x^1) = q(x^1) - q(x^0)$ for all $x^0, x^1 \in X$, $x^1 - x^0 \in J$. The restriction $q_J$ of $q$ to $J$ is uniquely defined.

**Proposition 2.** A $LQC$ system $(A, B, c)$ is non-degenerate if and only if $J$ does not contain any non-trivial $(A, J \cap \text{Image } B)$-controllability subspace.

## 2. Statement of the main results about optimal trajectories

The next definition states states precisely what an optimal trajectory is.

**Definition 2.** A trajectory $(\hat{x}, \hat{u}) : [0, \hat{T}] \to X \times U$ of the system $\frac{du}{dt} = Ax + Bu$ is called optimal if the cost $\int_0^{\hat{T}} c(\hat{x}(t), \hat{u}(t))dt$ of $(\hat{x}, \hat{u})$ is equal to $v(\hat{x}(0), \hat{x}(\hat{T}), \hat{T})$.

**Theorem 1.** If the $LQC$ system satisfies (C), (E), (U), for any two affine spaces $J^0, J^1$ parallel to $J$ and any time $T > 0$, there exists a unique optimal trajectory $\Gamma(J^0, J^1, T)$ : $[0, T] \to X \times U$ starting in $J^0$ and ending in $J^1$. In other words given $J^0, J^1, T$ as above there exists a unique optimal trajectory $(\hat{x}, \hat{u}) : [0, T] \to X \times U$ such that $\hat{x}(0) \in J^0$, $\hat{x}(T) \in J^1$. $\Gamma(J^0, J^1, T)$ is an analytic function in $J^0, J^1, T$. It is an affine function in the first two variables $J^0, J^1$.

This shows in particular that if $J$ is not reduced to 0, there does not exist in general an optimal trajectory from a point $x^0$ to a point $x^1$ in time $T > 0$.

This indicates that we cannot expect in general a regular synthesis of the problem in the classical case. But we have a substitute.

## 3. Generalised optimal control synthesis

In order to get a synthesis we are going to generalise the concept of trajectory of the system. There is an excellent justification for our generalisation based on the concept of a Lie saturate we have introduced in [JK1] and [JK2]. We are going to explain it in section 5.

**Definition 3.** Consider a *LQC* system $(A, B, c)$ satisfying (C) and (E) having a vector space $J$ of jump directions not reduced to 0. A generalised trajectory of the system is a curve $(x, u) : [0.T] \to X \times U$ such that:

(i) $u$ is square integrable.

(ii) $x : [0, T] \to X$ is a *rcll* as the probabilists say, that is, it is continuous on the right and at any discontinuity point the left limit exists.

(iii) If $t$ is any discontinuity point of $x$, the jump $x(t + 0) - x(t - 0)$ belongs to $J$.

(iv) The closure of the set of discontinuity points of $x$ has an empty interior. On its open and dense complement, $x$ is absolutely continuous and $\frac{dx}{dt}(t) = Ax(t) + Bu(t)$ for almost every $t$.

The cost of such a generalised trajectory can also be defined:

**Definition 4.** Let $(x, u) : [0, T] \to X \times U$ be a generalized trajectory of a *LQC* system $(A, B, c)$ satisfying (C) and (E). Denote by $D$ the set of discontinuity points of $x$ and by $\{\Delta_n | n \in \mathbb{N}\}$ the set of all connected components of the complement of the closure of $D$ in $[0, T]$. Then the cost $C(x, u)$ of $(x, u)$ is

$$C(x, u) = \sum_{t \in D} v_\infty\big(x(t - 0), x(t)\big) + \sum_n \int_{\Delta_n} c\big(x(t), u(t)\big) dt$$

if the righthand side has a meaning and $+\infty$ otherwise.

The next theorem states the generalised optimal synthesis

**Theorem 2.** Consider a $LQC$ system satisfying assumptions (C), (E), (U) and having jump space $J$.

(i) For any triple $(x^0, x^1, T) \in X \times X \times ]0, +\infty[$, the optimum cost to go from $x^0$ to $x^1$ in $T$ units of time using generalised trajectories is $v(x^0, x^1, T)$, the same as using only ordinary trajectories:

$$v(x^0, x^1, T) = \inf\Big\{ C(x,u) \Big| (x,u) : [0,T] \to X \times U \text{ generalised trajectory}$$
$$x(0-0) = x^0,\ x(T) = x^1 . \Big\}$$

(ii) A generalised trajectory $(x,u) : [0,T] \to X \times U$ is optimal (that is, its cost $C(x,u)$ is equal to $v(x^0, x^1, T)$) if and only if:

(a) $x$ has at most two discontinuity points located at 0 and $T$.

(b) One the time interval $[0,T[ = \{t | 0 \le t < T\}$, $(x,u)$ coincides with the optimal trajectory $\Gamma(J^0, J^1, T)$ defined in Theorem 1, where $J^0 = x^0 + J$, $J^1 = x^1 + J$

(iii) Given any two points $x^0, x^1$ in $X$ and any time $T > 0$, there exists a unique optimal generalized trajectory $(\hat{x}, \hat{u}) : [0,T] \to X \times U$ going from $x^0$ to $x^1$ in time $T$. It is defined as follows: let $J^0 = x^0 + J, J^1 = x^1 + J$ and denote by $\bar{x}^0$ (resp. $\bar{x}^1$) the origin (resp. end point) of $\Gamma(J^0, J^1, T)$

(a) If $x^0 \ne \bar{x}^0$ $\hat{x}$ starts with a jump $x^0 \to \bar{x}^0$: $\hat{x}(0-0) = x$ $\hat{x}(0) = \bar{x}^0$. Otherwise $\hat{x}(0-0) = \hat{x}(0) = x^0 = \bar{x}^0$

(b) On the interval $[0,T[, (\hat{x}, \hat{u})$ coincides with $\Gamma(J^0, J^1, T)$

(c) If $\bar{x}^1 \ne x^1$, $\hat{x}$ ends with a jump $\bar{x}^1 \to x^1$, $x(T-0) = \bar{x}^1$, $\hat{x}(T) = x^1$. If $x^1 = \bar{x}^1$, $\hat{x}(T-0) = \hat{x}(T) = \bar{x}^1 = \bar{x}^1$.

## 4. Indications on the proof of theorem

Let us recall that $X \times X'$ carries a well known flat symplectic structure $\sigma$. A subspace $Z \subset X \times X'$ is called symplectic if the restriction of $\sigma$ to $Z$ is non degenerate. A subspace $Z$ is called Lagrangian if the restriction of $\sigma$ to $Z$ is zero and $Z$ is maximal for this property. The dual $M'$ of $M = X \times X'$ carries a symplectic structure dual of $\sigma$. It is given by the Poisson bracket $\{f, g\}$.

If $x^1, \ldots, x^d : X \to \mathbf{R}$ $d = \dim X$ is a system of linear coordinates on $X$ and $p', \ldots, p^d$ the dual system of coordinates on $X'$ then $\sigma = \sum_{i=1}^{d} x^i \wedge p^i$ and for any two functions $f, g$ on $M$: $\{f, g\} = \sum_{i=1}^{d} \left[\frac{\partial f}{\partial x^i}\frac{\partial y}{\partial p^i} - \frac{\partial f}{\partial p^i}\frac{\partial y}{\partial x^i}\right]$

Using the maximum principle it is fairly easy to show that any optimal trajectory $(\hat{x}, \hat{u}) : [0, T] \to X \times U$ satisfies the conditions: let $H : X \times X' \times U \to \mathbf{R}$ be the function $H(x, p, u) = \langle Ax + Bu, p\rangle - c(x, u)$. Then there exists an absolutely continuous function $\hat{p} : [0, T] \to X'$ such that

(i) $-\frac{d\hat{p}}{dt}(t) = \frac{\partial H}{\partial x}(\hat{x}(t), \hat{p}(t), \hat{u}(t))$ for almost every $t$

(ii) $\frac{\partial H}{\partial u}(\hat{x}(t), \hat{p}(t), \hat{u}(t)) = 0$ for almost every $t$.

The curve $(\hat{x}, \hat{p}, \hat{u})$ is an extremal of the system. If $P$ is degenerate there will exist vectors $q \in U$ such that the linear form $f_q \in M'$, $f_q(x, p) = \langle \frac{\partial H}{\partial u}(x, p, u), q\rangle$ does not depend on $U$. Hence it is what Dirac called a first class constraint that is a linear form $f : X \times X' \to \mathbf{R}$ which is zero in any extremal that is any curve $(\hat{x}, \hat{p}, \hat{u}) : [0, T] \to X \times X' \times U$ satisfying (i) and (ii) above and:

$$\frac{d\hat{x}}{dt}(t) = \frac{\partial H}{\partial x}(\hat{x}(t), \hat{p}(t), \hat{u}(t)) \quad \text{for almost every} \quad t\,. \tag{0}$$

These first class constraints form a vector subspace $CI$ of $M'$. Theorem 1 is a consequence of the following proposition.

**Proposition 3.** If the $LQC$ system $(A, B, c)$ satisfies (C), (E), (U) then:

(i) the space of all first class constraints $CI$ is symplectic. As a consequence the subspace $\Omega$ of $X \times X'$, annihilator of $CI$, is also symplectic.

(ii) There exists a quadratic form $H_\Omega$ on $\Omega$ such that a trajectory $(\hat{x}, \hat{u}) : [0, T] \to X \times U$ is optimal if and only if the curve $\hat{x}$ is the projection on $X$ of a trajectory of the

Hamiltonian field $\vec{H}_\Omega$ of $H_\Omega$ in $\Omega$.

(iii) Let $\Lambda$ denote the vector subspace of $\Omega$ of all vectors in $\Omega$ whose projection on $X$ belongs to $J$, and let $\Lambda_t$ be the vector space image of $\Lambda$ at time $t \in \mathbf{R}$ under the flow of $\vec{H}_\Omega$. Then, for any $t \in \mathbf{R}$, $\Lambda_t \cap \Lambda = 0$.

Condition (iii) implies that given any two affine plane $\Lambda^0, \Lambda^1$ parallel to $\Lambda$ in $\Omega$, for any $t \in \mathbf{R}$ the image $\Lambda^0$ of $\Lambda^0$ at time under the flow of $\vec{H}_\Omega$ cuts $\Lambda^1$ in a unique point. By pprojection on $X$ we get the existence and uniqueness of the optimal trajectory $\Gamma(J^0, J^1, T)$ going from $J^0$ to $J^1$ in time $T$.

## 5. The jump directions $J$

To get a good picture of what the jump directions are and their meaning for the problem, it is convenient to extend our system to a bigger state space.

The new state space $\tilde{x}$ will be $\mathbf{R} \times \mathbf{R} \times X$. If $\tilde{x} \in \tilde{X}$, then $\tilde{x} = (x_{-1}, x_0, x)$ where $x_{-1} \in \mathbf{R}, x_0 \in \mathbf{R}$, $x \in X$. The variable $x_{-1}$ will register the cost, the variable $x_0$ the time. Now we extend our linear system, $\frac{dx}{dt} = Ax + Bu$ to a system $\frac{d\tilde{x}}{dt} = \tilde{F}(\tilde{x}, u)$ on $\tilde{X}$ as follows:

$$\tilde{F}(\tilde{x}, u) = \big(c(x,u), 1, Ax + Bu\big), \qquad \tilde{x} = (x_{-1}, x_0, x) \; .$$

In [JK1] we have introduced the concept of Lie saturate. Let this Lie $(\tilde{F})$ be the Lie algebra of vector-fields generated by the family of vector fields $\tilde{F}_u \,:\, \tilde{F}_u(\tilde{x}) = \tilde{F}(\tilde{x}, u)$. Thus a finite dimensional Lie algebra. All the elements $\tilde{G}$ of Lie $(\tilde{F})$ are of the form $G(\tilde{x}) = \big(q(x), \alpha, Ax + \beta\big)$ where $q$ is a polynomial of degree $\leq 2$, $\alpha \in \mathbf{R}$ and $\beta$ is a vector in $\sum_{n=0}^{\infty} A^n\,(\text{Image}\, B)$ (This last space is $X$ if the system $\frac{dx}{dt} = Ax + Bu$ is controllable.) Now the definition of Lie saturate.

**Definition 4.** The Lie saturate $LS(\tilde{F})$ of the extended system $\tilde{F}$ is the set of all vector fields $\tilde{G}$ in the Lie algebra, Lie $(\tilde{F})$, of $\tilde{F}$ such that for any $\tilde{a} \in \tilde{X}$, the positive semi-trajectory of $\tilde{G}$ starting at $\tilde{a}$ is contained in the closure $\mathcal{A}(\tilde{a}, \tilde{F})$ of the accessibility set of $\tilde{a}$ under the action of the system $\tilde{F}$.

The Lie saturate $LS(\tilde{F})$ contain a subset $LS_0(\tilde{F})$ of vector fields $\tilde{G}$ along which the time $x_0$ is constant. They are of the form $\big(\ell(x), 0, d\big)$ where $\ell : X \to \mathbf{R}$ is a linear form and

$d$ is a constant vector in $X$. Now the relation of $LS_0(\tilde{F})$ with $J$ is the following: a vector $d \in X$ belongs to $J$ if and only if the vector field $\tilde{G}, \tilde{G}(\tilde{x}) = \left(\langle \frac{\partial q}{\partial x}(x), d\rangle, 0, d\right)$ belongs to $LS_0(\tilde{F})$. $q$ has the same meaning as in Proposition 1 (iii). This means precisely that for any $x^0 \in X$ any $d \in J$, any $T > 0$ *as small as we want*, to points as near as we want to $x^1$, the system can steer $x^0$ in time $T$ with cost approximately equal to $q(x^0 + d) - q(x^0)$. Hence this gives a precise meaning and a good justification for our heuristic discussion of the jump lines following their Definition 2.

**Conclusion.** The preceding sketch of our work gives a complete picture for $LQC$ systems satisfying the assumptions (C), (E), and (U). For degenerate systems, that is those not satisfying (U), we get analogous results except we lose the uniqueness of optimal trajectories between $J^0, J^1$.

## References


[AC] B.D.O. Anderson-D.J. Clements, "Singular optimal control: The linear-quadratic problem", Lecture notes in Control and Information Sciences, no. 5, Springer-Verlag 1978.

[BJ] D.J. Bell-D.H. Jacobson, "Singular optimal control problems", Academic Press 1975.

[HS] M.L.J. Hautus-L.M. Silverman, "System structure and singular control", *Linear Algebra and Its Application* **50** (1983) 369-402.

[JK1] V. Jurdjevic-I. Kupka, "Control systems on semi simple Lie groups and their homogeneous spaces", *Annales de l'Institut Fourier* 31, 4 (1981) 151-179.

[JK2] V. Jurdjevic-I. Kupka, "Linear systems with quadratic costs", to appear.

[KE] H. Kelly-T. Edelbaum, "Energy climbs, energy turns and asymptotic expansions", *J. Aircraft* **7** (1970) 93-95.

[K01] J. Kogan, "Bifurcation of extremals in optimal control", Lecture Notes in Math. no. 1216, Springer-Verlag (1986).

[K02] J. Kogan, "Structure of minimizers in linear quadratic Bolza problems of optimal control", preprint, Purdue University 1987.

[KSW] A. Kitapcu-L.M. Silverman-J.C. Willems, "Singular optimal control: a geometric approach", *SIAM J. on Contr. and Optim.* **24** (1986) 369-402.

[T] L. Tonelli, "Fondamenti di calolo elle variazioni", vol 1-2, Zanichelli editors 1923.

[Wi] J. Willems, "Least square stationary optimal control and the algebraic Riccati equation", IEEE trans. Aut. Control, vol. AC15, 6 (1971) 621-634.

[Wo] M. Wohnam, "Linear multivariable control - A geometric approach", Springer-Verlag (1983).

[Y] V.A. Yakubovich, "Optimization and invariance of linear stationary control systems", Automatika and Telemekhamka no. 8 (1984) 5-44.

# IV - STABILITY, ROBUSTNESS AND SINGULAR PERTURBATIONS TECHNIQUES

## STABILIZATION TECHNIQUES

# STABILIZABILITY AND ASYMPTOTIC STABILIZABILITY OF THE ANGULAR VELOCITY OF A RIGID BODY

Dirk Aeyels

Department of Systems Dynamics
University of Gent
Grotesteenweg Noord 2
9710 Gent
Belgium

ABSTRACT

*The stabilizability and asymptotic stabilizability of Euler's angular velocity equations of the rigid body are investigated. The results presented follow either as an application of Lyapunov theory or the center manifold approach. The proposed controls are discussed with respect to their robustness properties.*

## 1. INTRODUCTION

Our goal in this paper is to discuss the asymptotic stabilization and the stabilization of the angular velocity of a rigid body by means of a smooth feedback control. The results presented follow either as an application of Lyapunov theory or the so-called center manifold approach, introduced in [ 1 ]. In section 2 we recall the center manifold approach. Then, in section 3 Euler's angular velocity equations of a rigid body are considered where two inputs along principal axes are available. We show by means of center manifold theory how the angular velocity can be asymptotically stabilized and this in a robust manner with respect to the parameters defining the control law. The problem whether asymptotic stabilization of the angular velocity is possible if there is only one control available is treated in Section 4. The main result is that asymptotic stabilizability is possible if the control axis has components with respect to all three principal axes. This is derived by means of Lyapunov theory. However the proposed control is not robust. Whether a scalar robust asymptotically stabilizing control exists is unknown at present. Also in this section an interesting result is derived stating that asymptotic stabilizability is impossible with scalar control if the control axis is aligned with one of the principal axes. This fact leads us to the problem of stabilizability by means of a scalar control with the control axis along a principal axis. This problem is developed in

Section 5, where we concentrate on robust stabilization. The techniques used in this section are center manifold theory as well as Lyapunov theory.

## 2. THE CENTER MANIFOLD APPROACH

Consider a system of the form

$$\dot{x} = f(x) + bu \tag{2.1}$$

with scalar input u and constant control vector b, and with $f(0) = 0$. The case with multivariable input is treated similarly and will be considered in section 3 where we study the angular velocity equations. After performing appropriate changes of basis ([1]) the system (2.1) is written in the form

$$\begin{aligned} \dot{x}_1 &= A_{11}x_1 + f_1(x_1,x_2,x_3) \\ \dot{x}_2 &= A_{22}x_2 + f_2(x_1,x_2,x_3) \\ \dot{x}_3 &= A_{33}x_3 + f_3(x_1,x_2,x_3) + bu \end{aligned} \tag{2.2}$$

with $A_{11}$ critical, $A_{22}$ and $A_{33}$ having eigenvalues with negative real part and $A_{33}$ such that $(A_{33},b)$ is a controllable pair. The functions $f_1$, $f_2$ and $f_3$ are the higher order terms; they vanish at the origin.

The problem is to construct a feedback control law $u=F(x_1,x_2,x_3)$ such that the origin of (2.2) is asymptotically stable.

Consider the (generalized) eigenspace $E^c$ corresponding to the eigenvalues of $A_{11}$. Then, by the <u>center manifold theorem</u> [2] there exists a "center manifold" $W^c$ <u>tangent to</u> $E^c$ at the origin. This center manifold has sufficiently high degree of smoothness and is <u>invariant</u> with respect to the flow.

Since $W^c$ is tangent to $E^c$ (the space $x_2=x_3=0$) it can be represented as a (local) graph in the $(x_1,x_2,x_3)$-space by

$$W^c = \{(x_1,h(x_1));\ h(0) = Dh(0) = 0\}$$

with $h = (h_1,h_2)$ defined in some neighborhood of the origin. Consider the projection of the vectorfield on $W^c$ onto $E^c$

$$\dot{x} = A_{11}x_1 + f_1(x_1,h_1(x_1),\ h_2(x_1)) \tag{2.3}$$

Since $h(x_1)$ is tangent to $x_2 = x_3 = 0$ and since the $\dot{x}_2$ and $\dot{x}_3$ equations are asymptotically stable up to first order, it is reasonable to expect that the the stability behavior of (2.2) is determined by that of (2.3). We state this as a theorem; for a proof we refer to the literature.

Theorem 1 [ 2 ]. If the origin of (2.3) is locally asymptotically stable, then the origin of (2.2) is also locally asumptotically stable.

The computation of $h(x_1)$ follows from the invariance of the flow on the center manifold. This leads to a partial differential equation. For purpose of stability investigation, approximations of $h(x)$ will do. This will be illustrated in what follows. For details, see [ 2 ].

## 3. ASYMPTOTIC STABILIZATION OF THE ANGULAR VELOCITY OF A RIGID BODY WITH TWO CONTROLS

Consider a rigid body in an intertial reference frame. Let $\omega_1$, $\omega_2$, $\omega_3$ be the angular velocity components with respect to a body-fixed reference frame with origin at the center of gravity and consisting of the principal axes. Let $I_1$, $I_2$, $I_3$ denote the principal moments of inertia (positive real numbers). The Euler equations are

$$I\dot{\omega} = S(\omega)I\omega + Bu$$

with $I = \text{diag}\,(I_1, I_2, I_3)$ the inertia matrix $\omega \in \mathbb{R}^3$

$$B = (b_1, b_2, b_3) \in \mathbb{R}^{3\times 3}$$

$$S(\omega) = \begin{bmatrix} 0 & \omega_3 & -\omega_2 \\ -\omega_3 & 0 & \omega_1 \\ \omega_2 & -\omega_1 & 0 \end{bmatrix}$$

$$u^T = (u_1, u_2, u_3) \in \mathbb{R}^3$$

For later reference:

$$I_{23} = (I_2-I_3)/I_1,\ I_{31} = (I_3-I_1)/I_2,\ I_{12} = (I_1-I_2)/I_3.$$

Notice that each $u_i$ represents an external torque aligned with $b_i$. Assume now that only two controls are available aligned with a principal axis:i.e.

$$b_1 = 0,\ b_2 = \begin{bmatrix} 0 \\ 1 \\ 0 \end{bmatrix},\ b_3 = \begin{bmatrix} 0 \\ 0 \\ 1 \end{bmatrix}$$

then with $u:=u_2$ and $v:=u_3$ the angular velocity equations are

$$\dot{\omega}_1 = I_{23}\omega_2\omega_3$$

$$\dot{\omega}_2 = I_{31}\omega_3\omega_1 + u$$

$$\dot{\omega}_3 = I_{12}\omega_1\omega_2 + v$$

or, since u and v are state feedback controls, equivalently, without changing notation

$$\dot\omega_1 = I_{23}\omega_2\omega_3$$
$$\dot\omega_2 = u$$
$$\dot\omega_3 = v$$

Consider the feedback law $u = -\omega_2 + \alpha\omega_1^2$, $v = -\omega_3 + \beta\omega_1^3$.

Then

$$\begin{bmatrix}\dot\omega_1\\ \dot\omega_2\\ \dot\omega_3\end{bmatrix} = \begin{bmatrix}0 & 0 & 0\\ 0 & -1 & 0\\ 0 & 0 & -1\end{bmatrix}\begin{bmatrix}\omega_1\\ \omega_2\\ \omega_3\end{bmatrix} + \begin{bmatrix}I_{23}\omega_2\omega_3\\ \alpha\omega_1^2\\ \beta\omega_1^3\end{bmatrix}$$

which has a linear part in diagonal form with a zero eigenvalue and two eigenvalues with negative real part. Therefore there exists a center manifold described by $\omega_2 = h_2(\omega_1)$, $\omega_3 = h_3(\omega_1)$ which has the property that the stability behavior of the system is given by the stability behavior of the reduced sysstem $\dot\omega_1 = I_{23}h_2(\omega_1)h_3(\omega_1)$.

Let $h_2(\omega_1) = h_{21}\omega_1^2+h_{22}\omega_1^3+\ldots$ and $h_3(\omega_1) = h_{31}\omega_1^2+h_{32}\omega_1^3+\ldots$ then since the flow is invariant on the center manifold, i.e. the curve described by $W^c = \{(\omega_1,h_2(\omega_1),h_3(\omega_1)\}$,

we obtain for the solutions for which $\omega_2(t)= h_2(\omega_1(t))$, $\omega_3(t)=h_3(\omega_1(t))$, for all t, that

$$\dot\omega_2(t) = \frac{\partial h_2}{\partial\omega_1}(\omega_1(t))\dot\omega_1(t)$$

$$\dot\omega_3(t) = \frac{\partial h_3}{\partial\omega_1}(\omega_1(t))\dot\omega_1(t)$$

for all t, where $\omega_2(t) = h_2(\omega_1(t))$ and $\omega_3(t) = h_3(\omega_1(t))$. Therefore, $\forall t$

$$-\omega_2(t) + \alpha\omega_1^2(t) = (2h_{21}\omega_1(t) + 3h_{22}\omega_1^2(t)+\ldots)I_{23}\omega_2(t)\omega_3(t)$$
$$-\omega_3(t) + \beta\omega_1^3(t) = (2h_{31}\omega_1(t) + 3h_{32}\omega_1^2(t)+\ldots)I_{23}\omega_2(t)\omega_3(t)$$

and after substitution of $\omega_2(t)$ and $\omega_3(t)$ one obtains

$$h_{21} = \alpha,\ h_{31} = 0,\ h_{32} = \beta.$$

The reduced system then is of the form

$$\dot\omega_1 = I_{23}\alpha\beta\omega_1^5 + \text{higher order terms.}$$

This system is locally asymptotically stable if $I_{23}\alpha\beta < 0$. Therefore the control $u = -\omega_2 + \alpha\omega_1^2$, $v = -\omega_3 + \beta\omega_1^3$ with $\alpha$ and $\beta$ such that $I_{23}\alpha\beta < 0$, is an asymptotically stabilizing feedback.

It is also easy to recover the asymptotically stabilizing control (assume without loss of generality that $I_{23} = 1$)

$$u = -\omega_2 + \omega_1$$

$$v = -\omega_3 - \omega_1^2$$

proposed by Brockett [ 3 ] .Indeed with this control the system becomes

$$\begin{bmatrix} \dot{\omega}_1 \\ \dot{\omega}_2 \\ \dot{\omega}_3 \end{bmatrix} = \begin{bmatrix} 0 & 0 & 0 \\ 1 & -1 & 0 \\ 0 & 0 & -1 \end{bmatrix} \begin{bmatrix} \omega_1 \\ \omega_2 \\ \omega_3 \end{bmatrix} + \begin{bmatrix} \omega_2\omega_3 \\ 0 \\ -\omega_1^2 \end{bmatrix}$$

With a diagonalizing linear transformation

$$\begin{bmatrix} \omega_1 \\ \omega_2 \\ \omega_3 \end{bmatrix} = \begin{bmatrix} 1 & 0 & 0 \\ 1 & 1 & 0 \\ 0 & 0 & 1 \end{bmatrix} \begin{bmatrix} x_1 \\ x_2 \\ x_3 \end{bmatrix}$$

one obtains

$$\begin{bmatrix} \dot{x}_1 \\ \dot{x}_2 \\ \dot{x}_3 \end{bmatrix} = \begin{bmatrix} 0 & 0 & 0 \\ 0 & -1 & 0 \\ 0 & 0 & -1 \end{bmatrix} \begin{bmatrix} x_1 \\ x_2 \\ x_3 \end{bmatrix} + \begin{bmatrix} (x_1+x_2)x_3 \\ -(x_1+x_2)x_3 \\ -x_1^2 \end{bmatrix}$$

which is in diagonal form. Therefore there exists a center manifold described by

$$x_2 = h_2(x_1) = h_{21}x_1^2 + h_{22}x_1^3 + \ldots$$

$$x_3 = h_3(x_1) = h_{31}x_1^2 + h_{32}x_1^3 + \ldots$$

From the theory of section 2 it follows that

$$h_{21} = 0 \ , \ h_{31} = -1$$

The reduced equation is

$$\dot{x}_1 = (x_1 + 0.x_1^2 + \ldots)(-x_1^2 + \ldots)$$

or $\quad \dot{x}_1 = -x_1^3$ + higher order terms

which is locally asymptotically stable at the origin.

A remark on robustness is in order. In fact the asymptotically stabilizing control that we have proposed is (with respect to the original system description)

$$u = -I_{31}\omega_3\omega_1 - \omega_2 + \alpha\omega_1^2$$
$$v = -I_{12}\omega_1\omega_2 - \omega_3 + \beta\omega_1^3$$

It is clear that small changes in the coefficients of $\omega_2, \omega_3, \omega_1^2$ or $\omega_1^3$ will not affect the stabilizing potential of the control. The coefficients of $\omega_3\omega_1$ and $\omega_1\omega_2$ cancel. We will now show that even if this canceling is not exact, then we still end up with an asymptotically stable system.

For the control

$$u = -(I_{31} - \varepsilon)\omega_3\omega_1 - \omega_2 + \alpha\omega_1^2$$
$$v = -(I_{12} - \delta)\omega_1\omega_2 - \omega_3 + \beta\omega_1^2$$

with $\varepsilon, \delta$ small, we obtain

$$\begin{bmatrix} \dot\omega_1 \\ \dot\omega_2 \\ \dot\omega_3 \end{bmatrix} = \begin{bmatrix} 0 & 0 & 0 \\ 0 & -1 & 0 \\ 0 & 0 & -1 \end{bmatrix} \begin{bmatrix} \omega_1 \\ \omega_2 \\ \omega_3 \end{bmatrix} + \begin{bmatrix} I_{23}\omega_2\omega_3 \\ \varepsilon\omega_3\omega_1 + \alpha\omega_1^2 \\ \delta\omega_1\omega_2 + \beta\omega_1^3 \end{bmatrix}$$

The stability properties of this system are given by the first order system

$$\dot\omega_1 = I_{23}h_2(\omega_1)h_3(\omega_1).$$

From the invariance of the flow on the center manifold we obtain that

$$h_{21} = \alpha \quad , \quad h_{31} = 0$$
$$h_{32} = \delta\alpha + \beta$$

Therefore the reduced system is of the form

$$\dot\omega_1 = I_{23}(\alpha\omega_1^2 + \ldots)((\delta\alpha + \beta)\omega_1^3 + \ldots)$$

or

$$\dot\omega_1 = I_{23}\alpha(\delta\alpha + \beta)\omega_1^5 + \text{higher order terms.}$$

Therefore, with $I_{23}\alpha\beta < 0$ and $\delta$ small the reduced system and therefore the original system remains asymptotically stable.

## 4. ASYMPTOTIC STABILIZABILITY OF THE ANGULAR VELOCITY EQUATIONS, WITH JUST ONE CONTROL AVAILABLE

The problem whether asymptotic stabilizability can be achieved with only one control comes up naturally after the treatment in section 3. We will show that there exists a scalar linear asymptotically stabilizing feedback to Euler's angular velocity equations. However the control is not robust with respect to parameter changes.

Consider the angular velocity equations with scalar control aligned with an axis having components along <u>all three</u> principal axes

$$I_1\dot{\omega}_1 = (I_2-I_3)\omega_2\omega_3 + a_1u$$
$$I_2\dot{\omega}_2 = (I_3-I_1)\omega_3\omega_1 + a_2u$$
$$I_3\dot{\omega}_3 = (I_1-I_2)\omega_1\omega_2 + a_3u$$

with $a_i \neq 0$, $i = 1,2,3$.

This is rewritten as

$$\dot{\omega}_1 = I_{23}\omega_2\omega_3 + b_1u$$
$$\dot{\omega}_2 = I_{31}\omega_3\omega_1 + b_2u$$
$$\dot{\omega}_3 = I_{12}\omega_1\omega_2 + b_3u$$

where $b_i = a_i/I_i$, $i = 1, 2,3$.

Consider the energy function

$$V(\omega_1,\omega_2,\omega_3) = \frac{1}{2}(I_1\omega_1^2 + I_2\omega_2^2 + I_3\omega_3^2)$$

Then the derivative $\dot{V}$ of V along solutions of the differential equations is

$$\dot{V} = I_1\omega_1\dot{\omega}_1 + I_1\omega_2\dot{\omega}_2 + I_3\omega_3\dot{\omega}_3$$
$$= (I_1b_1\omega_1 + I_2b_2\omega_2 + I_3b_3\omega_3)u$$

When we take

$$u^* = -(I_1b_1\omega_1 + I_2b_2\omega_2 + I_3b_3\omega_3)$$

then

$$\dot{V} \leqslant 0.$$

With this control, the origin is stable. By Lasalle's theorem u* would make the origin asymptotically stable if the larges invariant set M in $\dot{V} = 0$ is the origin. We will now show that M is the origin.

Let $\omega_1(t)$, $\omega_2(t)$, $\omega_3(t)$ be a trajectory in $\dot{V} = 0$, i.e. $\dot{V}(\omega(t)) \equiv 0, \forall t$. Then it is also true that $V(\omega(t)) = L$ for some $L \geqslant 0$, and that $\ddot{V}(\omega(t)) = 0$.

Therefore the trajectory $\omega_1(t)$, $\omega_2(t)$, $\omega_3(t)$ belongs to the set $\{\omega_1, \omega_2, \omega_3 : V(\omega) = L,\ \dot{V}(\omega) = 0,\ \ddot{V}(\omega) = 0\}$, i.e.

(α) $I_1\omega_1^2 + I_2\omega_2^2 + I_3\omega_3^2 = L$

(β) $I_1b_1\omega_1 + I_2b_2\omega_2 + I_3b_3\omega_3 = 0$

(γ) $b_1(I_2-I_3)\omega_2\omega_3 + b_2(I_3-I_1)\omega_3\omega_1 + b_3(I_1-I_2)\omega_1\omega_2 = 0$

Notice that $\alpha$ is an ellipsoid, that $\beta$ is a plane and that $\gamma$ is a cone with top at the origin.
The intersection $\beta \cap \gamma$ is represented by a finite number of lines through the origin. And thus $\alpha \cap \beta \cap \gamma$ is represented by a finite number of isolated points different from the origin (if $L > 0$). This implies that $(\omega_1(t), \omega_2(t), \omega_3(t))$ is an equilibrium point different from the origin. However it is immediate that with the control proposed and with all $b_i \neq 0$, the controlled angular velocity equations have no equilibrium points other than the origin. Therefore L must be zero, which implies that M is identical to the origin. Thus u* globally asymptotically stabilizes the angular velocity equations.

Notice that the choice of the coefficients defining the control law u* is critical. Small changes in these coefficients results in a $\dot{V}$ which is indefinite. The control is not robustly asymptotically stabilizing.

Again consider the system with scalar control

$$\dot{\omega}_1 = I_{23}\omega_2\omega_3 + b_1 u$$
$$\dot{\omega}_2 = I_{31}\omega_3\omega_1 + b_2 u$$
$$\dot{\omega}_3 = I_{12}\omega_1\omega_2 + b_3 u$$

but assume that say $b_3 \neq 0$ and $b_1 = b_2 = 0$.
Then the control axis is aligned with a principal axis. We claim that the system is not asymptotically stabilizable. Indeed, consider $V = I_{31}\omega_1^2 - I_{23}\omega_2^2$. This is a first integral of the system, independent of the control. Therefore the level surfaces $V_c = \{\omega : V(\omega) = c\}$ (i.e. elliptic or hyperbolic cylinders, depending on the sign of $I_{31}I_{23}$) are invariant manifolds, precluding asymptotic stabilizability.

The question of asymptotic stabilizability with one torque perpendicular to a principal axis and not along a principal axis (e.g. $b_1 \neq 0$, $b_2 \neq 0$, $b_3 = 0$) remains open.

## 5. STABILIZABILITY OF THE ANGULAR VELOCITY EQUATIONS OF THE RIGID BODY

Again consider the system

$$\dot{\omega}_1 = I_{23}\omega_2\omega_3 + b_1 u$$
$$\dot{\omega}_2 = I_{31}\omega_3\omega_1 + b_2 u$$
$$\dot{\omega}_3 = I_{12}\omega_1\omega_2 + b_3 u$$

and let the control torque be aligned with the principal axis defined by $b_1 = b_2 = 0$, $b_3 \neq 0$. Since for this case, asymptotic stabilizability is impossible we relax our goal to an implementation of a stabilizing feedback.

First notice that there is an obvious stabilizing control: $u = 0$. Then the system is stable, since $V(\omega) = \frac{1}{2}\,\omega^T I\omega$ is a positive definite function with $\frac{d}{dt} V(\omega(t)) \equiv 0$. However a small deviation from the zero input makes the system unstable. Indeed, let $u = \varepsilon\omega_3$ (with sign $\varepsilon$ = sign $b_3$); then $\frac{d}{dt} V(\omega(t)) = I_3\omega_3 b_3.\varepsilon\omega_3 = \varepsilon b_3 I_3 \omega_3^2 \geqslant 0$.
We must therefore look for robust stabilizing control feedbacks: i.e. such that the addition of small feedback control terms does not annihilate its stabilizing potential.

By allowing a polynomial type of feedback the angular velocity equations are of the form

$$\dot\omega_1 = I_{23}\omega_2\omega_3$$

$$\dot\omega_2 = I_{31}\omega_3\omega_1$$

$$\dot\omega_3 = k\omega_1 + \ell\omega_2 + m\omega_3 + F(\omega_1, \omega_2, \omega_3)$$

where $F(\omega_1,\omega_2,\omega_3)$ contains the terms of order two or higher. Notice that for this system, the level surfaces (cylinders) defined by $H(\omega_1,\omega_2,\omega_3) \equiv I_{31}\omega_1^2 - I_{23}\omega_2^2 = c$ are invariant with respect to the flow, and independent of the choice of the input. By the center manifold approach the problem will be reduced to a two-dimensional problem which allows for a nice discussion of the stabilizing properties of the proposed feedbacks. Assume $m = -1$.
By change of basis

$$\begin{bmatrix}\omega_1\\ \omega_2\\ \omega_3\end{bmatrix} = \begin{bmatrix}1 & 0 & 0\\ 0 & 1 & 0\\ k & \ell & 1\end{bmatrix}\begin{bmatrix}x_1\\ x_2\\ y\end{bmatrix}$$

one obtains the equations

$$\begin{bmatrix}\dot x_1\\ \dot x_2\\ \dot y\end{bmatrix} = \begin{bmatrix}0 & 0 & 0\\ 0 & 0 & 0\\ 0 & 0 & -1\end{bmatrix}\begin{bmatrix}x_1\\ x_2\\ y\end{bmatrix} + \begin{bmatrix}I_{23}x_2(y+kx_1+\ell x_2)\\ I_{31}x_1(y+kx_1+\ell x_2)\\ G(x_1,x_2,y)\end{bmatrix}$$

with G in one-to-one correspondence with F and of the same order. The system fits into the center manifold approach. There exists a center manifold $W^c$, invariant under the flow and locally represented by

$$W^c = \{((x_1, x_2), y) : y = h(x_1, x_2);\ h(0) = 0,\ Dh(0) = 0\}$$

with $h: U \to \mathbb{R}$ defined on some neighborhood $U \subset \mathbb{R}^2$ of the origin. The stability properties of the system are the same as the stability properties of the reduced two-dimensional system

$$\dot{x}_1 = I_{23} x_2 (h(x_1, x_2) + k x_1 + \ell x_2)$$

$$\dot{x}_2 = I_{31} x_1 (h(x_1, x_2) + k x_1 + \ell x_2).$$

For this system the level curves defined by

$$H(x_1, x_2) \equiv I_{31} x_1^2 - I_{23} x_2^2 = c$$

are invariant under the flow.

If $I_{31}/I_{23} < 0$, then the planar system and therefore the original system is stable, since the level curves are <u>ellipses</u>. In fact we have shown that each control

$$u = -\omega_3 + k\omega_1 + \ell\omega_2 + F(\omega_1, \omega_2, \omega_3)$$

stabilizes the system, (and obviously does so in a robust manner) if $I_3$ is larger or smaller than $I_1$ and $I_2$.

Consider now the control $u = -b_3\omega_3$ (i.e. $k = \ell = 0$ and $F \equiv 0$), then the system is of the form

$$\dot{\omega}_1 = I_{23} \omega_2 \omega_3$$

$$\dot{\omega}_2 = I_{31} \omega_3 \omega_1$$

$$\dot{\omega}_3 = I_{12} \omega_1 \omega_2 - b_3^2 \omega_3$$

Consider $V(\omega) = \frac{1}{2}\omega^T I \omega$, $\omega \in \mathbb{R}^3$; then $\dot{V}(\omega(t)) = -I_3 b_3^2 \omega_3^2$. By LaSalle's theorem, each trajectory converges towards the largest invariant subset $M$ of the plane $\omega_3 = 0$, at $t \to \infty$. A simple argument shows that

$$M = \{\omega : \omega = (\omega_1, 0, 0) \text{ or } \omega = (0, \omega_2, 0);\ \omega_1, \omega_2 \in \mathbb{R}\}$$

with each point in $M$ an equilibrium point. Notice now that the sets defined by $\{\omega : I_{23}\omega_2^2 - I_{31}\omega_1^2 = \text{constant}\}$, are invariant cylinders. When $I_{23}/I_{31} < 0$, these cylinders are elliptic; when $I_{23}/I_{31} > 0$, these cylinders are hyperbolic. In <u>both cases</u> the control $u = -b_3\omega_3$ is stabilizing. In fact for the elliptic case the control is robust as shown above by means of center manifold theory. Also in the hyperbolic case we have found a stabilizing control $u = -b_3\omega_3$ : the trajectories remain on a patch of the hyperbolic cylinder and converge towards the <u>single equilibrium on this patch</u>. However this control is <u>not robust</u>. This again can in fact best be seen from center manifold theory which says that the stability properties of the original system are those of a reduced two-dimensional system. From above we know that this reduced system has level curves of the form $I_{23}\omega_2^2 - I_{31}\omega_1^2 = \text{constant}$. These curves are hyperbolas. Therefore a point starting on a branch of a hyperbola stays on it and travels off to infinity as $t \to \infty$ unless it is attracted by an equilibrium point on the branch. This is exactly what happens in case $u = -b_3\omega_3$. But if this control is slightly changed (e.g.by adding $\varepsilon\omega_1 + \delta\omega_2$) then these equilibrium points disappear The system is no longer stable and therefore $u = -b_2\omega_3$ is no longer robustly stabilizing.

## 6. REFERENCES


1. D. Aeyels, Stabilization of a class of nonlinear systems by a smooth feedback control, *Systems Control Lett.* 5 (1985), 289-294
2. J. Carr, *Applications of Centre Manifold Theory* (Springer, New York, 1981)
3. R.W. Brockett, Asymptotic stability and feedback stabilization, in : R.W. Brockett, R.S. Millmann and H.J.Sussmann, Eds., *Differential Geometric Control Theory*, Progress in Mathematics, Vol.27 (Birkhaüser, Boston, 1983), 181-191
4. D. Aeyels and M. Szafranski, Comments on the stabilizability of the angular velocity of a rigid body, *Systems Control Lett.*, 10(1988) 35-39

# ASYMPTOTIC PROPERTIES OF NONLINEAR MINIMUM PHASE SYSTEMS


C. Byrnes*, A.Isidori**

*Department of Mathematics and
Department of Electrical and Computer Engineering
Arizona State University
Tempe, Arizona 85224

**Dipartimento di Informatica e Sistemistica
Università di Roma "La Sapienza"
18 Via Eudossiana , 00184 Roma


## 1. Introduction.

In this paper, we consider nonlinear control system described by equations of the form:

$$\dot{x} = f(x) + g(x)u \qquad u \in \mathbf{R}^m \tag{1.1a}$$
$$y = h(x) \qquad y \in \mathbf{R}^m \tag{1.1b}$$

with state x defined on an open set U of $\mathbf{R}^n$. In these equations, f and the m columns of g are smooth vector fields of U, and h is a smooth mapping.

Our purpose is to illustrate and discuss the properties of a broad class of system, which can be considered as the extension, in the nonlinear setting, of the class of linear system having all transmission zeroes in the (open) left-half plane. In particular, we show that for any system in this class certain control problems of paramount importance - like, e.g., asymptotic stabilization by smooth state feedback, or asymptotic tracking of a fixed reference output trajectory - can be solved.

In Section 2, we review the notion of "zero dynamics", namely the internal dynamics of the system that are consistent with the constraint that the output is zero for all times, and we show how - under mild regularity assumptions - these dynamics can be identified by means of a suitable alogorithm. Then, we call "minimum phase" systems those nonlinear systems whose "zero dynamics" are asymptotically stable at a given equilibrium point. In Section 3, we show that from the "zero dynamics algorithm" it is possible to deduce, as a by product, a special set of new local coordinates - for the state space - in which the equations describing the system assume a "normal form" of special interest. This normal form, whose existence is not tied to the (rather common) assumption that the system can be rendered noninteractive by means of static state feedback, includes a large number of particular cases already present in the literature. Once the system has been put in this form, it is particularly easy to understand - using e.g. Center Manifold Theory - how an asymptotically stabilizing state feedback can be designed, as shown in Section 4, or how problems of asymptotic tracking can be solved, as done in Section 5. An appropriate application of Center Manifold Theory is instrumental in our analysis, because the only restriction we impose on the zero dynamics is that of asymptotic stability, thus allowing critical cases, as well as cases in which the state response is explicitly dependent on higher order derivatives of the output.

## 2. The $Z^*$ algorithm.

Consider the following problem. Given the nonlinear system (1.1) find, if possible, a smooth submanifold $M^*$ of the state space satisfying:

(i) for each $x \in M^*$, $h(x) = 0$;

(ii) for each $x^\circ$ in $M^*$, there exists a smooth input function $u_{x^\circ} : \mathbf{R} \to \mathbf{R}^m$ such that the integral curve of $\dot{x}(t)= f(x(t))+g(x(t))u_{x^\circ}(t)$, satisfying $x(0) = x^\circ$, is such that $x(t) \in M^*$ for all $t \in \mathbf{R}$;

(iii) $M^*$ is maximal, with respect to (i) and (ii).

This problem is defined as the *problem of zeroing the output*. In fact, once such a problem has been solved, the set $\{(x^\circ, u_{x^\circ}): x^\circ \in M^*\}$ characterizes the set of all possible pairs (initial state, input function) yielding an output function which is zero for all times.

In general, it is not clear whether or not the manifold $M^*$ exists. However, under some mild regularity assumptions, in a neighorhood U of a reference point $x^\circ$ a manifold satifying requirements (i)-(iii) can be found rather easily. To this end consider the following recursive construction.

**$Z^*$ algorithm** ( the Zero Dynamics Algorithm) [4].
Step 0: set $M_0 = h^{-1}(0)$.
Step k: suppose $M_{k-1}$ is a smooth manifold and define $M_k$ as

$$M_k = \{ x \in M_{k-1} : f(x) \in \text{span}\{g_1(x),...,g_m(x)\} + T_xM_{k-1} \} \blacksquare$$

**Proposition 2.1**. Suppose $M_{k^*+1} = M_{k^*}$ for some $k^*$, and $\text{span}\{g_1(x),...,g_m(x)\} \cap T_xM_{k^*} = 0$ for all $x \in M_{k^*}$. Then $M_{k^*} = M^*$.

*Proof.* A straightforward consequence of the assumptions is the existence a unique smooth mapping $u^*: M_{k^*} \to \mathbf{R}^m$, such that:

$$f(x) + g(x)u^*(x) \in T_xM_{k^*}$$

for all $x \in M_{k^*}$. If $x(t)$ is an integral curve of $\dot{x} = f(x)+g(x)u^*(x)$, satisfying $x(0) = x^\circ$, the input function:

$$u_{x^\circ}(t) = u^*(x(t))$$

satsfies (ii). $M_{k^*}$ also satisfies (i), by construction. Observe now that if $M^*$ exists, then necessarily $M^* \subset M_k$ for all $k \geq 0$. This is proved, by induction, showing that $M^* \subset M_{k-1}$ implies $M^* \subset M_k$. In fact:

$$\begin{aligned} x \in M^* \quad & \Rightarrow \; f(x) \in \text{span}\,\{g_1(x),\dots,g_m(x)\} \; + \; T_xM^* \\ & \Rightarrow \; f(x) \in \text{span}\,\{g_1(x),\dots,g_m(x)\} \; + \; T_xM_{k-1} \\ & \Rightarrow \; x \in M_k \end{aligned}$$

From this we deduce that $M_{k^*}$ satisfies also (iii), i.e. that $M_{k^*} = M^*$.∎

In local coordinates, the algorithm thus described essentially coincides with the so-called Silverman's "Structure Algorithm"output, as shown in [14]. The submanifold $M^*$ is called the *zero dynamics submanifold*, and the vector field of $M^*$:

$$f^*(x) = f(x) + g(x)u^*(x) \tag{2.1}$$

is called the zero dynamics vector field or, simply, the *zero dynamics*.

In the following definition we characterize the asymptotic properties of the zero dynamics at an equilibrium point $x°$. In analogy with an appealing terminology, we refer to systems with a "stable" zero dynamics as to "minimum-phase systems".

**Definition**. A system is *minimum-phase* at $x°$ if the latter is an asymptotically stable equilibrium of the zero dynamics. A system is *hyperbolically* minimum-phase (respectively, *critically* minimum-phase) if it is minimum-phase and the eigenvalues of the jacobian matrix of (2.1) at $x°$ are in the open left-half complex plane (respectively, closed left-half complex plane).∎

### 3. Generalized Normal Forms.

A by-product the zero dynamics algorithm is the possibility of defining a local coordinates transformation in the state space, which induces an structure of special interest (Normal Form) on the equations describing the system. We present in this section a generalized structure which incorporates a number of special cases already considered in the literature. As a matter of fact, after such a change of coordinates and possibly a reordering of the outputs, the equations describing a system with m inputs and m outputs exhibit a form of the type:

$$\begin{aligned} \dot{z} &= f_0(z, \xi^1, \dots, \xi^m) + g_0(z, \xi^1, \dots, \xi^m)u \\ \dot{\xi}^1_1 &= \xi^1_2 \\ &\cdots \\ \dot{\xi}^1_{n_1-1} &= \xi^1_{n_1} \\ \dot{\xi}^1_{n_1} &= b^1(x) + a^1(x)u \end{aligned}$$

$$\dot{\xi}_1^2 = \xi_2^2 + \delta_{11}^2(x)(b^1(x) + a^1(x)u) + \sigma_1^2(x)u$$

...

$$\dot{\xi}_{n_2-1}^2 = \xi_{n_2}^2 + \delta_{n_2-1,1}^2(x)(b^1(x) + a^1(x)u) + \sigma_{n_2-1}^2(x)u$$

$$\dot{\xi}_{n_2}^2 = b^2(x) + a^2(x)u$$

...

$$\dot{\xi}_1^i = \xi_2^i + \sum_{j=1}^{i-1} \delta_{1j}^i(x)(b^j(x) + a^j(x)u) + \sigma_1^i(x)u$$

...

$$\dot{\xi}_{n_i-1}^i = \xi_{n_i}^i + \sum_{j=1}^{i-1} \delta_{n_i-1,j}^i(x)(b^j(x) + a^j(x)u) + \sigma_{n_i-1}^i(x)u$$

$$\dot{\xi}_{n_i}^i = b^i(x) + a^i(x)u$$

and:

$$y_i = \xi_1^i$$

for $i = 1, \dots, m$. In these equations:

$$x = (z, \xi^1, \dots, \xi^m)$$

$$\xi^i = (\xi_1^i, \xi_2^i, \dots, \xi_{n_i}^i)$$

$$n_1 \le n_2 \le \dots \le n_m$$

and:

$$a^i(x) = L_g\xi_{n_i}^i(x)$$

$$b^i(x) = L_f\xi_{n_i}^i(x)$$

The coordinate functions $\xi^i_k(x)$, the coefficients $\delta^i_{kj}(x)$ and $\sigma^i_k(x)$ are such that:

$$\xi_{k+1}^i(x) = -\sum_{j=1}^{i-1} \delta_{kj}^i(x)b^j(x) + L_f\xi_k^i(x) \qquad 1 \le k \le n_i-1, \quad 2 \le i \le m$$

$$L_g\xi_k^i(x) = \sum_{j=1}^{i-1} \delta_{kj}^i(x)a^j(x) + \sigma_k^i(x) \qquad 1 \le k \le n_i-1, \quad 2 \le i \le m$$

In the new coordinates, the submanifold M* is described as:

$$M^* = \{x : \xi^i = 0,\ 1 \le i \le m\}$$

and the functions $\sigma^i_k(x)$ vanish on $M^*$. The input $u^*(x)$ which imposes the zero dynamics is the unique solution of:

$$b^i(x) + a^i(x)u^*(x) = 0 \qquad 1 \le i \le m$$

Thus, the zero dynamics of the system is given by:

$$\dot{z} = f^*(z) = f_0(z, 0, \dots, 0) + g_0(z, 0, \dots, 0)u^*(z, 0, \dots, 0)$$

## 4. Asymptotic stabilization via state feedback

A system which is minimum phase at some equilibrium point, can be locally stabilized - at this point - by smooth state feedback. For, consider again the generalized normal form introduced in the previous section and set $u = \alpha(x)$, a state feedback, with $\alpha(x)$ satisfying:

$$\begin{aligned} b^i(x) + a^i(x)\alpha(x) &= - c^i_0\xi^i_1(x) - c^i_1\xi^i_2(x) - \dots - c^i_{n_i-1}\xi^i_{n_i}(x) \\ &= c^i\xi^i(x) \qquad 1 \le i \le m \end{aligned} \tag{4.1}$$

where:

$$c^i = [\, -c^i_0 \quad -c^i_1 \quad \dots \quad -c^i_{n_i-1} \,]$$

Note that these equations define a unique and smooth $\alpha(x)$, and this $\alpha(x)$ is related to the input $u^*(x)$ by the following:

$$\alpha(x) = u^*(x) + M^{-1}(x)\begin{bmatrix} c^1\xi^1(x) \\ \dots \\ c^m\xi^m(x) \end{bmatrix}$$

Moreover $\alpha(0) = 0$.

The effect of this feedback is to change the system into one of this type:

$$\dot{z} = f(z,\xi)$$

$$\dot{\xi} = A\xi + p(z,\xi)$$

with:

$$f(z,0) = f^*(z)$$

i.e. the zero dynamics vector field, and:

$$p(z,0) = 0 \qquad \frac{\partial p}{\partial \xi}(0,0) = 0$$

Note also that the eigenvalues of the matrix A are assignable, by proper choice of the coefficients $c^i_j$, $1 \le j \le n_i$, $1 \le i \le m$.

At this point, in order to prove that the system in question is asymptotically stable (provided that the system is minimum phase and the eigenvalues of the matrix A are in the left-half complex plane), we only need the following result.

**Lemma**. Consider a system:

$$\dot{z} = f(z,y) \tag{4.2a}$$
$$\dot{y} = Ay + p(z,y) \tag{4.2b}$$

and suppose that $p(z,0) = 0$ for all z near 0 and:

$$\frac{\partial p}{\partial y}(0,0) = 0$$

If $f(z,0)$ is asymptotically stable at $z=0$ and the eigenvalues of A are in the left-half complex plane, then the system (4.2) is asymptotically stable.∎

*Proof*. Use Center Manifold Theorem∎

**Remark**. We stress that the result of this Lemma does not require the dynamics of:

$$\dot{z} = f(z,0)$$

to be hyperbolically stable, but just asymptotically stable.∎

Merging the results of the previous discussion with that of this Lemma, we arrive at the main result of this section.

**Theorem**. Suppose the nonlinear system (1.1) is minimum phase at a given equilibrium point. Then, the smooth state feedback (4.1) asymptotically stabilizes (1.1) at this point, provided that the roots of the polynomials:

$$d^i(\lambda) = c^i_0 + c^i_1\lambda + \ldots + c^i_{n_i-1}\lambda^{n_i-1} + \lambda^{n_i}$$

for $1 \le i \le m$, are in the (open) left-half complex plane.∎

In order to explain the usefulness of this result, a few remarks are in order.

**Remark.** We do not need the zero dynamics of (1.1) to be hyperbolically minimum phase, because this is not required in Lemma 4.1. As a result, we see that the control (4.1) stabilizes any (*critically* as well as *hyperbolically*) minimum phase system. It might also be important to observe that the problem of locally stabilizing an hyperbolically minimum phase nonlinear system (see also [7]) is not quite a problem in a nonlinear setting, because all the data needed to solve that problem refer to the linear approximation of the system. For, observe that the composition of linear approximation of system (1.1) with the linear approximation of the feedback (4.1) yields a system having the triagular structure:

$$\dot{z} = Qz + Py$$
$$\dot{y} = Ay$$

If the nonlinear system was hyperbolically stable, the matrix Q (that characterizes the linear approximation of the zero dynamics at the equilibrium point) had all the eigenvalues in the (open) left-half-plane and the asymptotic stability of the closed loop is a trivial consequence of the well-known "principle of stability in the first approximation".

**Remark.** It is also quite known that, in the problem of *locally* asymptotically stabilize a nonlinear system, the only *nontrivial* case is one in which the linear approximation of the system in question has uncontrollable modes associated with eigenvalues on the imaginary axis (see e.g. [1]).An elementary calculation (see e.g. [4]) shows that *if a system has a linear approximation with uncontrollable modes and a zero dynamics is defined, then the eigenvalues associated with the uncontrollable modes arenecessarily part of the eigenvalues of the linear approximation of the zero dynamics.* Thus, if a system has a linear approximation with uncontrollable modes associated with eigenvalues having nonpositive real part and an asymptotically stable zero dynamics can be defined, then this zero dynamics is necessarily *critically minimum phase.* Our Theorem shows that any system whose linear approximation has uncontrollable modes associated with eigenvalues on the imaginary axis and in the left-half complex plane - i.e. any system in which the stabilization problem is nontrivial - can be locally stabilized by smooth state-feedback, *provided* that for some choice of an "output" map h(x) the system itself is minimum phase.

**Remark.** We stress that the result of the previous Theorem does not require any special form for dependence of the right-hand side of the equation:

$$\dot{z} = f_0(z, \xi^1, \dots, \xi^m) + g_0(z, \xi^1, \dots, \xi^m)u$$

on $(\xi^1, \dots, \xi^m)$, as postulated e.g. in [16] and [20], where only dependence on the first component $\xi^1_i$ of the $\xi^i$'s was allowed. The reason of such a restriction is to be found in the concern about "peaking" fenomena due to the presence of eigenvalues of A - in eq. (4.2a) - with "very large" (negative) real part. However, our construction shows that no restriction whatsoever is to be imposed on the eigenvalues of A, provided they are in the (open) left-half complex plane.

## 5. Asymptotic output tracking

As we have seen in section 4, one of the main features of the concept of zero dynamics is the possibility of asymptotically stabilizing minimum phase systems by means of a smooth state feedback. In this section, we briefly indicate a number of other relevant control problems in which the concept of zero dynamics plays an important role.

*Output reproduction.* Consider a control system of the form (1.1) and suppose a smooth reference output function:

$$y_R = \mathbf{R} \to \mathbf{R}^m$$

is given. This output function is said to be *reproducible*, for (1.1), if there exists an initial state $x^\circ$, an initial time $t^\circ$, and an input function:

$$u_R = \mathbf{R} \to \mathbf{R}^m$$

such that the output of (1.1), initialized in $x^\circ$ at some time $t = t^\circ$ and subject to the input $u_R$, is exactly $y_R$.

The question of deciding whether or not a fixed output $y_R$ is reproducible, can be easily answered on the basis of the zero dynamics alogorithm. For, consider the "error" system:

$$\dot{x} = f(x) + g(x)u \tag{5.1a}$$

$$\dot{x}_{n+1} = 1 \tag{5.1b}$$

with "output":

$$e = h(x) - y_R(x_{n+1}) \tag{5.1c}$$

The output $y_R$ is reproducible for (1.1) if and only if the problem of zeroing the output is solvable for the "error" system (6.1). Infact, if there exists a pair $z^\circ = (x^\circ, x_{n+1}{}^\circ)$ and an input $u^\circ\colon t \to u^\circ(t)$ yielding $e(t) = 0$ for all $t \in \mathbf{R}$, then the output of (1.1), initialized in $x^\circ$ at time $t^\circ = x_{n+1}{}^\circ$ and subject to the input $u_R\colon t \to u^\circ(t - x_{n+1}{}^\circ)$, is exactly $y_R$. Therefore, a convenient way to check whether or not $y_R$ is reproducible is to perform the zero dynamics algorithm on the triplet:

$$\bar{f}(z) = \begin{bmatrix} f(x) \\ 1 \end{bmatrix}, \quad \bar{g}(z) = \begin{bmatrix} g(x) \\ 0 \end{bmatrix}, \quad \bar{h}(z) = h(x) - y_R(x_{n+1})$$

*Asymptotic tracking.* Suppose the zero dynamics algorithm, performed on $\{\bar{f}, \bar{g}, \bar{h}\}$ has provided a positive answer to the problem of reproducing $y_R$. More precisely, suppose the algorithm in question has a point $z^\circ$ of regularity. Then, a (local) zero dynamics manifold, noted $\bar{M}^*$, can be found, with a smooth feedback $\bar{u}^*(z)$ defined on $\bar{M}^*$, rendering the vector field:

$$\bar{f}^*(z) = \bar{f}(z) + \bar{g}(z)\bar{u}^*(z)$$

tangent to $\bar{M}^*$. Let Z be a nonempty subset of $\bar{M}^*$, invariant under the flow of $\bar{f}^*$. Then, for any pair $(x^\circ, t^\circ) \in Z$, there is an input $u^\circ$ such that the integral curve of

$$\dot{x} = f(x) + g(x)u^\circ(t) \qquad\qquad x(t^\circ) = x^\circ$$

yields $h(x(t)) = y_R(t)$ for all $t \in \mathbf{R}$.

This input, if the initial condition is properly chosen, provides exact reproduction of the output $y_R$. If the initial condition cannot be arbitrarily preset, it is still possible to achieve at least *asymptotic tracking* of $y_R$, provided that on the "error" system $\{\overline{f}, \overline{g}, \overline{h}\}$ a stabilizing feedback similar to the one developed in Section 4 is imposed. We illustrate how this result can be pursued on the simple example of a single-input single-output system.

This system has normal form of the type:

$$\dot{z} = f_0(z, \xi)$$
$$\dot{\xi}_1 = \xi_2$$
$$\cdots$$
$$\dot{\xi}_{r-1} = \xi_r$$
$$\dot{\xi}_r = b(z, \xi) + a(z, \xi)u$$
$$y = \xi_1$$

The set $\overline{M}^*$ is the set of pairs $(x^\circ,t^\circ)$ such that

$$\xi_i(x^\circ) = y_R^{(i-1)}(t^\circ) \qquad 1 \leq i \leq r$$

and $\overline{u}^*(x,t)$ is given by:

$$\overline{u}^*((z,\xi),t) = \frac{1}{a(z,\xi)}\,(-\,b(z,\xi) + y_R^{(i-1)}(t))$$

Define now:

$$e(t) = h(x(t)) - y_R(t)$$

and set:

$$\phi_i\,(t) = e^{(i-1)}(t) = \xi_i(t) - y_R^{(i-1)}(t) \qquad\qquad 1 \leq i \leq r$$

Then, choosing:

$$\overline{u}^*((z,\xi),t) = \frac{1}{a(z,\xi)}\,(-\,b(z,\xi) + y_R^{(r)}(t) - \sum_{i=0}^{r-1} c_i\phi_i)$$

one obtains, for the "error" system :

$$\dot{z} = f_0(z, \phi_1 + y_R(t), \phi_2 + y_R^{(1)}(t), \ldots, \phi_r + y_R^{(r-1)}(t)\,)$$
$$\dot{t} = 1$$

$$\dot{\phi}_1 = \phi_2$$
$$\ldots$$
$$\dot{\phi}_{r-1} = \phi_r$$
$$\dot{\phi}_r = - c_0\phi_1 - \ldots - c_{r-1}\phi_r$$
$$e = \phi_1$$

Thus, it is clear that, by a proper choice of $c_0, \ldots, c_{r-1}$, one can have the error e(t) converging to 0 as $t \to \infty$. In the genaral case the construction is analogous. One obtains for the "error" system equations of the form:

$$\dot{z} = f(z, t, y)$$
$$\dot{t} = 1$$
$$\dot{y} = Ay + p(z, t, y)$$
$$e = Py$$

where P is a matrix selecting certain components of y. The convergence of y to 0, and thus that of the error e, is guaranteed by an extension of Lemma given in Section 4.

## References

[1] D.Aeyels: Stabilization of a class of nonlinear systems by a smooth feedback control, *Systems and Control Letters* , 5 (1985), pp. 289-294.

[2] R.W.Brockett: Asymptotic stability and feedback stabilizability, Differential Geometric Control Theory (R.W.Brockett, R.S.Millman, H.J.Sussmann, eds.) Birkhauser (1983), pp. 181-191.

[3] C.Byrnes, A.Isidori: A frequency domain philosophy for nonlinear systems, with applications to stabilization and adaptive control, *23rd IEEE Conf. Decision and Control* , (1984), pp. 1569-1573.

[4] C.I.Byrnes, A.Isidori: Local stabilization of minimum phase systems, *Systems and Control Letters* , 11 (1988), to appear.

[5] C.I.Byrnes, A.Isidori: Analysis and design of nonlinear feedback systems, I°: zero dynamics and global normal forms, submitted for publication.

[6] C.I.Byrnes, A.Isidori: Analysis and design of nonlinear feedback systems, II°: global stabilization of minimum phase systems, submitted for publication.

[7] B. D'Andrea, L. Praly: About finite nonlinear zeros for decouplable systems, *Systems and Control Letters* , 10 (1988), pp. 103-108.

[8] J.Descusse, C.H.Moog: Decoupling with dynamic compensation for strong invertible affine nonlinear systems, *Int. J. of Control*, 42 (1985), pp. 1387-1398.

[9] M.Fliess, A note on the invertibility of nonlinear input-output differential systems, *Systems and Control Letters*, 8 (1986), pp. 147-151.

[10] R.M.Hirschorn: Invertibility of multivariable nonlinear control systems, *IEEE Trans. Automatic Control*, AC-24 (1979), pp. 855-865.

[11] A.Isidori, *Nonlinear control systems: an introduction*, Springer Verlag, Lecture Notes in Control and Information Sciences, 72 (1985).

[12] A.Isidori, Control of nonlinear systems via dynamic state-feedback, *Algebraic and geometric methods in nonlinear control theory*, M.Fliess and M.Hazewinkel (eds.), D.Reidel (1986), pp. 121-145.

[13] A.Isidori, A.J.Krener, C.Gori-Giorgi, S.Monaco, Nonlinear decoupling via feedbac: a differential-geometric approach, *IEEE Trans. Automatic Control*, AC-26 (1981), pp. 331-345.

[14] A.Isidori, C.Moog: On the nonlinear equivalent o the notion of transmission zeros, *Modeling and Adaptive Control* (C.I.Byrnes and A.H.Kurszanski, eds.), Springer Verlag, Lecture Notes in Control and Information Sciences, 105 (1988), to appear.

[15] A.Isidori, C.Moog, A.DeLuca: A sufficient condition for full linearization via dynamic state-feedback, *25th IEEE Conf. Decision and Control*, (1986), pp. 203-208.

[16] H.Khalil, A.Saberi: Adaptive stabilization of a class of nonlinear systems using high-gain feedback, *IEEE Trans. Automatic Control*, AC-32 (1987), pp. 270-276.

[17] P.Kokotovic: Applications of singular perturbation techniques to control problems, SIAM Review, (1984), pp. 501-550

[18] A.J.Krener, $(Ad_f,g)$, $(ad_fg)$ and locally $(ad_fg)$ invariant and controllability distributions, *SIAM J. Control Optim.*, 30 (1985), pp. 566-573.

[19] A.J.Krener, A.Isidori: Nonlinear zero distributions, *19th IEEE Conf. Decision and Control*, (1980), pp.

[20] R.Marino: Feedback stabilization of single input nonlinear systems, *Systems and Control Letters*, 11 (1988), pp. 201-206.

[21] S.Monaco, D.Normand-Cyrot: Zero dynamics of sampled nonlinear systems, *Systems and Control Letters*, 11 (1988),

[22] L.M.Silverman: Inversion of multivariable linear systems, *IEEE Trans. Automatic Control*, AC-14 (1969), pp. 270-276.

[23] A.J. van der Schaft: On clamped dynamics of nonlinear systems, *Math. Theory of Network and Systems* (1987), to appear.

# FRACTION REPRESENTATIONS AND ROBUST STABILIZATION OF NONLINEAR SYSTEMS

Jacob Hammer

Center for Mathematical System Theory, Department of Electrical Engineering
University of Florida, Gainesville, FL 32611, USA

## ABSTRACT

The purpose of this note is to provide a survey of the theory of fraction representation and robust stabilization of nonlinear systems developed by the author over the last few years. The note contains an exposition of the main results obtained so far and some examples, but no proofs are included. The results are all explicit and implementable.

## 1. INTRODUCTION

Over the last few years, the author has been engaged in the development of a theory of stabilization for nonlinear systems (HAMMER [1984a,b, 1985a,b, 1986, 1987a,b, and 1988]). The basic mathematical notion on which this theory rests is the notion of fraction representations of nonlinear systems. Generally speaking, a fraction representation of a nonlinear system is a factorization of the system into a composition of two nonlinear systems, one of which is stable and the other is the inverse of a stable system. More specifically, one distinguishes between two kinds of fraction representations - a right fraction representation and a left fraction representation. A right fraction representation of a nonlinear system $\Sigma$ is a representation of the form $\Sigma = PQ^{-1}$, where $P$ and $Q$ are stable systems, with $Q$ being invertible (i.e., a set isomorphism). A left fraction representation of the system $\Sigma$ is of the form $\Sigma = G^{-1}T$, where $G$ and $T$ are stable systems, with $G$ being invertible. As it turns out, and as we manifest throughout the present note, fraction representations play a fundamental role in the theory of stabilization for nonlinear systems, and their construction is instrumental for the computation of compensators that stabilize a given system.

The general appearance of the stabilization theory we develop resembles very closely the transfer matrix theory of linear systems. The mathematical techniques we use for the nonlinear case are, of course, of a totally different nature, and no transforms are involved. In our presentation, we limit ourselves to the use of common mathematical techniques, and the general mathematical background we use can be found in any basic book on topology (e.g., KURATOWSKI [1961]). We shall discuss the robust stabilization of discrete-time nonlinear recursive systems, and the results we obtain are all explicit and can be directly implemented on digital computers. The purpose of this note is to provide a brief survey of the status of our theory at the present time. For proofs and detailed technical discussions of the results we survey here, see the appropriate full text papers.

We mention briefly the literature background. As we said, the results surveyed in this note are taken from HAMMER [1984a,b, 1985a,b, 1986, 1987a,b, and 1988]. Alternative recent studies on the stabilization of nonlinear systems can be found in VIDYASAGAR [1980], SONTAG [1981], DESOER and LIN [1984], ISIDORI [1985], the references cited in these papers, and others. Studies on the effect of feedback on system uncertainties appeared in BLACK [1934], BODE [1945], NEWTON, GOULD, and KAISER [1957], ZAMES [1966 and 1981], ROSENBROCK [1970 and 1974], DESOER and VIDYASAGAR [1975], KIMURA [1984], the references cited in these papers, and others.

## 2. MOTIVATION AND GENERALITIES

The basic control configuration that we use in our study of the stabilization of nonlinear systems is the following classical one.

(2.1)

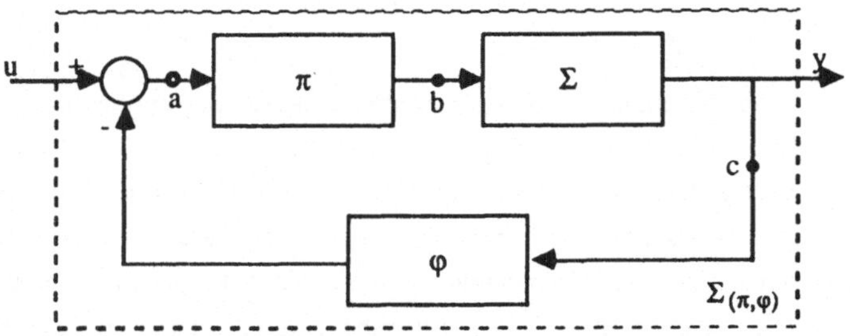


Here, $\Sigma$ is the given system that needs to be stabilized, $\pi$ is a dynamic precompensator, $\varphi$ is a dynamic feedback compensator, and $\Sigma_{(\pi,\varphi)}$ denotes the closed loop system. We have repeatedly concluded in our studies of the nonlinear stabilization problem that it is of particular advantage to choose the precompensator $\pi$ and the feedback compensator $\varphi$ in the form

(2.2) $\pi = B^{-1}, \ \varphi = A,$

where $A$ and $B$ are stable systems with $B$ being invertible, and where $A$ and $B^{-1}$ are causal. The advantage of using this particular form of the compensators is twofold. First, this form of the compensators leads to particularly simple and transparent conditions for input/output stabilization, as we show in a moment. Second, the conditions for internal stability become substantially simplified when this form of compensators is used, to the point where internal stability is almost implied by input/output stability, and an explicit derivation of compensators that internally stabilize the system becomes possible. The way these advantages come about will become clear from our ensuing discussion. It is not less important to note that the price that we pay for restricting ourselves to compensators of the form (2.2) is rather low. With this choice of compensators we can achieve virtually arbitrary dynamics assignment for the internally stable closed loop (HAMMER [1987b]), and we can design the closed loop to be robustly stable (HAMMER [1988]). Thus, this configuration allows us to do more or less everything we would like to do from a stabilization point of view, with a minimal amount of complication.

Throughout our discussion, we make the basic assumption that the amplitudes of the input sequences to any of the systems we consider are bounded by a fixed bound, namely, that there is a real number $\alpha > 0$ such that all the input sequences to our systems are of amplitude not exceeding $\alpha$. In most practical situations, this does not really amount to an assumption, but rather to a description of the actual physical reality. The input sequences, being generated by a physical device, are naturally of bounded amplitudes, the bound being determined, for instance, by saturation phenomena.

Let us now turn to a preliminary analysis of the control cofiguration (2.1). Assume that the system $\Sigma$ has a right fraction representation $\Sigma = PQ^{-1}$, and that the compensators $\pi$ and $\varphi$ are given by (2.2). Then, it can be readily seen (e.g., HAMMER [1984a]) that, under some standard mild assumptions, the input/output relationship induced by the closed loop system $\Sigma_{(\pi,\varphi)}$ is given by

(2.3) $$\Sigma_{(\pi,\varphi)} = \Sigma\pi[I + \varphi\Sigma\pi]^{-1} = PQ^{-1}B^{-1}[I + APQ^{-1}B^{-1}]^{-1} = P[AP + BQ]^{-1}.$$

Denoting

(2.4) $$M := AP + BQ,$$

we obtain that

(2.5) $$\Sigma_{(\pi,\varphi)} = PM^{-1},$$

in close analogy to the linear situation. Clearly, if the stable systems $A$ and $B$ are selected so that the stable system $M$ also has a stable inverse $M^{-1}$, then the closed-loop system $\Sigma_{(\pi,\varphi)}$ becomes input/output stable. In fact, as we discuss later, $\Sigma_{(\pi,\varphi)}$ will also be internally stable under these circumstances if the systems $A$ and $B$ satisfy some additional mild requirements (HAMMER [1986b and 1987b]). A stable system $M$ which is invertible and whose inverse $M^{-1}$ is also stable is called a*unimodular* system. For the existence of stable systems $A, B$ satisfying the equation $AP + BQ = M$ with $M$ unimodular, we need $P$ and $Q$ to be right comprime, as we elaborate in a later section.

It is rather obvious from our discussion so far that the problem of finding stable systems $A$ and $B$ which satisfy the equation $AP + BQ = M$, where $P, Q$, and $M$ are given, is of central importance to our discussion. In order to be able to choose the compensators $\pi$ and $\varphi$ most convenient for implementation, we would in fact like to know all pairs of stable systems $A, B$ satisfying that equation. It is comforting to know that in order to obtain all such pairs of stable systems, all we need is one pair, and, given one pair, all other pairs can be obtained in a straightforward way from transparent parametrization equations. For this purpose we need left fraction representations of nonlinear systems.

Let $\Sigma = G^{-1}T$, where $G$ and $T$ are stable systems, be a left fraction representation of the given system $\Sigma$. Recalling the right fraction representation $\Sigma = PQ^{-1}$ from before, we obtain $G^{-1}T = PQ^{-1}$, or

$$TQ = GP.$$

Now, assume we have one pair of stable systems A, B satisfying $AP + BQ = M$. To obtain other pairs of such systems, we can proceed simply as follows. Choose an arbitrary stable system h, and define the stable systems

**(2.6)** $A' := A - hG, \ B' := B + hT.$

Then, using the fact that $GP = TQ$, we obtain $A'P + B'Q = AP - hGP + BQ + hTQ = AP + BQ = M$, and A', B' satisfy our equation. Thus, for every choice of h we obtain a new pair of solutions, and we see that left fraction representations allow us to parametrize solutions of our basic equation in a rather transparent way. In fact, for the systems we consider in this note, (2.6) provides all pairs of solutions A', B' of the equation $A'P + B'Q = M$, when one such pair A, B is known. We conclude that left fraction representations also are of crucial importance to the theory of stabilization of nonlinear systems.

Returning now to equation (2.5), we see that the unimodular transformation M controls the dynamical properties of the closed loop system $\Sigma_{(\pi,\varphi)}$. By choosing M appropriatly, we can achieve dynamics assignment for our systems. Of course, detailed attention has to be given to the problem of internal stability of the closed loop, and we shall describe in a later section how internal stability of the configuration can be guarantied.

The final topic we would like to review in this note is the question of robust stabilization of nonlinear systems. Suppose the accurate description of the system $\Sigma$ that needs to be stabilized is not known, and that only a nominal description of $\Sigma$ is given. We denote the nominal description of the system that has to be stabilized by $\Sigma_n$, and we allow the actual system $\Sigma$ to deviate from its nominal description. The central question in the theory of robust stabilization is the following. Is it possible to use the nominal description $\Sigma_n$ to design the control configuration (2.1) in such a way that it will preserve its stability when the actual system $\Sigma$ is inserted in it, instead of the nominal system $\Sigma_n$ for which it was designed. We will describe in a later section a solution to the robust stabilization problem obtained in HAMMER [1988]. The underlying ideas on which this solution is based can be qualitatively (and quite inaccurately) described as follows. Let $\Sigma = PQ^{-1}$ be a fraction representation of the given system. Suppose we have one approppriate pair of systems A and B for which $M := AP + BQ$ is unimodular. The systems P and Q, which arise from a right fraction representation of the system $\Sigma$, depend, of course, on $\Sigma$. Consequently, deviations of $\Sigma$ from its nominal value $\Sigma_n$ will cause deviations of P and of Q from their nominal values. Let $\Sigma_n = P_nQ_n^{-1}$ be a fraction representation of the nominal system. Let $\Sigma$ be the actual system with the deviation, and suppose we can construct for it a fraction representation $\Sigma = PQ^{-1}$ in which the numerator P satisfies $P = P_n$, where $P_n$ is the numerator of the fraction representation of the nominal system $\Sigma_n$. Namely, assume that the effect of the deviation can be completely described by a deviation of the denominator system Q from its nominal value $Q_n$. Denote $\omega := Q - Q_n$, and notice that $\omega$ is a stable system, and that $Q = Q_n + \omega$. Suppose further that, for every real number $\varepsilon > 0$, there is a causal and stable system $A_\varepsilon$ satisfying the equation $A_\varepsilon P_n + \varepsilon Q_n = M$. Notice that when the latter holds and (2.2) is used, the system $\Sigma$ can be stabilized using $B = \varepsilon I$ (and $A = A_\varepsilon$), in which case the precompensator $\pi = (1/\varepsilon)I$ is a simple amplifier. By taking $\varepsilon$ arbitrarily small, we can arbitrarily increase the gain of this amplifier, and thus arbitrarily increase the forward path gain. Finally, suppose there is a real number $\delta > 0$ such that the system $\mathcal{M} := M + \mu$ stays unimodular for every stable system $\mu$ with 'magnitude' not exceeding $\delta$, so that a deviation of ' less than $\delta$ ' does not destroy the unimodularity of M. The existence of $\delta$ as well as its value depend, of course, on the nature of the particular unimodular system M.

Now, assume that the nominal system $\Sigma_n$ is stabilized using the compensators induced by $A = A_\varepsilon$ and $B = \varepsilon I$, via (2.2). Then, when the system $\Sigma$ is inserted in the loop instead of the nominal system $\Sigma_n$ for which the loop was designed, we obtain, recalling the fraction representation $\Sigma = PQ^{-1}$, that $AP + BQ = A_\varepsilon P_n + \varepsilon(Q_n + \omega) = A_\varepsilon P_n + \varepsilon Q_n + \varepsilon\omega = M + \varepsilon\omega =: \mathcal{M}$. Whence, if we choose $\varepsilon$ small enough so that the 'magnitude' of $\mu := \varepsilon\omega$ is smaller than $\delta$, the system $\mathcal{M}$ will still be unimodular, and the input/output relationship $\Sigma_{(\pi,\varphi)} = P\mathcal{M}^{-1}$ induced by the closed loop will remain stable, despite the deviation in the system $\Sigma$. Thus, the deviation will not destroy stability. Basically, our discussion in this paragraph is but a restatement of the qualitative principle that, in a closed feedback loop, high gain in the forward path can counteract deviations in the parameters of the forward path systems, a principle which has been widely accepted on an intuitive level ever since the classical work of BLACK [1934] on linear feedback systems. The main advantage of the particular form in which we formulate this principle here is that, in this formulation, the principle can be readily applied to nonlinear situations, and incorporated within the requirements of the theory of internal stability.

The qualitative ideas that we presented in this section form the crude material for the theory of fraction representations and robust stabilization that we survey in the remaining parts of this note. As we shall see, the theory is rather general in its scope, and the results it provides are explicit and implementable.

## 3. THE BASIC FRAMEWORK AND FRACTION REPRESENTATIONS

The systems we consider are discrete-time systems, accepting sequences of m-dimensional real vectors as their input, and generating sequences of p-dimensional real vectors as their output. To introduce our notation, we let R be the set of real numbers, and, for an integer $m > 0$, we let $R^m$ be the set of all m-dimensional real vectors. By $R^0$ we simply mean the zero element 0. We denote by $S(R^m)$ the set of all sequences of the form $u_0,u_1,u_2, \ldots$, with each element $u_i$ belonging to $R^m$. Given a sequence $u \in S(R^m)$ and an integer $i \geq 0$, we denote by $u_i$ the i-th element of the sequence, and we interpret the integer i as the time marker. For two integers $i \geq j \geq 0$, we denote by $u_j^i$ the set of elements $u_j, u_{j+1}, \ldots,u_i$. In the set $S(R^m)$ we induce the usual operation of addition elementwise, so that, given a pair of sequences $u, v \in S(R^m)$, the sum $w := u + v$ is again a sequence in $S(R^m)$, with each one of its elements being given by $w_i := u_i + v_i$, $i = 0,1,2, \ldots$ .

Adopting the input/output point of view, we conceive a system as a device that transforms input sequences into output sequences. In accurate terms, a system $\Sigma$ is simply a map $\Sigma : S(R^m) \to S(R^p)$, transforming input sequences from $S(R^m)$ into output sequences from $S(R^p)$, where m and p are arbitrary positive integers. In order to be able to obtain results which are simple, explicit, computable, and implementable, we shall not discuss here systems on this level of generality. Instead, we shall restrict ourselves to recursive systems which have their state as output, namely, to systems $\Sigma$ possessing a recursive representation of the form

(3.1) $x_{k+1} = f(x_k,u_k)$,

where $f : R^p \times R^m \to R^p$ is a function, which we shall usually assume to be continuous. Here, the input sequence of the system is $\{u_k\}$ and its output sequence is $\{x_k\}$, and we assume that the initial condition $x_0$ is specified. The function $f$ is called a *recursion function* of the system $\Sigma$. Given a subspace $S \subset S(R^m)$, we denote by $\Sigma[S]$ the image of the set $S$ through $\Sigma$, namely, the set of all output sequences that $\Sigma$ generates from input sequences belonging to $S$. Also, given a subspace $S' \subset S(R^p)$, we denote by $\Sigma^*[S']$ the inverse image of the set $S'$ through $\Sigma$, namely, the set of all input sequences that generate output sequences belonging to the set $S'$.

We review now briefly some notions related to causality of systems. A system $\Sigma : S(R^m) \to S(R^p)$ is causal (respectively, strictly causal) if the following holds for every pair of input sequences $u, v \in S(R^m)$: for all integers $i \geq 0$ for which $u_0^i = v_0^i$, one also has $\Sigma u]_0^i = \Sigma v]_0^i$ (respectively, $\Sigma u]_0^{i+1} = \Sigma v]_0^{i+1}$). A system $M : S(R^m) \to S(R^m)$ is a bicausal system if it is invertible, and if $M$ and $M^{-1}$ are both causal systems.

Most of our discussion is related, of course, to the stability of systems. The notion of stability that we adopt is in the spirit of the Lyapunov notion of stability, and thus is related to the continuity of the system as a map. For the purpose of introducing the notion of stability, we need to induce some norms on the space of sequences $S(R^m)$. First, let $w = (w^1, ..., w^m)$ be a vector in $R^m$. We denote $|w| := \max \{|w^i|, i = 1, ...,m\}$, the maximal absolute value of the coordinates of $w$. Next, we define a norm on the space $S(R^m)$ given, for any element $u \in S(R^m)$, by $\rho(u) := \sup \{2^{-i}|u_i|, i = 0,1,2, ... \}$, and we note that this is simply a weighted $L^\infty$-norm. We use this norm to define a metric $\rho(u,v)$ on $S(R^m)$, by letting $\rho(u,v) := \rho(u-v)$ for every pair of elements $u, v \in S(R^m)$. Whenever referring to continuity, we shall always mean continuity with respect to the topology induced by the metric $\rho$, unless explicitly stated otherwise. It will also be convenient for us to use the notation $|u| := \sup \{|u_i|, i = 0,1,2, ...\}$ for an element $u \in S(R^m)$, so that $|\cdot|$ is simply the $L^\infty$-norm. Then, for a real number $\theta > 0$, we denote by $S(\theta^m)$ the set of all elements $u \in S(R^m)$ satisfying $|u| \leq \theta$, namely, the set of all sequences bounded by $\theta$. A system $\Sigma : S(R^m) \to S(R^p)$ is BIBO (Bounded-Input Bounded-Output)-stable if, for every real number $\theta > 0$, there is a real number $D > 0$ such that $\Sigma[S(\theta^m)] \subset S(D^p)$. Finally, we say that a system $\Sigma : S(R^m) \to S(R^p)$ is *stable* if it is BIBO-stable, and if, for every real number $\theta > 0$, the restriction $\Sigma : S(\theta^m) \to S(R^p)$ is a continuous map.

Before turning to a review of our theory of fraction representations for nonlinear systems, we wish to discuss two basic assumptions that we make in the development of our framework. The first assumption is that all the systems we consider are operated by bounded input sequences, namely, that there is a fixed real number $\alpha > 0$ such that all our systems have $S(\alpha^m)$ as their domain. As we have remarked already in an earlier section, this is hardly a restrictive assumption from the practical point of view. In practice, input sequences are generated by a physical device, and their maximal amplitude is limited by the physical characteristics of that device. The second assumption we make is that the system $\Sigma$ that needs to be stabilized is an injective (one to one) map. At first glance, this looks like a restrictive assumption, since many systems of practical interest are, of course, not injective systems. However, further reflection shows that the assumption that the system that needs to be stabilized is an injective system is not really restrictive, for the following reason. Assume that the system $\Sigma$ that needs to be stabilized is a strictly causal system.

This is always true for systems having recursive representations of the form (3.1). Then, instead of stabilizing the system $\Sigma$ directly, consider the stabilization of the system $I+\Sigma$, the sum of $\Sigma$ and the identity system $I$, ignoring for a second the fact that this sum might not be well defined due to different input space and output space dimensionalities. Then, the strict causality of $\Sigma$ implies that $I+\Sigma$ is bicausal, and hence injective. Moreover, if we stabilize the system $I+\Sigma$, we shall also obtain stabilization of the original system $\Sigma$ (in a somewhat different control configuration), as we now show.

Let $\Sigma : S(R^m) \to S(R^q)$ be a strictly causal system. Let $p := \max\{m, q\}$, and define the identity injection maps $\mathfrak{I}_1 : S(R^m) \to S(R^p)$ and $\mathfrak{I}_2 : S(R^q) \to S(R^p)$ as follows. If $q \geq m$, write $S(R^p) = S(R^q) = S(R^m) \times S(R^{q-m})$, let $\mathfrak{I}_1 : S(R^m) \to S(R^p) : \mathfrak{I}_1[S(R^m)] = S(R^m) \times 0$ be the obvious identity injection, and let $\mathfrak{I}_2 : S(R^q) \to S(R^p)$ ( $=S(R^q)$ ) be the identity map. If $q < m$, write $S(R^p) = S(R^m) = S(R^q) \times S(R^{m-q})$, let $\mathfrak{I}_2 : S(R^q) \to S(R^p) : \mathfrak{I}_2[S(R^q)] = S(R^q) \times 0$ be the obvious identity injection, and let $\mathfrak{I}_1 : S(R^m) \to S(R^p)$ ( $=S(R^m)$ ) be the identity map. Then, as we show in a minute, the system

(3.2) $$\Sigma_\gamma := \gamma\mathfrak{I}_1 + \mathfrak{I}_2\Sigma \ : \ S(R^m) \to S(R^p),$$

where $\gamma$ is a $p \times p$ constant nonsingular matrix, is injective by the strict causality of the system $\Sigma$. The implementation of the injections $\mathfrak{I}_1$ and $\mathfrak{I}_2$ is very simple - it just amounts to increasing the dimension of some vectors through augmentation by entries of zeros (see HAMMER [1987b] for details). To simplify our notation, we shall usually abbreviate and denote $\mathfrak{I}_1 u$ by $u$ and $\mathfrak{I}_2 y$ by $y$. It can be seen that, when stabilizing the system $\Sigma_\gamma$ in the configuration (2.1), we in fact obtain stabilization of the original system $\Sigma$ in the following configuration.

(3.3)

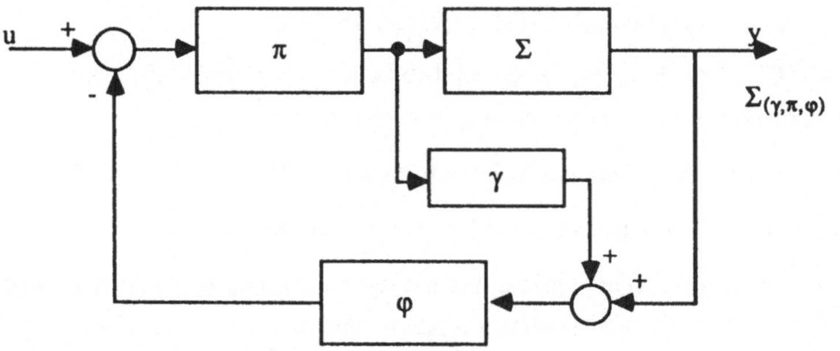


(Note that in the configuration (3.2), $\gamma$ is to be interpreted as $\gamma\mathfrak{I}_1$, in consistency with our notational convention.)

Let $\Sigma : S(R^m) \to S(R^p)$ be a strictly causal system. Then, the system $\Sigma_\gamma$ of (3.2) is an injective system whenever the $p \times p$ matrix $\gamma$ is nonsingular, and thus $\Sigma_\gamma$ possesses a left inverse. Moreover, when the original system $\Sigma$ is recursive, the left inverse of $\Sigma_\gamma$ is very easy to compute. Indeed, assume that $\Sigma$ has a recursive representation $x_{k+1} = f(x_k, u_k)$. Let $u \in S(R^m)$ be an input sequence, and let $x := \Sigma u$ be the corresponding output sequence. Denoting $z := \Sigma_\gamma u$, and using the abbreviated notation mentioned in the previous paragraph, we obtain $z = x + \gamma u$, so that $z_i = x_i + \gamma u_i$ for all integers $i \geq 0$. Therefore, $z_{k+1} = x_{k+1} + \gamma u_{k+1} = f(x_k, u_k) + \gamma u_{k+1} = f((z-\gamma u)_k, u_k) + \gamma u_{k+1}$, and, invoking the invertibility of $\gamma$, we obtain

(3.4) $u_{k+1} = \gamma^{-1}\{z_{k+1} - f((z-\gamma u)_k, u_k)\}$ , $k = 0,1,2, \ldots,$ $u_0 = \gamma^{-1}(z_0 - x_0)$,

where $x_0$ is the given initial condition of the system $\Sigma$, and where the relations are valid for any sequence $z \in \operatorname{Im} \Sigma_\gamma$. Thus, the input sequence $u$ of $\Sigma_\gamma$ can be readily computed from the output sequence $z$ of $\Sigma_\gamma$ in a recursive manner, using the given recursion function and initial conditions of the system $\Sigma$. This evidently amounts to a left inversion of the system $\Sigma_\gamma$, and we shall use these formulas repeatedly in the sequel. It is also clear from (3.4) that this left inverse is causal, and we have the following

**(3.5) PROPOSITION.** Let $\Sigma : S(R^m) \to S(R^q)$ be a strictly causal recursive system having a recursive representation $x_{k+1} = f(x_k,u_k)$. Let $p := \max\{m, q\}$, and let $\gamma$ be a $p\times p$ constant invertible matrix. Then, the system $\Sigma_\gamma : S(R^m) \to \operatorname{Im} \Sigma_\gamma$ defined by (3.2) is a bicausal system.

We can summarize our discussion in the last few paragraphs by saying that we can always transform our situation into one where the system that needs to be stabilized is injective, even if the original system $\Sigma$ is not injective. Consequently, from a stabilization point of view, it is not overly restrictive to limit our attention to the discussion of injective systems.

We provide now a brief survey of the theory of right and of left fraction representations for an injective system $\Sigma : S(\alpha^m) \to S(R^p)$, where $\alpha > 0$ is a fixed, but otherwise arbitrary, real number. As we shall see, the theory is surprisingly simple.

A right fraction representation of a system $\Sigma : S(\alpha^m) \to S(R^p)$ involves an integer $q > 0$, a subspace $S \subset S(R^q)$, called the *factorization space*, and a pair of stable systems $P : S \to S(R^p)$ and $Q : S \to S(\alpha^m)$, where $Q$ is invertible, so that $\Sigma = PQ^{-1}$. Of particular importance to us are coprime right fraction representations, which are fraction representations in which the systems $P$ and $Q$ are right coprime according to the following definition (HAMMER [1985a, 1987a]).

**(3.6) DEFINITION.** Let $S \subset S(R^q)$ be a subspace. A pair of stable systems $P : S \to S(R^p)$ and $Q : S \to S(R^m)$ are *right coprime* if the following two conditions are satisfied.

(i) For every real $\tau > 0$, there is a real $\theta > 0$ such that $P^*[S(\tau^p)] \cap Q^*[S(\tau^m)] \subset S(\theta^q)$.

(ii) For every real $\tau > 0$, the set $S \cap S(\tau^q)$ is a closed subset of $S(\tau^q)$.

It is quite easy to see why right coprime fraction representations are important to our discussion. In (2.4) we saw that the solution of the stabilization problem involves the search for a pair of stable systems $A$ and $B$ satisfying the equation $AP + BQ = M$, where $P$ and $Q$ arise from a fraction representation $\Sigma = PQ^{-1}$ of the given system $\Sigma$, and where $M$ is a unimodular system. The existence of such systems $A$ and $B$ is guarantied whenever $P$ and $Q$ are right coprime, as follows (HAMMER [1987a]).

**(3.7) THEOREM.** Let $\Sigma : S(\alpha^m) \to S(R^p)$ be an injective system, and assume it has a right coprime fraction representation $\Sigma = PQ^{-1}$, where $P : S \to S(R^p)$ and $Q : S \to S(\alpha^m)$, and where $S \subset S(R^q)$ for some integer $q > 0$. Then, for every unimodular system $M : S \to S$, there exists a pair of stable systems $A : S(R^p) \to S(R^q)$ and $B : S(\alpha^m) \to S(R^q)$ such that $AP + BQ = M$.

Theorem (3.7) underscores the importance of right coprime fraction representations to our discussion. At the same time, it opens a new question - what systems possess right coprime fraction representations.

The existence of right coprime fraction representations is related in a fundamental way to the concept of a homogeneous system, which is defined as follows (HAMMER [1985a, 1987a]).

**(3.8) DEFINITION.** A system $\Sigma : S(R^m) \to S(R^p)$ is a *homogeneous system* if the following holds for every real number $\alpha > 0$ : for every subspace $S \subset S(\alpha^m)$ for which there exists a real number $\tau > 0$ satisfying $\Sigma[S] \subset S(\tau^p)$, the restriction of $\Sigma$ to the closure $\bar{S}$ of $S$ in $S(\alpha^m)$ is a continuous map $\Sigma : \bar{S} \to S(\tau^p)$.

As the next statement shows, (injective) homogeneous systems possess right coprime fraction representations, and they are the only systems possessing such representations. Thus, the concept of a homogeneous system provides a complete characterization of the existence of right coprime fraction representations, in terms of input/output properties of the system (HAMMER [1985a, 1987a]).

**(3.9) THEOREM.** An injective system $\Sigma : S(\alpha^m) \to S(R^p)$ has a right coprime fraction representation if and only if it is a homogeneous system.

Of course, the obvious question now is - how common are homogeneous systems in practical applications. A partial answer to this question is given by the following statement, which shows that all the systems we consider in our present note are homogeneous. More general classes of homogeneous systems are described in the references (HAMMER [1987a]).

**(3.10) PROPOSITION.** Let $\Sigma : S(R^m) \to S(R^p)$ be a recursive system. If $\Sigma$ has a recursive representation $x_{k+1} = f(x_k,u_k)$ with a continuous recursion function f, then $\Sigma$ is a homogeneous system.

As we have discussed in detail earlier in this section, we usually prefer to study the stabilization of the system $\Sigma_\gamma$ of (3.2) instead of studying directly the stabilization of the given system $\Sigma$. The reasons for this are twofold. First, the theory of fraction representations for the case of injective systems is simpler and more transparent in its appearance, and $\Sigma_\gamma$ is always injective when $\Sigma$ is strictly causal. Secondly, the solution to the stabilization problem becomes simpler if the system $\Sigma_\gamma$ is used instead of $\Sigma$, even in the case where $\Sigma$ is itself injective, due to the simplicity of the inversion formulas (3.4) for $\Sigma_\gamma$. It is therefore of interest to know that when the system $\Sigma$ is homogeneous, so also is the system $\Sigma_\gamma$ (HAMMER [1987a]).

**(3.11) PROPOSITION.** Let $\Sigma : S(R^m) \to S(R^p)$ be a homogeneous system, and let $\Sigma_\gamma$ be defined as in (3.2). Then, $\Sigma_\gamma : S(R^m) \to S(R^p)$ is a homogeneous system.

It is quite easy to construct a right coprime fraction representation for an injective homogeneous system $\Sigma : S(\alpha^m) \to S(R^p)$. Indeed, since $\Sigma$ is injective, its restriction $\Sigma : S(\alpha^m) \to \mathrm{Im}\,\Sigma$ is a set isomorphism, and, consequently, it possesses an inverse $\Sigma^{-1} : \mathrm{Im}\,\Sigma \to S(\alpha^m)$. We have shown in HAMMER [1987a, section 3] that $\Sigma^{-1}$ is a stable system. This is a significant departure from the situation in the case of linear systems, where only very special systems possess stable inverses. The root of this departure is the fact that the domain $S(\alpha^m)$ of our systems here is compact in our topology, a fact that originates from the realistic assumption that all systems are operated by bounded input sequences. Thus, we see that the nonlinear framework allows us to take advantage of inherent restrictions in the physical operation of practical systems to simplify the mathematical structure of the problem. Defining the systems

$$P := I : \mathrm{Im}\,\Sigma \to \mathrm{Im}\,\Sigma, \quad Q := \Sigma^{-1} : \mathrm{Im}\,\Sigma \to S(\alpha^m), \tag{3.12}$$

where I denotes the identity system, we obtain a right fraction representation $\Sigma = PQ^{-1}$, which, as one can readily see, is right coprime. Once we have one right coprime fraction representation $\Sigma = PQ^{-1}$ of the system $\Sigma$, any other right coprime fraction representation of $\Sigma$ is of the form $\Sigma = P_1Q_1^{-1}$ where $P_1 = PM$ and $Q_1 = QM$, and where M is a unimodular system (HAMMER [1985a, 1987a]).

We turn now to left fraction representations. A left fraction representation of a nonlinear system $\Sigma : S(\alpha^m) \to S(R^p)$ involves an integer $q > 0$, a subspace $S \subset S(R^q)$, and a pair of stable systems $G : \text{Im}\,\Sigma \to S$ and $T : S(\alpha^m) \to S$, where G is invertible, such that $\Sigma = G^{-1}T$. The main use of left fraction representations in our context is for the purpose of parametrizing the set of pairs of stable systems A, B which satisfy an equation of the form $AP + BQ = M$. Here, P and Q originate from a right coprime fraction representation $\Sigma = PQ^{-1}$ of the same system $\Sigma$, and M is a fixed unimodular system. We have already indicated in (2.6) how such a parametrization may be obtained. The only questions that we still have to deal with in this context are the questions of the existence and of the construction of left fraction representations. The existence of left fraction representations for the systems we consider is guarantied by the following result, which we reproduce from HAMMER [1987a].

**(3.13) THEOREM.** An injective homogeneous system $\Sigma : S(\alpha^m) \to S(R^p)$ has a left fraction representation.

It is also quite easy to construct a left fraction representation for an injective homogeneous system $\Sigma : S(\alpha^m) \to S(R^p)$. Indeed, using the above mentioned fact that $\Sigma^{-1} : \text{Im}\,\Sigma \to S(\alpha^m)$ is a stable system, and letting $I : S(\alpha^m) \to S(\alpha^m)$ be the identity map, the pair of stable systems

**(3.14)** $$G := \Sigma^{-1} : \text{Im}\,\Sigma \to S(\alpha^m),\ \ T := I : S(\alpha^m) \to S(\alpha^m),$$

induces a left fraction representation $\Sigma = G^{-1}T$ (HAMMER [1987, section 4]).

To summarize, we see that a theory of fraction representations can be developed for nonlinear systems. This theory bears, in its external appearance, a close resemblance to the theory of fraction representations for transfer matrices of linear systems. The computations involved in the construction of fraction representations in the nonlinear case are relatively simple, and they become particularly simple for systems of the form $\Sigma_\gamma$, due to the simplicity of the inversion formula (3.4) for these systems. In our next section we discuss the problem of robust stabilization of nonlinear systems. The derivation of the results presented in the next section depends heavily on the theory of fraction representations that we have described here.

## 4. ROBUST STABILIZATION AND DYNAMICS ASSIGNMENT

In the present section we provide a survey of our results on the stabilization of nonlinear systems, following HAMMER [1987b and 1988]. In general, when considering stabilization of a system, one has to pay attention to three main issues - internal stability, dynamics assignment, and robustness.

Internal stability is a strong notion of stability, which is essential when the stability of composite systems is considered. In our case, internal stability of the configuration (3.3) means (i) that the configuration is input/output stable, (ii) that all the internal signals of the configuration are bounded, and (iii) that (i) and (ii) continue to hold when small noise signals are added to the signals at the points of entry of the subsystems $\Sigma$, $\pi$, $\varphi$, and $\gamma$ of which the configuration consists. Only internally stable systems possess stable physical implementations. All composite systems that we construct below are internally stable.

The issue of dynamics assignment deals with the characterization of the dynamical properties that can be assigned to the internally stable closed loop (3.3), through proper choice of the compensators $\pi$, $\varphi$ and $\gamma$. It provides the designer with the methodology to achieve a desired dynamical behviour for the final stabilized closed loop system. We show that, except for some obvious limitations, the stabilized closed loop system can be designed to have any desired dynamical behaviour, and we provide explicit constructions for compensators achieving that dynamical behaviour. In a qualitative way, the situation here is similar to the well known situation in the case of pole assignment for linear time invariant systems.

The issue of robustness deals with the stabilization of systems whose descriptions are not accurately known. Specifically, the situation in our case is as follows. Recall that the systems $\Sigma$ whose stabilization we consider are given by recursive representations of the form $x_{k+1} = f(x_k,u_k)$, $k = 0,1,2, ...,$ where the initial condition $x_0$ is specified. We shall consider the case where the recursion function $f$ of the system that needs to be stabilized is not accurately known. Rather, a nominal recursion function $f_n$ is given for the system $\Sigma$, and the actual recursion function $f$ of the system may deviate from its nominal description. We assume that the actual recursion function is of the form

$$(4.1) \qquad f(x_k,u_k) = f_n(x_k,u_k) + \upsilon(x_k,u_k),$$

where the function $\upsilon$ describes the deviation from nominality. Of course, the function $\upsilon$ is not known, and, qualitatively speaking, we assume only that a bound on the magnitude of its parameters is given. We shall make the last statement more precise in the sequel. The fundamental question in the theory of robust stabilization can then be stated as follows. Assume that the nominal recursion function $f_n$ of the system is given. Is it possible to design an internally stable control configuration that will stabilize the actual system $\Sigma$, irrespectively of the deviation function $\upsilon$, as long as the latter is continuous and its parameters do not exceed a prespecified bound. If such a design is possible, how is it done.

We describe now our design procedure for the robust stabilization of nonlinear systems. The procedure allows dynamics assignment. At the end of the section we provide an explicit example on the computation of robustly stabilizing compensators, using our procedure. Throughout our review here we shall assume that the system $\Sigma : S(\alpha^m) \to S(R^p)$ that needs to be stabilized has an input space which is of the same dimension as its output space, namely, that $m = p$. This assumption simplifies the presentation, but is of no fundamental consequence in our framework. The general case where $m \neq p$ is treated in HAMMER [1987b, 1988].

Let then $\Sigma : S(\alpha^m) \to S(R^m)$ be the system that needs to be stabilized, and assume it is strictly causal and homogeneous. As we have discussed before, the basic system whose stabilization we shall consider is the system $\Sigma_\gamma$, which here takes the form

$$(4.2) \qquad \Sigma_\gamma = \gamma + \Sigma \ : S(\alpha^m) \to S(R^m),$$

where $\gamma$ is an m×m nonsingular matrix. We recall that the system $\Sigma_\gamma$ is bicausal. One of the basic steps in our stabilization procedure is to find an m×m nonsigular matrix $\gamma$ for which the following condition is satisfied.

**(4.3) CONDITION.** There is a real number $\delta > 0$ such that $S(\delta^m) \subset \Sigma_\gamma[S(\alpha^m)]$ for some real number $\alpha > 0$.

In qualitative terms, the matrix $\gamma$ shifts the image of the system so as to include a subspace of the form $S(\delta^m)$; the restriction of the inverse system $\Sigma_\gamma^{-1} : S(\delta^m) \to S(\alpha^m)$ becomes then a stable system. The justification of these statements and explicit methods for the computation of $\gamma$ for some common classes of recursive systems are given in HAMMER [1987b and 1988]. We comment here that stronger results can be obtained when $\gamma$ is allowed to be a nonlinear system, as we discuss in a separate report. However, quite general results on robust stabilization and dynamics assignment for nonlinear systems can be obtained even when $\gamma$ is restricted to be an m×m nonsingular matrix, as we assume throughout our discussion here.

In our present context, the system $\Sigma$ is not accurately known, and we have to study the effects of the uncertainty in the description of $\Sigma$ on Condition (4.3). For this purpose we need to describe more accurately the nature of the deviation functions $\upsilon$ in (4.1). We do so by defining a 'neighbourhood' $\mathcal{N}(\Sigma_n,\Delta)$ of radius $\Delta$ around the nominal system $\Sigma_n$, which consists of all systems whose deviation from $\Sigma_n$ is permissible. We describe $\mathcal{N}(\Sigma_n,\Delta)$ in terms of quantities directly related to the recursion functions, distinguishing between two different classes of recursion functions, as follows.

The first class of recursion functions we consider is the class of recursion functions with bounded nonlinearities. Some notation. Given an m×m matrix A with entries $a_{ij}$, we denote $\|A\| := \max\{|a_{ij}|, i,j = 1, ...,m\}$. Let $\Sigma_n : S(R^m) \to S(R^m)$ be our nominal system, having a recursive representation $x_{k+1} = f_n(x_k,u_k)$ with $x_0 = 0$. Assume the nominal recursion function is of the form $f_n(x,u) = Fx + Gu + \psi(x,u)$, where F and G are m×m matrices, and where the function $\psi : R^m \times R^m \to R^m$ is continuous and bounded, say $|\psi(x,u)| \le D$ for all $x, u \in R^m$. Now, for a real number $\Delta > 0$, we define a class $\mathcal{N}(F,G,\Delta)$ of systems that deviate 'by $\Delta$' from the nominal system $\Sigma_n$. Specifically, $\mathcal{N}(F,G,\Delta)$ consists of all systems $\Sigma : S(R^m) \to S(R^m)$ having recursive representations $x_{k+1} = f_n(x_k,u_k) + \upsilon(x_k,u_k)$, $x_0 = 0$, with the deviation function $\upsilon : R^m \times R^m \to R^m$ being of the form $\upsilon(x,u) = \Gamma x + \Lambda u + \psi_\upsilon(x,u)$, where $\Gamma$ and $\Lambda$ are m×m matrices satisfying $\|\Gamma\| \le \Delta$ and $\|\Lambda\| \le \Delta$, and where $\psi_\upsilon : R^m \times R^m \to R^m$ is a bounded continuous function, say $|\psi_\upsilon(x,u)| \le D$ for all $x, u \in R^m$. As usual, we say that a linear system is stabilizable if all its unreachable modes correspond to eigenvalues having absolute value strictly less than one. We then have the following result, which guaranties the existence of an m×m nonsingular matrix $\gamma$ satisfying Condition (4.3) for all our deviated systems.

**(4.4) PROPOSITION.** Let $\mathcal{N}(F,G,\Delta)$ be the class of systems $\Sigma : S(R^m) \to S(R^m)$ defined in the previous paragraph, and assume that the pair F, G is stabilizable. Then, there is a real number $\Delta > 0$ and an m×m nonsingular matrix $\gamma$ such that the following holds true. For every real number $\delta > 0$, there is a real number $\alpha > 0$ satisfying $S(\delta^m) \subset \Sigma_\gamma[S(\alpha^m)]$ for all systems $\Sigma \in \mathcal{N}(F,G,\Delta)$.

The second class of systems we consider is more general than the one considered in Proposition (4.4), and it consists of systems having recursion functions which are differentiable. The results for this more general class of systems are somewhat weaker in the sense that the real number $\delta$ can no longer be chosen arbitrarily large when $\gamma$ is restricted to be a constant m×m matrix. Nevertheless, robust stabilization with dynamics assignment can still be achieved for this rather general class of systems. When $\gamma$ is allowed to be a nonlinear system, then the restriction on $\delta$ is removed, and $\delta$ can again be chosen arbitrarily large. However, we do not elaborate on this point here. Let $\Sigma_n : S(R^m) \to S(R^m)$ be our nominal system, having a recursive representation $x_{k+1} = f_n(x_k,u_k)$ with $x_0 = 0$. Assume that the nominal recursion function $f_n$ is differentiable at the origin and that $f(0,0) = 0$, and let (F,G), where F and G are m×m matrices, be the Jacobian matrix of the partial derivatives of $f_n$ at the origin. Now, given a real number $\Delta > 0$, we define a class $\mathcal{N}(f_n,\Delta)$ of systems that deviate 'by $\Delta$' from the nominal system $\Sigma_n$. First, we fix a neighbourhood $\mathcal{N}_0$ of the origin and a real number $D > 0$. Then, $\mathcal{N}(f_n,\Delta)$ consists of all systems $\Sigma : S(R^m) \to S(R^m)$ having recursive representations of the form $x_{k+1} = f_n(x_k,u_k) + \upsilon(x_k,u_k)$, $x_0 = 0$, where the deviation function $\upsilon : R^m \times R^m \to R^m$ satisfies the following conditions. (i) $\upsilon$ is twice continuously differentiable over $\mathcal{N}_0$, and all its second order partial derivatives there are bounded in absolute value by D; (ii) $\upsilon(0,0) = 0$; and (iii) the Jacobian matrix $(\Gamma,\Lambda)$ of the partial derivatives of $\upsilon$ at the origin, partitioned into the m×m matrices $\Gamma$ and $\Lambda$, satisfies $\|\Gamma\| < \Delta$ and $\|\Lambda\| < \Delta$.

**(4.5) PROPOSITION.** Let $\mathcal{N}(f_n,\Delta)$ be the class of systems $\Sigma : S(R^m) \to S(R^m)$ defined in the previous paragraph. Let (F,G), where F and G are m×m matrices, be the Jacobian matrix of the partial derivatives of the nominal recursion function $f_n$ at the origin, and assume that the pair F, G is stabilizable. Then, there are real numbers $\Delta, \delta, \alpha > 0$ and an m×m nonsingular matrix $\gamma$ such that $S(\delta^m) \subset \Sigma_\gamma[S(\alpha^m)]$ for all systems $\Sigma \in \mathcal{N}(f_n,\Delta)$.

As an example of a class of systems satisfying the conditions of Proposition(4.5), consider the following single-input single-output case. Let the nominal system $\Sigma_n : S(R) \to S(R)$ be given by the recursive representation $x_{k+1} = 2\mathrm{Exp}(x_k + u_k) - 2 =: f_n(x_k,u_k)$. Now, fix some real number $D > 0$. Then, the class of systems $\Sigma : S(R) \to S(R)$ having recursive representations of the form $x_{k+1} = 2\mathrm{Exp}(x_k + u_k) - 2 + ax_k + bx_k^2 + cu_k + du_k^2 + gx_ku_k$, where $|a|, |c| < \Delta$ and $|b|, |d|, |g| < D/2$, is a class of systems contained in $\mathcal{N}(f_n,\Delta)$, and hence the Proposition applies to it.

We remark that in HAMMER [1988] we described explicit ways for the computation of nonsingular matrices $\gamma$ satisfying the conditions of Propositions (4.4) and (4.5). Once the matrix $\gamma$ is at our disposition, we can directly proceed to the construction of the stabilizing compensators $\pi$ and $\varphi$ of configuration (3.2). We provide now a step by step description of the construction of compensators that robustly stabilize our system, and allow for assignment of dynamical properties for the final internally stable closed loop.

Let $\Sigma_n : S(R^m) \to S(R^m)$ be the given nominal system, and let $x_{k+1} = f_n(x_k,u_k)$ be its recursive representation, with the initial condition $x_0 = 0$. As before, we use the notation $\mathcal{N}(\Sigma_n,\Delta)$ for a 'neighbourhood' of 'radius' $\Delta$ of the system $\Sigma$, by which we simply mean a generic notation, referring to one of the sets $\mathcal{N}(F,G,\Delta)$ or $\mathcal{N}(f_n,\Delta)$ mentioned in Propositions (4.4) or (4.5). We shall assume that the given nominal recursion function $f_n$ satisfies all the conditions involved in the use of these sets of systems, so that, when $\mathcal{N}(\Sigma_n,\Delta)$ is $\mathcal{N}(F,G,\Delta)$, the pair F, G is stabilizable; and when $\mathcal{N}(\Sigma_n,\Delta)$ is $\mathcal{N}(f_n,\Delta)$, the Jacobian matrix (F,G) of $f_n$ at the origin, when partitioned into the pair of m×m matrices F and G , yields a stabilizable pair.

Our stabilization procedure consists then of the following steps.

**Step 1.** Choose a real number $\theta > 0$. This number will serve as the bound on the amplitude of the input sequences of the final stabilized closed-loop system. The choice of the number $\theta$ is usually determined by practical considerations, and there are no theoretical restrictions on its choice.

**Step 2.** Find a constant $m \times m$ nonsingular matrix $\gamma$ for which there are three real numbers $\Delta, \delta, \alpha > 0$ such that the condition $S(\delta^p) \subset \Sigma_\gamma[S(\alpha^p)]$ holds for all systems $\Sigma \in \mathcal{N}(\Sigma_n,\Delta)$. The existence of such a matrix $\gamma$ is guarantied by Propositions (4.4) and (4.5). Explicit methods for the computation of the matrix $\gamma$ are described in HAMMER [1988].

**Step 3.** Choose a positive number $\xi$, and, using the numbers $\theta$, $\delta$, and $\alpha$ of the previous steps, choose constant positive numbers $\zeta < \delta$ and $\varepsilon < \min\{\theta/\alpha, \xi/(2\alpha)\}$.

**Step 4.** Choose a recursive, unimodular, bicausal, and uniformly $L^\infty$-continuous system $M : S(R^m) \to S(R^m)$. The system $M$ will determine the dynamical behavior of the closed loop system, as in (2.5) (see also HAMMER [1987b]). An elementary possible choice for $M$ is $M := \beta I$, where $I : S(R^m) \to S(R^m)$ is the identity system and $\beta$ is a nonzero constant.

**Step 5.** Find a real number $c > 0$ so that the system $\mathcal{M} := Mc$ satisfies the condition $\mathcal{M}^{-1}[S((5\theta + \xi)^m)] \subset S(\zeta^m)$. This is simply a scaling operation which has no dynamical implications, and is performed as follows. In view of the fact that $M$ is unimodular, the system $M^{-1}$ is stable, and, consequently, there is a real number $\lambda > 0$ satisfying $M^{-1}[S((5\theta + \xi)^m)] \subset S(\lambda^m)$. But then, taking $c := (\lambda/\zeta)$, we obtain that the system $\mathcal{M} := Mc$ satisfies $\mathcal{M}^{-1}[S((5\theta + \xi)^m)] \subset S(\zeta^m)$. If we use the choice $M = \beta I$ mentioned in Step 4, then, for $\beta \geq (5\theta+\xi)/\zeta$, we obtain directly $M^{-1}[S((5\theta+\xi)^m)] \subset S(\zeta^m)$.

**Step 6.** Construct the static system $E : S(R^m) \to S(\zeta^m)$ given by the representation

$$(4.6) \qquad E: \; y_k := e(u_k), \; k = 0, 1, 2, \ldots\ldots, \; y = Eu,$$

where $e$ is a function $R^m \to [-\zeta,\zeta]^m$ defined as follows. For every vector $x = (x^1, \ldots, x^m) \in R^m$, it takes the value $e(x^1, \ldots, x^m) := (\alpha^1, \ldots, \alpha^m)$, where $\alpha^i := x^i$ if $|x^i| \leq \zeta$ and $\alpha^i := \zeta \mathrm{sign}(x^i)$ if $|x^i| > \zeta$, and where $\mathrm{sign}(\cdot)$ is $\pm 1$, depending on the sign of the argument. It is clear that the system $E$ is recursive, causal, stable, and uniformly $L^\infty$-continuous, and it is in fact an extension of the identity system $I : S(\zeta^m) \to S(\zeta^m)$.

**Step 7.** Using the nominal system $\Sigma_n$, we construct the system $\Sigma_{n\gamma} := \gamma + \Sigma_n$ and its inverse $\Sigma_{n\gamma}^{-1}$, which exists by virtue of the bicausality of $\Sigma_{n\gamma}$. Using (3.4) and the fact that $\gamma$ is invertible, we obtain an explicit recursive representation for $\Sigma_{n\gamma}^{-1}$, given by

$$u_{k+1} = \gamma^{-1}\{z_{k+1} - f_n((z-\gamma u)_k, u_k)\}, \; k = 0,1,2, \ldots, \; u_0 = \gamma^{-1}(z_0 - x_0),$$

where $x_0$ is the given initial condition of the system $\Sigma_n$; $z$ is the input sequence of $\Sigma_{n\gamma}^{-1}$; and $u$ is the output sequence of $\Sigma_{n\gamma}^{-1}$. We shall only be interested in the restriction $\Sigma_{n\gamma}^{-1} : S(\delta^m) \to S(\alpha^m)$, which, as we mentioned earlier, is a stable and bicausal system.

**Step 8.** Combining the results of Steps 6 and 7, we construct the system

$$Q_n^E := \Sigma_{n\gamma}^{-1} E \; : S(R^m) \to S(\alpha^m)$$

as a composition of the two recursive systems $\Sigma_{n\gamma}^{-1}$ and E. The system $Q_n^E$ is stable and causal, and it can be readily implemented on a digital computer.

**Step 9.** Construct the two systems

$$A := M' - \varepsilon Q_n^E : S(R^m) \to S(R^m), \quad B := \varepsilon I : S(R^m) \to S(R^m),$$

where $I : S(R^m) \to S(R^m)$ is the identity system and $\varepsilon$ is from Step 3. From these, using (2.2), construct the compensators

(4.7) $$\pi = (1/\varepsilon)I : S(R^m) \to S(R^m), \quad \varphi = \mathcal{M} - \varepsilon Q_n^E : S(R^m) \to S(R^m).$$

Notice that the precompensator $\pi$ here is simply an amplifier with amplification factor of $1/\varepsilon$.

Steps 1 to 9 complete the construction of compensators $\pi$ and $\varphi$ which, when connected in the closed loop $\Sigma_{(\gamma,\pi,\varphi)}$, yield robust stabilization of the system $\Sigma$. The closed loop configuration $\Sigma_{(\gamma,\pi,\varphi)}$ will be internally stable for any system $\Sigma \in \mathcal{N}(\Sigma_n,\Delta)$. According to our selection in Step 1, the input sequences to $\Sigma_{(\gamma,\pi,\varphi)}$ must be taken from $S(\theta^m)$. In our construction of the compensators $\pi$ and $\varphi$ we have used only the given nominal recursion function $f_n$, and the compensators we derived are given in explicit form and are implementable. We can achieve desirable dynamics assignment for the stabilized closed loop system through the selection of the unimodular system M in Step 4. Of course, the exact input/output relation induced by the closed loop configuration depends on the particular system $\Sigma$ inserted in it, but internal stability holds for any system $\Sigma \in \mathcal{N}(\Sigma_n,\Delta)$. Detailed proofs and justification for the design procedure we have outlined here are given in HAMMER [1987b and 1988], where more general forms of stabilizing compensators are also described.

We conclude this note with a rather simple example on the computation of compensators $\pi$ and $\varphi$ that yield robust stabilization of a given nominal system. The example is reproduced from HAMMER [1988], where all the computations are described in detail. Here, we only exhibit the class of systems that is stabilized and the final form of the compensators, in order to provide a feeling of the explicit form of the solutions provided by procedure described before. We emphasize that the procedure is valid for multivariable systems as well, and the stabilizing compensators in the multivariable case are of a similar form.

**(4.8) EXAMPLE.** We consider the design of a robust stabilization scheme for the following class of single-input single-output systems. The nominal system $\Sigma_n : S(R) \to S(R)$ is given by the recursive representation $\Sigma_n :\ x_{k+1} = 2x_k + 2u_k + \sin(x_k u_k),\ x_0 = 0$. The disturbed system $\Sigma$ belongs to the class of systems having recursive representations of the form $\Sigma :\ x_{k+1} = (2+\kappa)x_k + (2+\chi)u_k + \sigma\sin(x_k u_k)$, $x_0 = 0$, where $\kappa$, $\chi$, and $\sigma$ are real numbers in the intervals $-\frac{1}{3} \le \kappa \le \frac{1}{3}$, $-\frac{1}{3} \le \chi \le \frac{1}{3}$, and $-1 \le \sigma \le 1$. Our objective is to design an internally stable configuration that will stabilize any system of this class. We use only the parameters of the nominal recursion function. We note that $\gamma$ is a scalar here. In HAMMER [1988] we have used the following values for the design parameters : $\theta = 2$, $\gamma = 1$, $\delta = 1$, $\alpha = 19$, $\xi = 2$, $\zeta = 1/2$, and $\varepsilon = 1/20$. We take the simple choice $M = \beta I$ for our unimodular system, with $\beta = 25$, so that, recalling that we have a scalar system here, $\mathcal{M} = 25$.

The system $Q_n^E$ here can be readily computed, and, denoting by $\{x_k\}$ the input sequence of $Q_n^E$ and by $\{z_k\}$ the output sequence of $Q_n^E$, the representation of $Q_n^E$ is given by the relations

$$Q_n^E : \begin{cases} y_k := e(x_k) \\ z_{k+1} = y_{k+1} - 2y_k - \sin[(y_k - z_k)z_k] \\ z_0 = y_0 \end{cases}$$

$k = 0,1,2, \ldots$ . Here, the function $e : R \to [-\frac{1}{2}, \frac{1}{2}]$ is defined by $e(x) = x$ if $|x| \leq \frac{1}{2}$ and $e(x) = \frac{1}{2}\text{sign}(x)$ if $|x| > \frac{1}{2}$, where $\text{sign}(x) = \pm 1$, depending on the sign of $x$. The compensators become

$$\varphi = 25 - \frac{1}{20}Q_n^E, \quad \pi = 20.$$

Then, for any system $\Sigma : S(R) \to S(R)$ having a recursive representation of the form $x_{k+1} = (2+\kappa)x_k + (2+\chi)u_k + \sigma\sin(x_k u_k)$, where $-\frac{1}{3} \leq \kappa \leq \frac{1}{3}$, $-\frac{1}{3} \leq \chi \leq \frac{1}{3}$, and $-1 \leq \sigma \leq 1$, the closed loop $\Sigma_{(1,\pi,\varphi)}$ around $\Sigma$ will be internally stable for all input sequences from $S(2)$. As we see, the compensators $\pi$ and $\varphi$ that we obtained can be readily implemented.

## 5. REFERENCES


H.S. BLACK [1934], "Stabilized feedback amplifiers", Bell Systems Tech. J., Vol 13, pg. 1.

H.W. BODE [1945], "Network analysis and feedback amplifier design", D. Van Nostrand, N.Y.

C.A. DESOER and C.A. LIN [1984], "Nonlinear unity feedback systems and Q-parametrization", Int. J. Control, Vol. 40, pp. 37-51.

C.A. DESOER and M. VIDYASAGAR [1975], "Feedback systems: input-output properties", Academic Press, N.Y..

J. HAMMER [1984a], "Nonlinear systems: stability and rationality", Int. J. Control, Vol. 40, pp. 1-35.

[1984b], "On nonlinear systems, additive feedback, and rationality", Int. J. Control, Vol. 40, pp. 953-969.

[1985a], "Nonlinear systems, stabilization, and coprimeness", Int. J. Control, Vol. 42, pp. 1-20.

[1985b], "The concept of rationality and stabilization of nonlinear systems", Proceedings of the symposium on the Mathematical Theory of Networks and Systems, Stockholm, Sweden, June 1985 (C. Byrnes and A. Lindquist editors), North Holland Publishers.

[1986], "Stabilization of nonlinear systems", Int. J. Control, Vol. 44, pp. 1349-1381.

[1987a], "Fraction representations of nonlinear systems: a simplified approach", Int. J. Control, Vol. 46, pp. 455-472.

[1987b], "Assignment of dynamics for nonlinear recursive feedback systems", Int. J. Control, to appear.

[1988], "On robust stabilization of nonlinear systems", Preprint, Center for Mathematical System Theory, University of Florida, Gainesville, Florida 32611, USA.

A. ISIDORY [1985], "Nonlinear control systems: an introduction", Lecture Notes in Control and Information Sciences, Vol. 72, Springer Verlag, Berlin.

H. KIMURA [1984], "Robust stabilizability for a class of transfer functions", Trans. IEEE, Vol. AC-29, pp. 788-793.

K. KURATOWSKI [1961], "Introduction to set theory and topology", Pergamon Press, N.Y..

G.C. NEWTON, Jr., L.A. GOULD, and J.F. KAISER [1957], "Analitical design of linear feedback controls", Wiley, N.Y..

H.H. ROSENBROCK [1970], "State space and multivariable theory", Nelson, London.

[1974], "Computer-aided control system design", Academic Press, London.

E.D. SONTAG [1981], "Conditions for abstract nonlinear regulation", Information and Control, Vol. 51, pp. 105-127.

M. VIDYASAGAR [1980], "On the stabilization of nonlinear systems using state detection", Trans. IEEE, Vol. AC-25, pp. 504-509.

G. ZAMES [1966], "On the input-output stability of time varying nonlinear feedback systems - I: Conditions derived using concepts of loop gain, conicity, and positivity", Trans. IEEE, Vol. AC-11, pp. 228-239.

[1981], "Feedback and optimal sensitivity: model reference transformations, multiplicative seminorms, and approximate inverses", Trans. IEEE, Vol. AC-26, pp. 301-320.


## ACKNOWLEDGEMENT

This research was supported in part by the National Science Foundation, USA, Grants Number 8501536 and 8896182. Starting August 1987, this research was also supported in part by US Air Force Grant AFOSR-85-0186, by office of Naval Research Grant N00014-86-K-0538, and by US Army Grant DAAG29-85-K-0099, through the Center for Mathematical System Theory, University of Florida, Gainesville, Florida, 32611.

# STABILIZATION OF NONLINEAR SYSTEMS BY MEANS OF LINEAR FEEDBACKS

A. ANDREINI (1), A. BACCIOTTI (2),
P. BOIERI (2) and G. STEFANI (3)

Abstract - Solutions of the stabilizability problem for nonlinear systems are usually supposed to be nonlinear state feedback functions. Despite this common opinion, several examples show that linear stabilizing feedbacks may exist also in the case of nonlinear systems whose linearization is not stabilizable. The aim of this paper is to give a summary of recent results concerning certain classes of systems admitting linear stabilizing feedbacks.

---

(1) Istituto di Matematica Applicata, Facoltà di Ingegneria
Via S. Marta 3, 53134 Firenze (Italy)
(2) Dipartimento Matematico del Politecnico,
Corso Duca degli Abruzzi 24, 10129 Torino (Italy)
(3) Dipartimento di Matematica e Applicazioni
Via Mezzocannone 8, 80134 Napoli, (Italy)

## 1. Introduction

In this paper we shall review some results concerning smooth stabilizability of nonlinear systems recently obtained by the authors. The general statement of the problem is the following (see [B1] for more details and general references). Consider a control system

(1.1) $x' = f(x,u)$ , $x \in R^n$ , $u \in R^m$

where $f \in C^1$ and $f(0,0)=0$. We shall say that (1.1) is (smoothly) stabilizable if there exists a smooth function u: $R^n \longrightarrow R^m$ such that $u(0)=0$ and the origin is a (locally) asymptotically stable equilibrium point for the closed-loop system

(1.2) $x' = f(x,u(x))$ .

Such a function u is called a stabilizing feedback. We want to find conditions for the existence of stabilizing feedbacks (note that stabilizing feedbacks, when they exist, are not unique).

A natural approach to the problem is to consider the linear system

(1.3) $x' = Ax + Bu$

where $A=D_xf(0,0)$ and $B=D_uf(0,0)$. Indeed, conditions for stabilizability of (1.3) are well known. Furthermore, the theorem of stability at the first approximation tells us that if (1.3) is stabilizable, then there exists a linear stabilizing feedback $u(x)=Fx$ for (1.1), which provides also exponential decay for trajectories of (1.2) in a neighborhood of zero. This result admits a partial converse. Indeed, it is proven in [Z] that if there exists a stabilizing feedback for (1.1) providing exponential decay for trajectories of (1.2) in a neighborhood of zero, then the linearized system (1.3) must be stabilizable and hence, stabilization can be achieved by means of linear feedbacks. On the contrary, simple examples show that (1.1) can be stabilized by means of linear feedbacks even if (1.3) is not stabilizable. It is noteworthy that linear feedbacks work also in cases of practical interest. Consider for instance the stabilization problem for the angular velocity equations of a rigid body. It has been recently proved that when only one control torque (not aligned with a principal axis) is available, stabilization can be achieved by means of a linear feedback ([AS]). Moreover, although never noticed before, it is possible to prove that a linear stabilizing feedback exists also in the case of two independent control torques. This remarks motivate the following definition.

Definition 1.1 - The system (1.1) is said to be linearly stabilizable if there exists a linear stabilizing feedback $u(x)=Fx$ . #

When the existence of a stabilizing feedback cannot be stated (or excluded) by looking at (1.3), the system is called critical. A computer aided approach to linear stabilizability of critical systems has been developed in [BB] and [Bo] for low dimensions and simple criticalities. It will be described in Sec. 2 where, in particular, we shall focus on the so-called affine systems.

Of course, from a practical point of view, local stabilizability is not sufficient: informations about the size of the region of attraction are also needed. Neither existence results obtained on the base of (1.3) nor the approach of [BB], [Bo] give such information. Actually, simple examples show that in general global stabilizability of nonlinear systems cannot be achieved by using only linear feedbacks. Thus, we are led to the following definition.

Definition 1.2 - We shall say that (1.1) is linearly potentially globally stabilizable if for each $\rho>0$ there is a linear function $u_\rho(x)=F_\rho x$ such that:

(i) $u_\rho$ is a stabilizing feedback;

(ii) the region of attraction of (1.2) with $u=u_\rho$ contains the ball of radius $\rho$ and centre zero. #

In Sec. 3 we shall state a sufficient condition for the existence of a linear potentially global stabilizing feedback; it is applicable to systems of the form

$$x' = f(x) + Bu \tag{1.4}$$

where B is a (n,m)-matrix and f is a homogeneous vector field of odd degree. This condition is an immediate consequence of the theory developed in [ABS1] and [ABS2]. In Sec. 4, the result will be extended to systems of the form (1.4) where f is a polynomial vector field (see also [B2]).

## 2. Local stabilizability of planar single-input systems

A very general and very natural approach to linear stabilizability of critical systems can be based on certain geometric methods, recently introduced in control theory by D. Aeyels and other authors (see [B1]). It consists of the following steps:

a) put $u(x)=Fx$ in (1.1), where F is considered an unknown matrix;
b) if necessary, make a linear change of coordinates, in order to put $D_x[f(x,Fx)]_{x=0}$ in Jordan form;
c) perform a centre manifold reduction;
d) determine the elements of F in order to ensure asymptotic stability of the reduced equation.

In principle, step d) enables us to obtain conditions which guarantee the solvability of the problem. As a rule, such conditions involve the coefficients of the first few terms of the Taylor expansion of f.

In practice, the procedure is successful only for low dimensions and simple criticalities. Several results have been obtained in [BB] for analytic, planar and single-input systems, performing computations with the aid of a symbolic manipulation package. A sample of those results is reported below. In order to avoid complicated expressions, we shall limit ourselves to (analytic, planar, single-input) affine systems here, namely, systems of the form

$$(2.1) \qquad \begin{cases} x' = f_o(x,y) + u\, f_1(x,y) \\ y' = g_o(x,y) + u\, g_1(x,y) \end{cases} .$$

Moreover, we shall assume $f_1(0,0)=g_1(0,0)=0$. Since, by assumption, also $f_o(0,0)=g_o(0,0)=0$, we have the expansions

$$f_o(x,y)= \sum_{k=1}^{\infty} \sum_{i=0}^{k} a_{i,k-i,0}x^i y^{k-i} , \qquad g_o(x,y)= \sum_{k=1}^{\infty} \sum_{i=0}^{k} b_{i,k-i,0}x^i y^{k-i}$$

$$f_1(x,y)= \sum_{k=1}^{\infty} \sum_{i=0}^{k} a_{i,k-i,1}x^i y^{k-i} , \qquad g_1(x,y)= \sum_{k=1}^{\infty} \sum_{i=0}^{k} b_{i,k-i,1}x^i y^{k-i}$$

in a neighborhood of zero.

<u>Theorem 2.1</u> - Let $a_{100}=a_{010}=b_{100}=0$ and $b_{010}=-1$ (modulo linear changes of coordinates, this corresponds to the critical case where the linearized system has one zero uncontrollable eigenvalue). System (2.1) is linearly stabilizable if either

(i) $a_{101}\neq 0$ and $a_{101}b_{200}\neq a_{200}b_{101}$, or

(ii) $a_{101}\neq 0$, $a_{101}b_{200}=a_{200}b_{101}$, but $a_{300}<a_{200}a_{201}/a_{101}$.

The stabilizing feedback has the form $u(x)=-(a_{200}/a_{101})x+qy$, for suitable values of q (in case (ii), it can be taken q=0).

System (2.1) is linearly stabilizable even if

(iii) $a_{101}=a_{200}=0$ but $a_{011}b_{101}<0$.

In this case, the stabilizing feedback takes the form $u(x)=kx$, for $|k|>>0$. Moreover, if

(iv) $a_{101}\neq 0$, $a_{101}b_{200}=a_{200}b_{101}$ and $a_{300}>a_{200}a_{201}/a_{101}$, or

(v) $a_{101}=0$ and $a_{200}\neq 0$,

there is no linear stabilizing feedback. #

Assume now that $a_{100}=b_{010}=0$, $-a_{010}=b_{100}=1$ (this corresponds to the critical case where the linearized system has a pair of imaginary uncontrollable eigenvalues). Using a normal form expansion, it can be seen that asymptotic stability of system (2.1) is related to the sign of the expression

$$H(p,q)=H_{20}p^2+H_{02}q^2+H_{11}pq+H_{10}p+H_{01}q+H_{00}$$

where

$$H_{00}=a_{120}+3a_{300}+b_{210}+3b_{030}+a_{200}a_{110}-2a_{200}\,b_{200}$$
$$+a_{110}a_{020}+2a_{020}b_{020}-b_{110}b_{200}-b_{110}\,b_{020}$$

$$H_{10}=a_{021}+3a_{201}+b_{111}+a_{101}a_{110}-2a_{101}b_{200}$$
$$+a_{200}a_{011}-2a_{200}b_{101}+a_{011}a_{020}-b_{011}b_{200}$$
$$-b_{011}b_{020}-b_{101}b_{110}$$

$$H_{01}=a_{111}+3b_{021}+b_{201}+a_{101}a_{200}+a_{101}a_{020}+a_{011}a_{110}$$
$$+2a_{011}b_{020}+2a_{020}b_{011}-b_{011}b_{110}-b_{101}b_{200}-b_{101}b_{020}$$

$$H_{20}=a_{101}a_{011}-2a_{101}b_{101}-b_{011}b_{101}$$

$$H_{11}=a_{101}^2+a_{011}^2-b_{011}^2-b_{101}^2$$

$$H_{02}=a_{101}a_{011}+2a_{011}b_{011}-b_{011}b_{101} \quad .$$

More precisely,

<u>Theorem 2.2</u> - Let $a_{100}=b_{010}=0$, $-a_{010}=b_{100}=1$. If $H(p,q)<0$ for some $p,q\in R$, then the function $u(x)=px+qy$ is a stabilizing feedback. On the contrary, if $H(p,q)>0$ for every $p,q\in R$, the system is not linearly stabilizable. #

From Theorem 2.2, it is clearly possible to derive sufficient or necessary conditions for the existence of linear stabilizing feedbacks. For instance, if either $H_{20}$ or $H_{02}$ (or both) are negative, then $H(p,q)$ obviously takes negative values for some p and q: thus, the condition of the theorem is fulfilled.

The reader is referred to [BB] and [Bo] for proofs and details.

## 3. Stabilizability of homogeneous systems

For homogeneous system we mean a system of the form (1.4) where $f(\lambda x)=\lambda^k f(x)$ for each $x\in R^n$ and $\lambda\in R$. We shall assume that the degree of homogeneity k is an odd integer, $k\geq 1$.

For each real symmetric definite positive matrix P, let

$$C^-(P) = \{x\colon {}^t xPf(x)<0\}\cup\{0\}$$

(as before, $^t$ denotes transposition). Since k is odd, we see that $C^-(P)$ is a symmetric cone. This means that if $x\in C^-(P)$ and $\lambda\in R$, then $\lambda x\in C^-(P)$. The following theorem holds (see [ABS1]).

<u>Theorem</u> <u>3.1</u> - Let us assume that k is odd and that

(3.1) $\mathrm{Ker}\ {}^tBP \subset C^-(P)$

for some symmetric definite positive matrix P. Then, (1.4) is linearly potentially globally stabilizable. #

More precisely, there exists a positive number $\mu°$ such that for each $\mu>\mu°$ and each $\rho>0$, the function

(3.2) $u_\rho(x) = -\mu\rho^{k-1}\ {}^tBPx$

is a stabilizing feedback. Moreover, the region of attraction of the origin for the closed-loop system obtained substituting (3.2) in (1.4) contains the ball of radius $\rho$.

Condition (3.1) can be stated in an equivalent form.

<u>Theorem</u> <u>3.2</u> - There exists a symmetric positive definite matrix P satisfying (3.1) if and only if there exist a subspace V and a scalar product g on V such that

(i) $V \oplus \mathrm{ImB} = R^n$ ;

(ii) $g(x,\Phi(x))<0$ for $x\in V$ $(x\neq 0)$, where $\Phi$ is the projection of f on V along ImB. #

On the basis of this theorem (cf. [ABS2]), it is easy to see what happens near the origin when (3.2) is used in (1.4). Indeed, after substitution, the subspace V mentioned in Theorem 3.2 becomes the centre manifold for the closed-loop system. Thus, we see that the system presents exponential decay along the directions in ImB (fast dynamic) and weak attraction along the directions in V (slow dynamic).

Inspired by Theorem 3.2, we can give a slight generalization of the local content of Theorem 3.1.

<u>Theorem</u> <u>3.3</u> - Assume that there exists a subspace V such that:

(i) $V\oplus \mathrm{ImB} = R^n$ ;

(ii) the projection $\Phi(x)$ of $f(x)$ on V along ImB defines a vector field on V, for which the origin is asymptotically stable.

Then, the system is (locally) linearly stabilizable.

Sketch of the proof - By a linear change of coordinates, the system takes the form

$$(*)\qquad \begin{cases} x_1' = f_1(x_1,x_2) + u \\ x_2' = f_2(x_1,x_2) \end{cases}$$

where $V=\{x_1=0\}$. By (ii), the origin is asymptotically stable for the equation $x_2'= f_2(0,x_2)$.

Let $U(x_2)$ be a Liapunov function for this equation. Put $u=-\mu x_1$ in (*). By reasoning as in the proof of Theorem 3.1 (see [ABS1]), it is easy to see that for sufficiently large $\mu$,

$$W(x_1,x_2)=({}^t x_1 x_1/2) + U(x_2)$$

is a Liapunov function for (*). #

4. Stabilizability of polynomial systems

Let us consider again a system of the form (1.4), now, we assume that

$$f(x) = f_k(x)+f_{k+1}(x)+...+f_{k+s}(x)$$

where each $f_{k+i}(x)$ is a homogeneous polynomial vector field, whose homogeneity degree is $k+i$ ($i=0,1,...,s$). Let us define

$$C_k^-(P) = \{x: {}^t xPf_k(x)<0\}\cup\{0\}$$

and

$$Z_i(P) = \{x: {}^t xPf_{k+i}(x)\le 0\}\ , \quad i=1,2,...,s\ .$$

We know that $C_k^-(P)$ is a symmetric cone if $k$ is odd. The same is true for $Z_i(P)$ if $k+i$ is odd. On the contrary, when $k+i$ is even, a straight line $\{\lambda x\ ,\ \lambda\in R\}$ may be contained in $Z_i(P)$ only if ${}^t xPf_{k+i}(x)=0$. It is not difficult to prove that if

$$\mathrm{Ker}\ {}^t BP \subset C_k^-(P)$$

for some symmetric positive definite matrix $P$, then $u(x)=-\mu\, {}^t BPx$

is a locally stabilizing feedback for each sufficiently large $\mu$.

However, there is no hope for getting global or potentially global stabilizability without assumptions on the terms of degree greater than k. The following theorem holds (see [B2]).

Theorem 4.1 - Let $k$ be odd and let

$$\operatorname{Ker} {}^tBP \subset C_k^-(P) \cap Z_1(P) \cap \ldots \cap Z_s(P)$$

for some symmetric definite positive matrix P. Then, the system is linearly potentially globally stabilizable. #

More precisely, for each $\rho>0$ there exists a positive number $\mu(\rho)$ such that the function $u_\rho(x)=-\mu(\rho)\,{}^tBPx$ stabilizes the system, and the region of attraction of the origin for the closed-loop system contains the ball of radius $\rho$.

## References

[ABS1] A. Andreini, A. Bacciotti, G. Stefani, Global Stabilizability of Homogeneous Vector Fields of Odd Degree, to appear in Systems and Control Letters

[ABS2] A. Andreini, B. Bacciotti, G. Stefani, On the Stabilizability of Homogeneous Control Systems, 8th Int. Conf. Analysis and Optim. of Systems, Antibes 1988, INRIA, Lect. Notes in Control & Inf., Springer Verlag

[B1] A. Bacciotti, The Local Stabilizability Problem for Nonlinear Systems, IMA J. Math. Control and Inf., 5 (1988) pp. 27-39

[B2] A. Bacciotti, Further Remarks on Potentially Global Stabilizability, submitted

[BB] A. Bacciotti, P. Boieri, Linear Stabilizability of Nonlinear Systems, submitted

[Bo] P. Boieri, Linear Stabilizability of Tridimensional Nonlinear systems, in progress

[KD] D.E. Koditschek, K.S. Narendra, Stabilizability of Second-Order Bilinear Systems, IEEE Trans. AC-28 (1983) pp. 987-989

[Z] J. Zabczyk, Some Comments on Stabilizability, to appear in Appl. Math. Optim.

# ROBUSTNESS AND SINGULAR PERTURBATIONS TECHNIQUES

A NEW APPROACH TO IDENTIFICATION
PROBLEMS USING SINGULAR PERTURBATIONS.

H.W.Knobloch*)
Mathematisches Institut
Am Hubland
D-87oo Würzburg

## 1. Introduction. Review of previous results.

Singular perturbation theory is a well established tool for treating various types of control problems. Among others it leads to an understanding of the dynamics which is induced by means of so called "high-gain-feedgack". To give an example of this type of control law let us consider an affine single input/output system which is written in the form

$$\dot{x} = a(x) + ub(x), \quad y = h(x). \tag{1.1}$$

u ( = input), y (= output) are scalars, the vector $x = (x_1,\ldots,x_n)^T$ represents the state. The proportional control law $u = y/\varepsilon = h(x)/\varepsilon$ where $\varepsilon$ is a small positive parameter changes (1.1) to a closed loop system. The evolution of this system is governed by the ordinary d.e. ( = differential equation)

$$\dot{x} = a(x) + \frac{h(x)}{\varepsilon}b(x). \qquad (1.1'_\varepsilon)$$

The geometry of trajectories in the neighborhood of the set

$$M_o := \{x : h(x) = 0\} \tag{1.2}$$

can be analyzed for $(1.1'_\varepsilon)$ rather completely under the proviso

$$(L_b h)(x) := (\text{grad } h\cdot b)(x) < 0 \text{ for } x \in M_o . \tag{1.3}$$

This has been carried out in [1]-[3], the results which are relevant for our purposes will be summarized in the subsequent Theorem 1.1. What is actually used in the present paper is a special case of this theorem which is discussed in Section 2. It concerns the existence of a sufficiently smooth solution of the first order partial d.e. $(2.2_\varepsilon)$. Here $\varepsilon$ is again a small positive parameter, f a sufficiently smooth

*) Supported by Deutsche Forschungsgemeinschaft-Kn 164/3-1

vector field, $X$ a bounded open set in the $(x_2,\dots,x_n)$-space. No further conditions on f, $X$ are required in order to be sure that there exists a smooth solution. It is this generality which makes Theorem 2.1 useful for the subsequent considerations. On the other hand if it comes to the proof of this theorem one has to pay a certain prize for the freedom which one has in applying it to concrete situations. As it appears there exists no simple alternative way to establish existence of a global solution for the p.d.e. $(2.2_\varepsilon)$. In particular one cannot resort to the usual method of characteristics. Instead we pursue the following way. We set up the equations for the characteristics - these are the o.d.e.'s $(2.4_\varepsilon)$. Next we construct the global center manifold for $(2.4_\varepsilon)$ whose existence has been established in the literature quoted before. Then we take a suitable analytic representation of this manifold.

As a first application in the direction of control theory we present in Section 3 a recipe for solving - at least in principle - the following special reconstruction problem. Given a solution $x(t) = (x_1(t),\dots,x_n(t))^T$ of an o.d.e.

$$\dot{x} = f(x),\ x = (x_1,\dots,x_n)^T. \tag{1.4}$$

Assume that the component $x_1$ of the state has been measured on some time interval I up to an error with a known bound $\delta > 0$, i.e. assume that there is given a function $\xi(t)$ on I such that

$$|x_1(t) - \xi(t)| < \delta,\ t \in I. \tag{1.5}$$

What we wish to demonstrate is that a similar approximation for $\dot{x}_1(t)$ is available, namely

$$|\dot{x}_1(t) - \eta_\varepsilon(t)| < \rho \tag{1.6}$$

where

$$\eta_\varepsilon(t) := \frac{1}{\varepsilon^2}\{\varepsilon\xi(t) - e^{-t/\varepsilon}\int_0^t e^{\tau/\varepsilon}\xi(\tau)d\tau\}. \tag{1.7}$$

It is valid in this precise sense. Given $\rho > 0$. Then one can find positive $\delta$ and $\varepsilon$ such that (1.6) is implied by (1.5). $\delta$ and $\varepsilon$ depend upon $\rho$ and upon bounds for the vectorfield f and its partial derivatives up to a certain order. These bounds need not be valid in the whole state space but only on some bounded subset $\hat{X}$ which is such that $x(t) \in \hat{X}$ for all $t \in I$.

At first glance this result seems not to be too far from what can be gained by elementary arguments. In fact, $\eta_\varepsilon(t)$, as given by (1.7), provides an approximation for $\dot{\xi}(t)$ if the function $\xi(\cdot)$ is twice differentiable. This follows simply by integration by parts, see formula (4.1) below. Note however that our problem is to set up approximations for $\dot{x}_1(t)$ and not for $\dot{\xi}(t)$! Hence we cannot simply adopt standard techniques say for carrying out numerical differentiation. What is needed are arguments which exploit these facts: $x(\cdot)$ is solution of the d.e. (1.4), $x(t) \in \hat{X}$ for $t \in I$, certain information about the vector field f and its derivatives on $\hat{X}$ is available. Note also that we do not assume differentiability of $\xi(\cdot)$; the results of Sec.3 are valid under the only proviso that $\xi(\cdot)$ is an integrable function satisfying the condition (1.5). Hence the relation between $\eta_\varepsilon$ and $\dot{x}_1$ remains correct even if the question whether $\eta_\varepsilon$ is an approximation for $\dot{\xi}(\cdot)$ is meaningless.

On the other hand if we assume that $\xi(\cdot)$ is sufficiently smooth one can extract from the results of Sec. 1-3 a bound for

$$|\dot{x}_1(t) - \dot{\xi}(t)| . \tag{1.8}$$

This bound however is rather crude and probably of no interest from a practical viewpoint. A method which improves the results is outlined in Sec. 4. It has an interesting aspect which may help in the following situation. One starts out with some - say non-smooth-approximation $\xi$ for $x_1$. Then one wants to smooth out $\xi$ in such a way that $\dot{\xi}$ yields a good approximation for $\dot{x}_1$. Some hints for the actual performance of this procedure can be obtained from a careful analysis of the error terms in (4.6). The first one is a global one, i.e. it depends upon the behaviour of $\dot{\xi}, \ddot{\xi}$ on the whole time interval, cf. (4.4), (4.5). In contrast the second one - which estimates the error bound in the asymptotic expansion of s, cf. Theorem 2.1 - is local one. So in principle one can evaluate errors which arise by making $\dot{\xi}$ smooth at one point on the expense large amplitudes at some other place.

As announced earlier we conclude this section by refereeing the necessary background material about invariant manifolds for o.d.e.'s. By a manifold $M$ we always mean one which is compact and properly embedded in the state space $\mathbb{R}^n$. It may have a boundary which is denoted by $\partial M$. $M$ is called invariant with respect to a d.e. if solutions which have a point in common with $M$ can leave the manifold through boundary points only.

The d.e. which underlies the subsequent considerations is given by $(1.1'_\varepsilon)$ and we assume that a,b,h are defined and of class $C^N$ on a neighborhood of some compact subset $\hat{X}'$ of the state space $\mathbb{R}^n$. We introduce the set

$$M'_0 := \{x \in \hat{X}' : h(x) = 0\}$$

and assume that

$$L_b h(x) < 0 \quad \text{for} \quad x \in M'_0 \tag{1.9}$$

(for the definition of the Lie-derivative cf. (1.3)). Note that (1.9) implies that $\operatorname{grad}h(x) \neq 0$ for $x \in M_o'$. Hence $M_o'$ is a (n-1)-dimensional manifold, possibly with boundary.

It is convenient to state our main result using some terminology from [2], in particular the notion of "local coordinates near a point $\bar{x} \in M_o'$". This is a pair (y,z) where y is of dimension 1 and z of dimension n-1 which is related to the state variable x through a linear transformation

$$x-\bar{x} = P(y,z)^T . \qquad (1.10)$$

P is a nonsingular matrix depending upon $\bar{x}$ which has to satisfy the condition

$$P^{-1}J(\bar{x})P = \operatorname{diag}(a(\bar{x}),0) \qquad (1.11)$$

where

$$J(\bar{x}) := b(\bar{x})\cdot(\operatorname{grad}h)(\bar{x}),\ a(\bar{x}) := (L_b h)(\bar{x}). \qquad (1.12)$$

Note that P is not uniquely determined by these relations. However one can select P from a finite set $\mathcal{P} = \{P_1,\dots,P_k\}$ of smooth matrix functions in a certain way which has been explained in detail in [2].

Let now $M_o$ be a compact submanifold of $M_o'$ which is such that

$$M_o \cap \partial M_o' = \phi .$$

There exists then a fixed positive $\rho$ such that $M_o$ is contained in the union of local neighborhoods $N_{i,\rho}(\bar{x})$, $\bar{x} \in M_o$. These are open sets which are defined in local coordinates (1.10) around $\bar{x}$ simply by the inequalities $|y| < \rho$, $\|z\|<\rho$. The subscript i indicates the member of the set $\mathcal{P}$ whose value at $x=\bar{x}$ has been taken for P. Furthermore if $\rho$ is sufficiently small the intersection of $N_{i,\rho}(\bar{x})$ with $M_o$ is given in the form

$$\{y,z : y = y_o(z),\ \|z\|< \rho\} \qquad (1.13)$$

where $y_o(\cdot)$ is a $C^N$-function of z.

<u>Theorem 1.1.</u> Let (1.9) be satisfied. Then there exists an invariant manifold $M(\varepsilon)$ for the d.e. $(1.1_\varepsilon')$ whenever $\varepsilon \neq 0$, $|\varepsilon|$ sufficiently small, with this property

$$M(\varepsilon) \cap N_{i,\rho}(\bar{x}) = \{y,z : y=y(\varepsilon,z),\ \|z\|<\rho\} . \qquad (1.14)$$

$y(\varepsilon,z)$ is of class (at least) $C^{N-1}$ on the set

$$\{z,\varepsilon : \|z\|<\rho,\ |\varepsilon|<\varepsilon_o\}$$

where $\varepsilon_o$ is a fixed positive number. Furthermore we have $y(0,z)=y_o(z)$ (cf. (1.13)). All this holds true for every $\bar{x} \in M_o$ and sufficiently small fixed $\rho,\varepsilon_o$.

<u>Proof.</u> The usual rescaling of time changes $(1.1_\varepsilon')$ to a d.e. which formally is non-singular with respect to $\varepsilon$:

$$\dot{x} = \varepsilon a(x) + h(x)b(x) =: f(x,\varepsilon). \tag{1.15}$$

The right hand side can be viewed upon as a perturbation of the vectorfield $f(x,0) = h(x)b(x)$ and the relevant results from [2], in particular Theorem 2.1, (i)-(iii), can be applied. This remark proves our theorem. One gap remains: Smooth dependence upon x and $\varepsilon$ is not explicitly stated in [2]. It can be easily closed simply by updating $\varepsilon$ to a state variable. □

## 2. A first order partial d.e.

We assume throughout this section that the subset of the state space appearing in Section 1 is a cylinder

$$\hat{X} = \{x = (x_1,x_2,\dots,x_n)^T : |x_1| \le \kappa, (x_2,\dots,x_n) \in X'\} \tag{2.1}$$

where $\kappa$ is a positive number and $X'$ a compact subset of the $(x_2,\dots,x_n)$-space. Furthermore we assume that there is given a vectorfield $f = (f_1,\dots,f_n)$ which is of class $C^{N+1}$ in a neighborhood of $\hat{X}$. We wish to study solutions $s = s(\varepsilon,x_2,\dots,x_n)$ of the first order p.d.e.

$$f_1(s,x_2,\dots,x_n) - \frac{s}{\varepsilon} = \sum_{i=2}^{n} (\partial s/\partial x_i) f_i(s,x_2,\dots,x_n). \tag{2.2$_\varepsilon$}$$

Theorem 2.1. Let f be of class $C^{N+1}$ on some neighborhood of $\hat{X}$ and let $X$ be a subset of $X'$ which is open relative to the $(x_2,\dots,x_n)$-space. Then there exists a function $s = s(\varepsilon,x_2,\dots,x_n)$ which is defined and of class $C^N$ on the set

$$\{\varepsilon,x_2,\dots,x_n : |\varepsilon|<\varepsilon_o, (x_2,\dots,x_n)\in X\} \tag{2.3}$$

and which has these properties.

(i) $s(0,x_2,\dots,x_n) = 0$ identically in $x_2,\dots,x_n$,

(ii) s is solution of the p.d.e. (2.2$_\varepsilon$) for $\varepsilon \ne 0$, $|\varepsilon|<\varepsilon_o$.

Proof. We wish to apply Theorem 1.1 to the o.d.e.

$$\dot{x}_1 = f_1(x_1,x_2,\dots,x_n) - \frac{x_1}{\varepsilon}, \quad \dot{x}_i = f_i(x_1,\dots,x_n), \quad i=2,\dots,n. \tag{2.4$_\varepsilon$}$$

Note that it is of the form (1.1$'_\varepsilon$) with $a(x) = f(x)$, $b(x) = (-1,0,\dots,o)^T$, $h(x) = x_1$. Hence $(L_b h)(x) = -1$ everywhere, and (1.9) as well as the other conditions stated in Section 1 are satisfied if we put

$$M'_o = \{(x_1,\dots,x_n)^T \in \hat{X} : x_1 = 0\},$$

$$M_o = \{(x_1,\dots,x_n)^T : x_1 = 0, (x_2,\dots,x_n) \in X\}.$$

Since $J(\bar{x}) = \mathrm{diag}(-1,0,\dots,0)$ we can take as matrix P (cf. (1.11)) simply the identity. Local coordinates near a point $\bar{x} = (0,\bar{x}_2,\dots,\bar{x}_n) \in M_o$ are given as $y = x_1$, $z = (x_2,\dots,x_n)^T - (\bar{x}_2,\dots,\bar{x}_n)^T$ and the neighborhood $N_{i,\rho}(\bar{x})$, $\bar{x} \in M_o$, is simply the box $\{x : \| x-\bar{x}\| < \rho\}$ (N.B: $\|...\|$ always means maximum norm). The union of these neighborhoods contains the open set

$$\tilde{N}_\rho := \{ (x_1,x_2,\ldots,x_n) : |x_1|<\rho, (x_2,\ldots,x_n) \in X\}.$$

Let $M(\varepsilon)$ be the invariant manifold for $(2.4_\varepsilon)$ which has all properties as stated in Theorem 1.1. It follows then from this theorem that we have an analytic representation

$$M(\varepsilon)\cap\tilde{N}_\rho = \{(x_1,\ldots,x_n)^T : x_1 = s(\varepsilon,x_2,\ldots,x_n), (x_2,\ldots,x_n) \in X\} \tag{2.5}$$

in terms of a function s which is of class $C^{N-1}$ and vanishes for $\varepsilon=0$. Since $M(\varepsilon)$ is invariant with respect to the d.e. $(2.4_\varepsilon)$ the set (2.5) is locally invariant. In other words: The relation $x_1=s(\varepsilon,x_2,\ldots,x_n)$ is preserved under the flow defined by $(2.4_\varepsilon)$ as long as $(x_2,\ldots,x_n)\in X$. This statement is nothing else than the geometric version of statement (ii) of the theorem. □

<u>Corollary.</u> There exist functions $s_\nu(x_2,\ldots,x_n)$, $\nu = 1,\ldots,N$, which are defined on $X$ but depend on f only such that

$$s(\varepsilon,x_2,\ldots,x_n) = \sum_{\nu=1}^{N-1} \varepsilon^\nu s_\nu(x_2,\ldots,x_n) + \mathcal{O}(\varepsilon^N). \tag{2.6}$$

In particular

$$\begin{aligned} s_1(x_2,\ldots,x_n) &= f_1(0,x_2,\ldots,x_n), \\ s_2(x_2,\ldots,x_n) &= (\partial f_1/\partial x_1)(0,x_2,\ldots,x_n)s_1(x_2,\ldots,x_n) - \\ &\quad - \sum_{i=2}^{n} (\partial s_1/\partial x_i)(x_2,\ldots,x_n)f_i(0,x_2,\ldots,x_n) . \end{aligned} \tag{2.7}$$

The Landau-symbol $\mathcal{O}$ indicates a bound for the remainder term and its derivatives which depend upon bounds for the derivatives of f (up to a certain order) on $\hat{X}$ only.

<u>Proof.</u> The first statement is clear. $\sum_{\nu=1}^{N-1} \varepsilon^\nu s_\nu$ is simply the Taylor-polynomial in $\varepsilon$ of order N-1 for s. It satisfies the p.d.e. $(2.2_\varepsilon)$ up to an error of order $\varepsilon^N$ and this formal property leads to recursive relations for the coefficients $s_\nu$ which determine the latter ones uniquely. More substantial arguments are required if one wants to convince oneself that the remainder term can be estimated using nothing else than bounds for f and its partial derivatives on $\hat{X}$. The source of this estimate is the formula (5.7) given in [1], Theorem 5.1, and the subsequent explanations. If this theorem is applied to $(2.4_\varepsilon)$ or rather to its rescaled version

$$\dot{x} = \varepsilon f(x) - x_1(1,0,\ldots,0)^T \tag{2.8}$$

one sees that the constants $K_1$, $K_2$ appearing in [1] can be expressed in terms of bounds for f and Df whereas the numbers $\lambda_i$, i=1,4, can be taken as 1 in view of the simple type of local coordinates which has been introduced in the proof of Theorem 2.1. □

One last remark concerning an elementary proof of Theorem 2.1 in a special case: If every $f_i$ does not depend explicitly upon $x_1$ one can treat ($2.4_\varepsilon$) as a linear d.e. for $x_1$ and write down a solution for the p.d.e. ($2.2_\varepsilon$) in closed form, simply applying the variation-of-constants formula.

## 3. First applications.

For the remaining part of the paper we will tacitly assume that the integer N is as large as needed (what this means will become clear from the context). We use the Landau-symbol to indicate estimates which depend upon

(i) bounds on the set $\hat{X}$ for $\|f\|$, $\|Df\|$, D a differential operator of order $\leq N$,

(ii) bounds for Sup $|\xi(t)|$, $t \in I$, this function is explained below.

Given a solution $x(t) = (x_1(t),\ldots,x_n(t))$ of the d.e.

$$\dot{x} = f(x) \tag{3.1}$$

which satisfies $x(t) \in \hat{X}$ for t in some interval $I = [0,\bar{t})$, $\bar{t}<\infty$. Given also an integrable bounded function $\xi(t)$ such that

$$|x_1(t) - \xi(t)| \leq \varepsilon^2\lambda \tag{3.2}$$

$\varepsilon$ and $\lambda$ being positive numbers to be specified later. Now x(t) can be viewed upon as solution of an o.d.e which can be written in this form

$$\dot{x}_1 = f_1(x_1,x_2,\ldots,x_n) - \frac{x_1}{\varepsilon} + \frac{\xi(t)}{\varepsilon} + \varepsilon\pi(t);\dot{x}_i=f_i(x_1,x_2,\ldots,x_n), i=2,\ldots,n,$$

where

$$|\pi(t)| \leq \lambda \quad , \quad t \in I. \tag{3.3}$$

We introduce the solution $s(\varepsilon,x_2,\ldots,x_n)$ of the p.d.e. ($2.2_\varepsilon$) and thereby assume that $|\varepsilon|$ is as small as it is required in the considerations of Sec. 2.

The function

$$\Delta(t) := x_1(t) - \sigma(t) \text{ where } \sigma(t) := s(\varepsilon,x_2(t),\ldots,s_n(t)) \tag{3.4}$$

is then defined and differentiable for $t \in I$. We have

$$\dot{\Delta}(t) = f_1(x_1(t),\ldots,x_n(t)) - \frac{1}{\varepsilon}x_1(t) - \sum_{i=2}^{n} s_{x_i}(\varepsilon,x_2(t),\ldots,x_n(t))f_i(x_1(t),\ldots,x_n(t)) + \xi(t)/\varepsilon + \varepsilon\pi(t) \tag{3.5}$$

for all $t \in I$.

On the other hand, if $(2.2_\varepsilon)$, (3.4) and (3.5) are combined we arrive at these identities

$$0 = f_1(\sigma(t),x_2(t),\ldots,x_n(t)) - \frac{1}{\varepsilon}\sigma(t) - \sum_{i=2}^{n} s_{x_i}(\varepsilon,x_2(t),\ldots,x_n(t))f_i(\sigma(t),x_2(t)\ldots,x_n(t))$$

and

$$\dot{\Delta}(t) = (\alpha(t,\varepsilon) + \varepsilon\beta(t,\varepsilon) - \frac{1}{\varepsilon})\Delta(t) + \frac{\xi(t)}{\varepsilon} + \varepsilon\pi(t). \qquad (3.6)$$

The second one is obtained by subtracting the first one from (3.5) and factorizing terms as much as possible. That the coefficients $\alpha,\beta$ appearing in (3.6) are of order $\mathcal{O}(1)$ together with their derivatives (with respect to t and $\varepsilon$) can easily be verified if one exploits two facts, namely

(i) $s_{x_i}$ is of order $\mathcal{O}(\varepsilon)$, cf. the corollary to Theorem 2.1,

(ii) the quotient $(f_i(x_1,x_2,\ldots,x_n)-f_i(\sigma,x_2,\ldots,x_n))(x_1-\sigma)^{-1}$ can be regarded as a differentiable function of $x_1,x_2,\ldots,x_n,\sigma$ and the derivatives can be estimated in terms of the derivatives of $f_i$.

At a later occasion we need two relations which become a straightforward consequence of (3.6) if one uses (2.7) and observes that $s = \mathcal{O}(\varepsilon)$ :

$$\beta x_1 = \beta(x_1-s)+ \mathcal{O}(\varepsilon) = -\sum_{i=2}^{n}(s_1)_{x_i}[f_i(x_1,\ldots,x_n)-f_i(0,\ldots,x_n)] + \mathcal{O}(\varepsilon)$$

and $\qquad (3.7)$

$$\beta x_1 + s_2 = (f_1)_{x_1}(0,\ldots,x_n)f_1(0,x_2,\ldots,x_n) - \sum_{i=2}^{n}(f_1)_{x_i}(0,\ldots,x_n)f_i(x_1,\ldots,x_n)+\mathcal{O}(\varepsilon).$$

We now put

$$A(t,\varepsilon) := \int_0^t [\alpha(u,\varepsilon) + \varepsilon\beta(u,\varepsilon)]du.$$

Since the integrand is of order $\mathcal{O}(1)$ the following estimates - which hold for sufficiently small positive $\varepsilon$ - can be derived by standard arguments:

$$e^{-t/\varepsilon+A(t,\varepsilon)} = \mathcal{O}(e^{-t/2\varepsilon}), e^{-t/\varepsilon+A(t,\varepsilon)}\int_0^t e^{\tau/\varepsilon-A(\tau,\varepsilon)}d\tau = \mathcal{O}(\varepsilon). \qquad (3.8)$$

We now regard (3.6) as a d.e. for $\Delta$ and integrate by the variation - of - constants formula. Afterwards certain terms are neglected using (3.3), (3.8) and $\Delta(0) = \mathcal{O}(1)$. We arrive then at this representation for $\Delta$:

$$\Delta(t) = \frac{1}{\varepsilon} e^{-t/\varepsilon+A(t,\varepsilon)} \int_0^t e^{\tau/\varepsilon-A(\tau,\varepsilon)}\xi(\tau)d\tau + (\lambda\varepsilon^2+e^{-t/2\varepsilon}).$$

Comparison with the definition (3.4) and application of the corollary to Theorem 2.1 yields a proof for

<u>Lemma 3.1</u> under the proviso that $x(t)\in\hat{X}$ for $t\in I$ this relation holds true:

$$x_1(t)-\varepsilon s_1(t)-\varepsilon^2 s_2(t)+\mathcal{O}(\varepsilon^3) = \frac{1}{\varepsilon} e^{-t/\varepsilon+A(t,\varepsilon)} \int_0^t e^{\tau/\varepsilon-A(\tau,\varepsilon)} \xi(\tau)d\tau + \mathcal{O}(\lambda\varepsilon^2 + e^{-t/2\varepsilon})$$

where $s_i(t) := s_i(\ x_2(t),\dots,x_n(t))$.

Remark: The statement is valid for an unbounded region $\hat{X}$ also, provided one can find finite bounds for $\|f\|$, $\|Df\|$ .

Let us apply the lemma to the d.e.

$$\dot{x}_o = f_1(x_1,\dots,x_n),\ \dot{x}_1=f_1(x_1,\dots,x_n),\ \dot{x}_i=f_i(x_1,\dots,x_n), i=2,\dots,n. \qquad (3.9)$$

which arises from (3.1) by "doubling" the first component. In fact $x_o(t) = x_1(t)$ for all t if $x_o(0) = x_1(0)$ and (3.2) holds with $x_o(t)$ instead of $x_1(t)$. On the other hand the right hand sides of (3. ) do not depend explicitly upon $x_o$. So if we let $x_o$ play the role of $x_1$ the lemma can be applied and we have

$$\alpha(t,\varepsilon) = 0,\ \beta(t,\varepsilon) = 0,\ A(t,\varepsilon) = 0,$$

$$s_1(t) = f_1(x(t)) = \dot{x}_1(t),\ x_o(t) = x_1(t) = \xi(t) + \mathcal{O}(\varepsilon^2), \qquad (3.10)$$

$$s_2(t) = -\sum_{i=1}^{n} \partial f_1/\partial x_i f_i = -\ddot{x}(t).$$

Because of (3.10) the statement can be phrased in this form

$$\xi(t) - \varepsilon\dot{x}_1(t) = \frac{1}{\varepsilon} e^{-t/\varepsilon} \int_0^t e^{\tau/\varepsilon} \xi(\tau)d\tau + \mathcal{O}(\lambda\varepsilon^2 + \varepsilon^2 + e^{t/2\varepsilon}).$$

This is a type of result as announced in Sec. 1 (cf. (1.6), (1.7)).

## 4. Discussion of Lemma 3.1

We wish to analyse the statement of the lemma further. The procedure in this section will be rather formal, some arguments as $t, x_1,\dots,x_n$ which appear in symbols for functions are in general omitted. We write for shortness $F(x_1)$, $F'(x_1)$ respectively instead of $f_1(x_1,\dots,x_n)$, $(\partial f_1/\partial x_1)(x_1,\dots,x_n)$ respectively. Here and there it is assumed that $x_1(t) = \xi(t)$ for all t, hence $\lambda = 0$. The complete formula of Lemma 3.1 reads then for the updated system (3.9) (cf. also (3.10))

$$x_1-\varepsilon\dot{x}_1 + \varepsilon^2\ddot{x}_1 = \frac{1}{\varepsilon}\zeta + \mathcal{O}(\varepsilon^3+e^{-t/2\varepsilon}),$$

$$\zeta(t) := e^{-t/\varepsilon}\int_0^t e^{\tau/\varepsilon} x_1(\tau)d\tau. \qquad (4.1)$$

We wish to examine the function $\alpha$ which was in introduced earlier (3.6)). Using the present notation the relation defining $\alpha$ can be written in this form

$$F(x_1)-F(s) = \alpha\cdot(x_1-s). \tag{4.2}$$

If we take as s the first partial sum of the asymptotic series (i.e. we put $s = \varepsilon s_1 + \mathcal{O}(\varepsilon^2)$) one finds by inspection an $\alpha$ which is linear in $\varepsilon$ and satisfies (4.2) up to terms of order $\mathcal{O}(\varepsilon^2)$. Hence

$$\alpha = \frac{1}{x_1}\{F(x_1)-F(0)+\varepsilon s_1[\frac{F(x_1)-F(0)}{x_1} - F'(0)]\} + \mathcal{O}(\varepsilon^2). \tag{4.3}$$

Assume now that $\xi$ is twice differentiable and that we have

$$\ddot{\xi} = \mathcal{O}_1(1). \tag{4.4}$$

Integrating by parts twice one obtains

$$\begin{aligned} &e^{-t/\varepsilon+A(t,\varepsilon)}\int_0^t e^{\tau/\varepsilon-A(\tau,\varepsilon)}\xi(\tau)d\tau = \\ &= \varepsilon\xi + \varepsilon^2[(\alpha+\varepsilon\beta)\xi-\dot{\xi}] + \mathcal{O}_1(\varepsilon^3). \end{aligned} \tag{4.5}$$

From (4.3) we have (notation is now as in the previous section)

$$\alpha = \frac{1}{x_1}[\dot{x}_1-f_1(0,x_2,\ldots,x_n)] + \mathcal{O}_1(\varepsilon).$$

If we use this relation together with Lemma 3.1 we arrive at an approximation for $\dot{x}_1$ in terms of $\dot{\xi}$, namely

$$\frac{x_1-\xi}{\varepsilon} - s_1 - \frac{\dot{x}_1}{x_1}\xi + \frac{\xi}{x_1}f_1(0,x_2,\ldots,x_n) + \dot{\xi} = \mathcal{O}_1(\varepsilon).$$

We rewrite it as follows and distinguish more carefully the various contributions to the error bound;

$$\begin{aligned} &(\xi-x_1)\{\frac{1}{\varepsilon} - \frac{\dot{x}_1}{x_1} + \frac{1}{x_1}f_1(0,x_2,\ldots,x_n)\} - \dot{x}_1 + \dot{\xi} = \\ &= \mathcal{O}_1(\varepsilon) + \mathcal{O}(\varepsilon) + \mathcal{O}(\lambda\varepsilon) + \mathcal{O}(\frac{1}{\varepsilon}e^{-t/2\varepsilon}). \end{aligned} \tag{4.6}$$

The first term is the same as in the relation (4.5). The second one is the remainder term in the expansion of s.:

$$s = \varepsilon s_1 + \mathcal{O}(\varepsilon^2).$$

The third is obtained by combining (3.2) and (3.8).

We now wish to sketch a formal procedure which leads to the same type of relation as in (4.6), with the same precise description of the error terms, but with a left hand side $\dot{\xi}-\dot{x}_1$.

We augment the given d.e. (3.1) by adding the equations $\dot{t} = 1$ and

$$\dot{\sigma} = (\xi(t)-x_1-\sigma)\{\frac{1}{\varepsilon} - \frac{\dot{x}_1}{x_1+\sigma} + \frac{1}{x_1+\sigma} f_1(-\sigma,x_2,\ldots,x_n)\}. \qquad (4.7)$$

Furthermore we introduce $x_1+\sigma$ instead of $x_1$ as first component of the state. If we write down (4.6) for the modified system - taking as $\xi(t)$ the same function as before and $x_1+\sigma$, $f_1+\dot{\sigma}$ respectively instead of $x_1, f_1$ respectively - one sees that the left hand side assumes the simple form $-\dot{x}_1+\dot{\xi}$ .

To handle - in the case of the modified d.e. - the error terms on the right hand side of (4.6) one needs above all an estimate of the type (3.2), i.e.

$$|x_1(t) + \sigma(t) - \xi(t)| < \lambda'\varepsilon^2.$$

We now claim: One can simply take $\lambda'=2\lambda$ on grounds of the estimate

$$0 \leq \sigma(t) \leq \lambda\varepsilon^2. \qquad (4.8)$$

To justify (4.8) we assume first of all and without loss of generality (replace $\xi(t)$ by $\xi(t)-c$, c a sufficiently small constant!) that $\xi(t) > x_1(t)$ for all t. Secondly we take $\varepsilon$ so small that the expression in {} on the right hand side of (4.7) is positive. It follows then from this d.e for $\sigma$ that we have

$$\dot{\sigma} < 0 \quad \text{if} \quad \sigma > \lambda\varepsilon^2, \text{ since } \quad \xi(t) - x_1(t) < \lambda\varepsilon^2,$$

and

$$\dot{\sigma} > 0 \quad \text{if} \quad \sigma = 0 \quad \text{since} \quad \xi(t) > x_1(t).$$

(4.8) is therefore true for all $t \geq 0$ if it is satisfied for the initial value which is at our disposal.

References

[1] Knobloch,H.W.: Dichotomy and integral manifolds. Part I: General Principles. *Resultate Mathe.* 14 (1988), pp. 93-124.

[2] Knobloch,H.W.: Stabilization of Control Systems by Means of High-Gain feedback. In: Optimal Control Theory and Economic Analysis 3, G.Feichtinger ed., North-Holland-Amsterdam, New York, Oxford, Tokyo, 1988. pp. 153-173.

[3] Knobloch,H.W.: Invariant manifolds and singular perturbation. In: Proceedings of the eleventh international conference on non-linear oscillations, M.Farkas, V.Kertész, G.Stépan eds. Janos Bolyai Mathematical Society, Budapest 1988. pp. 109-118.

## Erratum

in the preceding article "A new approach to identification..." by H.W.Knobloch.

The expression on the left hand side of (4.6) should be corrected and reads then

$$(x_1-\xi)\{\frac{1}{\varepsilon} + \frac{\dot{x}_1}{x_1} - \frac{1}{x_1}f_1(0,x_2,\dots,x_n)\} - \dot{x}_1+\dot{\xi}. \qquad (4.6')$$

The arguments used in the last portion of Section 4 - starting with (4.7) - should be modified as follows. Assume that $x_1(t)$ is positive and bounded away from zero, say $x_1(t) > c > 0$ for all t. Choose a constant $K_o$ such that the function (of t and $\sigma$)

$$K(t,\sigma) := K_o + \frac{\dot{x}_1}{x_1+\sigma} - \frac{1}{x_1+\sigma} f_1(-\sigma,x_2,\dots,x_n),\ x_i \to x_i(t), \dot{x}_1 \to \dot{x}_1(t),$$

has constant sign on the set $\{t,\sigma : t\in I,\ 0 \le \sigma \le \lambda\varepsilon^2\}$ . Assume as before that $\xi(t) > x_1(t)$ for all t. This implies that the function

$$x_1(t) + \sigma - \xi(t)$$

has opposite constant sign for $\sigma = 0$ and $\sigma = \lambda\varepsilon^2$. Hence one can find a solution $\sigma(t)$ of the first order o.d.e.

$$\dot{\sigma} = (x_1(t) + \sigma - \xi(t))K(t,\sigma)$$

satisfying the inequality $0 < \sigma(t) < \lambda\varepsilon^2$. Consider now the o.d.e.

$$\dot{x}_1 = f_1(x_1-\sigma(t),x_2,\dots,x_n)+\dot{\sigma}(t), \dot{x}_i=f_i(x_1-\sigma(t),x_2,\dots,x_n), i=2,..,n, \dot{t}=1 \quad (*)$$

which has the solution $(x_1(t)+\sigma(t),x_2(t),\dots,x_n(t),t)^T$. Evaluate the expression (4.6') along this solution, taking as underlieing d.e. (*) instead of (3.1) but as $\xi$ the same function as before. The expression (4.6') assumes then the form

$$(x_1(t) - \xi(t))(\frac{1}{\varepsilon} - K_o) + \dot{\xi}(t) - \dot{x}_1(t).$$

The first term is of order $\mathcal{O}(\lambda\varepsilon)$ and hence can be added to a similar error term on the right hand side of (4.6). Thereby the approximation result for $\dot{x}_1$ - as it was claimed in the introduction - is justified.

GENERATING SERIES AND SINGULARLY PERTURBED BILINEAR SYSTEMS

**F. Rotella, G. Dauphin-Tanguy**

Laboratoire d'Automatique et d'Informatique Industrielle (L.A.I.I.)
INSTITUT INDUSTRIEL DU NORD (I.D.N.)
B.P. 48
59651 VILLENEUVE D'ASCQ CEDEX - FRANCE

**KEYWORDS** : Bilinear Systems, Generating Series, Singular Perturbations, Reduced Order Model.

**ABSTRACT** : In this paper, we study the interconnection between singularly perturbed bilinear systems and generating series. Special attention is focused on the expansion of the generating series of the initial system into a fast one and a slow one.

## INTRODUCTION

For some years, particular attention has been paid to bilinear systems |1| and some works have dealt with the problem of singularly perturbed bilinear systems |2||3||4|. But, due to the presence of input variables in the state matrix, the construction of reduced order models is not as simple as in the linear case, and some singularities may occur. Except in some particular cases |2||3|, the slow reduced order model appears as a non-linear-in-the-inputs model ; several approximating techniques have been proposed |5| to replace this exact slow reduced model by a bilinear one, but in each case attention must be paid to the inputs variations domains.

In this paper, we are going to give another construction of the reduced order models based on the generating series |6| of non-linear models. The paper is organized as follows : after a short review of the generating series, we will determine a decomposition of the generating series of a singularly perturbed bilinear model into a fast series and a slow one.

## I - GENERATING SERIES |6||7|

Let us consider the bilinear system :

$$\mathring{x}(t) = \left[ \sum_{i=0}^{m} A_i \, u_i(t) \right] x(t) + B\, u(t) \tag{1.1}$$

$$y(t) = C\, x(t) \tag{1.2}$$

where : $\forall t \in \mathbb{R}^+$,

$x(t) = \left[ x_1(t), \ldots, x_q(t) \right]^T \in \mathbb{R}^q$ is the state vector,

$u(t) = \left[ u_1(t), \ldots, u_m(t) \right]^T \in \mathbb{R}^m$ is the input vector,

$u_o(t) \equiv 1$, for notational compactness,

$y(t) \in \mathbb{R}^n$ is the output vector, and

$A_i$, B, C are constant matrices of adapted dimensions.

By the use of the Fundamental Formula |7| for control-affine non-linear systems, we can associate to the system a generating series :

$$y = h|_{x(0)} + \sum_{\nu \geq 0} \; \sum_{j0,\ldots,j\nu=0} \mathcal{A}_{j0}\, \mathcal{A}_{j1} \ldots \mathcal{A}_{j\nu}\, h|_{x(0)}\, \alpha_{j\nu} \ldots \alpha_{j1}\, \alpha_{j0} \tag{2}$$

where, here : $x(0)$ is the initial condition at $t=0$,

$h = C\, x$,

$\mathcal{A}_0 = A_0$,

$i = [1,\ldots,m] : \mathcal{A}_i = (A_i\, x + B_i)$,

$\mathcal{A}_{j0}\, \mathcal{A}_{j1} \ldots \mathcal{A}_{j\nu}\, h$ means the iterated Lie derivative of h with respect to $\mathcal{A}_{j\nu}, \ldots, \mathcal{A}_{j0}$,

the bar $|_{x(0)}$ indicates evaluation at $x(0)$, and

$\alpha_{j\nu} \ldots \alpha_{j0}$ is a word of a free monoïd generated by the alphabet of symbols $\{\alpha_0, \alpha_1, \ldots, \alpha_m\}$.

The output of the system is then obtained by replacing the words $\alpha_{j\nu} \ldots \alpha_{j0}$ in (2) by the iterated integral $\int_0^t d\xi_{j\nu} \ldots d\xi_{j0}$ defined recursively on its length by :

$$\xi_0(t) = t, \; \xi_i(t) = \int_0^t u_i(\tau)\, d\tau, \; i = [1,\ldots,m] \tag{3.1}$$

$$\int_0^t d\xi_j = \xi_j(t), \; i = [0,\ldots,m] \tag{3.2}$$

$$\int_0^t d\xi_{j\nu} \cdots d\xi_{j0} = \int_0^t d\xi_{j\nu}(\tau) \int_0^\tau d\xi_{j,\nu-1} \cdots d\xi_{j0} \tag{3.3}$$

The use of Noncommutative Symbolic Calculus |7||8| enables the generating series (2) to be written in a more compact form. The integration of (1.1) leads to :

$$x(t) = x(0) + \sum_{i=0}^{m} A_i \int_0^t u_i(\tau)\, x(\tau)\, d\tau + B \int_0^t u(\tau)\, d\tau \tag{4}$$

If we denote by $X$ the generating series associated to the state vector of (1), we have with the foregoing notations :

$$X = x(0) + \left\{ \sum_{i=0}^{m} A_i\, \alpha_i \right\} X + B\,\alpha \tag{5}$$

where : $\alpha = \left[ \alpha_1, \alpha_2, \ldots, \alpha_m \right]^T$.

Thus, the generating series of the system (1) can be written on a transfer function form :

$$Y = C \left[ I_q - \sum_{i=0}^{m} A_i\, \alpha_i \right]^{-1} \{ x(0) + B\,\alpha \} \tag{6}$$

In all the following we will use this form, where, for simplicity's sake, we will suppose $x(0) = 0$.

## II - SINGULARLY PERTURBED BILINEAR SYSTEMS

The bilinear system (1) is singularly perturbed if the matrices $A_i$ and B have the following structure :

$$i = \left[ 0, \ldots, m \right] \qquad A_i = \begin{bmatrix} A_{11}^{i} & A_{12}^{i} \\ \frac{A_{21}^{i}}{\varepsilon} & \frac{A_{22}^{i}}{\varepsilon} \end{bmatrix} \tag{7.1}$$

$$B = \begin{bmatrix} B_1 \\ \frac{B_2}{\varepsilon} \end{bmatrix} \tag{7.2}$$

where : $\forall\ i \in \{0,\ldots,m\}$ · $A^i_{k\ell} \in \mathbb{R}^{q_k \times q_\ell}$, $(k,\ell) \in \{1,2\}^2$

$B_1 \in \mathbb{R}^{q_1 \times m}$, $B_2 \in \mathbb{R}^{q_2 \times m}$

$q_1 + q_2 = q$

Under these conditions, the state vector, $x(t)$, can be decomposed as :

$$x(t) = \left[\, v(t)^T\ ,\ w(t)^T \,\right]^T \tag{8}$$

where the $q_1$-dimensional vector $v(t)$ is predominantly slow and the $q_2$-vector $w(t)$ contains fast transients surimposed on a slowly varying "quasi-steady-state".

These variables are defined by the following state-space equations :

$$\mathring{v}(t) = \left[ \sum_{i=0}^{m} u_i(t)\ A^i_{11} \right] v(t) + \left[ \sum_{i=0}^{m} u_i(t)\ A^i_{12} \right] w(t) + B_1\ u(t) \tag{9.1}$$

$$\varepsilon\ \mathring{w}(t) = \left[ \sum_{i=0}^{m} u_i(t)\ A^i_{21} \right] v(t) + \left[ \sum_{i=0}^{m} u_i(t)\ A^i_{22} \right] w(t) + B_2\ u(t) \tag{9.2}$$

$$y(t) = C_1\ v(t) + C_2\ w(t) \tag{9.3}$$

where : $C_1 \in \mathbb{R}^{n \times q_1}$, and $C_2 \in \mathbb{R}^{n \times q_2}$.

Following the singular perturbations method |9|, a reduced order model can be constructed by setting $\varepsilon$ equal to zero in (9.2). Let us denote :

$$\forall\ (k,\ell) \in \{1,2\}^2 \qquad A_{k\ell}(u) = \sum_{i=0}^{m} A^i_{k\ell}\ u_i(t) \tag{10}$$

If, $\forall\ t \in \mathbb{R}^+$, $A_{22}(u)$ is not singular, we obtain the reduced slow model :

$$\mathring{v}_s(t) = A_s(u)\ v_s(t) + B_s(u)\ u \tag{11.1}$$

$$y_s(t) = C_s(u)\ v_s(t) + D_s(u)\ u \tag{11.2}$$

where :

$$A_s(u) = A_{11}(u) - A_{12}(u)\, A_{22}(u)^{-1}\, A_{21}(u) \tag{11.3}$$

$$B_s(u) = B_1 - A_{12}(u)\, A_{22}(u)^{-1}\, B_2 \tag{11.4}$$

$$C_s(u) = C_1 - C_2\, A_{22}(u)^{-1}\, A_{21}(u) \tag{11.5}$$

$$D_s(u) = - C_2\, A_{22}(u)^{-1}\, B_2 \tag{11.6}$$

and if moreover, all the eigenvalues of $A_{22}(u)$ have real parts less than a fixed negative number, the following approximation holds for all $t \in \mathbb{R}^+$ |10| :

$$v(t) = v_s(t) + \sigma(\varepsilon) \tag{12}$$

where $\sigma(\varepsilon)$ indicates a function of $\varepsilon$ of norm less than $c\varepsilon$ in $[0, \varepsilon^*]$, $c > 0$, $\varepsilon^* > 0$.

Except in particular cases, like single input strictly bilinear systems ($A_o = 0$ and $m = 1$) |3| or when structural constraints are imposed on the matrices $A_i$ and B |2|, the reduced order slow model is not a bilinear one and moreover it is not a control-affine non-linear model. In order to solve the problem of analysis and control of such systems, several methods have been proposed in |5| in order to bilinearize this model. But all these methods take in account the input variables, and the bilinear reduced order model is valid in a certain input-domain. To overcome this problem, we are going to propose a method based on the analytic expression of the generating series of the singularly perturbed system (9).

## III - ASYMPTOTIC DEVELOPMENT OF THE GENERATING SERIES

To obtain this development we first rewrite (9) in the fast time scale, $\tau = t/\varepsilon$ :

$$\underline{v}'(\tau) = \varepsilon \left[ A_{11}(\underline{u})\, \underline{v}(\tau) + A_{12}(\underline{u})\, \underline{w}(\tau) + B_1\, \underline{u}(\tau) \right] \tag{13.1}$$

$$\underline{w}'(\tau) = A_{21}(\underline{u})\, \underline{v}(\tau) + A_{22}(\underline{u})\, \underline{w}(\tau) + B_2\, \underline{u}(\tau) \tag{13.2}$$

where : $v' = dv/d\tau$ and $\underline{v}(\tau) = v(\varepsilon\tau)$.

For notational compactness, the argument of the functions $\underline{u}$, $\underline{v}$ and $\underline{w}$ will be omitted. If we apply the principle of asymptotic expansions in singular perturbations problems |11|, we can write v and w as analytic forms in $\varepsilon$ :

$$\underline{v} = \sum_{i=0}^{\infty} \underline{v}_{(i)} \varepsilon^i \tag{14.1}$$

$$\underline{w} = \sum_{i=0}^{\infty} \underline{w}_{(i)} \varepsilon^i \tag{14.2}$$

where $\underline{v}_{(i)}$ and $\underline{w}_{(i)}$ are time-dependent vectors.

If we replace (14) in (13) the identity of the coefficients of $\varepsilon^k$ in the two members of (13.1) and (13.2) leads to a set of equations defining $\underline{v}_{(i)}$ and $\underline{w}_{(i)}$, $i \in \mathbb{N}$ :

$$\Sigma_0 \left| \begin{array}{ll} \underline{v}'_{(0)} = 0 & (15.1) \\ \underline{w}'_{(0)} = A_{21}(\underline{u})\, \underline{v}_{(0)} + A_{22}(\underline{u})\, \underline{w}_{(0)} + B_2\, \underline{u} & (15.2) \end{array} \right.$$

$$\Sigma_1 \left| \begin{array}{ll} \underline{v}'_{(1)} = A_{11}(\underline{u})\, \underline{v}_{(0)} + A_{12}(\underline{u})\, \underline{w}_{(0)} + B_1\, \underline{u} & (15.3) \\ \underline{w}'_{(1)} = A_{21}(\underline{u})\, \underline{v}_{(1)} + A_{22}(\underline{u})\, \underline{w}_{(1)} & (15.4) \end{array} \right.$$

$i \geqq 2$

$$\Sigma_i \left| \begin{array}{ll} \underline{v}'_{(i)} = A_{11}(\underline{u})\, \underline{v}_{(i-1)} + A_{12}(\underline{u})\, \underline{w}_{(i-1)} & (15.5) \\ \underline{w}'_{(i)} = A_{21}(\underline{u})\, \underline{v}_{(i)} + A_{22}(\underline{u})\, \underline{w}_{(i)} & (15.6) \end{array} \right.$$

The output is then defined, from (9.3) by :

$$\underline{y} = \sum_{i=0}^{\infty} \underline{y}_{(i)} \varepsilon^i \tag{16.1}$$

where :

$$\forall\, i \in \mathbb{N},\ \underline{y}_{(i)} = C_1\, \underline{v}_{(i)} + C_2\, \underline{w}_{(i)} \tag{16.2}$$

Let us denote by $\underline{\mathcal{V}}_{(i)}$, $\underline{\mathcal{W}}_{(i)}$ and $\underline{\mathcal{Y}}_{(i)}$ the generating series associated respectively to $\underline{v}_{(i)}$, $\underline{w}_{(i)}$ and $\underline{y}_{(i)}$, $i \in \mathbb{N}$, and by $\underline{\alpha}_i$, $i \in \{0,\ldots,m\}$, the symbols of integration in the fast time scale. It can be deduced from (15), where the initial conditions are taken equal to zero :

$$\underline{V}_{(0)} = 0 \qquad (17.1)$$

$$\underline{W}_{(0)} = \underline{\Delta}_{22}\,\underline{W}_{(0)} + B_2\,\underline{\alpha} \qquad (17.2)$$

$$\underline{V}_{(1)} = \underline{\Delta}_{11}\,\underline{V}_{(0)} + \underline{\Delta}_{12}\,\underline{W}_{(0)} + B_1\,\underline{\alpha} \qquad (17.3)$$

$$\underline{W}_{(1)} = \underline{\Delta}_{21}\,\underline{V}_{(1)} + \underline{\Delta}_{22}\,\underline{W}_{(1)} \qquad (17.4)$$

$i \geqq 2$

$$\underline{V}_{(i)} = \underline{\Delta}_{11}\,\underline{V}_{(i-1)} + \underline{\Delta}_{12}\,\underline{W}_{(i-1)} \qquad (17.5)$$

$$\underline{W}_{(i)} = \underline{\Delta}_{21}\,\underline{V}_{(i)} + \underline{\Delta}_{22}\,\underline{W}_{(i)} \qquad (17.6)$$

where : $\underline{\alpha} = \left[\underline{\alpha}_1, \ldots, \underline{\alpha}_m\right]^T$

$$(k,\ell) \in \{1,2\}^2 \qquad \underline{\Delta}_{k\ell} = \sum_{j=0}^{m} \underline{\alpha}_j\, A^j_{k\ell}$$

With some manipulations, these expressions can be written as :

$$\begin{bmatrix} \underline{V}_{(0)} \\ \underline{W}_{(0)} \end{bmatrix} = G_0\,\underline{\alpha} \qquad (18.1)$$

$$\begin{bmatrix} \underline{V}_{(1)} \\ \underline{W}_{(1)} \end{bmatrix} = G_1\,\underline{\alpha} \qquad (18.2)$$

$i \geqq 2$

$$\begin{bmatrix} \underline{V}_{(i)} \\ \underline{W}_{(i)} \end{bmatrix} = M \begin{bmatrix} \underline{V}_{(i-1)} \\ \underline{W}_{(i-1)} \end{bmatrix} \qquad (18.3)$$

where :

$$G_0 = \begin{bmatrix} 0 \\ \underline{\Delta}_f\, B_2 \end{bmatrix} \qquad (18.4)$$

$$G_1 = \begin{bmatrix} I_{q_1} \\ \underline{\Delta}_f \; \underline{\Delta}_{21} \end{bmatrix} \underline{B} \tag{18.5}$$

$$M = \begin{bmatrix} I_{q_1} \\ \underline{\Delta}_f \; \underline{\Delta}_{21} \end{bmatrix} \left[ \underline{\Delta}_{11} \; \underline{\Delta}_{12} \right] \tag{18.6}$$

$$\underline{\Delta}_f = \left[ I_{q_2} - \underline{\Delta}_{22} \right]^{-1}$$

$$\underline{B} = B_1 + \underline{\Delta}_{12} \, \underline{\Delta}_f \, B_2$$

From which we deduce :

$$\forall \, i \geqq 1 \qquad \begin{bmatrix} \underline{V}_{(i)} \\ \underline{W}_{(i)} \end{bmatrix} = M^{i-1} \, G_1 \, \underline{\alpha} \tag{19}$$

The generating series of the output is given by :

$$\underline{Y} = \sum_{i=0}^{\infty} \varepsilon^i \, \underline{Y}_{(i)} \tag{20}$$

where :

$$\underline{Y}_{(i)} = C \begin{bmatrix} V_{(i)} \\ W_{(i)} \end{bmatrix}$$

We obtain thus, from (19) :

$$\underline{Y} = \left[ C \, G_0 + \varepsilon \, C \left\{ \sum_{i=0}^{\infty} \varepsilon^i \, M^i \right\} G_1 \right] \underline{\alpha} \tag{21}$$

With the use of the identity :

$$\sum_{i=0}^{\infty} \varepsilon^i \, M^i = \left[ I_q - \varepsilon \, M \right]^{-1} \tag{22}$$

and the matric inversion lemma :

$$(A + B C D)^{-1} = A^{-1} - A^{-1} B (C^{-1} + D A^{-1} B)^{-1} D A^{-1} \tag{23}$$

we obtain the following decomposition :

$$\left[ I_q - \varepsilon M \right]^{-1} = I_q + \varepsilon \begin{bmatrix} I_{q_1} \\ \underline{\Delta}_f \underline{\Delta}_{21} \end{bmatrix} \left[ I_{q_1} - \varepsilon \underline{\Delta}_s \right]^{-1} \left[ \underline{\Delta}_{11} \; \underline{\Delta}_{12} \right] \tag{24}$$

where : $\underline{\Delta}_s = \underline{\Delta}_{11} + \underline{\Delta}_{12} \underline{\Delta}_f \underline{\Delta}_{21}$

Substituting (24) in (21), and taking into account the relationships (18.4) and (18.5), the generating series can be rewritten as :

$$\underline{y} = \left[ C_2 \underline{\Delta}_f B_2 + \varepsilon \underline{C} \underline{B} + \varepsilon^2 \underline{C} \left[ I_{q_1} - \varepsilon \underline{\Delta}_s \right]^{-1} \underline{\Delta}_s \underline{B} \right] \underline{\alpha} \tag{25}$$

where : $\underline{C} = C_1 + C_2 \underline{\Delta}_f \underline{\Delta}_{21}$

But the following equalities hold :

$$\underline{C} \underline{B} + \varepsilon \underline{C} \left[ I_{q_1} - \varepsilon \underline{\Delta}_s \right]^{-1} \underline{\Delta}_s \underline{B}$$

$$= \underline{C} \left[ I_{q_1} + \varepsilon \left[ I_{q_1} - \varepsilon \underline{\Delta}_s \right]^{-1} \underline{\Delta}_s \right] \underline{B} \tag{26.1}$$

$$= \underline{C} \left[ I_{q_1} - \varepsilon \underline{\Delta}_s \right]^{-1} \left[ I_{q_1} - \varepsilon \underline{\Delta}_s + \varepsilon \underline{\Delta}_s \right] \underline{B} \tag{26.2}$$

Then, finally, the generating series takes the following form :

$$\underline{y} = \left[ C_2 (I_{q_2} - \underline{\dot{\Delta}}_{22})^{-1} B_2 + \varepsilon \underline{C} (I_{q_1} - \varepsilon \underline{\Delta}_s)^{-1} \underline{B} \right] \underline{\alpha} \tag{27}$$

## IV - CONCLUSION

In the decomposition (27) we have obtained two generating series :

- the fast generating series :

$$\underline{y}_f = C_2 (I_{q_2} - \underline{\Delta}_{22})^{-1} B_2 \underline{\alpha} \tag{28.1}$$

- the slow generating series :

$$\underline{y}_s = \varepsilon \, \underline{C} \, (I_{q_1} - \varepsilon \, \underline{\Delta}_s)^{-1} \, \underline{B} \, \underline{\alpha} \qquad (28.2)$$

and the realization of every one of these series will lead to reduced order models : a slow one and a fast bilinear one.

The treatment and the interpretation of the generating series $\underline{y}_s$ can be implemented in two different ways : in one hand, we can study the beginning of the motion, and in another hand, we can see the asymptotic behaviour of the slow reduced order model. In every case we can realize a bilinear approximation model of the non-linear slow reduced model by the use of the algorithm proposed in |12|.

From the decomposition of the generating series of a singularly perturbed bilinear system, we have proposed ways to construct reduced order models. This aim is not achieved because we have not looked for nonlinear realizations of the obtained generating series but we have restricted our study to bilinear realizations, especially for the slow generating series.

The study presented, as we have shown it, brings out that the use of generating series enables some previous results of singular perturbations methods to be extended.

With the reduced order models it is possible to construct suboptimal control laws by the use, in one hand, the principle of composite control (used for singularly perturbed systems) |13|, and in another hand, the analytic control of nonlinear systems |5|. The application of this mixed method of control is carried out in |5| on singularly bilinear models.

## REFERENCES

|1| R.R. MOHLER, W.J. KOLODZIEJ
"An overview of bilinear systems : theory and applications"
IEEE-SMC, Vol. 10, n° 10, pp. 683-688, 1980.

|2| J.M. GUILLEN, M.A. ARMADA
"A singular perturbation method for order reduction of large scale bilinear dynamical systems"
in "Large Scale Systems : Theory and Applications", A. Titli, M.G. Singh Ed., Pergamon Press, pp. 229-236, 1981.

|3| S.G. TZAFESTAS, K.E. ANAGNOSTOU
"Stabilization of singularly perturbed strictly bilinear systems"
IEEE-AC, Vol. 29, n° 10, pp. 943-946, 1984.

|4| F. ROTELLA, G. DAUPHIN-TANGUY
"Multi-model representation for singularly perturbed bilinear systems"
in "Applied Modelling and Simulation of Technological Systems", P. Borne, S.G. Tzafestas Ed.,.North-Holland, pp. 139-145, 1987.

|5| F. ROTELLA
"Méthodes algébriques et analytiques pour la simplification et la commande de systèmes bilinéaires à deux dynamiques"
Thèse d'Etat, Lille, n° 732, 1987.

|6| M. FLIESS
"Un outil algébrique : les séries formelles non commutatives"
in "Mathematical System Theory", Lect. Notes Econom. Math. Syst., Springer-Verlag, Vol. 131, pp. 122-148, 1976.

|7| M. FLIESS, M. LAMNABHI, F. LAMNABHI-LAGARRIGUE
"An algebraic approach to nonlinear functional expansions"
IEEE-CAS, Vol. 30, n° 8, pp. 554-570, 1983.

|8| M. LAMNABHI
"Series de Volterra et séries génératrices non commutatives"
Thèse de Docteur-Ingénieur, Orsay, n° 432, 1980.

|9| V.R. SAKSENA, J. O'REILLY, P.V. KOKOTOVIC
"Singular perturbations and time-scale methods in control theory : survey 1976-1983"
Automatica, Vol. 20, n° 3, pp. 273-293, 1984.

|10| A.N. TIHONOV
"Systems of differential equations containing a small parameter multiplying the derivative"
Mat. Sb., Vol. 31, n° 73, pp. 575-586, 1952.

|11| G. DAUPHIN-TANGUY, P. BORNE
"Singular perturbations : boundary-layer problem"
Systems and Control Encyclopedia, M.G. Singh Ed., Pergamon Press, Vol. 7, pp. 4425-4429, 1987.

|12| C. HESPEL, G. JACOB
"Approximation of nonlinear systems by bilinear ones"
in "Algebraic and Geometric Methods in Nonlinear Control Theory", M. Fliess, M. Hazewinkel Ed., D. Reidel pub. Comp., pp. 511-520, 1986.

|13| J.H. CHOW, P.V. KOKOTOVIC
"A decomposition of near-optimum regulators for systems with slow and fast modes"
IEEE-AC, Vol. 21, pp. 701-705, 1976.

# FROM THE ROBUSTNESS OF STABILITY DEGREE IN NATURE TO THE CONTROL OF HIGHLY NON LINEAR MANIPULATORS

A. OUSTALOUP
Equipe Systèmes et Commande d'Ordre Non Entier
L.A.R.F.R.A. - E.N.S.E.R.B. - Université de Bordeaux I
351, cours de la Libération - 33405 TALENCE CEDEX - FRANCE

## SUMMARY

This paper deals with the robustenss as far as damping is concerned, and more particularly the robustness as for control damping versus the parameters of the plant.

After defining robustness in time domain, it presents the non integer approach of the CRONE control, a french abbreviation of "Commande Robuste d'Ordre Non Entier", namely "Non Integer Order Robust Control". This approach uses the mathematical principle which insures the robustness of stability degree in nature, namely non integer derivation.

An open loop frequency template is deduced from the non integer order differential equation which describes the relaxation of the ebb and flow on a porous dyke, this phenomenon being robust as for stability degree since the damping factor is independent of motion water mass. This template illustrates robustness in frequency domain.

In the last part, the CRONE control which uses this template is applied in manipulator control, through an inclining polar table the inertia of which very much varies because of a direct drive by the motorization ; moreover, it presents strong dynamic couplings and a large number of non-linearities.

## I - INTRODUCTION

For many years, it is common to speak of robustness. But this concept is very wide, even in a domain such as the automatic control one. In fact, robustness is a notion which always translates the same idea, namely insensitivity.

In automatic control, it is frequent to consider the robustness as far as stability is concerned.

In the non integer approach, the considered robustness is much stricter, that is to say the robustness concerning stability degree. More precisely, the robustness which is at stake translates the insensitivity of the damping factor or the stability degree of the control to the plant parameters ; at least, in so far as they remain within given ranges.

Although time domain makes it possible to illustrate the definition of robustness, particularly from the transient of the step response, it is not a priviliged domain for specifying robustness, not in terms of response performances in closed loop, but in terms of control performances in open loop.

It is true that frequency domain is a domain in which robustness can be illustrated by a characteristic transfer of the control in open loop.

The approach we propose is based on the concept of non integer derivation, in so far as it uses the mathematical principle which constitutes the origin of the robustness of stability degree in nature, that is to say non integer derivation. It is true that the relaxation of the ebb and flow on a porous dyke, described by a non integer order differential equation, is characterized by a damping which is independent of the motion water mass.

From such a differential equation it is possible to determine an "open loop frequency template" (or more simply "template") which illustrates robustness in frequency domain, in this case, a vertical straight line segment lying between the abscissae $-\pi/2$ and $-\pi$ in the Black plane.

The synthesis of the template is carried out in the case of a polar table, the configuration of which insures large inertia variations, strong dynamic couplings and a great number of non-linearities.

The performances obtained are very remarkable ; it is true that, for an inertia variation by a factor of 50, which is very large, and without mentioning the other variations (Coriolis and centrifugal effects) which are also very large, the transients of the step responses of the freedom degrees keep their forms with or without a time scale changing.

Although these performances are remarkable, they are easily explained. Indeed, if the control is robust versus the coefficients of a non stationary linear plant, even if these vary as fast as the variables, one understands why the control may be robust versus the non linearities which can be interpreted in term of non stationarity.

## II - REPRESENTATION OF ROBUSTNESS IN TIME DOMAIN

In time domain, the principle of robustness is translated by a step response which presents the same overshoots independently of the parameters of the plant ; only the natural frequency changes ; so, the transient keeps its form with only a time scale changing (figure 1).

## III - FROM THE ROBUSTENSS OF STABILITY DEGREE IN NATURE TO A NEW ROBUST CONTROL STRATEGY : THE NON INTEGER APPROACH OF THE CRONE CONTROL

Our approach, the aim of which is the conception and the application of a new robust control strategy, makes use of the observation of a natural phenomenon, that of the ebb and flow on a porous dyke. Already in the 17th century, the constructors of dykes had noted the damping properties of the very disturbed dykes and particularly

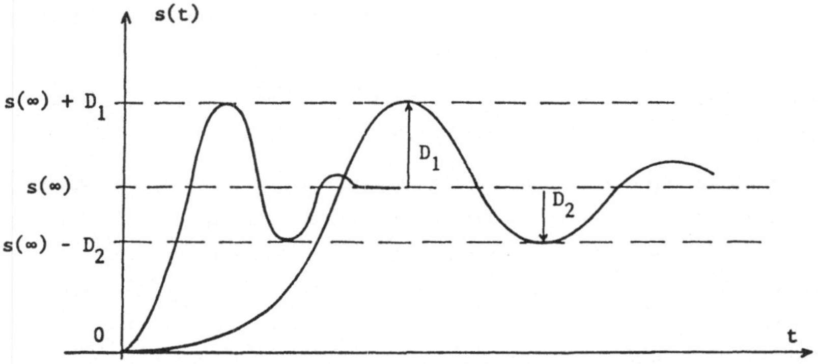


Figure 1 - Illustration of robustness in time domain

those forming air pockets which can be compressed by the advance of water. Otherwise, an attentive observation of the ebb and flow phenomenon consecutive to the damping of water on fluvial or coastal dykes, shows that in the case of very damping (or absorbing) dykes through a porous volumic structure and a rough* surfacic structure :

- the natural frequency of the relaxation is different whether the dyke is fluvial or coastal ;

- the damping of the relaxation seems to be independent of the dyke, whether it is fluvial or coastal.

Given that the fluvial and coastal tests can be distinguished by very different motion water masses, the observation seems to show that the relaxation is characterized by a natural frequency which depends on the motion water mass and by a damping which is independent of it. Although it should be paradoxical when one knows the properties of a pendular relaxation, this result is as remarkable as fundamental in so far as it reveals the insensitivity of the damping factor to a parameter of the process, in this case the motion water mass ; in automatic language, this translates the phenomenon robustness as for stability degree.

After trying to determine the mathematical origin of this type of natural robustness, it appears that it resides in non integer derivation. Indeed, by taking into account the fractality of porosity and the corresponding recursivity, we show (4) that the process is described by a differential equation of non integer order n' between 1 and 2, namely :

$$\tau^{n'}(d/dt)^{n'}\ P(t) + P(t) = 0, \qquad (1)$$

P(t) designating the dynamic pressure at the water-dyke interface.

The corresponding characteristic equation is of the form :

---

* The consideration of a rough surfacic structure (or very disturbed in the sense of B. MANDELBROT), permits a minimization of the reflections on the dyke faces and so, frees oneself from stationary wave phenomena which stems from them ; that is to say that the observation turns, not on water motions consecutive to reflections, but on the motion of the water which rushes into the dykes through their faces.

$$(\tau s)^{n'} + 1 = 0. \quad (2)$$

Finally, the purpose is to obtain the same thing in automatic control, that is to say a control which should be characterized by such a characteristic equation. Indeed, it seems interesting to use such a fundamental result for synthesizing a robust control strategy : it is the approach (said non integer) that the CRONE control uses.

## IV - REPRESENTATION OF ROBUSTNESS IN FREQUENCY DOMAIN : OPEN LOOP FREQUENCY TEMPLATE

### IV.1 - Transfer in closed loop

As "synthesis transmittance in closed loop", one considers a transfer function of the form :

$$F(s) = (1 + (\tau s)^{n'})^{-1}, \quad (3)$$

whose characteristic equation is indeed that given by relation (2).

### IV.2 - Transfer in open loop

Let us designate by E(s) and S(s) the Laplace transforms of the input and output of the control. Relation (3) permits then to write :

$$S(s)/E(s) = (1 + (\tau s)^{n'})^{-1}, \quad (4)$$

from where one draws :

$$S(s) = [E(s) - S(s)]/(\tau s)^{n'}, \quad (5)$$

a symbolic equation which defines an open loop transfer fucntion of the form :

$$\beta(s) = (1/\tau s)^{n'} \; ; \quad (6)$$

this one can be considered as a "synthesis transmittance in open loop" of a robust control.

The corresponding open loop frequency response, namely

$$\beta(j\omega) = (1/j\tau\omega)^{n'}, \quad (7)$$

admits, as Black locus, a vertical straight line of abscissa between -90° and -180°.

### IV.3 - Open loop frequency template

Given that the dynamic behaviour in closed loop is essentially linked to the behaviour in open loop close to the unit gain frequency $\omega_u$, a vertical straight line segment is sufficient to insure the robustness of damping. This segment, called "open loop frequency template" (or more simply "template"), illustrates robustness in frequency domain (figure 2) ; the longer the segment, the greater the robustness.

If the parameters of the plant vary, the segment AB slides vertically on itself. This insures a constant phase margin (independent of the plant parametric state) and, consequently, the invariance of the corresponding damping factor in time domain.

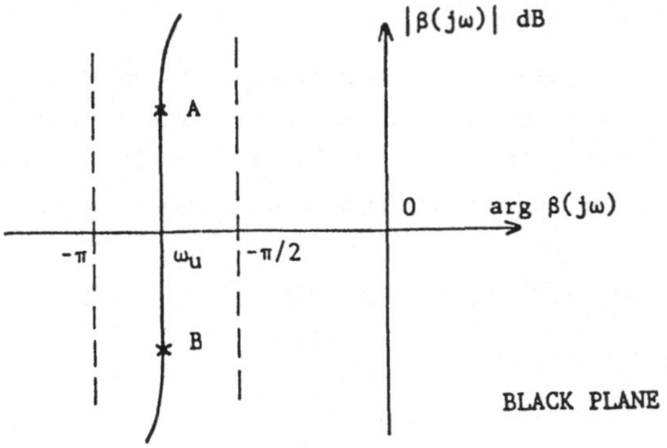


Figure 2 - Illustration of robustness in frequency domain : AB is the segment to be synthesized

## V - IDEA OF THE SYNTHESIS OF THE TEMPLATE IN THE CASE OF AN ASYMPTOTIC FREQUENCY PLACEMENT

### V.1 - Asymptotic frequency behaviour of the plant

An "asymptotic frequency behaviour" (or more simply "asymptotic behaviour") is characterized by a locked phase which is defined by a phase independent of frequency ; it corresponds to a range of frequencies in which the phase diagram is comparable to the corresponding asymptotic diagram. A plant presents one or several asymptotic behaviours. An order n asymptotic behaviour corresponds to a locked phase at $-n\pi/2$ (figure 3).

### V.2 - Asymptotic frequency placement of the template

In such a placement, the template belongs to a frequency range which corresponds to an asymptotic behaviour of the plant (figure 3).

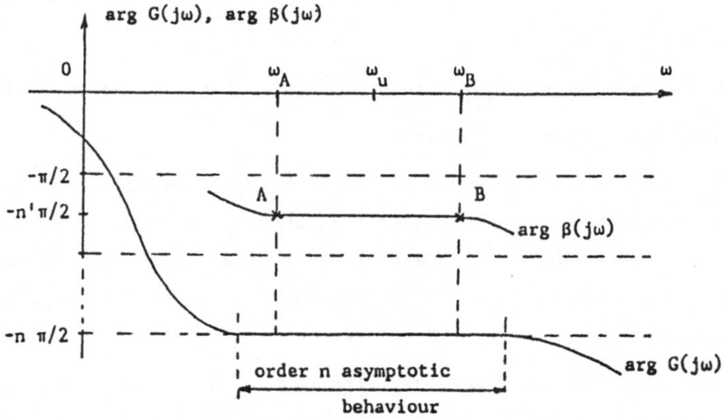


Figure 3 - Illustration of an asymptotic frequency placement of the template in the Bode plane : $G(j\omega)$ designates the frequency response of a plant which presents an order n asymptotic behaviour.

To pass from the argument of $G(j\omega)$ to the argument of $\beta(j\omega)$ for $\omega_A < \omega < \omega_B$, the observation of figure 3 shows that the regulator placed in cascade with the plant, must provide an advance of phase equal to $m'\pi/2$, with $m' = n - n'$, in the frequency range corresponding to the template, namely $[\omega_A, \omega_B]$.

Such a phase advance can be obtained with an order m' frequency response of the form :

$$C_{m'}(j\omega) = C_0 \left(\frac{1 + j\omega/\omega_b}{1 + j\omega/\omega_h}\right)^{m'}, \tag{8}$$

in which the transitional frequencies $\omega_b$ and $\omega_h$ satisfy the conditions :

$$\omega_b \ll \omega_A \qquad \text{and} \qquad \omega_h \gg \omega_B. \tag{9}$$

From ( 5 ), m' is given by :

$$m' = n - 2(1 - \Phi_m/\pi) \tag{10}$$

or

$$m' = n - \frac{2}{\pi} \text{ arc sin } 1/Q \tag{11}$$

or more

$$m' = n - \pi/\text{arc cos } (-\zeta), \tag{12}$$

where $\Phi_m$ is the phase margin, Q the resonance factor and $\zeta$ the damping factor.

As far as $C_0$ is concerned, it is given by the relation :

$$C_0 = \frac{1}{|G(j\omega_u)|} \left(\frac{1 + (\omega_u/\omega_h)^2}{1 + (\omega_u/\omega_b)^2}\right)^{m'/2}. \tag{13}$$

Always from ( 5), one knows that $C_{m'}(j\omega)$ can be approximated by a physically achievable frequency response. This one is of finite integer order N, namely :

$$C_N(j\omega) = C_0 \prod_{i=1}^{N} \frac{1 + j\omega/\omega'_i}{1 + j\omega/\omega_i}, \tag{14}$$

in which

$$\omega_i/\omega'_i = \alpha \qquad \text{and} \qquad \omega'_{i+1}/\omega_i = \eta, \tag{15}$$

where $\alpha$ and $\eta$ are superior to unity and called "recursive factors". Moreover, m' is given by :

$$m' = (1 + \log \eta/\log \alpha)^{-1}. \tag{16}$$

In practice, the ratio between two consecutive zeros or poles must make a compromise between two conflicting requirements :

- it should be close enough to unity to limit the undulations associated to the order m' asymptotic behaviour, due to the frequency connections of the various steps and crenels of the asymptotic diagrams
- it must be sufficiently greater than unity to cover a wide range of frequencies with a reasonable number of zeros and poles.

This compromise is achieved by taking a ratio between 5 and 10 and a number of zeros or poles also between 5 and 10. Satisfactory results are obtained with a ratio of 6 and 8 zeros and 8 poles ; in robotics, or more precisely in manipulator control, 4 zeros and 4 poles are often sufficient.

## VI - APPLICATION IN MANIPULATOR CONTROL

### VI.1 - Study plant : inclining polar table

The study plant is achieved by an inclining polar table which constitutes an elementary manipulator with three degrees of freedom $q_1$, $q_2$ and $q_3$ (figure 4). It is achieved from an electrical motor $m_1$ driving in rotation a ruler ($m_1$, $l_1$) on which a guided lead $m_2$ is moved in translation by a motor $M_2$ ; the whole is driven in rotation by a motor $M_3$.

Given that the phenomena are sufficiently interesting without taking into account the gravitational forces, this paper exclusively deals with the case of the horizontality of the table. In this configuration, the equations describing the dynamic behaviour of the table are the following ones :

$$(I_1 + m_2 q_2{}^2)\ddot{q}_1 + (f_1 + \alpha_1\beta_1/R_1 + 2m_2 q_2 \dot{q}_2)\dot{q}_1 = (\alpha_1/R_1)u_1 \qquad (17)$$

and

$$(I_2/R^2)\ddot{q}_2 + (f_2 + \alpha_2\beta_2/R^2R_2)\dot{q}_2 - m_2 \dot{q}_1{}^2 q_2 = (\alpha_2/R\, R_2)u_2 , \qquad (18)$$

in which the different magnitudes are defined as follows :

- $I_1$ : inertia in relation to $q_1$ of the ruler, the motor $M_1$ rotor and the motor $M_2$
- $I_2$ : inertia of the motor $M_2$ and the lead in relation to $q_2$
- $f_1$ and $f_2$ : viscous friction coefficients associated to $q_1$ and $q_2$
- $\alpha_1$ and $\alpha_2$ : proportionality coefficients between motor torque and armature current
- $\beta_1$ and $\beta_2$ : proportionality coefficients between f.c.e.m. and rotor angular speed
- $R_1$ and $R_2$ : armature resistances
- $u_1$ and $u_2$ : armature voltages (controls)
- R : radius of driving pulleys of the lead

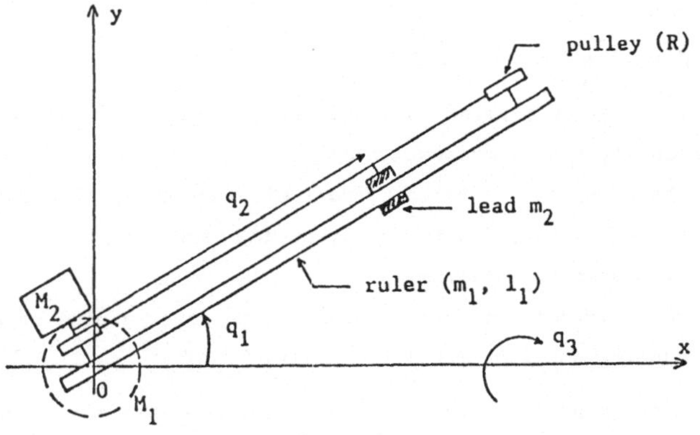


Figure 4 - Configuration of the table

VI.2 - Setting of the problem and adopted control scheme

- *Setting of the problem* : The control of such a table is very difficult, particularly if one wants to have mastery of the damping of each degree of freedom $q_i$, independently of the temporal configuration of the table. Nevertheless, the aim of the non integer approach is to insure this mastery.

- *Adopted control scheme* : The corresponding control strategy consists in elaborating the control $u_i$ of the freedom degree $q_i$ from the error signal $\varepsilon_i = q_{c_i} - q_i$ through a non integer order robust regulator, $q_{c_i}$ designating the input reference of rank i.

VI.3 - Interpretation of the non linear model of the table

The non linear mdoel of the table given by relations (17) and (18) may be interpreted as a linear non stationnary model whose coefficients can vary as fast as the variables, namely :

$$a_1(t)\,\ddot{q}_1(t) + b_1(t)\,\dot{q}_1(t) = d_1\,u_1(t) \qquad (19)$$

and

$$a_2\,\ddot{q}_2(t) + b_2\,\dot{q}_2(t) + c_2\,q_2(t) = d_2\,u_2(t), \qquad (20)$$

the corresponding coefficients being functions of time.

VI.4 - Frequency approach in the case of a weak non-stationarity of the table : local (or instantaneous frequency representation)

In the case of a weak-stationarity, the coefficients of the dynamic model of the plant vary relatively slowly in comparison with the variables :

* Their variations are not sufficiently slow for continuing to apply the results obtained in stationary state, that is to say for justifying the approximation of the near stationary states.

* Their variations are sufficiently slow in order that the response to a sinusoidal input keeps a near sinusoidal form, or more precisely that the response to a sinusoidal stationary oscillation should be a sinusoidal non-stationary oscillation (amplitude and phase angle depending on time).

In the particular case of the table, a weak non-stationarity correspond to a slow evolution of $q_2$ in comparison with that of $q_1$ when the dynamic behaviour in relation to $q_1$ is at stake, and vice versa when the dynamic behaviour in relation to $q_2$ is at stake.

At the view of equations (19) and (20), it appears that the differential equation which describes the dynamic behaviour of the table in relation to the freedom degree $q_i$, admits a general expression of the form :

$$a_i(t)\,\ddot{q}_i(t) + b_i(t)\,\dot{q}_i(t) + c_i(t)\,q_i(t) = d_i\,u_i(t). \qquad (21)$$

As it's the case in sinusoidal oscillation modulation, one assumes that the response to a stationary sinusoidal oscillation is a sinusoidal oscillation modulated both in amplitude and in phase. To that effect, if the concrete functions

$$u_i(t) = u_{m_i} \cos \omega t \quad \text{and} \quad q_i(t) = q_{m_i}(\omega,t) \cos\left(\omega t + \phi_i(\omega,t)\right) \tag{22}$$

are the solutions of equation (21), this one is verified by the complex functions

$$U_i(j\omega,t) = u_{m_i} \exp j\omega t$$

and

$$Q_i(j\omega,t) = q_{m_i}(\omega,t) \exp j\left(\omega t + \phi_i(\omega,t)\right), \tag{23}$$

which permits to write :

$$a_i(t)\, \ddot{Q}_i(j\omega,t) + b_i(t)\, \dot{Q}_i(j\omega,t) + c_i(t)\, Q_i(j\omega,t) = d_i\, U_i(j\omega,t). \tag{24}$$

It is then possible to define a frequency response, namely

$$G_i(\omega,t) = \frac{Q_i(j\omega,t)}{U_i(j\omega,t)} = \frac{q_{m_i}(\omega,t)}{u_{m_i}} \exp j\, \phi_i(\omega,t), \tag{25}$$

called "generalized frequency response".

Its determination ( 5 ) shows that it presents an order 2 asymptotic behaviour which is not modified by non-stationarity, which constitutes a fundamental property. In these conditions, the problem can be solved like the previous, described by figure 3.

VI.5 - Simulation of the control

We have simulated the dynamic behaviour of the control through a numerical processing of the differential equations of the table and the regulators. The used numerical integration method is that of Runge-Kutta (order 4). The simulation has been executed from the following data.

A - First kind of data : features of the motorized table

- ruler : $l_1$ = 0.5 m - lead : $m_2$ = 0.5 kg - driving pulleys : R = 0.5 cm - motor $M_1$ : CEM-AXEM-F9M4 ; $\alpha_1$ = 5.92 N cm/A ; $\beta_1$ = 6.2 V/1000 tr/mn ; $R_1$ = 1.1 Ω ; rotor inertia : 350 g cm² ; motor mass : 2.3 kg - motor $M_2$ : CEM-AXEM-UGP MEE 09B12 ; $\alpha_2$ = 3.1 N cm/A ; $\beta_2$ = 3.3 V/1000 tr/mn ; $R_2$ = 1.1 Ω ; rotor inertia : 340 g cm² ; motor mass : 0.6 kg - $I_1 = 2.5.10^{-3}$ kg m² ; $I_1 = 5.10^{-5}$ kg m²

B - Second kind of data : features of the regulators

- Regulator corresponding to $q_1$ : it is defined by the frequency response

$$C_{2_1}(j\omega) = 6433 \frac{(1 + j\omega/38.7)(1 + j\omega/477)}{(1 + j\omega/190.7)(1 + j\omega/2350.6)} \tag{26}$$

- Regulator corresponding to $q_2$ : it is defined by the frequency response

$$C_{2_2}(j\omega) = 5.888.10^5 \frac{(1 + j\omega/316)(1 + j\omega/2966.5)}{(1 + j\omega/1348)(1 + j\omega/12644.9)} \tag{27}$$

C - Performances

The recorded step responses corresponding to the freedom degrees $q_1$ and $q_2$ are shown in figure 5.

- Figure 5.a shows the influence of $q_2$ on the dynamics of $q_1$. In each case, the input reference of the control of $q_1$ is submitted to a step whose amplitude is equal to 90°. $q_2$ is successively defined as follows : (a) $q_2 = 0$ ; (b) $q_2 = 25$ cm ; (c) $q_2 = 50$ cm ; (d) $q_2$ increases from 0 to 50 cm ; (e) $q_2$ decreases from 50 cm to 0. Although the inertia in relation to $q_1$ varies by a factor of 50, the first overshoot of $q_1$ remains practically constant ; only the natural frequency changes.

- Figure 5.b shows the influence of $q_1$ on the dynamics of $q_2$. In each case, the input reference of the control of $q_2$ is a step whose amplitude equals 25 cm. $q_1$ is sucessively defined as follows : (a) $q_1$ increases from 0 to 270° in steady state (response to a step of 270°) ; (b) $q_1$ decreases from 270° to 0 in steady state (response to a step of -270°). One notes that all the dynamics of $q_2$ is independent of the variations of $q_1$, or practically independent in so far as the graphic resolution does not permit to distiguish between two very close curves.

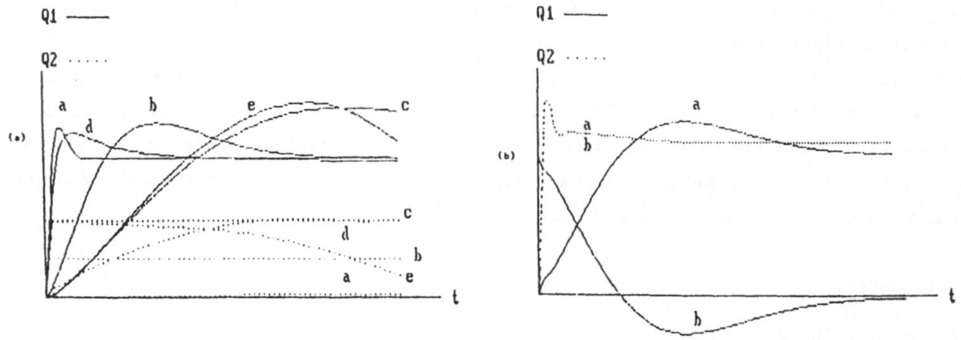


Figure 5 - Step responses of freedom degrees $q_1$ and $q_2$

## VII - CONCLUSION

We show how the natural robustness of the damping of water on a porous dyke can be used in Automatic Control so as to synthesize a robust control. The idea consists in synthesizing an open loop frequency template defined by a straight line segment around the unit gain frequency.

The non integer order robust control which uses this template is applied in manipulator control through an inclining polar table. The interest of the choice of such a table comes from its structure ; it is true that this one permits to insure large inertia variations, strong dynamic couplings and a great number of non-linearities. In fact, the very satisfactory damping performances of the control recorded with this table, allow to expect better performances in the case of manipulators which are less constraining.

In the consideration of the horizontality of the table and after establishing the model which describes its dynamic behaviour, the first stage of the approach consists in interpreting the non linear plant that the table constitutes as a non stationary linear plant whose coefficients vary as fast as the variables.

A local frequency representation associated to each freedom degree through a generalized frequency response, permits to show that non-stationarity does not modify the asymptotic frequency behaviours of the table, defined by a locked phase. So, the templates retain their forms despite non-stationarity. This phenomenon is fundamental for explaining the remarkable robustness performances obtained in non linear domain. Indeed, if the control is robust versus the coefficients of a linear non stationary plant, even if these one vary as fast as the variables, one understands why the control may be robust versus the non-linearities which can be interpreted in term of non-stationarity. In fact, what does it matter if the cause of the coefficient variations is different in non linear domain ?

A numerical simulation in time domain shows that our frequency approach is correct. Indeed, the synthesized control presents a very robust damping versus the table temporal configuration on which the coefficients of the model depend, so verifying the frequency approach. In fact, the performances obtained are very remarkable : it is true that, for an inertia variation by a factor of 50, which is very large, and without mentioning the other variations (Coriolis and centrifugal effects) which are also very large, the transients of the step responses of the freedom degrees keep their forms with or without a time scale changing.

BIBLIOGRAPHY

(1) - A. OUSTALOUP - IEEE, Chicago, April 27-29, 1981 - IEEE Transactions on Circuits and Systems, vol. cas 28, n°10, 1981

(2) - A. OUSTALOUP - Syst. asserv. liné. d'ordre fract. : théor. et prat., MASSON 1983

(3) - A. OUSTALOUP and B. BERGEON - IFAC'87, Munich, July 27-31, 1987

(4) - A. OUSTALOUP - From fractality to non integer derivation. 8th Internat. Conf. "Analysis and Optimization of Systems", INRIA, Antibes (France), June 8-10, 1988

(5) - A. OUSTALOUP - From the robustness of stability degree in nature to the control of highly non linear manipulators - Preprints Colloque International Automatique Non Linéaire, Nantes (France), 13-17 Juin 1988

(6) - A. OUSTALOUP - From fractality to non integer derivation through recursivity - Survey, session "Fractality and non integer derivation", 12th IMACS World Congress on Scientific Computation, Paris, July 18-22, 1988

# AFFINE REALIZATIONS OF MULTIMODELS
## characterization, stability, identification

**Guy BORNARD, Li-Ping HU**

Laboratoire d'Automatique de Grenoble - UA 228 CNRS-INPG

BP 46 - 38402 Saint-Martin-d'Hères - FRANCE

**SUMMARY** : Knowing a set of linear models indexed by an independent parameter - a multimodel - there exists an infinite class of state-affine models compatible with this multimodel - the realizations of the multimodel. The set of the realizations of a given multimodel is characterized. Stability properties are studied as well as the identification problem.

**KEY-WORDS** : nonlinear systems, multimodels, realization, identification, stability.

## I - INTRODUCTION

It is not always possible to derive a model of a plant to be controlled from the theoretical knowledge available on the process. In such a case, an alternate approach consists in performing experiments on the plant and extracting some "black box" model from the information so obtained. For linear systems, this procedure of identification has been widely used and is a current tool of the control engineer.

For the nonlinear case, the difficulty arises from the number of possible structures for the model, and from the large amount of information that would be required from the plant in order to obtain a model available for every admissible input.

Due to this difficulty and to the fact that the problem of control is often restricted to the regulation of the plant near a certain number of operating points, most of the general purpose nonlinear identification studies achieved up to now are restricted to the following scheme:

- analyse the behavior of the plant near a certain number of steady state operating points,
- get a linear or state affine model for each one,
- construct a unique nonlinear, generally state affine, model by interpolation or regression from the preceding ones.

Note that the construction of such a model is not necessarily achieved by a two-stages procedure and better results can be obtained by direct identification.

For such works, see for instance [PONCET, 1982], [NORMAND-CYROT, 1983], [MEIZEL, 1984], [THOMASSET, 1987], [NEYRAND, 1987].

The basic information contained in such models can only predict the behavior for constant or "slowly variable" values of the parameter which indexes the operating points.

In this paper, we study the problem of representing the behavior of the process for a certain class of fast change in this parameter, i.e. for large transcients. We shall restrict the study to the case where a linear behavior was first obtained for a certain set of operating points. This set of linear models, indexed by an independant parameter representing the operating points, is called here a multimodel, by reference to the terminology also used in so-called multimodel control techniques. See for instance [MONNIER, 1977].

It should be noted that the term "operating point" does not necessarily mean a small region near a steady state.

When achieving the interpolation process in order to get a state-affine model, it appears that the solution is not unique, but depends on an arbitrary affine change of coordinates when representing each linear model, so that there is an infinite equivalence class of state-affine global models -the realizations of the multimodel- which are compatible with the multimodel.

It will be shown that the behavior of two distinct realizations of the same multimodel may be quite far one from the other. Clearly, it is highly desirable to choose the most adequate realization of a multimodel, on the basis of extra information to be got on the plant.

The paper is organized as follows:

- The Section II is devoted to the definition and notations and to the characterization of the set of realizations of a multimodel.
- In the Section III, the stability of subclasses of realizations is studied.
- In the Section IV, the identification problem is considered for certain classes of time evolution of the indexing parameter.

All proofs are omitted. See for instance [BORNARD, 1988].

## II - Realizations of multimodels

### Definition II.1

A multimodel is the set M defined as follows:

(II.1) $$M = \{ \Lambda_\varphi , \mathring{u}_\varphi , \mathring{y}_\varphi \mid \varphi \in \Phi \}$$

where: $\mathring{u}_\varphi \in \mathbb{R}^m$, $\mathring{y}_\varphi \in \mathbb{R}^p$, $\Lambda_\varphi : u \longrightarrow y$ is a linear input-output mapping with the input function u (measurable, bounded) taking its values in $\mathbb{R}^m$, the output function y taking its values in $\mathbb{R}^p$, $\Lambda_\varphi$ being moreover realizable by a system on $\mathbb{R}^n$, and $\Phi$ is the set indexing the elements of M.

If, for every $\varphi \in \Phi$, the realization of $\Lambda_\varphi$ in $\mathbb{R}^n$ is minimal, the multimodel is said regular.

**Definition II.2**

A realization of the multimodel M is a system of the form:

$$(\text{II.2}) \qquad \sigma \begin{cases} \dot{x}(t) = A(\varphi(t))\,(x(t) - f(\varphi(t))) + B(\varphi(t))\,(u(t) - u^\circ(\varphi(t))) \\ y(t) = C(\varphi(t))\,(x(t) - f(\varphi(t))) + y^\circ(\varphi(t)) \end{cases}$$

where: $u : \mathbb{R} \longrightarrow \mathbb{R}^m$ is the measurable bounded input function

$x(t) \in \mathbb{R}^n$ is the state

$y(t) \in \mathbb{R}^p$ is the output

$\varphi : \mathbb{R} \longrightarrow \Phi$

$A : \Phi \longrightarrow \mathcal{M}_{n\times n}$, bounded

$B : \Phi \longrightarrow \mathcal{M}_{n\times m}$, bounded

$C : \Phi \longrightarrow \mathcal{M}_{p\times n}$, bounded

$f : \Phi \longrightarrow \mathbb{R}^n$, bounded

$\varphi$ and A, B, C are such that $A\circ\varphi$, $B\circ\varphi$, $C\circ\varphi$, $f\circ\varphi$ are measurable and, for each constant function $\varphi$, the input-output mapping generated by the system $\sigma$ is exactly:

$$\mathcal{P}\,(A(\varphi), B(\varphi), C(\varphi) = \Lambda_\varphi$$

The system $\sigma$ defined by (II.2) associates to each function $\varphi$ a system $\sigma_\varphi$ which is, with u as input, a time varying linear system.

In (II.2), $\varphi$ might also be considered as an input function. However, in order to distinguish its role from that of u, it will be regarded as a "time varying parameter".

In the following, in order to shorten the notations, we shall denote x for x(t), $A_\varphi$ for $A(\varphi(t))$, etc., and use indifferently $\varphi$ for the function or its value.

Let G be the set of the mappings from $\Phi$ to $Gl(n) \times_s \mathbb{R}^n$, the semi-direct product of the general linear group Gl(n) by the additive group $\mathbb{R}^n$. G is given naturally a group structure from that of $Gl(n) \times_s \mathbb{R}^n$.

Consider the action of $g=(T, z) \in G$ on the set $\Sigma$ of systems of the form (II.2) defined by:

$$(\text{II.3}) \qquad \begin{matrix} A_\varphi \\ B_\varphi \\ C_\varphi \\ f_\varphi \end{matrix} \quad \xrightarrow{(T_\varphi,\, z_\varphi)} \quad \begin{matrix} T_\varphi A_\varphi T_\varphi^{-1} \\ T_\varphi B_\varphi \\ C_\varphi T_\varphi^{-1} \\ T_\varphi f_\varphi + z_\varphi \end{matrix}$$

**Proposition II.1**

Let $\sigma$ be a realization of a multimodel M. Then, every element of the orbit $G.\sigma$ in $\Sigma$ is a realization of M. Moreover, if M is regular, then there exists no other realization of M.

**Definition II.3**

Consider a subset $\mathfrak{D}$ of the admissible functions for $\varphi$. Two realizations $\sigma$ and $\sigma'$ of a multimodel M are said $\mathfrak{D}$-equivalent iff, for each $\varphi \in \mathfrak{D}$, the input-output mappings $\mathcal{P}_\sigma$ and $\mathcal{P}_{\sigma'}$ generated by $\sigma$ and $\sigma'$ are identical. If they are $\mathfrak{D}$-equivalent for every admissible $\mathfrak{D}$, they are said equivalent.

**Proposition II.2**

Let $\sigma$ be a realization of a multimodel M. Assume that $g_c$ is a constant element of G. Then $\sigma$ and $g_c.\sigma$ are equivalent. If moreover M is regular, then the converse is true.

The set of realizations of a regular multimodel M is then parametrized, up to equivalence, by the elements of G modulo the constant elements.

## III - Stability

The behavior of two distinct realizations of the same multimodel may be quite different, as it will be shown in the following example.

**Example 1**

Let $\Phi = \{1, 2\}$, and consider two realizations of the same multimodel:

$$\sigma \begin{cases} A_1 = \begin{pmatrix} -.1 & -1 \\ 1 & -.1 \end{pmatrix} & B_1 = \begin{pmatrix} 0 \\ 1 \end{pmatrix} & A_2 = \begin{pmatrix} -.2 & -1 \\ 1. & -.2 \end{pmatrix} & B_2 = \begin{pmatrix} 0 \\ 1 \end{pmatrix} \\ C_1 = (1\ 0) & & C_2 = (1\ 0) & \end{cases}$$

$$\sigma' \begin{cases} A'_1 = \begin{pmatrix} -.1 & -.1 \\ 10 & -.1 \end{pmatrix} & B'_1 = \begin{pmatrix} 0 \\ 10 \end{pmatrix} & A'_2 = \begin{pmatrix} -.2 & -10 \\ .1 & -.2 \end{pmatrix} & B'_2 = \begin{pmatrix} 0 \\ .1 \end{pmatrix} \\ C'_1 = (1\ 0) & & C'_2 = (1\ 0) & \end{cases}$$

Let $u \equiv 0$ and consider $\sigma$. For every function $\varphi$, every $x(0) \in \mathbb{R}^n$, $x(t) \to 0$ when $t \to +\infty$. The same does not apply for $\sigma'$ : let $x'(0) = (1, 0)$, $\varphi(t) = \varphi_1$ if $2k\, t_0 \leq t < (2k+1)\, t_0$, $\varphi = \varphi_2$ if $(2k+1)\, t_0 \leq t < (2k+2)\, t_0$, $k \in N$, where $t_0$ is the smallest positive time for which the trajectory crosses the vertical axis ($t_0 \sim \frac{\Pi}{2}$).

Then $\| x(t) \| \to \infty$ when $t \to +\infty$.

The situation is the following: the multimodel under consideration has all its elements stable. However, the first realization is stable in some sense while the second one is not. This problem is discussed in the present section.

**Definition III.1**

A multimodel is said stable iff the two following conditions are fulfilled:

i) for every $\varphi \in \Phi$, the linear input-output mapping $\Lambda_\varphi$ is asymptotically stable,

ii) their exist a lower bound $\lambda_0 > 0$ for the absolute value of the real part of the eigenvalues of $A_\varphi$ in some realization of M for every $\varphi \in \Phi$.

**Remarks** :

- The notion of stability used here is the classical one for linear systems (see [KALMAN, 1960] for instance).
- The second condition can be relaxed if $\Phi$ is a finite set.

**Definition III.2**

Let $\mathfrak{D}$ be some subset of admissible functions $\varphi$. A realization $\sigma$ is said $\mathfrak{D}$-stable if, for every $\varphi \in \mathfrak{D}$, the linear time-varying system generated by $\sigma$ is stable in the bounded-input bounded-state sense, i.e. to every $\alpha > 0$ corresponds $\beta > 0$, such that for $\forall x \in \mathbb{R}^n$ there exists $t_0 > 0$ such that $\|u\| \leq \alpha$ implies $\|x(t)\| \leq \beta$ for $\forall t \geq t_0$.

If $\sigma$ is $\mathfrak{D}$-stable for every $\mathfrak{D}$, it is said stable.

**Remark** : Let us recall that A, B, C, f are bounded mappings.

We can give two stability results.

**Theorem III.1**

Let M be a stable regular multimodel. Then there exists a stable realization of M.

Let A be an endomorphism of $\mathbb{R}^n$. We shall call, by abuse, real Jordan form of type 1 and real Jordan form of type 2, the matrices $A_1$ and $A_2$ representing A in bases such that :

( III1.a) $A_1 =$

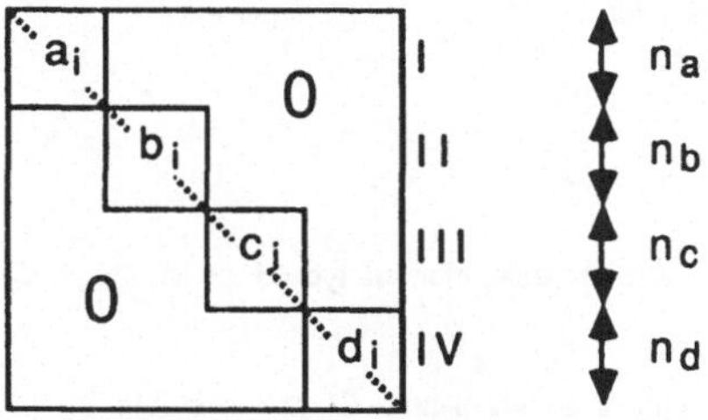


(III1.b) $A_2 =$

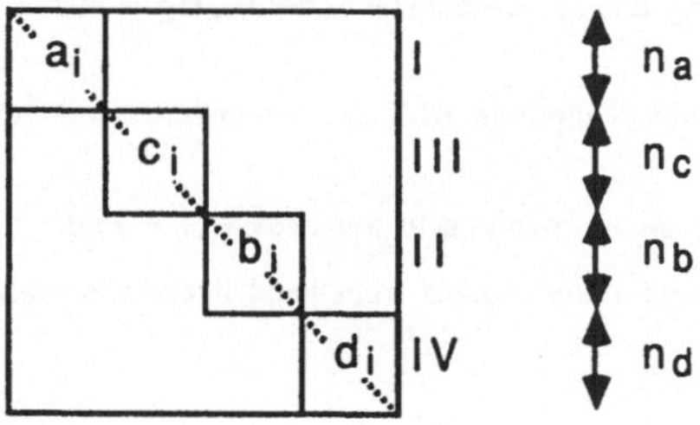

where the four blocks are respectively of dimensions $n_a$, $n_b$, $n_c$, $n_d$, each block being itself block diagonal, with $a_i$, $b_i$, $c_i$, $d_i$ of the form :

$a_i$ = scalar element (single real eigenvalue)

$$b_i = \begin{pmatrix} \lambda_i & & 0 \\ 1 & \ddots & \\ 0 & 1 & \lambda_i \end{pmatrix}$$ (multiple real eigenvalue)

$$c_i = \begin{pmatrix} \lambda_i & \gamma_i \\ -\gamma_i & \lambda_i \end{pmatrix}$$ (single complex conjugate pair of eigenvalues)

$$d_i = \begin{bmatrix} \begin{matrix} \lambda_i & \gamma_i \\ -\gamma_i & \lambda_i \end{matrix} & & \\ I & \ddots & \\ & I & \begin{matrix} \lambda_i & \gamma_i \\ -\gamma_i & \lambda_i \end{matrix} \end{bmatrix}$$ (multiple complex conjugate pair of eigenvalues)

Let us call Jordan realization of type 1 (resp. type 2) of a multimodel M a realization such that, for every $\varphi \in \Phi$, the matrix $A_\varphi$ is under the real Jordan form of type 1 (resp. type 2).

<u>**Theorem III.2**</u>

Let M be a stable multimodel, $\sigma_1$ a Jordan realization of type 1 of M, $\sigma_2$ a Jordan realization of type 2 of M. Then :

i) If $\sigma_1$ is such that the quantity $n_a+n_b$ is constant over $\Phi$, $\sigma_1$ is stable.

ii) If $\sigma_2$ is such that the quantities $n_b$ and $n_d$ are constant over $\Phi$, $\sigma_2$ is stable.

**Remark** : If $\Phi$ is a finite set, the proof of theorem III.1 can be initiated with any realization.

The result of theorem III.2 seems to give arguments to the choice of a Jordan form, which is sometimes made, without justification, when building state-affine models from local information available near operating points.

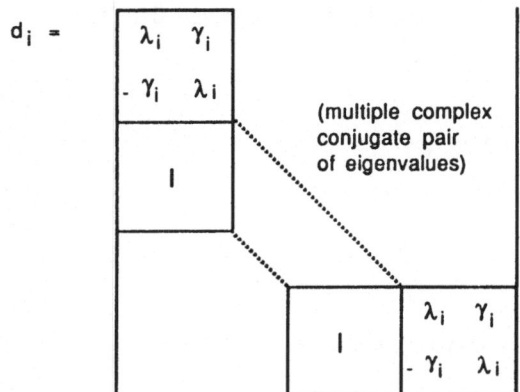

Looking towards the slowly varying parameter case, it would perhaps have been expected that, when $\varphi$ remains constant during sufficiently long time intervals, all the realizations of a multimodel would be stable. This is not true.

Denote by $\mathfrak{D}_\theta^c$ the set of the function $\varphi$ which are constant on time intervals greater or equal to $\theta$, $\theta>0$. Consider again the example III.1. It is clear that, for any given $\theta>0$, one can make "sharper" the second realization so that it is not $\mathfrak{D}_\theta$-stable.

## IV - Identification

Assume now that in addition to the multimodel M, extra information is known from the plant through some input-output data obtained for large changes on $\varphi$. The identification problem is then to choose in the orbit associated to M the realization which fits best with the experimental data. This problem will be discussed under additional conditions which make it tractable through least squares techniques.

### A one to one parametrization

The section II gives a natural parametrization to the class of realizations of a given stable regular multimodel M. Choose first some realization $\sigma = (A, B, C, f\ i\ 0)$. Any realization of M is of the form $\sigma' = g.\sigma$, $g \in G$, and can be writen:

$$(IV.1)\quad \begin{cases} \dot{x}(t) = T\varphi(t)\ A\varphi(t)\ T\varphi(t)^{-1}\ (x(t) - z\varphi(t)) + T\varphi(t)\ B\varphi(t)\ (u(t) - u\mathring{\varphi}(t)) \\ y(t) = C\varphi(t)\ T\varphi(t)^{-1}\ (x(t) - z\varphi(t)) + y\mathring{\varphi}(t) \end{cases}$$

This parametrization by G is not one to one, because of the equivalence under constant elements of G. However, we can obtain another one to one parametrization in the following way:

For $\varphi \in \Phi$, let $x_\varphi$ be defined by :

$$(IV.2)\qquad x(t) = T_\varphi\ x_\varphi(t) + z_\varphi(t)$$

One has then, for any $\varphi, \varphi_1, \varphi_2, \varphi_3 \in \Phi$ :

$$x_{\varphi_2}(t) = H_{\varphi_2\varphi_1} \cdot x_{\varphi_1}(t) + h_{\varphi_2\varphi_1}$$

where: $$H_{\varphi_2\varphi_1} = T_{\varphi_2}^{-1}\ T_{\varphi_1}$$

$$h_{\varphi_2\varphi_1} = T_{\varphi_2}^{-1}\ (z_{\varphi_1} - z_{\varphi_2})$$

$$(IV.3)\quad \begin{aligned} &(H_{\varphi_3\varphi_1}, h_{\varphi_3\varphi_1}) = (H_{\varphi_3\varphi_2}, h_{\varphi_3\varphi_2})\ (H_{\varphi_2\varphi_1}, h_{\varphi_2\varphi_1}) \\ &(H_{\varphi\varphi}, h_{\varphi\varphi}) = (I, 0) \end{aligned}$$

In (IV.3) the couples $(H_{\varphi_i\varphi_j}, h_{\varphi_i\varphi_j})$ are considered as elements of $Gl(n) \times_s \mathbb{R}^n$.

Let $\mathfrak{D}^c$ be the set of the piecewise constant functions from $\mathbb{R}$ to $\Phi$, with a finite number of commutations on every compact interval.

Let $\varphi \in \mathfrak{D}^c$. On the interior of each time interval on which $\varphi$ is constant, (IV.1) can be rewriten as:

$$(IV4.a) \begin{cases} \dot{x}_{\varphi(t)}(t) = A_{\varphi(t)}\, x_{\varphi(t)}(t) + B_{\varphi(t)}\,(\,u(t) - \mathring{u}_{\varphi(t)}(t)\,) \\ y(t) = C_{\varphi(t)}\, x_{\varphi(t)}(t) + \mathring{y}_{\varphi(t)}(t) \end{cases}$$

At the commutation instants $t_k$, one has:

$$(IV.4.b) \qquad x_{\varphi(t_k+)}(t_k) = H_{\varphi(t_k+)\varphi(t_k-)}\, x_{\varphi(t_k-)}(t_k) + h_{\varphi(t_k+)\varphi(t_k-)}$$

(IV.4) is a "system" with affine jumps on the state at the commutation instants. It is parametrized by the group $\tilde{G}$, whose elements are:

$$(IV.5) \qquad \tilde{g} : \begin{array}{l} \Phi \times \Phi \longrightarrow Gl(n) \times_s \mathbb{R}^n \\ (\varphi_1, \varphi_2) \longrightarrow (H_{\varphi_2\varphi_1}, h_{\varphi_2\varphi_1}) \end{array}$$

From (IV.3), it is clear that to every element of an orbit in G of a constant element $g_c \in G$ corresponds a unique $\tilde{g} \in \tilde{G}$.

However $\tilde{G}$ is a much larger set than G and the elements $\tilde{g} \in \tilde{G}$ which represent a realization of the multimodel M are those satisfying (IV.3). This defines a subset $\tilde{G}^a \in \tilde{G}$, which is in one to one correspondance with the set of the classes of equivalent realizations of M.

Remark that, for $\tilde{g} \notin \tilde{G}^a$, it would be possible to go back to the form (IV.1), but T and z would no longer be indexed by the function $\varphi$, but by the semigroup generated by the functions $\varphi \in \mathfrak{D}^c$ (for the exact definition of this semigroup, see [GAUTHIER, 1986].

**Discrete-time multimodels**

At this point, one may have some doubt about the practical tractability of the identification problem. Let us first consider the discrete-time case, for which all the properties of the continuous case apply, but which exhibits some particularities.

M is now a discrete-time (stable, regular) multimodel, i.e. $\Lambda\varphi$ is a discrete-time linear input-output mapping in the definition (II.1).

Copying the continuous-time case, a realization can be writen either:

$$(\text{IV.6})\quad \begin{cases} x(t+1) = T\varphi_{(t)}\, A\varphi_{(t)}\, T\varphi_{(t)}^{-1}\,(x(t) - z\varphi_{(t)}) + T\varphi_{(t)}\, B\varphi_{(t)}\,\big(u(t) - u\mathring{\varphi}_{(t)}\big) \\ y(t) = C\varphi_{(t)}\, T\varphi_{(t)}^{-1}\,\big(x(t) - z\varphi_{(t)}\big) + y\mathring{\varphi}_{(t)} \end{cases} \qquad t\in\mathbb{Z}$$

or:

$$(\text{IV.7})\quad \begin{cases} x\varphi_{(t)}(t+1) = A\varphi_{(t)}\, x\varphi_{(t)} + B\varphi_{(t)}\,\big(u(t) - u\mathring{\varphi}_{(t)}\big) \\ x\varphi_{(t+1)}(t+1) = H\varphi_{(t+1)}\varphi_{(t)}\, x\varphi_{(t)}\,(t+1) + h\varphi_{(t+1)}\varphi_{(t)} \\ y(t) = C\varphi_{(t)}\, x\varphi_{(t)}\,(t) + y\mathring{\varphi}_{(t)} \end{cases} \qquad t\in\mathbb{Z}$$

and finally (IV.7) gives:

$$(\text{IV.8})\quad \begin{cases} x\varphi_{(t+1)}(t+1) = H\varphi_{(t+1)}\varphi_{(t)}\, A\varphi_{(t)}\, x\varphi_{(t)}\,(t) \\ \qquad + H\varphi_{(t+1)}\varphi_{(t)}\, B\varphi_{(t)}\,\big(u(t) - u\mathring{\varphi}_{(t)}\big) + h\varphi_{(t+1)}\varphi_{(t)} \\ y(t) = C\varphi_{(t)}\, x\varphi_{(t)}\,(t) + y\mathring{\varphi}_{(t)} \end{cases}$$

Let $\overline{x}\,(t) = x\varphi_{(t)}\,(t)$. (IV.8) becomes:

$$(\text{IV.9})\quad \begin{cases} \overline{x}\,(t+1) = H\varphi_{(t+1)}\varphi_{(t)}\, A\varphi_{(t)}\, \overline{x}\,(t) \\ \qquad + H\varphi_{(t+1)}\varphi_{(t)}\, B\varphi_{(t)}\,\big(u(t) - u\mathring{\varphi}_{(t)}\big) + h\varphi_{(t+1)}\varphi_{(t)} \\ y(t) = C\varphi_{(t)}\, \overline{x}\,(t) + y\mathring{\varphi}_{(t)} \end{cases}$$

One can remark that, for every $(H, h)\in\tilde{G}$ and every admissible function $\varphi$, (IV.9) is a linear time-varying discrete-time system. The notion of jump as encountered in (IV.4) has no meaning here.

(IV.9), with (H, h) exploring all $\tilde{G}$, can be interpreted as an extended definition of a realization of a discrete-time multimodel M. The set of realizations, in this sense, of M (regular) is exactly one orbit of $\tilde{G}$ for the action defined by (IV.9).

## The finite memory (discrete-time) case

It will be assumed now that every $\Lambda\varphi$ is a finite memory input-output mapping, i.e. $A_\varphi^{\,n}=0$ in any representation of $\Lambda\varphi$.

This could seem to be a very drastic assumption. However, one should note that, for a stable $\Lambda\varphi$ (continuous or discrete-time), there is a time $\theta$ -the "response time"- after which the effect of the initial state vanishes from a practical point of view. So that the finite memory assumption may be regarded in practice as a quite natural one for stable multimodels.

Let $\mathfrak{D}_k^d$ be the set of the functions from $\mathbb{Z}$ into $\Phi$ which are constant on intervals of at least k consecutive instants. Assume that $\varphi\in\mathfrak{D}_n^d$, and consider the commutation times $t_{12}$, $t_{23}$ such that:

$$\varphi\,(t) = \varphi_1 \quad \text{for} \quad t_{12} - n \le t \le t_{12} - 1$$
$$\varphi\,(t) = \varphi_2 \quad \text{for} \quad t_{12} \le t \le t_{23} - 1$$

Since M is finite memory, $x(t_{12})$ does not depend upon $x(t_{12}-n)$ but only upon $u(t_{12}-n), \ldots, u(t_{12}-1)$. In particular, it is independent of the elements $(H_{\varphi_i\varphi_j}, h_{\varphi_i\varphi_j})$ corresponding to the commutations occurred previously One has, from (IV.7):

$$
\text{(IV10)} \quad \begin{cases}
x_{\varphi_1}(t_{12}) & = \Psi^n_{\varphi_1} u^n_{\varphi_1}(t_{12}) \\
\overline{x}(t_{12}) & = H_{\varphi_2\varphi_1} \Psi^n_{\varphi_1} u^n_{\varphi_1}(t_{12}) + h_{\varphi_2\varphi_1} \\
y(t_{12}+k) & = C_{\varphi_2} \overline{x}(t_{12}+k) \\
& = C_{\varphi_2} \Psi^n_{\varphi_2} u^n_{\varphi_2}(t_{12}+k) + C_{\varphi_2} A^k_{\varphi_2} (H_{\varphi_2\varphi_1} \Psi^n_{\varphi_1} u^n_{\varphi_1}(t_{12}) + h_{\varphi_2\varphi_1}) \\
& \quad k = 0, \ldots, (t_{23} - t_{12} - 1)
\end{cases}
$$

where :
$$\Psi^0_\varphi = 0$$
$$\Psi^k_\varphi = (B_\varphi, A_\varphi B_\varphi, \ldots, A^{k-1}_\varphi B_\varphi)$$
$$u^k_\varphi(t) = ((u(t-1) - u^0_\varphi)^T, \ldots, (u(t-k) - u^0_\varphi)^T)^T$$

In (IV.10), y is of the form:

$$\text{(IV.11)} \qquad y(t_{12}+k) = \Omega^k_0 + \Omega^k_1 (H_{\varphi_2\varphi_1} \Omega_2 + h_{\varphi_2\varphi_1})$$

Clearly, the output $y(t)$, $t = t_{12}, \ldots, t_{23}-1$, depends linearly upon the coefficients of $H_{\varphi_2\varphi_1}$, $h_{\varphi_2\varphi_1}$ and is independent of the other elements $(H_{\varphi_i\varphi_j}, h_{\varphi_i\varphi_j})$. To determine completely $H_{\varphi_2\varphi_1}$, $h_{\varphi_2\varphi_1}$, it is required to have at least n+1 occurrences of a commutation of $\varphi$ from $\varphi_1$ to $\varphi_2$, and that the corresponding values of $x_{\varphi_1}(t_{12})$ span $\mathbb{R}^n$. If the available data is redondant, a least square solution can be used.

So, if M is a discrete-time finite-memory regular multimodel, and if $\varphi \in \mathcal{D}^c_n$, the elements $(H_{\varphi_i\varphi_j}, h_{\varphi_i\varphi_j})$ can be estimated independently one from each other, through a least square technique.

__Jumps from steady states__

Assume that M is again a discrete-time finite memory regular multimodel and that u and $\varphi$ are jointly restricted as follows:

- $\varphi \in \mathcal{D}^c_n$,
- at each commutation from $\varphi_i$ to $\varphi_j$ at time $t_{ij}$, one has $x_{\varphi_i}(t_{ij})=0$ (steady state corresponding to $u^0_{\varphi_i}$).

Then the $H_{\varphi_j\varphi_i}$'s don't play any role, and one occurrence of a commutation of $\varphi$ from $\varphi_i$ to $\varphi_j$, is sufficient to determine $h_{\varphi_i\varphi_j}$. Moreover, taking $H_{\varphi_i\varphi_j}=I$ for every i, $j \in \mathbb{Z}$, $\tilde{g} \in \tilde{G}^a$ becomes $h_{\varphi_j\varphi_i} = -h_{\varphi_i\varphi_j}$. Then this constraint can be easily introduced in the least square problem, if required.

There is an important difference between this case and the previous one: in the finite memory case, a limitation is imposed on $\varphi$ only for the identification data. On the contrary, what is identified here is a class of realizations which are equivalent only for the couples (u, $\varphi$) satisfying the imposed restriction.

This case may be interpreted as representing the situation where the plant is to be controlled near steady state operating points and is submitted to large step changes occurring at instants separated by at least the response time of the plant.

## V - Conclusion

We discussed in this paper several aspects of the problems arising when looking for a state-affine realization from a known multimodel - a case of interest in black box nonlinear modelling. A characterization of the equivalence classes of realizations was given. Stability properties and identification procedures were discussed.

Let us mention as potentially interesting extensions of the scope of this work the case where the elements of the multimodel are state-affine rather than linear, and the case where they are indexed by the input, for instance.

## REFERENCES


— GAUTHIER J.P., BORNARD G. : "Global realization of input-output mappings", SIAM J. of Control and optimization, Vol. 24, No 3, May 1986, pp. 509-521.

— KALMAN R.E., BELTRAM J.E. : "Control system analysis and design via the second method of Lyapunov-I continuous-time systems", Journal of Basic Engineering, Trans. of the ASME, June 1960, pp. 371-393.

— LASALLE J., LEFSCHETZ S. : "Stability by Lyapunov's direct method with applications", Mathematics in Science and Engineering, Academic Press, New York, 1961.

— MEIZEL D. : "Sur la synthèse paramétrique d'asservissements de processus non linéaires", Thèse d'état, Université de Lille, 21 sept. 1984.

— MONNIER B. : "Contribution à la commande d'une classe de procédés dynamiques industriels dans de grands domaines de fonctionnement", Thèse D. Ing., INP Grenoble, 1977.

— NEYRAN B. : "Identification et commande en temps discret des systèmes linéaires à paramètres variables en utilisant des modèles à état affine", Thèse Doct. INSA, Lyon, 8 juillet 1987.

— NORMAND-CYROT D. : "Théorie et pratique de systèmes non linéaires en temps discret", Thèse d'état, Université de Paris Sud, mars 1983.

— PONCET B. : "A new approach of nonlinear systems : homodynamic systems", IFAC Symposium on identification and system parameter estimation, Washington, June 1982, Vol. 1, pp. 80-86.

— THOMASSET D. : "Réalisation et commande en temps discret des systèmes continus retardés linéaires invariants et des systèmes linéaires à paramètres variables", Thèse d'état, INSA, Université CB Lyon, 8 juillet 1987.

— BORNARD G., HU L.P. : "Affine realizations of multimodels : characterization, stability, identification", Colloque International Automatique Non Linéaire, Nantes, juin 1988.

# ADAPTIVE TECHNIQUES

# STATE ESTIMATION AND ADAPTIVE CONTROL OF MULTILINEAR COMPARTMENTAL SYSTEMS: THEORETICAL FRAMEWORK AND APPLICATION TO (BIO)CHEMICAL PROCESSES.

**G. BASTIN**
Laboratoire d'Automatique, Dynamique et Analyse des Systèmes
Catholic University of Louvain,Bâtiment Maxwell, Place du Levant, 3,
B-1348 LOUVAIN LA NEUVE (Belgium).

Work supported by the CEC Biotechnology Action Programme (contract BAP-0032)

## ABSTRACT

A general theoretical framework for the design of state observers and adaptive controllers for a class of multilinear compartmental systems (arising from (bio)chemical applications) is presented.

## 0. MOTIVATION AND INTRODUCTION.

This paper is devoted to state estimation and adaptive control of a class of compartmental systems which arise from chemical and biochemical engineering applications. The motivation is twofold:

a) A critical issue in controlling (bio)chemical processes in stirred tank reactors is the lack (or the prohibitive cost) of reliable sensors for on line measurement of the main state variables (i.e. reactants and reaction products). The use of (possibly adaptive) observers as "software sensors" for some of these variables can therefore constitute a valuable alternative.

b) The dynamics of (bio)chemical processes is most often non linear and non stationary, and even unstable in some instances (due, for example to inhibition effects). Adaptive nonlinear control, in order to compensate for both non linearities and non stationarities, should therefore be a valuable tool for process stabilization and optimization.

The state estimation problem has received much attention in the last decade: the extended Kalman filter is used in numerous aplications (e.g.Takamatsu et al. (1981), Dekkers (1983), Bellgardt et al. (1983), Marsilli-Libelli (1983), Stephanopoulos and San (1984)) while dedicated algorithms for specific applications are presented by e.g. Aborhey and Williamson (1978), Holmberg and Ranta (1982), Holmberg (1983). On the other hand "classical" adaptive controllers (i.e. based on time varying approximate linear black box models) for biochemical processes are discussed by Bastin et al. (1983), Williams et al. (1984), Dekkers and Voetter (1986), Andersen and Joergensen (1988), Axelsson (1988) while several dedicated nonlinear algorithms have been recently published (Dochain and Bastin (1984,1985), Rundquist (1986), Chamilothoris et al. (1988)).

Our objective, in this paper, is to present a new general theoretical framework for the design of observers and adaptive controllers for (bio)chemical processes which includes most of the previous works in the field.

In section I, we present a general multilinear state space representation able to describe a wide class of (bio)chemical processes in stirred tank reactors. A particular structural property of this state space model is emphasized. The model is linearly parameterized by a set of "specific parameters" which are assumed unknown.

In section II, we show that, provided a suitable subset of the state is available from measurements, the remaining state variables can be estimated on line without the knowledge nor the estimation of the specific parameters being necessary. In addition a linear regression parameter estimator is presented for the on line estimation of the specific parameters. The stability and the robustness of this estimator are analysed.

In section III, state feedback control is considered. Two different situations are examined which cover a wide range of potential applications. In each case, it is shown how adaptive controllers can be designed by combining state feedback linearization with an adaptive "Luenberger-type" observer which is shown to be asymptotically convergent (the analysis is inspired by Taylor et al.(1988)).

Finally in section IV, we mention a number of experimental applications where these state estimators and adaptive controllers have proved to be effective.

## I. A CLASS OF MULTILINEAR COMPARTMENTAL MODELS.

In this section, we shall present a class of multilinear compartmental models which are able to represent (bio)chemical processes in stirred tank reactors and for which the estimation and control methods of this paper are developpped.

### I.1. State-space model.

A (bio)chemical process is defined as a set of M (bio)chemical reactions which take place simultaneously in a reactor and which involve N components.The N-vector of the concentrations of the components (i.e. reactants and reaction products) in the reactor is denoted:

$$x^T = (X_1, X_2, \dots , X_N)$$

When the process takes place in a continuous stirred tank reactor with a constant volumetric inflow/outflow rate Q, the mass balance dynamics of the components is described by the following state space model:

$$\dot{x} = -Qx + K\varphi(x,t) + v \qquad (1.1)$$

where: Qx is the rate of mass transfer of the components from the reactor in the effluent stream; K is the NxM matrix of yield (possibly stoichiometric) parameters; $\varphi^T(x,t) = (\varphi_1, \varphi_2, \dots, \varphi_M)$ is the M-vector of reaction rates which are usually time varying and depending on the state x; v is the N-vector of representing the balance between the rate of mass inflow of the components in the reactor feed stream and the rate of mass outflow of the components from the reactor in gazeous form.

***Example: Ethanolic fermentation.***

The following scheme is a plausible (and commonly used) scheme for the growth of yeasts (Saccharomyces Cerevisiae) on glucose with ethanol production.

$$X_1 + X_2 \xrightarrow{\varphi_1} X_3 + X_4$$

$$X_1 + X_3 \xrightarrow{\varphi_2} X_3 + X_4 + X_5 \qquad (1.2)$$

$$X_2 + X_5 \xrightarrow{\varphi_3} X_3 + X_4$$

Five components are involved: $X_1$: Glucose; $X_2$: Dissolved Oxygen; $X_3$: Yeasts; $X_4$: Carbon Dioxide; $X_5$: Ethanol

The first and the third reactions represent the yeasts growth on glucose and ethanol respectively. The second reaction represents the enzymatic synthesis of ethanol.

The state space model is written as follows:

$$\begin{pmatrix} \dot{X}_1 \\ \dot{X}_2 \\ \dot{X}_3 \\ \dot{X}_4 \\ \dot{X}_5 \end{pmatrix} = -Q\begin{pmatrix} X_1 \\ X_2 \\ X_3 \\ X_4 \\ X_5 \end{pmatrix} + \begin{pmatrix} -k_{11} & -k_{12} & 0 \\ -k_{21} & 0 & -k_{23} \\ k_{31} & 0 & k_{33} \\ k_{41} & k_{42} & k_{43} \\ 0 & k_{52} & -k_{53} \end{pmatrix}\begin{pmatrix} \varphi_1 \\ \varphi_2 \\ \varphi_3 \end{pmatrix} + \begin{pmatrix} V_1 \\ V_2 \\ 0 \\ V_4 \\ 0 \end{pmatrix} \quad (1.3)$$

The process "inputs" $V_i$ have the following meaning: $V_1$ is the glucose inflow (which is introduced in the reactor to maintain the reaction); $V_2$ is the balance between gazeous oxygen inflow and outflow rates (expressed in dissolved oxygen units); $V_4$ is the output flow rate of gazeous carbon dioxyde.

### I.2. A basic structural property of the model.

We define:

$p = \text{rank}(K)$; $K_1$ a (pxM) full rank arbitrary submatrix of K; $K_2$ the remaining [(N-p)xM] submatrix of K; $(x_1,x_2)$ and $(v_1,v_2)$ the partitions of x and v induced by $(K_1,K_2)$.

The state-space model (1.1) is rewritten:

$$\dot{x}_1 = -Qx_1 + K_1\varphi(x_1,x_2,t) + v_1$$
$$\dot{x}_2 = -Qx_2 + K_2\varphi(x_1,x_2,t) + v_2 \quad (1.4)$$

***Property.***

There exists a state transformation:

$$z = Ax_1 + x_2 \quad (1.5)$$

where A is solution of the matrix equation:

$$AK_1 + K_2 = 0 \quad (1.6)$$

such that the state-space model is equivalent to:

$$\dot{x}_1 = -Qx_1 + K_1\varphi(x_1, z - Ax_1, t) + v_1 \quad (1.7.a)$$
$$\dot{z} = -Qz + Av_1 + v_2 \quad (1.7.b)$$

### I.3. Modelling the reaction rates.

The reaction rate $\varphi(x,t)$ is most often a very complex function of the operating conditions and of the state of the process. The analytical modelling of this function is often cumbersome and a continuing subject of intensive investigations. The following fact is however undeniable: *the reaction can take place only if all the reactants are present in the reactor*. In other words, the reaction rate is necessarily zero whenever the concentration of one of the reactants is zero. This basic fact can be represented as follows:

$$\varphi_j(x,t) = \alpha_j(x,t)\left[\prod_{l(r)j} X_l\right] \quad (1.8)$$

$$0 \le \alpha_j(x,t) \le \alpha_{max} \quad (1.9)$$

The notation l(r)j means that the multiplication (Π) is taken on the components with index l which are reactants in the reaction j. $\alpha_j(x,t)$ is called the "specific reaction rate", since it is the reaction rate per unit of concentration of each reactant separately. It must be a bounded function for evident reasons of mathematical consistency.

Defining the vector $\alpha^T = (\alpha_1,...,\alpha_M)$ and the matrix F(x):

$$F(x) = \text{diag}\left\{\prod_{l(r)j} x_l\right\}$$

the state-space model (1.1) is rewritten:

$$\dot{x} = -Qx + KF(x)\alpha + v \qquad (1.10)$$

A plethora of analytical expressions have been suggested to describe the specific reaction rate. The simplest model (but also, probably, the most commonly used in (bio)chemical engineering) is to consider that the specific reaction rates are independent of the state x and depend only on the temperature (e.g. according to the Arrhenius law):

$$\alpha(x,t) = \alpha(T(t)) \qquad (1.11)$$

with T(t) the temperature.

In particular, when the temperature is regulated at a constant value, a multilinear state-space model with constant parameters is obtained. In terms of chemical kinetics theory, such a model corresponds to the assumption that all the reactions are governed by the "law of mass action" with a unit partial order with respect to each reactant.

## II. PARAMETER AND STATE ESTIMATION.

### II.1. Formulation of the estimation problems.

We consider a (bio)chemical process whose dynamics is described by the multilinear state-space model (1.10).

We assume that:

H1) The matrix K is known (rank(K)=p)

H2) The specific reaction rates $\alpha_j$ (j=1,...,M) are *unknown*.

H3) p state variables are measured on line. The vector of these measurements is denoted $x_1$. The corresponding matrix $K_1$ is full rank (see section I.2).

H4) The input variables $V_i$ (i=1,...,N) are known, either by measurement or by a choice of the user.

H5) The experimental conditions are such that all the state variables are strictly positive and bounded:

$$0 < \underline{X}_i \leq X_i(t) \leq \overline{X}_i \qquad \forall\, t \qquad (2.1)$$

Then we address the two following estimation problems:

Problem 1: On line estimation of the N-p non measured state variables (the vector of these variables is denoted $x_2$).

Problem 2: On line estimation of the specific reaction rates (vector $\alpha$).

**II.2. Problem 1: State estimation.**

The structural property of the model together with assumption H3 leads quite naturally to the following asymptotic observer for the estimation of $x_2$:

$$\dot{\hat{z}} = -Q\hat{z} + Av_1 + v_2$$

$$\hat{x}_2 = \hat{z} - Ax_1 \qquad (2.2)$$

where the matrix A is the solution of the matrix equation (1.6). The existence of this matrix is guaranteed by assumption H3.

The convergence of this observer is given by the following theorem.

***Theorem 1.*** Under assumptions H1 to H4,

$$\lim_{t\to\infty} \| x_2 - \hat{x}_2 \| = 0 \qquad (2.3)$$

*Proof:* From (2.2) and (1.7), it is readily shown that the dynamics of the estimation error is as follows:

$$\dot{\tilde{x}}_2 = -Q\tilde{x}_2 \qquad (2.4)$$

where $x_2$ denotes the estimation error:

$$\tilde{x}_2 = x_2 - \hat{x}_2$$

The theorem follows.

*Remark.* The main merit of this very simple observer is to allow the on line estimation of the state variables without the knowledge (nor the estimation) of the specific reaction rates being necessary. Furthemore, it was proved to be very efficient in practical applications (see e.g. Dochain et al.(1988)).

**II.3. Problem 2: On line estimation of the specific reaction rates.**

The dynamics of the measurement $x_1$ is as follows:

$$\dot{x}_1 = -Qx_1 + K_1F(x_1,x_2)\alpha + v_1 \qquad (2.11)$$

This equation, which is linear with respect to the unknown parameter $\alpha$, can be shown to be equivalent to the following "regressor form":

$$x_1 = \Psi^T\alpha + \Psi_0 \qquad (2.12)$$

where the "filtered regressor" $\Psi$ and the auxiliary quantity $\Psi_0$ are outputs of a linear filter:

$$\dot{\Psi}^T = -\omega\Psi^T + K_1F(x_1,x_2) \qquad (2.13.a)$$

$$\dot{\Psi}_0 = -\omega\Psi_0 + (\omega - Q)x_1 + v_1 \qquad (2.13.b)$$

The time constant $\omega^{-1}$ of this filter is arbitrary.
The regressor form (2.12) naturally suggests to use a regression technique for the estimation of $\alpha$. We consider the following continuous unnormalized least-squares algorithm:

$$\dot{\hat{\Psi}}^T = -\omega\Psi^T + K_1F(x_1,x_2) \quad (2.14.a)$$

$$\hat{x}_1 = \Psi^T\alpha + \Psi_0 \quad (2.14.b)$$

$$\dot{\hat{\alpha}} = \Gamma\Psi(x_1 - x_1) \quad (2.14.c)$$

$$\dot{\Gamma} = -\Gamma\Psi\Psi^T\Gamma \qquad \Gamma(0) > 0 \quad (2.14.d)$$

where the unknown state $x_2$ is substituted by its estimate (2.2).

The convergence of this algorithm, in the ideal situation where $\alpha$ is constant, is demonstrated in the following theorem:

***Theorem 2.*** Under assumptions H1 to H5,

$$\lim_{t\to\infty} \hat{\alpha} = \alpha \quad (2.15)$$

*Proof:*
i) From assumption H5, $F(x_1,x_2)$ is a persistently exciting (PE) signal.
ii) then, by theorem 1, $F(x_1,\hat{x}_2)$ is also asymptotically a PE signal.
iii) then $\hat{\Psi}$ is a PE signal since the pair $(\omega I_p,K_1)$ is reachable.
iv) we define:

$$e = \tilde{x}_1 - \hat{\Psi}^T\tilde{\alpha} \qquad \text{with:} \quad \tilde{x}_1 = x_1 - \hat{x}_1 \quad \text{and} \quad \tilde{\alpha} = \alpha - \hat{\alpha} \quad (2.16)$$

we have the following "error system":

$$\dot{e} = -\omega e + K_1\tilde{F}\alpha \quad \text{with} \quad \tilde{F} = F(x_1, x_2) - F(x_1, \hat{x}_2) \quad (2.17.a)$$

$$\dot{\tilde{\alpha}} = -\Gamma\hat{\Psi}\hat{\Psi}^T\tilde{\alpha} - \Gamma\hat{\Psi}e \quad (2.17.b)$$

v) From H5 and theorem 1, it can be shown that :

$$\lim_{t\to\infty} \tilde{F} = 0 \quad \text{and hence} \quad \lim_{t\to\infty} e = 0$$

vi) Then the theorem follows from iii).

In the non ideal case (but more realistic) where the parameter vector $\alpha$ is time-varying, the parameter estimation error remains bounded provided the parameter derivative is bounded:

$$\| \dot{\alpha} \| \le \alpha_0 \quad (2.18)$$

Indeed, in that case, the "error" system (2.17) is modified as follows:

$$\dot{e} = -\omega e + K_1\tilde{F}\alpha \quad (2.19.a)$$

$$\dot{\tilde{\alpha}} = -\Gamma\hat{\Psi}\hat{\Psi}^T\tilde{\alpha} - \Gamma\hat{\Psi}e + \dot{\alpha} \quad (2.19.b)$$

The BIBO stability of this system (and hence the boundedness of $\alpha$) follows from theorem 2.

*Comment.* It must be emphasized that assumption H5 which is completely realistic from an operating viewpoint and which is easily checked in practice, is sufficent to guarantee the persistency of the regressor.

## III. STATE FEEDBACK CONTROL.

### III.1. Statement of the control problem.

We consider (bio)chemical processes whose dynamics is described by the multilinear state space model (1.10), operating under assumptions H1 to H5.

The scalar controlled output y(t) is a linear combination of the measured state variables, i.e.:

$$y = C^T x_1 \tag{3.1}$$

whose dynamics is as follows:

$$\dot{y} = -Qy + C^T K_1 \varphi(x_1, x_2) + C^T v_1(t) \tag{3.2}$$

The control input (denoted "u" as usual) is the feedrate of one external reactant of the process (i.e. a reactant which is introduced from the outside in the reactor). We define the unit vector H and the measured disturbance input w(t) such that:

$$v(t) = Hu(t) + w(t) \tag{3.3}$$

The partitions of H and w induced by the partition $(v_1, v_2)$ of v are denoted $(H_1, H_2)$ and $(w_1, w_2)$ respectively.

We then consider the two following situations:

***First situation.***

We suppose that one of the components involved in the output y is precisely the reactant which is used as control input, i.e. $u = V_j$ and $X_j$ belongs to y.This implies that $C^T H_1 = 0$, and therefore that the input-output dynamics is rewritten:

$$\dot{y} = -Qy + C^T K_1 \varphi(x_1, x_2) + C^T H_1 u + C^T w_1 \tag{3.4}$$

***Second situation.***

We suppose that, in (3.2), $C^T v_1$ is identically zero: there is no "first order" connection between input $v_1(t)$, and hence u(t), and output y(t). But we assume in addition that at least one component involved in y appears in a reaction to which the control reactant also belongs. This implies that the following quantity is not identically zero:

$$C^T K_1 \frac{\partial \varphi}{\partial x} H \neq 0 \tag{3.5}$$

The meaning of both situations, which cover a wide majority of practical problems, is more clearly emphasized by the ethanolic fermentation example.

We shall present solutions to these control problems in the ideal case where the specific reaction rates α are unknown. Adaptive versions will be subsequently discussed, in the case where α is unknown.

*Notations.* The following notations will be used in the sequel:

reference input (set point): $y^*(t)$

control error: $\varepsilon = y^* - y$

### III.2. Linearizing control.

***First situation.***

The following linear reference model is considered:

$$\dot{\varepsilon} + \lambda\varepsilon = 0 \tag{3.9}$$

whith $\lambda$ a design parameter ($\lambda > 0$).

Then the following "linearizing control law" is readily shown to achieve the reference model:

$$u = (C^T H_1)^{-1}[\lambda\varepsilon + \dot{y}* + Qy - C^T K_1 \varphi(x_1,x_2) - C^T w_1(t)] \tag{3.10}$$

This control law contains a feedforward compensation of the measurable disturbance w(t).

***Second situation.***

Taking the derivative of (3.2), with $C^T v_1 = 0$, we obtain the following second order input output dynamics:

$$\ddot{y} = g_0(x_1,x_2) + g_1(x_1,x_2)u + g_2(x_1,x_2)w \tag{3.11}$$

with:

$$g_0(x_1,x_2) = C^T\Big[\, Q^2 x_1 - QK_1\varphi(x_1,x_2) + K_1\frac{\partial\varphi}{\partial x}\{K\varphi(x_1,x_2) - Qx\}\Big]$$

$$g_1(x_1,x_2) = C^T K_1 \frac{\partial\varphi}{\partial x} H$$

$$g_2(x_1,x_2) = C^T K_1 \frac{\partial\varphi}{\partial x}$$

The following linear reference model is adopted:

$$\ddot{\varepsilon} + \lambda_1\dot{\varepsilon} + \lambda_2\varepsilon = 0 \tag{3.12}$$

where $\lambda_1$ and $\lambda_2$ are design parameters chosen such that the following matrix is strictly stable:

$$\begin{pmatrix} 0 & 1 \\ -\lambda_1 & -\lambda_2 \end{pmatrix} \tag{3.13}$$

The linearizing control law which achieves the reference model is as follows:

$$u = g_1^{-1}(x_1,x_2)[\ddot{y}* + \lambda_1(\dot{y}* + Qy - C^T K_1\varphi(x_1,x_2)) + \lambda_2\varepsilon$$
$$-g_0(x_1,x_2) - g_2(x_1,x_2)w] \tag{3.14}$$

### III.3. Adaptive control.

Under assumptions H1 to H3, the control laws (3.10) and (3.14) cannot actually be applied since $\alpha$ is unknown and $x_2$ is not measured. In this section, we shall show that they can nevertheless be realized asymptotically by using adaptive certainty equivalence forms of these laws together with the asymptotic observer of section 2. We first present the parameter estimator.

***Parameter estimator.***

Many other parameter estimators than (2.14) can be defined for the model (2.11). As it will be apparent hereafter, the following adaptive "Luenberger-type" observer will be suitable for the design of the adaptive control laws:

$$\dot{\hat{x}}_1 = -Qx_1 + \Phi^T(x_1,\hat{x}_2)\hat{\alpha} + u_1 + \omega(x_1 - \hat{x}_1) \qquad (3.21.a)$$

$$\dot{\hat{\alpha}} = \gamma[\Phi(x_1,\hat{x}_2)](x_1-\hat{x}_1) \qquad (3.21.b)$$

where:

$$\Phi^T(x_1,\hat{x}_2) \equiv K_1F(x_1,\hat{x}_2)$$

In this algorithm, $\hat{x}_2$ is the estimate provided by the observer (2.2). We define the estimation errors:

$$\tilde{x}_1 = x_1 - \hat{x}_1 \qquad \tilde{x}_2 = x_2 - \hat{x}_2 \qquad \tilde{\alpha} = \alpha - \hat{\alpha} \qquad (3.22)$$

Then the following "error system" derives from (3.21):

$$\dot{\tilde{x}}_1 = -\omega\tilde{x}_1 + \tilde{\Phi}^T\alpha + \hat{\Phi}^T\tilde{\alpha} \qquad (3.23.a)$$

$$\dot{\tilde{\alpha}} = -\gamma\hat{\Phi}\tilde{x}_1 \qquad (3.23.b)$$

$$\dot{\tilde{x}}_2 = -Q\tilde{x}_2 \qquad (3.23.c)$$

with the compact notations:

$$\hat{\Phi} = \Phi(x_1,\hat{x}_2) \qquad \tilde{\Phi} = \Phi - \hat{\Phi} \qquad (3.24)$$

Since $\Phi$ is multilinear in the state x, we have:

$$\tilde{\Phi}^T\alpha = \Phi^*(x_1,x_2,\hat{x}_2,\alpha)\tilde{x}_2 = \Phi^*\tilde{x}_2 \qquad (3.25)$$

and the error system is rewritten:

$$\dot{\tilde{x}}_1 = -\omega\tilde{x}_1 + \hat{\Phi}^T\tilde{\alpha} + \Phi^*\tilde{x}_2 \qquad (3.26.a)$$

$$\dot{\tilde{\alpha}} = -\gamma\hat{\Phi}\tilde{x}_1 \qquad (3.26.b)$$

$$\dot{\tilde{x}}_2 = -Q\tilde{x}_2 \qquad (3.26.c)$$

The following theorem establishes the properties of the parameter estimator.

***Theorem 3.*** Under assumptions H1 to H5,

$\tilde{x}_1$, $\tilde{x}_2$, $\tilde{\alpha}$, $\hat{\Phi}$ and $\Phi^*$ are bounded

$\dot{\tilde{x}}_1$, $\dot{\tilde{\alpha}}$ are bounded

$\lim_{t\to\infty} \tilde{x}_1 = 0, \qquad \lim_{t\to\infty} \dot{\tilde{\alpha}} = 0.$

*Proof:*

1) by assumption H5 and theorem 1, $\tilde{x}_2$, $\Phi$, $\hat{\Phi}$ and $\Phi^*$ are bounded:

$$|\Phi^*| \leq \Phi^*_{max} \qquad (3.27)$$

2)We consider the following Lyapunov function:

$$V = \frac{1}{2}[\tilde{x}_1^T\tilde{x}_1 + \gamma^{-1}\tilde{\alpha}^T\tilde{\alpha} + b\tilde{x}_2^T\tilde{x}_2] \qquad (3.28)$$

where b is chosen such that:

$$b > \frac{(\Phi^*_{max})^2}{4\omega Q} \tag{3.29}$$

3) The time derivative of (3.28) evaluated along the trajectories of (3.23) is:

$$\dot{V} = -\omega\tilde{x}_1^T\tilde{x}_1 + \tilde{x}_1^T\tilde{\Phi}^*\tilde{x}_2 - bQ\tilde{x}_2^T\tilde{x}_2 \tag{3.30}$$

and hence:

$$\dot{V} \leq -\omega|\tilde{x}_1|^2 + \Phi^*_{max}|\tilde{x}_1|\ |\tilde{x}_2| - bQ|\tilde{x}_2|^2 = -(|\tilde{x}_1|\ |\tilde{x}_2|)\begin{pmatrix} \omega & \frac{\Phi^*_{max}}{2} \\ \frac{\Phi^*_{max}}{2} & bQ \end{pmatrix}\begin{bmatrix} |\tilde{x}_1| \\ |\tilde{x}_2| \end{bmatrix} \tag{3.31}$$

Clearly $\dot{V}$ is negative definite and we conclude sucessively that:
$\tilde{x}_1$ and $\tilde{\alpha}$ are bounded;
$\dot{\tilde{x}}_1$ and $\dot{\tilde{\alpha}}$ are bounded;

$\lim_{t\to\infty} \tilde{x}_1 = 0, \quad \lim_{t\to\infty} \dot{\tilde{\alpha}} = 0$

***Adaptive control: first situation.***

Involving the certainty equivalence principle, we consider the linearizing control law (3.10) but with the parameter $\alpha$ and the state $x_2$ replaced by their current estimates provided by the adaptive observer (3.21) and the asymptotic observer (2.2) respectively. This adaptive control law is denoted:

$$u(x_1,\hat{x}_2,\hat{\alpha}) = (C^TH_1)^{-1}[\lambda\varepsilon + \dot{y}^* + Qy - C^TK_1\Phi^T(x_1,\hat{x}_2)\hat{\alpha} - C^Tw_1] \tag{3.32}$$

It is then easily shown that the closed loop behavior can be described as follows:

$$\dot{\zeta} = -\lambda\zeta + (\omega - \lambda)\tilde{x}_1 \tag{3.33.a}$$

$$\varepsilon = C^T(\zeta + \tilde{x}_1) \tag{3.33.b}$$

This means that the control error $\varepsilon = y^* - y$ can be viewed as the output of a stable linear filter driven by the observation error $\tilde{x}_1$. Then, it is clear that, provided assumptions H1 to H5 are satisfied, theorem 3 implies the convergence of the control system, i.e.:

$$\lim_{t\to\infty} \varepsilon = 0 \tag{3.34}$$

It must be noticed, however, that this result is not a proof of closed loop stability whose analysis is beyond the scope of this paper.

***Adaptive control: second situation.***

By analogy with the first situation, the adaptive controller is designed here in order to obtain again a stable linear transfer between the observation error $x_1$ and the control error $\varepsilon$. A straightforward analysis reveals that the linearizing control law (3.14) has to be modified as follows:

$$u_{ad} = u(x_1,\hat{x}_2,\hat{\alpha}) - \gamma g_2^{-1}(x_1,\hat{x}_2,\hat{\alpha})C^T\hat{\Phi}^T\hat{\Phi}(x_1 - \hat{x}_1) \tag{3.35}$$

where $u(x_1,\hat{x}_2,\hat{\alpha})$ is the "certainty equivalence control law" (i.e. (3.14) with $x_2$ and $\alpha$ substituted by their estimates).

With this control law, the closed loop can be shown to be described as follows:

$$\dot{\zeta}_1 = -\lambda_1\zeta_1 + \zeta_2 + \omega\tilde{x}_1 \qquad (3.36.a)$$

$$\dot{\zeta}_2 = -\lambda_2\zeta_1 - \lambda_2\tilde{x}_1 \qquad (3.36.b)$$

$$\varepsilon = C^T(\zeta_1 + \tilde{x}_1) \qquad (3.36.c)$$

Hence the control error is the output of a linear filter driven by the observation error $\tilde{x}_1$ and, under assumptions H1 to H5, the convergence of the closed loop follows from theorem 3.

## IV. CONCLUSIONS.

Algorithms for state estimation and adaptive control of multilinear compartmental systems have been presented and theoretically discussed. These algorithms have proved to be very efficient in several practical applications, namely:

- on line estimation of microbial specific growth rates in ethanolic fermentations and anaerobic digestion processes (Bastin and Dochain, 1986)

- on line estimation of product concentration and specific growth rate in a PHB production bioprocess in collaboration with the Solvay Company (Dochain et al., 1988)

- on line estimation of biomass concentration in a yeasts fermentation process (in collaboration with the Smith Kline company)

- adaptive control of anaerobic digestion processes, in collaboration with the Laboratory of Bioengineering, Louvain University (Renard et al., 1988;Dochain et al., 1988; Van Breusegem et al., 1988)

- adaptive control of fedbatch fermentation processes (Dochain and Bastin, 1988)

In this paper, the design of *indirect* adaptive controllers (where the parameter adaptation is driven by an observation error) has been addressed, for a class of multilinear *continuous* time systems. Thee *discrete* time counterpart of this approach is presented in Bastin and Dochain (1988) while the design problem of direct adaptive controllers (Lyapunov design) is discussed in Bastin (1988).

## V. REFERENCES.

ABORHEY S. and WILLIAMSON D. (1978), "State and parameter estimation of microbial growth processes", Automatica, vol 14, pp.493-498.

AXELSSON J.P. (1988), "On the role of adaptive controllers in fed-batch yeast production", Proceedings IFAC ADCHEM 88, Copenhagen, August 1988, pp. 115-120.

BASTIN G. and DOCHAIN D. (1986), "On line estimation of microbial specific growth rates", Automatica, vol.22(6), pp.705-709.

BASTIN G. and DOCHAIN D. (1988), "Non linear adaptive control of biotechnological processes", IEEE-Aut. Cont. Conf., Atlanta (USA), June 1988, pp.1124-1128.

BASTIN G., DOCHAIN D., HAEST M., INSTALLE M., OPDENACKER P. (1983), "Modelling and adaptive control of a biomethanization process", in Modelling and Data Analysis in Biotechnology, Van Steenkiste and Youg ed., North-Holland Pub., pp.271-282.

BELLGARDT K.H., MEYER H.D., KUHLMANN W., SCHUGERL K. and THOMA M., (1984), " On line estimation of biomass and fermentation parameters by a Kalman filter during a cultivation of sacharomyces cerevisiae", 3rd Europ. Cong. Biotech., Verlag Chemie, Weinheim, vol.2, pp.607-616.

CHAMILOTHORIS G., RENAUD P.Y., SEVELY Y., VIGIE P. (1988), "Adaptive predictive control of a multistage fermentation process", Int. Journal of Control, in press.

DEKKERS R.M. (1983), State estimation of a fed-batch bakers yeast fermentation", in A.Halme ed., Modelling and Control of Biotechnical Processes, Pergamon, pp.201-212.

DEKKERS R.M. and VOETTER M.H. (1986), "Adaptive control of fed-batch bakers yeast fermentation", in Modelling and Control of Biotechnological Processes, A. Halme ed., Pergamon Press, pp.201-212.
DOCHAIN D. and BASTIN G. (1984), "Adaptive identification and control algorithms for non linear bacterial growth systems", Automatica, vol.20(5), pp.621-634.
DOCHAIN D. and BASTIN G. (1985), "Stable adaptive algorithms for estimation and control of fermentation processes", Proc.1st IFAC Symp. Mod. and Cont. Biotech. Proc., Noordwijckerhout, Dec. 1985, pp.1-6.
DOCHAIN D. and BASTIN G. (1988), "Adaptive control of fed-batch fermentation processes", IFAC Symposium on Adaptive Control of Chemical Processes, ADCHEM 88, August 1988, Copenhagen (Denmark), pp. 103-108.
DOCHAIN D., DE BUYL E. and BASTIN G. (1988), "Experimental validation of an algorithm for on line state estimation in bioreactors", 4th Int. Conf. Appl. Fermentation Technology, Cambridge (U-K), September 1988.
DOCHAIN D., BASTIN G., ROZZI A., PAUSS A. (1988), "Adaptive estimation and control of biotechnological processes", Proc. Int. Workshop on Adaptive Control Strategies for Industrial Use, June 1988, Banff, pp.126-141.
HOLMBERG A. (1983), "A micro processor based estimation and control system for the activated sludge process", in Halme A. ed., Modelling and Control of Biotechnical Processes, Pergamon, pp.111-120.
HOLMBERG A. and RANTA J. (1982), "Procedures for parameter and state estimation of microbial growth models", Automatica, vol.13, pp.181-193.
MARSILI LIBELLI S. (1983), "On line estimation of bioactivities in activated sludge processes", in Halme A. ed., Modelling and Control of Biotechnical Processes, Pergamon, pp.121-126.
RENARD P., DOCHAIN D., BASTIN G., NAVEAU H., NYNS E.J. (1988), "Adaptive control of anaerobic digestion processes: a pilot scale experiment", Biotechnology and Bioengineering, vol. 31, pp. 287-294.
RUNDQUIST L. (1986), "Self tuning control of the dissolved oxygen concentration in an activated sludge process", Ph.D. Thesis, Lund Inst. Techn., Sweden.
STEPHANOPOULOS G. and SAN K-Y (1984), "Studies on on line bioreactor identification", Biotechnology and Bioengineering, vol 26, pp.1176-1188.
TAKAMATSU T., Shoya S., Yokoyama K., Kurome Y and Morisaki K (1981), "On line monitoring and control of biochemical reaction processes", Proc. 8th IFAC World Congress, Kyoto, August 1981, Vol.22, pp.146-151.
TAYLOR D.G., KOKOTOVIC P.V., MARINO R., KANELLAKOPOULOS I. (1988), "Adaptive regulation of non linear systems with unmodelled dynamics", IEEE-ACC, Atlanta (USA), June 1988, pp. 360-365.
VAN BREUSEGEM V., ROZZI A., BASTIN G. (1988), "Feedback control of anaerobic digestion processes through adaptive bicarbonate regulation", 5th Int. Symp. on Anaerobic Digestion, Bologna, Italy, May 1988.
WILLIAMS D., YOUSEPFOUR P. and SWANICK B. (1984), "On-line adaptive control of a fermentation process"., IEE Proc. 133/D4, pp.117-125.

# ADAPTIVE NONLINEAR CONTROL
## an estimation-based algorithm

*Jean-Baptiste Pomet* *Laurent Praly*

Ecole Nationale Supérieure des Mines de Paris, C. A. I. Section d'Automatique.
35 rue Saint Honoré, 77305 FONTAINEBLEAU cédex, France
Tél.: (1) 64 22 48 21 Telex: Minefon 694 736 Fax: (1) 64 22 39 03

## 1. INTRODUCTION

We consider a family of non-linear systems with state $x$ in $\mathbf{R}^n$ and input $u$, indexed by the parameter vector $p$:

$$J(p,x)\,\dot{x} \;=\; f(p,x) + u\;g(p,x) \tag{1.1}$$

The maps involved in this equation are smooth with respect to $x$ and linear with respect to $p$. More precisely (with looseness in the notations):

$$\begin{aligned} J(p,x) &= J_s(x) + J_p(x).p \\ f(p,x) &= f_s(x) + f_p(x).p \quad , \; f(p,0) = 0 \\ g(p,x) &= g_s(x) + g_p(x).p \end{aligned} \tag{1.2}$$

The parameter vector $p$ lies in $\Pi$, a smooth closed convex subset of $\mathbf{R}^q$ with a non-empty interior such that:
- $J(p,x)$ is invertible for all $(p,x)$ in $\Pi\times\mathbf{R}^n$,
- For each system (1.1), i.e. each $p$ in $\Pi$, there exists a smooth state feedback control law $u = v(p,x)$ making the origin a global attractor.

The functions $J$, $f$, $g$, $v$ and the set $\Pi$ being known, our problem is to design a controller which guarentees state regulation and solution boundedness when placed in feedback with one of the systems (1.1) corresponding to some parameter vector denoted $p^*$, $p^*$ being unknown.

Usual adaptive controllers have the following form:

$$u = v(\hat{p},x) + w \tag{1.3}$$

$\hat{p}$ being an updated parameter vector, $v(p,x)$ the so-called "certainty equivalence" control law, and $w$ an additional corrective term eventually added to $v$.

This has been succesfully performed in the case of a robot arm with as many motors as axis. Two rather different methods have been used. For example:
- [Mi-Go] proposes an adaptive controller where the estimation of $p^*$ is performed without taking into account the control law to be used, the certainty equivalence law is feedback linearisation, and some additive terms are added to compensate for the effect of updating $\hat{p}$. This is an estimation-based approach.
- [Sl-Li] proposes to update $\hat{p}$ to make a global positive function decrease. The certainty equivalence control law is not feedback linearization there.

In [Ta-Ko-Ma-Ka], an adaptive controller is designed for a family of systems of the type (1.1), specified by the fact that $J$ is the identity matrix and there exist a Hurwitz matrix $A$ and a diffeomorphism $\Phi(x)$ such that:

$$\frac{\partial \Phi}{\partial x}(x)^{-1} A\, \Phi(x) - f(p,x) \in \text{range } g(p,x) \qquad \forall (p,x) \quad , \tag{1.4}$$

which means that all the systems (1.1) are linearizable via feedback and diffeomorphism and all the dynamic uncertainties (i.e. the part of the fields actually depending on $p$) are contained in the subspace spanned by the input fields. A global Lyapunov function is used to obtain the update law of $\hat{p}$.

In this paper, we propose an estimation-based adaptive controller, using the certainty equivalence feedback law. We derive its equations in section 2. We establish boundedness of the solutions and convergence of $x$ to zero in section 3. The assumptions we require, collected below, are discussed in section 4. Note that in [Po-Pr,88], we propose an algorithm with corrective terms, which works on a more particular class of systems, but requires far less restrictive assumptions on the growth of the fields at infinity (see assumption A2 below).

**ASSUMPTIONS:**

**Assumption A0 (invertibility):** $J(p,x)$ is uniformly invertible for $(p,x)$ in $\Pi \times \mathbf{R}^n$:

$$| J(p,x)^{-1} | \leq k \tag{1.5}$$

**Assumption A1 ("satisfactory" control law for all $p$):** There exist $C^1$ functions $V(p,x)$ and $v(p,x)$ defined on $\Pi \times \mathbf{R}^n$, such that:

1- $V(p,x)$ is non negative and is zero if and only if $x$=0.

2- For any positive $M$, the set $\{(p,x) \,/\, V(p,x) < M \;, p \in \Pi \text{ and } |p| \leq M\}$ is bounded.

3- Let the map $F$ be defined by:

$$F(p,x) = J(p,x)^{-1} \,[\, f(p,x) + v(p,x)\, g(p,x) \,] \quad . \tag{1.6}$$

There exists $a$, a strictly positive real number, such that, for any $(p,x)$ in $\Pi \times \mathbf{R}^n$,

$$\frac{\partial V}{\partial x}(p,x) \,.\, F(p,x) \leq - a \; V(p,x) \tag{1.7}$$

**Assumption A2 (growth at infinity):** There exist a positive continuous function $k$ and positive constants $d$, $\alpha$, $\beta$, $\gamma$, $\delta$ such that, for all $(p,x)$ in $\Pi \times \mathbf{R}^n$:

$$\alpha < 1 \tag{1.8}$$

$$\left| \frac{\partial V}{\partial x}(p,x) \right| \leq k(p) \text{ Sup } \left\{ 1 \,,\, V(p,x)^{\alpha} \right\} \tag{1.9}$$

$$\left| \frac{\partial V}{\partial p}(p,x) \right| \leq k(p) \text{ Sup } \left\{ 1 \,,\, V(p,x)^{\beta} \right\} \tag{1.10}$$

$$\begin{aligned} |f_s(x)| &\leq k(p) \text{ Sup } \left\{ 1 \,,\, V(p,x)^{\gamma} \right\} \quad ; \quad |v(p,x)\, g_s(x)| \leq k(p) \text{ Sup } \left\{ 1 \,,\, V(p,x)^{\gamma} \right\} \\ |f_p(x)| &\leq k(p) \text{ Sup } \left\{ 1 \,,\, V(p,x)^{\gamma} \right\} \quad ; \quad |v(p,x)\, g_p(x)| \leq k(p) \text{ Sup } \left\{ 1 \,,\, V(p,x)^{\gamma} \right\} \end{aligned} \tag{1.11}$$

$$\left| J(p,x)^{-1} J_p(x) \right| \leq k(p) \tag{1.12}$$

$$|F(p,x) - F(p,y)| \leq k(p) \text{ Sup} \left\{ 1 \,,\, V(p,x)^{\delta} \right\} \left[ 1 + |x-y|^{d} \right] |x-y| \tag{1.13}$$

**Assumption A3 (realizability):** There exist smooth functions $h_s(p,x)$ and $h_p(p,x)$ such that, for $x$ in $\mathbf{R}^n$ and $p,q$ in $\Pi$,

$$\left[ J(p,x)^{-1}\begin{bmatrix} J_s(x) \\ J_p(x) \end{bmatrix} - \begin{bmatrix} \frac{\partial h_s}{\partial x}(p,x) \\ \frac{\partial h_p}{\partial x}(p_1,x) \end{bmatrix} \right] \frac{\partial}{\partial q}\left[ J(q,x)^{-1}\left( f(q,x) + v(p,x)g(q,x) \right) \right] = 0 \tag{1.14}$$

**Assumption A4 :** $\Pi$ is closed, convex, and has a non-empty interior. $p^*$ is an interior point of $\Pi$; i.e.

$$\mathrm{dist}\left[ p^* \text{ , } boundary \ of \ \Pi \right] \geq D^* > 0 \tag{1.15}$$

## 2. THE ADAPTIVE CONTROLLER

As mentioned above, our controller is based on certainty equivalence principle. Namely, $u$ is given by

$$u = v(\hat{p},x) \tag{2.1}$$

where $\hat{p}$ is an estimate of $p^*$. To achieve this estimation, we remark that, regardless of the control law actually used, (1.1) can be seen as a linear observation equation for the actual parameter vector $p^*$:

$$J_s(x(t))\dot{x}(t) - f_s(x(t)) - u(t)g_s(x(t)) = \left[ -J_p(x(t))\,\dot{x}(t) + f_p(x(t)) + u(t)\,g_p(x(t)) \right] p^* \tag{2.2}$$

Hence, we can think of using a linear estimation algorithm. Unfortunately, $\dot{x}$ is involved in this equation and is not measured. To overcome this difficulty, we integrate (2.2). More precisely, with $\varepsilon$ a positive constant, we define the "filtered" quantities $z$ and $Z$, as is usually done in the linear case:

$$\varepsilon\dot{Z}(t) + Z(t) = J(\hat{p}(t),x(t))^{-1}\left[ -J_p(x(t))\,\dot{x}(t) + f_p(x(t)) + u(t)\,g_p(x(t)) \right] \tag{2.3}$$

$$\varepsilon\dot{z}(t) + z(t) = J(\hat{p}(t),x(t))^{-1}\left[ J_s(x(t))\,\dot{x}(t) - f_s(x(t)) - u(t)\,g_s(x(t)) \right] \tag{2.4}$$

This definition cannot be used per se to obtain Z and $z$ since $\dot{x}$ is still explicitely needed. But, if Assumption A3 holds and the control law $u$ is $v(\hat{p},x)$, (2.3) and (2.4) may be realized in:

$$\begin{aligned}
\varepsilon\dot{\omega} + \omega &= -\frac{1}{\varepsilon}\, h_s(\hat{p},x) - \frac{\partial h_s}{\partial p}(\hat{p},x)\dot{\hat{p}} + [J(\hat{p},x)^{-1}J_s(x) - \frac{\partial h_s}{\partial x}(\hat{p},x)]J(\hat{p},x)^{-1}(\,f_s(x) + v(\hat{p},x)g_s(x)\,) \\
z &= \frac{1}{\varepsilon}\, h_s(\hat{p},x) + \omega \\
\varepsilon\dot{\Omega} + \Omega &= -\frac{1}{\varepsilon}\, h_p(\hat{p},x) - \frac{\partial h_p}{\partial p}(\hat{p},x)\dot{\hat{p}} + [J(\hat{p},x)^{-1}J_p(x) - \frac{\partial h_p}{\partial x}(\hat{p},x)]J(\hat{p},x)^{-1}(\,f_s(x) + v(\hat{p},x)g_s(x)\,) \\
Z &= \frac{1}{\varepsilon}\, h_p(\hat{p},x) + \Omega
\end{aligned} \tag{2.5}$$

As seen later, an interesting choice for the initial conditions may be

$$\omega(0) + \frac{1}{\varepsilon}\, h_s(\,\hat{p}(0)\,,\,x(0)\,) = \frac{x(0)}{\varepsilon} \quad , \ \Omega(0) + \frac{1}{\varepsilon}\, h_p(\,\hat{p}(0)\,,\,x(0)\,) = 0 \tag{2.6}$$

This is always possible since $h_s$ and $h_p$, only constrained by (1.14), are defined up to (at least) a constant.

Now, comparing (2.3) and (2.4), we obtain:

$$z(t) = Z(t)p^* + [\, z(0) - Z(0)p^* \,]\, e^{-t/\varepsilon} \tag{2.7}$$

which gives, in the case when the initial conditions are given by (2.6),

$$z(t) = Z(t)p^* + \frac{x(0)}{\varepsilon} e^{-t/\varepsilon} \tag{2.8}$$

Without the exponentially decaying term, (2.7) would be again a linear observation equation of $p^*$. But now $z$ and $Z$ can be computed. The estimation algorithm we shall use is the following generalization of least square algorithm: $\hat{p}$ being the estimate of $p^*$:

$$e = z - Z\hat{p} \tag{2.9}$$

$$\left.\begin{array}{ll} \dot{\hat{p}} = |e|^{m-1}\, \pi_{\hat{p}} \left[PZ^T e\right] & ; \hat{p}(0) \in \Pi \\ \dot{P} = -\, |e|^{m-1} PZ^T ZP & ; P(0) = I \end{array}\right\} \tag{2.10}$$

where:

- $\pi_{\hat{p}}$ is the identity map if $\hat{p}$ lies inside $\Pi$ and the "$P^{-1}$-orthogonal projection" onto the boundary of $\Pi$ if $\hat{p}$ is on this boundary; namely, in this latter case, with $n(\hat{p})$ the inward unit normal vector to $\partial\Pi$ at $\hat{p}$, for $s$ in $\mathbf{R}^q$,

$$\pi_{\hat{p}}(s) = \begin{cases} s & \text{if } n(\hat{p})^T s \geq 0 \\ s - \dfrac{n(\hat{p})^T s}{n(\hat{p})^T P\, n(\hat{p})}\, P\, n(\hat{p}) & \text{if } n(\hat{p})^T s < 0 \ ; \end{cases} \tag{2.11}$$

- $m$, larger or equal to 1, is a degree of freedom of the algorithm.

Our complete adaptive controller is (2.1),(2.5),(2.9),(2.10).

## 3. BEHAVIOUR OF THE CLOSED-LOOP SYSTEM

The closed-loop system is made of the adaptive controller in feedback with:

$$\dot{x} = J(p^*,x)^{-1} \left[f(p^*,x) + u\, g(p^*,x)\right] \ . \tag{3.1}$$

Its state is $(x,\hat{p},P,\omega,\Omega)$. The right-hand side of this non-linear autonomous differential equation is not continuous (because of the projection in (2.10)), but, thanks to $\Pi$ 's convexity, existence, uniqueness and continuity of solutions hold for such a system (see [Br] prop 3.12 ).

### 3.1. A property of the estimation algorithm

As more or less well known in the case $m=1$, regardless of the control law effectively used, as long as the solution exists, the estimation algorithm (2.10)-(2.9) has the following property, established in Appendix A:

**Lemma 1:** Under assumptions A0, A3, A4 only, regardless of the control law actually used, i.e. regardless of the function $u(t)$, any solution of the closed-loop system which is defined on the time interval $[\,0\,,\,t_f\,)$ (maybe $t_f=\infty$ ) has the following properties:

- $P(t)$ is symmetric positive definite for any $t$ in $[0,t_f)$ and $P(t) \leq P(0)$. (3.2)
- $\hat{p}(t)$ is bounded and $\dot{\hat{p}}(t) \in L^1(0,t_f)$ [ hence $\hat{p}(t)$ has a limit when $t$ tends to $t_f$ ] (3.3)

  $e(t) \in L^{m+1}(0,t_f)$ (3.4)

and we have the following bounds ($W(0)$ is defined in (3.7) and $q$ is the dimension of the parameter $p$):

$$\hat{p}(t) \in \Pi \quad \text{and} \quad |p^* - \hat{p}(t)|^2 \leq 2\, W(0) \qquad \forall t \in [0, t_f) \tag{3.5}$$

$$\int_0^{t_f} |\dot{\hat{p}}|\, dt \;\leq\; (m+1)\, W(0) \quad ; \quad \int_0^{t_f} |e|^{m+1} dt \;\leq\; \sqrt{q\,(m+1)\, W(0)} + \frac{m+1}{D^*} W(0) \tag{3.6}$$

$$\text{where} \quad W(0) \;\stackrel{\Delta}{=}\; \frac{1}{2} |p^* - \hat{p}(0)|^2 + \frac{\varepsilon\, |z(0) - Z(0) p^*|^{m+1}}{(m+1)^2}\, ) \tag{3.7}$$

### 3.2. Boundedness and convergence

Lemma 1 gives us informations only on the $(\hat{p}, P)$-components of solutions of the closed-loop system and on $e$. The complete behaviour is precised in:

**Theorem 1:** Let the following index be defined from assumption A2:

$$\eta_1 = \text{Sup} \left\{ \alpha+\gamma\, ,\, \alpha+\delta\, ,\, \beta \right\} . \tag{3.9}$$

Suppose that assumptions A0, A1, A2, A3, A4 are met and, in the adaptation law (2.10), $m$ is chosen such that:

$$m+1 \geq \text{Sup} \left\{ 2\, ,\, d+1\, ,\, \gamma\, ,\, \frac{\gamma}{1-\alpha} \right\} \tag{3.8}$$

Under these conditions, the closed loop system satisfies:

- if $\eta_1 \leq 1$, all its solutions are defined and bounded on $[0, \infty)$, and $x$, as well as $e$ and $\dot{\hat{p}}$, tend to zero.
- if $\eta_1 > 1$, the above conclusion holds locally. Precisely, if $\omega(0)$ and $\Omega(0)$ are chosen to meet (2.6), it holds for solutions such that:

$$\frac{m+1}{2} |\, p^* - \hat{p}(0)\, |^2 + \varepsilon \frac{|x(0)|^{m+1}}{m+1} \leq C$$
$$\text{and} \quad P(0) = I \;;\; \hat{p}(0) \in \Pi \tag{3.10}$$

where $C$ is a constant only depending on the constants of the problem.

A complete proof of this theorem is in appendix B. Let us give a sketch of it.

With (2.9),(2.3),(2.4) defining $e$, $z$ and $Z$, we have:

$$\begin{aligned} \varepsilon \dot{e} + e &= \varepsilon \dot{z} + z - [\, \varepsilon \dot{Z} + Z\,]\hat{p} - \varepsilon Z \dot{\hat{p}} \\ &= \dot{x} - F(\hat{p}(t), x(t)) - \varepsilon Z \dot{\hat{p}} \end{aligned} \tag{3.11}$$

Hence, letting:

$$y = x - \varepsilon\, e \quad , \tag{3.12}$$

this equation and (2.2) allow us to write:

$$\begin{aligned} \dot{y} &= F(\hat{p}, y) + \Big[ F(\hat{p}, y+\varepsilon e) - F(\hat{p}, y) \Big] + \varepsilon Z \dot{\hat{p}} + e \\ \varepsilon \dot{Z} &= -Z + J(\hat{p}, x)^{-1} \Big[ -J_p(x)\, \dot{x} + f_p(x) + v(\hat{p}, x)\, g_p(x) \Big] \end{aligned} \tag{3.13}$$

Considering $\dot{\hat{p}}$ and $e$ as functions of time, (3.13) may be seen as a perturbation of the triangular system obtained for $\dot{\hat{p}} = 0$, $e = 0$. Without the terms involving $\dot{\hat{p}}$ and $e$, boundedness of $y$ would be guarenteed by assumption (1.7) and boundedness of $Z$ by the "growth estimates" of assumption A2. Boundedness of $(y, Z)$-solution of the actually perturbed system (3.13) is to be obtained via a robustness argument using the fact that, $e$ and $\dot{\hat{p}}$ are known to be $L^k$-functions (Lemma 1). All the conclusions of Theorem 1 follow from this point.

## 4. DISCUSSION OF THE ASSUMPTIONS

We have proposed an adaptive regulator based on estimation and certainty equivalence principle. It guarentees solution boundedness and asymptotic regulation when placed in feedback with an element of a linearly parameterized family of non-linear systems satisfying assumptions A0 to A3. Let us discuss these assumptions.

### 4.1. Linear parameterization and invertibility (A0)

Linear parameterization can be considered as a restrictive constraint. To enlarge the field of application, this assumption is invoked for an implicit form. This allows robot arms for example, or explicit homographic parametrization. The draw-back of allowing an implicit form is that we have to introduce an assumption ($J$'s invertibility ) to make sure that an explicit (nonlinearly parameterized) form can be obtained. Again, to relax this assumption, we ask for it to be satisfied not for all parameter vector $p$ but for those contained in a known convex set $\Pi$.

### 4.2. Satisfactory control law for all $p$ (A1)

Would $p$ be fixed, the possibility of meeting this condition is a non-adaptive control problem. An important aspect is that a global control law is required. Given the function $V(p,x)$, assumption (1.7) is met if the control $u$ satisfies (in the single input case, to simplify the notation):

$$u\ \frac{\partial V}{\partial x}(p,x)g(p,x) \leq -aV(p,x) - \frac{\partial V}{\partial x}(p,x)f(p,x)\ . \tag{4.1}$$

Clearly, this is possible for all $x$ such that $\frac{\partial V}{\partial x}(p,x)g(p,x)$ is invertible. Actually, the assumption means that we can find, for each $p$, a proper function $V(p,.)$ such that, whenever $\frac{\partial V}{\partial x}(p,x)g(p,x)$ is singular, $aV(p,x)+\frac{\partial V}{\partial x}(p,x)f(p,x)$ is non-positive and the control law (4.1) can be regularized in the neighborhoud of these singularities.

The parametric aspect enters in the fact that this holds for all $p$ in $\Pi$ and $V$ depends smoothly on $p$.

### 4.3. Realizability (A3)

We have mentioned that the parameter vector observation equation naturally given by the system equation involves the speed of $x$. To overcome the difficulty, we introduce a first order filter. Assumption A3 is one of the possible sufficient condition allowing us to obtain a realization.

We observe that, for second order systems (similar to robot arms) such as

$$J(p,\chi)\ddot{\chi} = f(p,\chi,\dot{\chi}) + ug(p,\chi,\dot{\chi}) \qquad \text{(the state is: } x = \begin{bmatrix}\chi\\ \dot{\chi}\end{bmatrix}\text{ )}\ ,$$

A3 is always satisfied with:

$$h(p,\chi,\dot{\chi}) = J(p,\chi)\dot{\chi}\ . \tag{4.3}$$

### 4.4. Growth at infinity (A2)

These assumptions ask, more or less, for the possibility of bounding all the functions given by the problem by a power of the function $V$. They seem to be inherent with the fact that the adaptation introduces in the closed-loop system the perturbing terms $e$ and $\dot{\hat{p}}$ not taken in account in the design. They are just hoped to be handled by the controller on its own via feedback. Our assumptions limits the strength of their affecting the closed-loop system. To obtain global convergence, we have to bound the values of the exponents $\alpha$, $\beta$, $\gamma$, $\delta$; this is restrictive since, for instance, $\alpha+\gamma\leq 1$ more or less asks the open loop fields to be globally lipschitz at infinity. In [Po-Pr,88], we mend the present algorithm, adding a term to the control law as in (1.3). This gives global convergence under far less restrictive conditions on the growth at infinity of the open-loop fields, but under some structural assumptions not needed here.

## REFERENCES

[Br] H. BREZIS , *Opérateurs maximaux monotones* . North holland , 1973.

[Mi-Go] R.H. MIDDLETON / G.C. GOODWIN , *Adaptive computed torque control for rigid links manipulators* . Systems & Control Letters, vol.10 No 1, pp. 9-16, 1988 .

[Po-Pr,88] J.-B. POMET / L. PRALY , *Estimation-based Non-linear Adaptive Control* . CDC proceedings, Austin, 1988.

[Sl-Li,86] J.-J. E. SLOTINE / W. LI , *On the adaptive control of robot manipulators* . M.I.T., 1986.

[Sl-Li,87] J.-J. E. SLOTINE / W. LI , *Theoretical issues in adaptive manipulator control* . M.I.T., 1987.

[Ta-Ko-Ma-Ka] D. G. TAYLOR / P. V. KOKOTOVIC / R. MARINO / I. KANELLAKOPOULOS , *Adaptive regulation of nonlinear systems with unmodeled dynamics* . Procced. of American Control Conference, June 1988.

## APPENDIX: THE PROOFS

### APPENDIX A: Proof of lemma 1

$P(t)$ clearly remains symmetric. In addition, it remains invertible because the matrix $I+\int_0^t |e(\tau)|^{m-1}\, Z^T(\tau)Z(\tau)d\tau$ is well defined for any $t$ in $[0,t_f)$ and it is not difficult to check (from (2.10)) that it is the inverse of $P(t)$. This proves (3.2).

Now, let (note that $W(0)$ defined in (3.7) is the same as the above $W(t)$ for $t=0$) :

$$W(t) = \frac{1}{2}\,(p^*-\hat{p})^T P^{-1}(p^*-\hat{p}) + \varepsilon\, \frac{|z(0) - Z(0)p^*|^{m+1}}{(m+1)^2}\, e^{-\frac{(m+1)t}{\varepsilon}} \quad . \tag{A.1}$$

From (2.7),(2.9),(2.10), we get the following inequality in which $\chi(t)$ is 1 if $\hat{p}$ is on the boundary of $\Pi$ and $PZ^Te$ points outward $\Pi$, 0 in the other cases:

$$\begin{aligned}\dot{W}(t) \;\leq & -\frac{1}{2}|e(t)|^{m+1} + \chi(t)\,\frac{(n^TPZ^Te)\,(n^T(p^*-\hat{p}))}{n^TPn} \\ & + \frac{1}{2}|e(t)|^{m-1}|z(0) - Z(0)p^*|^2\, e^{-\frac{2t}{\varepsilon}} - \frac{|z(0) - Z(0)p^*|^{m+1}}{m+1}\, e^{-\frac{(m+1)t}{\varepsilon}}\end{aligned} \tag{A.3}$$

since for any positive $E$ and $Y$, $-\frac{1}{2}E^{m+1} + \frac{1}{2}Y^2E^{m-1} \leq -\frac{1}{m+1}E^{m+1} + \frac{1}{m+1}Y^{m+1}$, and (from assumption A4) $n^T(p^*-\hat{p}) \leq D^*$, and $n^TPZ^Te \leq 0$ when $\chi$ is non-zero,

$$\dot{W}(t) \leq -\frac{1}{m+1}|e(t)|^{m+1} - \chi(t)\, D^* \frac{|n^TPZ^Te|}{n^TPn} \quad . \tag{A.4}$$

Therefore $W(t)$ is positive decreasing. This, together with (A.1) and (3.2), yields:

$$|p^*-\hat{p}(t)|^2 \leq 2W(t) \leq 2W(0) \quad \forall t \in [0,t_f) \; . \tag{A.5}$$

On the other hand, $\hat{p}$ remains in $\Pi$ because $\dot{\hat{p}}$ points inward $\Pi$ all along its boundary; this gives (3.5).

We shall now establish (3.3), (3.4), (3.6). (A.1) and (A.4) yield, for any $\tau$ in $[0,t_f)$:

$$\int_0^\tau |e|^{m+1} \le (m+1)\,[W(0) - W(\tau)\,] \le (m+1)\; W(0) \tag{A.7}$$

This gives (3.4) and the second part of (3.6). Now, from (2.10), with the same definition of $\chi$ as in (A.4),

$$\begin{aligned}\int_0^\tau |\dot{\hat{p}}| &\le \int_0^\tau |e|^{m-1} |PZ^T e| + \int_0^\tau \chi |e|^{m-1} \frac{|n^T PZ^T e|\,|Pn|}{n^T Pn} \\ &\le \left[\int_0^\tau |e|^{m+1}\right]^{\frac{1}{2}} \left[\int_0^\tau |e|^{m-3} e^T ZPPZ^T e\right]^{\frac{1}{2}} + \int_0^\tau \chi |e|^{m-1} \frac{|n^T PZ^T e|}{n^T Pn}\end{aligned} \tag{A.8}$$

(we use Shwartz inequality and the fact that $P \le I$. But

$$0 \le \int_0^\tau |e|^{m-3}\, e^T ZPPZ^T e \le \int_0^\tau \mathrm{tr}(PZ^T ZP)\; |e|^{m-1} = -\int_0^\tau \mathrm{tr}(\dot{P}) \le q$$

Then, since $|e|^{m+1} \le -(m+1)\dot{W}$ and $\dfrac{|n^T PZ^T e|}{n^T Pn} \le -\dfrac{1}{D^*}\dot{W}$, (A.8) implies the second part of (3.6).

**APPENDIX B: Proof of theorem 1**

Let $[0,t_f)$ be a maximal interval of definition of a solution of the closed-loop system. From (3.5) in lemma 3, we may use the inequalities in assumption A2 replacing $k(\hat{p}(t))$ with

$$\kappa(\,W(0)\,) \triangleq \mathrm{Sup}\left\{ k(p) \,\middle|\, |p^* - p|^2 \le 2W(0) \right\}. \tag{B.1}$$

$\kappa(.)$ is a positive increasing function. For the sake of simplicity, in the following, $\kappa$ will stand for any positive continuous increasing function of $W(0)$ (usually the original $\kappa$ mixed with some constants independant of $t$ and of the solution we consider).

Now, considering (3.13) and assumption A1, $V$ is clearly a good canditate as a commparison function for $y$. Hence, let:

$$W_1(t) = V(\hat{p}(t), y(t)) \quad , \tag{B.2}$$

Along (3.13), we have:

$$\dot{W}_1 = \frac{\partial V}{\partial x}(\hat{p},y).F(\hat{p},y) + \frac{\partial V}{\partial x}(\hat{p},y)\,[F(\hat{p},y+\varepsilon e) - F(\hat{p},y) + \varepsilon Z\dot{\hat{p}} + e\,] + \frac{\partial V}{\partial p}(\hat{p},y)\dot{\hat{p}} \tag{B.3}$$

which gives, considering (1.7),(1.9),(1.10) and (1.13),

$$\begin{aligned}\dot{W}_1 \le\; & -a\; W_1 + \kappa\; \mathrm{Sup}\left\{1\,,\, W_1^{\alpha+\delta}\right\}\left[1 + \varepsilon^d |e|^d\right]|e| \\ & + \kappa\; \mathrm{Sup}\left\{1\,,\, W_1^{\alpha}\right\}\left[\varepsilon |Z|\,|\dot{\hat{p}}| + |e|\right] \\ & + \kappa\; \mathrm{Sup}\left\{1\,,\, W_1^{\beta}\right\}|\dot{\hat{p}}|\end{aligned} \tag{B.4}$$

and finally

$$\dot W_1 \le -a\,W_1 + \kappa\,\mathrm{Sup}\left\{1\,,\,W_1^{\mathrm{Sup}(\alpha+\delta,\beta)}\right\}\left[|\dot{\hat p}| + |e|(1+|e|^d)\right] + \kappa\,\mathrm{Sup}\left\{1\,,\,W_1^{\alpha}\right\}|Z|\,|\dot{\hat p}| \tag{B.5}$$

Similarly, the autonomous part of the $Z$-equation in (3.13) being linear, we consider the following comparison function:

$$W_2(t) = \varepsilon|Z(t)|^2 \tag{B.6}$$

Along (3.13), we have:

$$\dot W_2 = 2\varepsilon Z^T\dot Z = -2|Z|^2 + 2Z^T J(\hat p, x)^{-1}\left[-J_p(x)\,\dot x + f_p(x) + v(\hat p, x)\,g_p(x)\right] \tag{B.7}$$

From (3.1) and inequalities (1.1) in assumption A1 and (1.11) to (1.12) in assumption A2, one gets:

$$\left|J(\hat p, x)^{-1}\left[-J_p(x)\,\dot x + f_p(x) + v(\hat p, x)\,g_p(x)\right]\right| \le \kappa\,\mathrm{Sup}\left\{1\,,\,V(\hat p, x)^{\gamma}\right\} \le \kappa\,\mathrm{Sup}\left\{1\,,\,V(\hat p, y)^{\gamma}\right\} + \kappa\,\mathrm{Sup}\left\{1\,,\,V(\hat p, y)^{\alpha\gamma}\right\}\left(|e|^{\gamma} + |e|^{\frac{\gamma}{1-\alpha}}\right) \tag{B.8}$$

where the last inequality is obtained from (1.9) in assumption A2, lemma 4 (given in appendix D), and the fact that, for some positive constant $k$,

$$(a+b)^{\gamma} \le k\,(a^{\gamma} + b^{\gamma}) \qquad \forall\, a, b \ge 0 \tag{B.9}$$

(B.8) and (B.7) yield:

$$\dot W_2 \le -\frac{2}{\varepsilon}W_2 + 2\frac{\kappa}{\sqrt{\varepsilon}}\sqrt{W_2}\,\mathrm{Sup}\left\{1, W_1^{\gamma}\right\} + 2\frac{\kappa}{\sqrt{\varepsilon}}\sqrt{W_2}\,\mathrm{Sup}\left\{1, W_1^{\alpha\gamma}\right\}\left[|e|^{\gamma} + |e|^{\frac{\gamma}{1-\alpha}}\right]\,. \tag{B.10}$$

To study, the "triangular system" (3.13), let us consider the following comparison function:

$$U(t) = W_1(t)^{2\gamma} + 2\,rW_1(t)^{\gamma}\sqrt{W_2(t)} + s^2\,W_2(t) \tag{B.11}$$

where $r$ and $s$ are two real positive numbers. We have the following majorations:

$$W_1 \le U^{\frac{1}{2\gamma}}\quad,\quad \sqrt{W_2}\,\mathrm{Sup}\{\,1\,,\,W_1^{\gamma}\,\} \le \frac{\mathrm{Sup}\{\,1\,,\,U\,\}}{s}$$
$$W_2 \le \frac{U}{s^2}\quad,\quad W_1^{\gamma}\,\mathrm{Sup}\{\,1\,,\,W_1^{\gamma}\,\} \le \mathrm{Sup}\{\,1\,,\,U\,\}\quad, \tag{B.12}$$

and the following equality (where no problem arises when $W_2$ vanishes, see (B.10) ) :

$$\dot U = 2\gamma W_1^{\gamma-1}(W_1^{\gamma} + r\sqrt{W_2})\,\dot W_1 + \left(r\frac{W_1^{\gamma}}{\sqrt{W_2}} + s^2\right)\dot W_2 \tag{B.13}$$

with (B.5), (B.10) and (B.12), this yields :

$$\dot U \le -2a\gamma W_1^{2\gamma} - 2r\left(a\gamma + \frac{1}{\varepsilon}\right)W_1^{\gamma}\sqrt{W_2} - \frac{2}{\varepsilon}s^2W_2 + 2\frac{\kappa}{\sqrt{\varepsilon}}(r+s)\,\mathrm{Sup}\{1,U\} + \kappa\,\mathrm{Sup}\{1,U^{\zeta}\}\left[|\dot{\hat p}| + P(|e|)\right] \tag{B.14}$$

where: $P(x) = x + x^{d+1} + x^{\gamma} + x^{\frac{\gamma}{1-\alpha}}$

$$\zeta = \mathrm{Sup}\{1+\frac{\eta_1-1}{2\gamma} , \frac{1+\alpha}{2}\} \tag{B.15}$$

Notice that, since $\alpha$ is smaller than 1,

$$\zeta \le 1 \iff \eta_1 \le 1 \tag{B.16}$$

Let us tune the positive numbers $r$ and $s$ so that $U$ is a good comparison function for the behaviour of $(y,Z)$ along (3.13):

**Lemma 2 :** Let $c, d, K$ be three strictly positive real numbers. There exists a positive number $\lambda$ such that, th and ch denoting the usual hyperbollic functions,

$$(1+\mathrm{th}\lambda)\, Min\ \{c,d\} = \frac{2}{3}[\ c+d-|c-d|\mathrm{ch}\lambda\ ] \tag{B.17}$$

then if we define $r, s, b$ by:

$$s = \frac{1}{2}\frac{Min\ \{c,d\}}{K\sqrt{d}} \quad , \quad r = s\,\mathrm{th}\lambda \quad , \quad b = \frac{1}{6}[\ c+d-|c-d|\mathrm{ch}\lambda\ ] \ , \tag{B.18}$$

then $b$ is stricly positive and:

$$X^2 + 2rXY + s^2Y^2 \quad \text{is positive definite.} \tag{B.19}$$

$$2\,[\ cX^2 + r(c+d)\,XY + s^2dY^2\ ] \ \ge \ \left[b + 2K\sqrt{d}\,(r+s)\right]\left[X^2 + 2rXY + s^2Y^2\right] \tag{B.20}$$

This lemma is not difficult to prove. Using it with $c=a\gamma$, $d=\frac{1}{\varepsilon}$, $K=\kappa$, (B.14) yields:

$$If \quad U(t) \ge 1 \ , \quad then \quad \dot{U} \le -bU + \kappa\, U^{\zeta}\left[|\dot{\hat{p}}| + P(|e|)\right] \tag{B.21}$$

with $b$ a strictly positive real number. Our conclusion will follow from the following lemma, established in Appendix C:

**Lemma 3:** Let $U(t)$ be a positive function defined on the time interval $[0,t_f)$ and such that

$$If \quad U(t) \ge 1 \ , \quad then \quad \dot{U}(t) \le -b\ U(t) + [\ \sum_i f_i(t)\ ]\ U(t)^{\zeta} \tag{B.22}$$

where $b$ and $\zeta$ are real positive numbers and $(f_i)$ is a (finite) family of real positive funtions such that:

$$\int_0^{t_f} [\ f_i(\tau)\ ]^{k_i} d\tau = S_i < \infty \quad ; \quad k_i \ge 1 \tag{B.23}$$

Then,

- If $\zeta \le 1$, $U$ is bounded on $[0,t_f)$ no matter the value of the $S_i$'s and the initial value.
- If $\zeta > 1$, $U$ is bounded on $[0,t_f)$ under the following condition:

$$\sum_i \frac{[(\zeta-1)\,S_i\,]^{1/k_i}}{(b\bar{k}_i)^{1/\bar{k}_i}} \le \mathrm{Inf}\ \{\ 1\ , \frac{1}{U(0)^{\zeta-1}}\ \} \quad \text{with} \quad \frac{1}{k_i} + \frac{1}{\bar{k}_i} = 1 \tag{B.24}$$

To use this lemma, let us first check that the needed hypotheses are met in our case:
(B.22) is the same as (B.21), and (B.23) follows from lemma 1, with:

$$\sum_i f_i = \kappa[\, P(|e|) + |\dot{\hat{p}}|\,]$$

$$k_1=m+1 \;\; ; \;\; k_2=\frac{m+1}{d+1} \;\; ; \;\; k_3=\frac{m+1}{\gamma} \;\; ; \;\; k_4=(m+1)\,\frac{1-\alpha}{\gamma} \;\; ; \;\; k_5=1 \tag{B.25}$$

$$S_i = \kappa^{k_i}\int_0^{t_f} |e|^{m+1} \quad i=1,2,3,4 \qquad ; S_5 = \kappa\int_0^{t_f} |\dot{\hat{p}}| \quad .$$

Let us come to (B.24). From (B.25), we have:

$$\sum_i \frac{[(\zeta-1)\, S_i\,]^{1/k_i}}{(b\bar{k}_i)^{1/\bar{k}_i}} = \kappa \sum_{i=1}^{4} \frac{1}{(b\bar{k}_i)^{1/\bar{k}_i}} \left[(\zeta-1)\int_0^{t_f} |e|^{m+1}\right]^{\frac{1}{k_i}} + \kappa\,(\zeta-1)\int_0^{t_f} |\dot{\hat{p}}| \;\cdot \tag{B.26}$$

On the other hand, if the initial conditions of the filters are chosen according to (2.6), with assumption A1, $U(0)$ is zero. It follows that (B.24) is satisfied if:

$$\sum_{i=1}^{4} \frac{1}{(b\bar{k}_i)^{1/\bar{k}_i}} \left[(\zeta-1)\int_0^{t_f} |e|^{m+1}\right]^{\frac{1}{k_i}} + (\zeta-1)\int_0^{t_f} |\dot{\hat{p}}| \leq \frac{1}{\kappa} \;\cdot \tag{B.27}$$

Since $\alpha$ is smaller than 1, $\zeta-1$ is equal to $(\eta_1-1)/2\gamma$. $b$, $\zeta$ and the $k_i$'s do not depend on the initial conditions, but $\kappa$ does (see (B.1)); as explained at the very beginning of this proof, it is a continuous function of $W(0)$. On the other hand, from (3.6), the integrals in (B.27) are upperbounded by a continuous function of $W(0)$ which tends to zero when $W(0)$ does. It follows that, if $C$ is chosen small enough, (3.10) implies (B.27) which implies (B.24).

Now, we may apply this lemma to our system, which gives the boundedness of $U(t)$, no matter the value of the initial conditions if $\eta_1 \leq 1$, and under condition (3.10) if $\eta_1 > 1$.

Since $U$ is a positive definite function of ( $W_1^{\gamma}$ , $\sqrt{W_2}$ ), and comparing (B.2), (B.6) and assumption A1, the boundedness of $U(t)$ implies this of $y(t)$ and $Z(t)$. It follows that $e$ is bounded because $\hat{p}$ is bounded (see (3.5)), and (2.7) and (2.9) yield:

$$e(t) = Z(t)[\, p^* - \hat{p}\,] + |z(0) - Z(0)p^*|\, e^{-\frac{t}{c}} \quad .$$

With (3.12), $x$ is bounded too. $P$ and $\dot{\hat{p}}$ are bounded (see (2.10),(3.2)). The state of the filters are bounded because they are states of strictly stable linear systems with bounded inputs (see (2.5)).

We have proved (under condition (3.10) if $\eta_1$ is larger than 1) that the solution of the closed-loop system we are following is bounded on $[0,t_f)$, which is its maximal definition interval. This implies that $t_f = \infty$. This would be very classical if the o.d.e. (2.10) were smooth; in spite of the discontinuity introduced by the projection in the estimation algorithm, this follows from the classical process of continuation thanks to the existence result of [Br] prop 3.12 and the fact that the right hand side of (2.10) is bounded by a continuous function.

We now check that $x(t)$ tends to zero:

$\dot{x}(t)$, given by (1), is bounded. Hence, considering (3.11), $\dot{e}(t)$ is bounded; this implies that $e$ tends to zero because $e^{m+1}$ belongs to $L^1$. It follows that (B.4) may be written

$$\dot{W}_1(t) \leq -aW_1(t) + \lambda(t) + \mu(t) \tag{B.29}$$

where $\lambda$ belongs to $L^1$ and $\mu$ tends to zero. Convergence to zero for $W_1$ follows. With the property of $V$ given by assumption A1, the same holds for $y$ and therefore $x$. □

**APPENDIX C: Proof of lemma 3**

**i) Case $\zeta$ no larger than 1:** Let $[t_1,t_2]$ be a maximum interval on which $U$ remains larger than 1. It is clear that either $t_1 = 0$ or $U(t_1) = 1$, and that, for any $t$ in $[t_1,t_2]$, we have

$$\dot U(t) \le [\, -b + \sum_i f_i(t)\,]\, U(t) \tag{C.1}$$

hence, for any t in $[t_1,t_2]$,

$$U(t) \le U(t_1)\, e^{-b(t-t_1) + \sum_i \int_{t_1}^{t} f_i(t)} \tag{C.2}$$

but $U(t_1) \le \mathrm{Sup}\{\, 1 \,,\, U(0)\,\}$, and, with Holder inequality,

$$-b(t-t_1) + \sum_i \int_{t_1}^{t} f_i(t) \;\le\; -b\,(t-t_1) + \sum_i S_i^{\frac{1}{k_i}} (t-t_1)^{\frac{1}{\bar k_i}} \;. \tag{C.4}$$

The $\frac{1}{\bar k_i}$'s being smaller than 1, the right hand side of (C.4) is bounded by a constant $\bar S$, depending only on b, the $k_i$'s and the $S_i$'s. We have proved that, for any t in $[0,t_f)$,

$$U(t) \le \mathrm{Sup}\left\{ 1 \,,\, e^{\bar S} \,,\, U(0)\, e^{\bar S} \right\} \tag{C.5}$$

**ii) Case $\zeta$ strictly larger than 1:** Let $\lambda$ be defined by

$$U(t) = \lambda(t)^{-\frac{1}{\zeta-1}}\, e^{-b(t-t_1)} \tag{C.6}$$

$$\text{so that:}\quad \dot U(t) = -bU(t) - \frac{1}{\zeta-1}\, \dot\lambda(t)\, \lambda(t)^{-\frac{\zeta}{\zeta-1}}\, e^{-b(t-t_1)} \tag{C.7}$$

Finally, (B.22) may be rewritten as follows when $U \ge 1$:

$$-\frac{1}{\zeta-1}\, \dot\lambda(t)\, e^{-b(t-t_1)} \;\le\; [\, \sum_i f_i(t)\,]\, e^{-b\zeta(t-t_1)} \tag{C.8}$$

$$\text{or}\qquad \dot\lambda(t) \;\ge\; -(\zeta-1)\, [\, \sum_i f_i(t)\,]\, e^{-b(\zeta-1)(t-t_1)} \tag{C.9}$$

$$\text{hence}\qquad \lambda(t) \;\ge\; \lambda(t_1) - (\zeta-1) \int_{t_1}^{t} [\, \sum_i f_i(\tau)\,] e^{-b(\zeta-1)(\tau-t_1)} d\tau \tag{C.10}$$

$$\text{but}\qquad (\zeta-1) \int_{t_1}^{t} f_i(\tau) e^{-b(\zeta-1)(\tau-t_1)} d\tau \;\le\; (\zeta-1)\, S_i^{\frac{1}{k_i}} \left[ \frac{1}{b\bar k_i(\zeta-1)} \right]^{\frac{1}{\bar k_i}} \;=\; S_i\, \frac{(\zeta-1)^{1/k_i}}{(b\bar k_i)^{1/\bar k_i}} \tag{C.11}$$

$$\text{and, finally,}\quad \lambda(t) \;\ge\; \lambda(t_1) - \sum_i \frac{[(\zeta-1)\, S_i\,]^{1/k_i}}{(b\bar k_i)^{1/\bar k_i}} \;. \tag{C.12}$$

Now, it is easy to conclude:

$\lambda(t_1)$ is equal to $U(t_1)^{-(\zeta-1)}$, which is either 1 (if $t_1 \ne 0$) or $U(0)^{-(\zeta-1)}$, so that, if (B.24) is met, we get

$$U(t)^{\zeta-1} \;\le\; \frac{1}{\mathrm{Inf}\{\; 1 \;,\; U(0)^{-(\zeta-1)} - \sum_i \frac{[(\zeta-1)\, S_i\,]^{1/k_i}}{(b\bar k_i)^{1/\bar k_i}} \;\}} \tag{C.13}$$

**APPENDIX D: Technical lemma.**

**Lemma 4:** Let $V$ be a smooth real positive function such that, for any $x$,

$$\left|\frac{\partial V}{\partial x}(x)\right| \leq k \ \mathrm{Sup}\ \{\ 1\ ,\ V(x)^{1-\frac{1}{q}}\ \} \tag{D.1}$$

then, there exists two positive real numbers $\lambda$ and $\mu$ such that, for any positive $t$,

$$\left|\ V(x) - V(y)\ \right| \leq \lambda \ \mathrm{Sup}\ \{\ 1\ ,\ V(x)^{1-\frac{1}{q}}\ \}\ |x-y| + \mu |x-y|^q \tag{D.2}$$

*Remark:* This technical result is used in the proof of theorem 1 with $1 - \dfrac{1}{q} = \alpha$.

*Proof:* Let $W = \mathrm{Sup}\{\ 1\ ,\ V^{\frac{1}{q}}\}$. With (D.1), we have, at all points where $V$ is larger than 1, $\left|\dfrac{\partial W}{\partial x}\right| \leq \dfrac{k}{q}$ ; so that, for all $x,y$:

$$\left|\ W(x) - W(y)\ \right| \leq \frac{k}{q}\ |x-y|\ . \tag{D.3}$$

It is not difficult to see that (D.3) yields:

$$\left|\mathrm{Sup}\{\ 1\ ,\ V(x)\ \} - \mathrm{Sup}\{\ 1\ ,\ V(y)\ \}\right| \leq (\ W(x) + \frac{k}{q}\ |x-y|\ )^q - W(x)^q \tag{D.4}$$

Now, with $L$ and $M$ such that, for any positive $a$ and $b$, $(\ a + b\ )^q \leq a^q + L\ a^{q-1}b + M\ b^q$, we have

$$\left|\mathrm{Sup}\{1,V(x)\} - \mathrm{Sup}\{1,V(y)\}\right| \leq L\ \mathrm{Sup}\{1,V(x)^{1-\frac{1}{q}}\}\ \frac{k}{q}|x-y| + M(\frac{k}{q}|x-y|)^q\ . \tag{D.5}$$

The lemma follows. □

# ADAPTIVE LINEAR CONTROL INTERPRETED AS NONLINEAR DYNAMICAL FEEDBACK FOR NONLINEAR SYSTEMS

David J. Hill

Changyun Wen

Department of Electrical and Computer Engineering,
University of Newcastle, New South Wales, 2308
Australia.

## ABSTRACT

The theory of stable parameter adaptive control has advanced to allow linear time-varying plants. However, a more honest view of such systems is that they are often derived from inexact linearization about a trajectory of a nonlinear system. Then adaptive control based on a linear model can be regarded as one way to realize a nonlinear controller for a nonlinear plant. The implications of this view are studied using a simple certainty equivalence adaptive scheme.

## 1. INTRODUCTION

Adaptive control is commonly presented as a technique whereby a self-tuning linear controller is used on a linear plant with unknown parameters. The theory and design principles for such controllers are becoming quite well understood. The literature is enormous; some attempts to cover recent developments are provided by [1-4]. The theory of stable parameter adaptive control has advanced to allow linear time-varying plants [5-8]. Generally speaking, the time-variations can only correspond to occasional jumps or slow trends (unless special features are built into the algorithms for classes of fast time-variation). In the latter case at least, a more honest view is that the linear time-varying systems are derived from inexact linearization about a trajectory of a nonlinear system. We then see that adaptive linear control is one approach to deriving a nonlinear dynamic controller which can cope with nonlinear dynamics. It is a continuation of the linearization tradition in the control field aided by techniques for parameter estimation. In this paper, we begin to study the implications of the nonlinear controller view and state some new results for stability.

Most system models emerge from the application of basic physical laws as a set of nonlinear differential (or difference) equations. Indeed these equations usually are defined in terms of smooth functions of some convenient analytical form. For example, power systems give trigonometric terms and flow systems give bilinear terms. In the absence of tools for designing nonlinear controllers (thirty years ago), the adaptive viewpoint offered a way to apply conventional linear control techniques. The nonlinear system can be viewed [9,10] as a time-varying linear part (with unknown parameters) and higher order terms which are ignored. Adaptive linear control is seen as a conceptually simple way to proceed towards a nonlinear controller. It is certainly a path taken in practice, e.g. power systems control, aircraft control. Knowing the nominal trajectory, adaptive control loops can be set up with signals measured relative to nominal values.

In [11] a complete stability analysis is given of adaptive linear control. Stability conditions are given on the nonlinear system functions and certain bias signals which ensure useful stability properties of the overall adaptive system. It is shown that it is not necessary to know the nominal trajectory exactly. If it is only approximately known (and the error can be large in principle), a fixed deadzone in the parameter estimator ensures stability. These results are briefly discussed here.

## II. NONLINEAR CONTROL PROBLEM

The class of plant to be considered in this paper is described by the difference equation

$$y(t) = f(y(t-1), y(t-2), \ldots, y(t-n), u(t-1), \ldots, u(t-n), \ldots, d(t-1), \ldots, d(t-p)) \tag{2.1}$$

Assumption 2.1. n is a known integer.
In simpler notation, we rewrite (2.1) as

$$y(t) = f(v(t-1), \omega(t-1)) \tag{2.2}$$

where

$$v^T(t-1) := [y(t-1), y(t-2), \ldots, u(t-1), \ldots, u(t-n)]$$

$$\omega^T(t-1) := [d(t-1), \ldots, d(t-p)]$$

For any function $f: \mathbb{R}^{2n} \times \mathbb{R}^p \to \mathbb{R}$, signals u, d and given initial conditions $v(0)$, $\omega(0)$, model (2.2) generates a unique solution y. We use $v_i$, $i=1, \ldots, 2n$, and

$\omega_i$, i=1, ..., p to denote the components of the vectors v and $\omega$.

The control problem is assumed to be regulation about a given desired output signal $y^*$. More precisely, we have

<u>Control Problem</u>: Synthesise a system with output u(t) and input v(t-1) which, when applied to system (2.2), ensures that $|y(t) - y^*(t)|$ is small in a chosen sense as $t \to \infty$.

There are of course many approaches to solving this problem. In many applications including power systems and aircraft control, function f is complicated, but known at least approximately. Otherwise, system behaviour may be sufficiently understood that there is some knowledge of a nominal trajectory. We establish this property precisely as

<u>Assumption 2.2</u>: For given $y^*$ and d, $\exists$ a unique input signal $u^*$ s.t.

$$y^*(t) = f(v^*(t-1), \omega(t-1)) \tag{2.3}$$

where $v^*$ is derived from $(u^*, y^*)$. Suppose that $u_1$ is known s.t.

$$\left\{ \sum_{t=i}^{n+i} (u_1(t) - u^*(t))^2 \right\}^{1/2} \leq U \text{ for all } i \tag{2.4}$$

and U is a constant.

At this stage, it is convenient to introduce the variables

$$\begin{aligned} \Delta u\ (t) &:= u(t) - u_1(t) \\ \Delta y(t) &:= y(t) - y^*(t) \end{aligned} \tag{2.5}$$

i.e. departures from the known nominal values. In order to relate our formulation to real-world control, consider the following scheme. The nonlinear system modelled by (2.2) is controlled by two loops. The high level control establishes the nominal trajectory. For instance, in power systems it represents the slower control actions due to a human operator, the governor etc.; in aircraft control it could represent the action of the pilot. This loop senses the system state x. If it is acting successfully, y will be somewhat close to $y^*$ under the action of $u_1$. Assumption 2.2 effectively says just this (although there is no constraint that U is small). The adaptive controller is the second loop based on departures from a nominal trajectory. Of course, $u_1$ and $y^*$ are known, so the departures $\Delta u$, $\Delta y$ are measurable.

Clearly, in order to analyse the adaptive regulator, we need to invoke an exact linearisation model, i.e. one which relates $\Delta y(t)$ to $\Delta v(t-1)$ via a linear system and whatever else is needed to make the model exact.

## III. EXACT LINEARIZATION MODEL

We consider the plant with representation (2.3). The following assumption imposes some additional smoothness restrictions to those implied by Assumption 2.2 on the function $f: \mathbb{R}^{2n} \times \mathbb{R}^{p} \to \mathbb{R}$.

Assumption 3.1

The partial derivatives $f_{v_i} := \frac{\partial f}{\partial v_i}$ $i = 1, \ldots 2n$ exist and are locally Lipschitz in $v$ and $\omega$.

We let $\mathcal{B}(x_0, r) := \{x \mid \|x - x_0\| \leq r\}$ where $\|\cdot\|$ denotes the Euclidean norm. Also let $\mathcal{B}_r = \mathcal{B}(0,r)$. An immediate consequence of Assumption 3.1 is that $f$ is locally Lipschitz in $v$ in the sense that $\exists$ $L \geq 0$ and $r > 0$ s.t.

$$|f(v,\omega) - f(\bar{v}, \omega)| \leq L \| v - \bar{v} \| \tag{3.1}$$

for a given $\omega$, $v - \bar{v} \in \mathcal{B}_r$.

Comment 3.1

The Lipschitz gain constant $L$ in (3.1) is dependent on $r$ and $\omega$.

We now rewrite model (2.2) in terms of the incremental variables (2.5). It is convenient to denote the signal vector $(y^*, u_1)$ by $v_1$. From Assumption 3.1, the system can be linearized about $v_1$. Note that

$$\begin{aligned}\Delta y(t) &= y(t) - y^*(t) && \text{from (2.5)}\\ &= f(v(t-1),\omega(t-1)) - f(v^*(t-1),\omega(t-1)) && \text{using (2.2) and (2.3)}\end{aligned}$$

$$= \phi^T(t-1)\,\theta(t-1) + R(t) \tag{3.2}$$

where

$$\theta^T(t) = [\theta_1(t), \ldots, \theta_{2n}(t)]$$

$$\phi^T(t-1) = [\Delta y(t-1), \ldots, \Delta y(t-n), \Delta u(t-1), \ldots, \Delta u(t-n)]$$

$$\theta_i(t) = f_{v_i}(v_1(t), \omega(t)) \qquad i=1, \ldots, 2n \tag{3.3}$$

and $R(t)$ is the remainder term. Note that (3.2) has the so-called regression form commonly used in analysis of parameter estimation algorithms. We refer to $\theta$ and $\phi$ as the parameter and regression vectors respectively. We now make a further assumption.

<u>Assumption 3.2</u> The reference signals, disturbance and derivatives $f_{v_i}$ are such that $\theta(t) \in \mathscr{C} \ \forall \ t$ where $\mathscr{C} \subset \mathbb{R}^{2n}$ is a known compact convex region.

This assumption restricts the slope of $f$ along $v_1$ to be within finite limits. As a consequence of Assumption 3.2, we note that

$$\|\theta_1 - \theta_2\| \le k_c \tag{3.4a}$$

$$\| \theta \| \le k_\theta \tag{3.4b}$$

$\forall \ \theta, \theta_1, \theta_2 \in \mathscr{C}$. $k_c$, $k_\theta$ are constants depending on the size of $\mathscr{C}$.

A key property of the linearization model (3.2) is that the remainder term $R$ is linearly bounded by $\phi$. It is convenient to describe $R$ by

$$R(t) = R_1(t) + f(v_1(t-1),\omega(t-1)) - f(v^*(t-1),\omega(t-1))$$

where

$$R_1(t) = f(v(t-1),\omega(t-1)) - f(v_1(t-1),\omega(t-1)) - \phi^T(t-1)\theta(t-1) \tag{3.5}$$

Then (3.1) and (3.4b) imply

$$|R_1(t)| \le (L + k_\theta)\| \phi(t-1)\| \tag{3.6}$$

for $\phi(t-1) \in \mathscr{B}_r$. This bound is very loose. There are some cancellations between the terms of (3.5). This can be demonstrated by example [11] and suggests that a tighter bound than (3.6) can be written in the form

$$|R_1(t)| \le \varepsilon_r \|\phi(t-1)\| \tag{3.7}$$

for $\phi(t-1) \in \mathcal{B}_r$ by taking $\varepsilon_r$ as a least upper bound over $\mathcal{B}_r$. In (3.7) there is no implication that L or $k_\theta$ should be small.

<u>Comments 3.1</u>

1. The value of parameter $\varepsilon_r$ reflects the degree of nonlinearity of the plant. Loosely speaking, a more nonlinear plant will have larger $\varepsilon_r$ for smaller r.
2. While being aware of its existence, we do not need to know the value of $\varepsilon_r$.

To complete the bound on R, we note

$$\begin{aligned} |R(t)| &\le |R_1(t)| + |f(v_1(t-1),\omega(t-1) - f(v^*(t-1),\omega(t-1)| \\ &\le \varepsilon_r \|\phi(t-1)\| + LU \end{aligned} \tag{3.8}$$

for $\phi(t-1) \in \mathcal{B}_r$ using (2.4), (3.1) and (3.7)

(Actually, L in (3.7) can be replaced by a smaller Lipschitz constant which governs changes in just u.)

To design the parameter estimator, we need to know a bound on the constant LU in (3.7).

<u>Assumption 3.3</u> We know a number $\ell$ s.t. $\ell \ge LU$.

<u>Comments 3.2</u>

1. In the stability analysis to follow, we see that $\ell$ is not as critical as $\varepsilon_r$. The bound in Assumption 3.3 can be loose. In fact if $f(\cdot, \omega)$ is globally Lipschitz, U and so $\ell$ can be arbitrarily large for stability. However, better performance is expected for smaller values of $\ell$.
2. If $u_1 = u^*$, then $\ell$ can be zero. This is the case if we use adaptive control as a means to design a controller for a known nonlinear system, i.e. f is known.
3. In the use of linear models [1], the system with unstructured disturbances is effectively described relative to signals $(0,u^*)$. In the design of adaptive control, we implicitly use $u_1 = 0$ to approximate $u^*$. The number U can be large. Anticipating later developments, we recall that it is standard to use a fixed deadzone in the estimator for this situation.

We now make some assumptions which restrict the speeds of the signals $u_1$, $y^*$ and d. These will be passed through the smoothness Assumption 3.1 into the time variations of the linearized form (3.2).

<u>Assumption 3.4</u> Suppose $y^*$, d satisfy

$$|y^*(t) - y^*(t-1)| \le \delta_1 \quad \forall t$$

$$|d(t) - d(t-1)| \leq \delta_1 \qquad \forall t$$

Further, the signal $u_1$ (Assumption 2.2) satisfies

$$|u_1(t) - u_1(t-1)| \leq \delta_1'' \quad \forall t$$

From Assumptions 3.1 and 3.4, we easily conclude [11] that there exists $\varepsilon_\theta \geq 0$ such that

$$\|\theta(t) - \theta(t-1)\| \leq \varepsilon_\theta \qquad \forall t \tag{3.9}$$

Comments 3.3

The essential parameters of the nonlinearity in (2.2) are $r$, $\varepsilon_r$ and $\varepsilon_\theta$. These are clearly related. In the bounds (3.8) and (3.9), $\varepsilon_r$ and $\varepsilon_\theta$ will typically be larger for larger $r$.

## IV. ADAPTIVE CONTROL SCHEME

We study a certainty equivalence adaptive controller along the lines studied previously in many situations [1]. The parameter estimator is taken as the simple gradient scheme

$$\hat{\theta}(t) = \mathcal{P}\left\{\hat{\theta}(t-1) + \frac{\phi(t-1)g(t)}{1+\phi^T(t-1)\phi(t-1)}\right\} \tag{4.1}$$

where $\hat{\theta}(t)$ denotes the estimate of $\theta(t)$ and $\mathcal{P}$ denotes the projection operator which ensures that $\hat{\theta}(t) \in \mathcal{C}\ \forall t$[1]. The signal $g(t)$ is given by

$$g(t) = \begin{cases} e(t) - \ell, & \text{if} \quad e(t) > \ell \\ 0 \quad , & \quad |e(t)| \leq \ell \\ e(t) + \ell, & \quad e(t) < -\ell \end{cases} \tag{4.2}$$

where

$$e(t) = \Delta y(t) - \phi^T(t-1)\,\hat{\theta}(t-1) \tag{4.3}$$

$e(t)$ is the prediction error and $g(t)$ is the output of a fixed deadzone.

The stability analysis relies on several key properties of the estimator [11].

Lemma 4.1. The estimator (4.1), (4.2), (4.3) applied to system (2.3) (or (3.6)) has the properties:

(a) $$|\tilde{g}(t)| \le k_c + \varepsilon_r \tag{4.4}$$

if $\phi(t-1) \in \mathcal{B}_r$ where

$$\tilde{g}(t) := \frac{g(t)}{(1+\|\phi(t-1)\|^2)^{1/2}}$$

(b) $$\|\hat{\theta}(t) - \hat{\theta}(t-1)\| \le |\tilde{g}(t)| \qquad \forall t \tag{4.5}$$

(c) $$\sum_{i=t_0+1}^{t} \tilde{g}^2(i) \le \alpha_1 + \alpha_2(t-t_0) \tag{4.6}$$

if $\phi(i) \in \mathcal{B}_r$, $i=t_0, t_0+1, \ldots, t-1$ where $\alpha_1$, $\alpha_2$ are given by

$$\alpha_1 = k_c^2 \tag{4.7a}$$

$$\alpha_2 = \varepsilon_\theta^2 + 2[(k_c+\varepsilon_r)(\varepsilon_\theta+\varepsilon_r) + k_c\varepsilon_\theta] \tag{4.7b}$$

<u>Comments 4.1</u>.

1. The form of (4.6) is special in that $\alpha_2$ can be made arbitrarily small by reducing $\varepsilon_r$ and $\varepsilon_\theta$.
2. The more commonly used (in practice) least squares estimator has similar properties, but the analysis is more tedious [1].

The estimator is combined with a conventional (non-adaptive) controller with the system parameters replaced by $\hat{\theta}(t)$. A wide variety of controllers are suitable [1]. Here we consider a pole assignment regulator of the form

$$\hat{L}(t-1)\Delta u(t) = -\hat{P}(t-1)\Delta y(t) \tag{4.8}$$

where $\hat{L}$, $\hat{P}$ are derived from

$$\hat{A}(t) = 1 - \hat{\theta}_1(t)q^{-1} - \ldots - \hat{\theta}_n(t)q^{-n} \tag{4.9a}$$

$$\hat{B}(t) = \hat{\theta}_{n+1}(t)q^{-1} + \ldots + \hat{\theta}_{2n}(t)q^{-n} \tag{4.9b}$$

$$\hat{A}(t-1)\hat{L}(t-1) + \hat{B}(t-1)\hat{P}(t-1) = A^* \tag{4.10}$$

$A^*$ is a given monic polynomial in shift operator $q^{-1}$ of degree 2n. Then $\hat{L}$ and $\hat{P}$ provide a strictly proper regulator. We express these in the form

$$\hat{L}(t) = 1 + \hat{\ell}_0(t)\, q^{-1} + \ldots + \hat{\ell}_{n-1}(t) q^{-n} \quad (4.11a)$$

$$\hat{P}(t) = \hat{p}_0(t)\, q^{-1} + \ldots + \hat{p}_{n-1}(t) q^{-n} \quad (4.11b)$$

Assumption 4.1. The polynomial $z^{2n}A^*$ is strictly (discrete-time) Hurwitz.

Just as in all previous discussions of indirect adaptive control, a technical difficulty (invariant under problem reformulation) arises in the solution of (4.10). We require $\|\hat{L}\|$, $\|\hat{P}\|$ to be bounded where $\|\cdot\|$ denotes the norm of the vector of polynomial coefficients. Thus we require a further restriction on the system model [4,7].

Assumption 4.2. The convex region $\mathscr{C}$ in Assumption 3.2 has the additional property that for all $\theta \in \mathscr{C}$, the linearized system model is uniformly controllable.

Comments 4.2.

1. Time-varying $A^*$ can easily be accommodated with a suitably modified version of Assumption 4.2 [4,7].
2. An arbitrarily large region of the parameter space can be covered by the use of multiple convex regions [4]. The analysis presented below can be extended to this situation.
3. Assumption 4.2 effectively introduces a local controllability requirement on the nonlinear plant.

## V. STABILITY ANALYSIS

The main stability results establish that the overall adaptive system is bounded-input bounded-state (BIBS) stable in the sense where all variables are departures from the nominal trajectory represented by $v_1$.

The establishment of a framework for analysis follows standard steps for certainty equivalence adaptive controllers [1]. Then the closed loop system equations (4.2) and (4.12) can be combined to give

$$\phi(t+1) = A(t)\, \phi(t) + b\, e(t+1) \quad (5.1)$$

where

$$A(t):=\left[\begin{array}{ccc|ccc}
\hat{\theta}_1(t) & \cdots & \hat{\theta}_n(t) & \hat{\theta}_{n+1}(t) & \cdots & \hat{\theta}_{2n}(t) \\
1 & \ddots & 0 & 0 & & 0 \\
0 & 1 & 0 & 0 & & 0 \\
\hline
-\hat{p}_0(t) & \cdots & -\hat{p}_{n-1}(t) & -\hat{\ell}_0(t) & \cdots & \hat{\ell}_{n-1}(t) \\
0 & & 0 & 1 & \ddots & 0 \\
0 & & 0 & 0 & 1 & 0
\end{array}\right] \qquad (5.2a)$$

$$b^T := [\ 1\ \ 0\ \ 0\ \ .\ .\ .\ \ 0\ ] \qquad (5.2b)$$

The stability analysis of (5.1) makes use of several standard tools in analysis of adaptive controllers [7] plus a novel induction procedure to handle the locally bounded error. The details are presented in [11].

The main stability result is as follows.

Theorem 5.1: Consider the adaptive scheme consisting of plant (2.2) (modelled by (3.2)), estimator (4.1), (4.3) and regulator (4.8) - (4.10). Under all above assumptions $\exists\ r^*$, $\varepsilon_r^*$, $\varepsilon_\theta^*$ and $r_0$ s.t. $r \geq r^*$, $\varepsilon_r \leq \varepsilon_r^*$, $\varepsilon_\theta \leq \varepsilon_\theta^*$ and $\phi(0) \in \mathfrak{B}_{r_0}$ ensures $\phi(t) \in \mathfrak{B}_r \forall t$. In particular, the control signal $\Delta u(t)$ and tracking error $\Delta y(t)$ remain bounded.

If the Lipschitz condition (3.1) is global, a simpler result is easily seen to hold.

Theorem 5.2: Suppose the conditions of Theorem 5.1 are altered to allow $f(\cdot,\omega)$ to be globally Lipschitz. Then $\exists\ \varepsilon_r^*$ and $\varepsilon_\theta^*$ s.t. $\varepsilon_r \leq \varepsilon_r^*$ and $\varepsilon_\theta \leq \varepsilon_\theta^*$ ensure the system is BIBS stable (about the nominal trajectory).

From the proof of theorem 5.1, we can see that $\ell$ does not affect stability in this case. This justifies our earlier Comment 3.2.1.

An appealing feature of the above results is that there has been no reliance on estimator modifications such as special normalization, persistence of excitation or so-called $\sigma$-modification. These have become common-place in recent results on robust adaptive control [3,4]. One key aspect of the problem treated here is that the 'modelling uncertainty' in (3.2) consists of a memoryless function of the data.

## 6.. CONCLUSIONS

The interpretation of adaptive linear control as a design technique for well-defined discrete-time nonlinear systems have been studied. The stability conditions essentially limit the gradient of the functional giving a difference equation description. For global stability, this functional is required to be globally Lipschitz.

## REFERENCES

[1] G.C. Goodwin and K.S. Sin, Adaptive Filtering Prediction and Control, Prentice-Hall, New Jersey, 1984.

[2] B.D.O. Anderson et.a., Stability of Adaptive Systems Passivity and Averaging Analysis, The MIT Press, Cambridge, Mass., 1986.

[3] K.J. Åström, Adaptive Feedback Control, *Proc. IEEE*, Vol.75, No.2, Feb. 1987.

[4] R.H. Middleton, G.C. Goodwin, D.J. Hill and D.Q. Mayne, "Design issues in adaptive control", *IEEE Trans. Auto. Control*, Vol. AC-33, No. 1, January 1988, pp 50-58.

[5] G. Kreisselmeir, "Adaptive control of a class of slowly time-varying plants", Systems and Control Letters, Vol. 8, 1986, pp.97-103.

[6] K. Tsakalis and P.A. Ioannou, "Adaptive control of linear time-varying plants", *Automatica*, Vol. 23, No. 4, July 1987, pp.459-468.

[7] R.H. Middleton and G.C. Goodwin, "Adaptive control of time varying linear systems", *IEEE Trans. Auto. Control*, Vol. AC-33, No.2, February 1988, pp 150-155.

[8] P. de Larminat and H.F. Raynaud, "A robust solution to the stabilizability problem in indirect passive adaptive control" *Proc. of 25th Conference on Decision and Control*, 1986, pp 462-467.

[9] M. Vidyasagar, Nonlinear Systems Analysis, Prentice-Hall, Englewood Cliffs, New Jersey, 1978.

[10] C.A. Desoer and K.K. Wong, "Small-signal behaviour of nonlinear lumped networks", *Proc. IEEE*, Vol.56, No.1, January 1968, pp.14-22.

[11] C. Wen and D.J. Hill, "Adaptive linear control of nonlinear systems" Dept. of Electrical and Computer Engineering, University of Newcastle, Technical Report EE8724, revised May 1988.

# V - STOCHASTIC SYSTEMS

# Approximation of a Stochastic Ergodic Control Problem

*Fabien Campillo, François Le Gland*

INRIA centre de Sophia Antipolis
Route des Lucioles
06565 Valbonne Cedex, FRANCE

*Etienne Pardoux*

INRIA centre de Sophia Antipolis
&
Mathématiques, UA 225
Université de Provence

**Abstract**

We study a degenerate non linear optimal stochastic control problem of ergodic type. We first prove that for each feedback control law, there exists a unique invariant measure which is equivalent to Lebesgue measure. This is proved using an accessibility property of the stochastic differential equation, after the discontinuous part of the drift has been removed via a change of probability measure. We then approximate the problem by ergodic control problems for finite state, continuous time Markov chains. We finally prove that the cost functionals of the approximate problems converge pointwise towards that of the continuous problem.

All the study is done for a particular problem introduced in [1], which is motivated by the optimal control of the shock–absorber of a road vehicle. The numerical results can be found in [1].

## 1 Introduction

The aim of this paper is the study and the approximation of a class of ergodic control problems. For clarity we will work on a particular problem already introduced in [1], which comes from a problem of optimization of controlled shock–absober. This involves three difficulties — which are met in most applied problems — : the diffusion we want to control is degenerate, some coefficients are discontinuous and the problem is strongly nonlinear.

Let us consider the following stochastic system

$$dX(t) = b(u(X(t)), X(t))\, dt + \begin{pmatrix} 0 \\ \sigma \end{pmatrix} dW(t) \; , \tag{1}$$

where $X$ is a process which takes values in $\mathbb{R}^2$, $W$ is a real standard Wiener process and $\sigma > 0$. $b$ maps $\mathbb{R} \times \mathbb{R}^2$ in $\mathbb{R}^2$ and is defined by

$$b(u,x) \triangleq \begin{pmatrix} b_1(u,x) \\ b_2(u,x) \end{pmatrix} \triangleq \begin{pmatrix} x_2 \\ -u\,x_2 - \beta\,x_1 - \gamma\,\mathrm{sign}(x_2) \end{pmatrix}, \quad x \triangleq \begin{pmatrix} x_1 \\ x_2 \end{pmatrix},$$

where $\beta$, $\gamma$ are strictly positive constants. In (1), $u$ is a feedback control which belongs to the class $\mathcal{U}$ of admissible controls defined by (fix $\underline{u}$, $\overline{u}$ such that $0 < \underline{u} < \overline{u} < \infty$)

$u \in \mathcal{U} \iff$ $u : \mathbb{R}^2 \to [\underline{u}, \overline{u}]$ and there exists a finite number of submanifolds of $\mathbb{R}^2$ with dimension less than or equal to 1 outside of which $u$ is continuous.

We are concerned with an ergodic type control problem, whose cost functional is

$$J(u) \triangleq \lim_{T\to\infty} \frac{1}{T} E \int_0^T f(u(X(t)), X(t))\,dt\,, \quad \forall u \in \mathcal{U}\,, \tag{2}$$

where the instantaneous cost function $f$ is defined by

$$f(u,x) \triangleq (u\,x_2 + \beta\,x_1 + \gamma\,\mathrm{sign}(x_2))^2\,. \tag{3}$$

From now on, we denote

$$b^u(x) \triangleq b(u(x),x)\,, \quad f^u(x) \triangleq f(u(x),x)\,, \quad \forall u \in \mathcal{U}\,.$$

The physical interpretation of this problem is the following: $y = X_1(t)$ is a solution of the equation

$$m\,\ddot{y} + v\,\dot{y} + K\,y + F\,\mathrm{sign}(\dot{y}) = m\,\ddot{e}\,. \tag{4}$$

($m$, $K$, $F > 0$) which describes a one–degree–of–freedom shock–absober system with dry friction. $y$ is the relative displacement, $v$ is the shock–absorber damping constant (the controlled parameter). $K\,y + F\,\mathrm{sign}(\dot{y})$ represents the restoring force (including the dry friction term). $\ddot{e}$ is the random input of the system (i.e. the road surface displacement) which is supposed to be a white noise. Taking $u = v/m$, $\beta = K/m$, $\gamma = F/m$, (4) can be rewritten as (1). The problem is to improve vehicle riding comfort by the choice of an adequate feedback $u$, i.e. to minimize

$$J(u) \triangleq \lim_{T\to\infty} \frac{1}{T} E \int_0^T f(u,y,\dot{y})\,dt\,, \tag{5}$$

where the instantaneous cost function $f(u,y,\dot{y})$ is the absolute acceleration squared, that is

$$f(u,y,\dot{y}) \triangleq |\ddot{a}|^2 = |\ddot{y} - \ddot{e}|^2 = \left|\frac{1}{m}(v\,\dot{y} + K\,y + F\,\mathrm{sign}(\dot{y}))\right|^2\,.$$

In [1], we present a numerical approach based on finite difference techniques [7,10]. For the discretized problem, we use the policy iteration algorithm for which we state a convergence property. In the present paper we give some results on the following properties

- existence and uniqueness of the invariant measure $\mu_u$ associated with (1),
- convergence result of the approximation when the discretizing parameter goes to 0.

The Hamilton–Jacobi–Bellman equation related to the ergodic control problem (1,2) can be formally stated as

$$\min_{u\in[\underline{u},\overline{u}]} (\mathcal{L}_u v(\cdot) + f(u,\cdot)) = \rho \quad \text{on } \mathbb{R}^2, \tag{6}$$

where $v : \mathbb{R}^2 \to \mathbb{R}$ is defined up to an additive constant. $\rho$ is a constant and $\mathcal{L}_u$ is the infinitesimal generator of (1)

$$\mathcal{L}_u \phi(x) \triangleq b_1^u(x)\frac{\partial\phi(x)}{\partial x_1} + b_2^u(x)\frac{\partial\phi(x)}{\partial x_2} + \frac{\sigma^2}{2}\frac{\partial^2\phi(x)}{\partial^2 x_2}\,. \tag{7}$$

Numerical approximation of the Hamilton–Jacobi–Bellman equation, in the nonergodic case, may be found in [4,13], as well as the study of the convergence of the approximation. Here, we are not studying directly the Hamilton–Jacobi–Bellman equation for which there seems to be no result proved in the present context. We want to study — using probabilistic techniques — the convergence of the approximation.

The ergodic control problem has been studied in [5] for discrete state space Markov processes, and in [2,3,9,11,15,16] for diffusion processes. Most of these last works are based on a strong ellipticity assumption, or establish a recurrence property with a different set of hypotheses than ours.

The bound $\underline{u} > 0$ is important both for mathematical and physical reasons in order to ensure the stability of the system (cf. the proof of lemma 2.1).

Because the system (1) is degenerate (the noise appears only in the second component), the uniqueness of the invariant measure is related to a controllability type property, but, due to the nonregularity of the coefficients, the standard techniques fail in proving this last property. However, this can be done via a change of probability law.

In section 2 we establish an existence and uniqueness result for the invariant measure corresponding to system (1), for any $u$ in $\mathcal{U}$. In 3 we present the approximation of the problem using finite difference techniques. The convergence of the approximate cost functionals to the original cost functional is studied in section 4.

## 2 The Invariant Probability Measure

The cost function (2) can be rewritten as

$$J(u) = \langle f^u\,,\,\mu_u\rangle\,, \quad \forall u \in \mathcal{U}\,, \tag{8}$$

where $\mu_u$ is the invariant probability measure associated with system (1). In this section we establish an existence and uniqueness property for $\mu_u$.

## 2.1 Existence

**Lemma 2.1** *There exists a constant $C$ such that*

$$E\,|X(t)|^2 \leq C\,, \quad \forall t \geq 0,\ \forall u \in \mathcal{U}\,. \tag{9}$$

**Proof** We define

$$\mathcal{V}(x) \triangleq \beta\, x_1^2 + \varepsilon\, x_1\, x_2 + x_2^2\,, \qquad \text{and} \quad V(t) \triangleq E\,\mathcal{V}(X(t))\,.$$

There exists $\varepsilon_0 > 0$ such that for any $\varepsilon_0 > \varepsilon > 0$

$$\mathcal{V}(x) \geq \frac{1}{2}\left(\beta\, x_1^2 + x_2^2\right)\,.$$

Hence, it is sufficient to show that $V(t) \leq$ Cte for any $t \geq 0$. From (1),

$$\begin{aligned} \frac{d}{dt}V(t) &= E\,\big[2\,\beta\, X_1(t)\, X_2(t) + \varepsilon\, X_2^2(t) - \varepsilon\,\beta\, X_1^2(t) - \varepsilon\, u(X(t))\, X_1(t)\, X_2(t) \\ &\quad -\varepsilon\,\gamma\, X_1(t)\,\mathrm{sign}(X_2(t)) - 2\,\beta\, X_1(t)\, X_2(t) - 2\,u(X(t))\, X_2^2(t) \\ &\quad -2\,\gamma\, X_2^2(t)\big] + \sigma^2\,. \end{aligned}$$

Using $\underline{u} \leq u(x) \leq \overline{u}$ and the following inequalities

$$\begin{aligned} -\varepsilon\, u(x)\, x_1\, x_2 &\leq \frac{\varepsilon\,\beta}{2}\, x_1^2 + \frac{\varepsilon\,\overline{u}^2}{2\,\beta}\, x_2^2\,, \\ -\varepsilon\,\gamma\, x_1\,\mathrm{sign}(x_2) &\leq \frac{\varepsilon\,\delta\,\gamma}{2}\, x_1^2 + \frac{\varepsilon\,\gamma}{2\,\delta}\,, \quad (\forall \delta > 0)\,, \end{aligned}$$

we get

$$\frac{d}{dt}V(t) \leq E\left[-\left(\frac{1}{2}\,\varepsilon\,\beta - \frac{1}{2}\,\varepsilon\,\delta\,\gamma\right) X_1^2(t) - \left(2\,\underline{u} + 2\,\gamma - \frac{1}{2\,\beta}\,\varepsilon\,\overline{u}^2\right) X_2^2(t)\right] + \frac{\varepsilon\,\gamma}{2\,\delta} + \sigma^2\,,$$

so there exists strictly positive constants $\varepsilon$ and $\delta$ such that

$$\frac{d}{dt}V(t) \leq -C(\varepsilon,\delta)\, V(t) + \frac{\varepsilon}{2\,\delta} + \sigma^2\,,$$

where $C(\varepsilon,\delta) > 0$. Applying Gronwall's lemma to this last inequality yields the conclusion. □

**Lemma 2.2** *The process $X(t)$ solution of (1) has the Feller property, i.e. for any $u \in \mathcal{U}$, $t \geq 0$ and $\phi \in C_b(\mathbb{R}^2)$, the function*

$$\mathbb{R}^2 \ni x \longrightarrow E\phi\,(X^x(t)) \tag{10}$$

*is continuous. $X^x(t)$ denotes the solution of (1) starting from $x$ at time $t = 0$.*

**Proof** In (1), the drift coefficient can be written as

$$b(u,x) = \overline{B}\,x + \begin{pmatrix} 0 \\ -u\,x_2 - \gamma\,\mathrm{sign}(x_2) \end{pmatrix} \triangleq \begin{pmatrix} 0 & 1 \\ \beta & 0 \end{pmatrix} \begin{pmatrix} x_1 \\ x_2 \end{pmatrix} + \begin{pmatrix} 0 \\ -u\,x_2 - \gamma\,\mathrm{sign}(x_2) \end{pmatrix}.$$

Let

$$\begin{aligned} \overline{W}(t) &\triangleq W(t) + \int_0^t \psi(X^x(s))\,ds\ , \\ \psi(x) &\triangleq -\frac{1}{\sigma}\,(u(x)\,x_2 + \gamma\,\mathrm{sign}(x_2))\ , \\ Z^x(t) &\triangleq \exp\left(\int_0^t \psi(X^x(s))\,d\overline{W}(s) - \frac{1}{2}\int_0^t \psi(X^x(s))^2\,ds\right)\ . \end{aligned} \tag{11}$$

We define a new probability law

$$\left.\frac{d\overline{P}}{dP}\right|_{\mathcal{F}_t} \triangleq (Z^x(t))^{-1}\ .$$

$X$ satisfies

$$dX(t) = \overline{B}\,X(t)\,dt + \begin{pmatrix} 0 \\ \sigma \end{pmatrix} d\overline{W}(t)\ , \tag{12}$$

where — from Girsanov's theorem — $\overline{W}(t)$ is a real standard Wiener process under the probability law $\overline{P}$.

For any sequence $x_n \to x$, we want to prove that

$$E\phi(X^{x_n}(t)) = \overline{E}\,[\phi(X^{x_n}(t))\,Z^{x_n}(t)] \underset{n\to\infty}{\longrightarrow} E\phi(X^x(t)) = \overline{E}\,[\phi(X^x(t))\,Z^x(t)]\ , \tag{13}$$

where $\overline{E}$ denotes the expectation with respect to $\overline{P}$. So, it is sufficient to check that

$$X^{x_n}(t) \underset{n\to\infty}{\longrightarrow} X^x(t) \quad \overline{P}\text{–a.s.}\ , \tag{14}$$

$$Z^{x_n}(t) \underset{n\to\infty}{\longrightarrow} Z^x(t) \quad \text{in } \overline{P}\text{–probability.} \tag{15}$$

Assume for a moment that (14) and (15) hold. Then, $\overline{E}Z^{x_n}(t) = \overline{E}Z^x(t) \equiv 1$ and (15) imply that $Z^{x_n}(t) \to Z^x(t)$ in $L^1(\overline{P})$, so

$$\begin{aligned} \left|\overline{E}\,(\phi(X^{x_n}(t))\,Z^{x_n}(t) - \phi(X^x(t))\,Z^x(t))\right| \ \le\ & \overline{E}\,(|\phi(X^{x_n}(t)) - \phi(X^x(t))|\,Z^x(t)) \\ & + C\,\overline{E}\,|Z^{x_n}(t) - Z^x(t)|\ , \end{aligned}$$

the second term tends to 0, the first one also by dominated convergence.

We now prove (14) and (15). Under the probability law $\overline{P}$, $X(t)$ is the solution of a *linear* stochastic differential system, so (14) is obvious. For (15), we show that

$$\overline{E}\int_0^t [u(X^{x_n}(s))\,X_2^{x_n}(s) - u(X^x(s))\,X_2^x(s)]^2\,ds \underset{n\to\infty}{\longrightarrow} 0\ , \tag{16}$$

$$\overline{E}\int_0^t [\mathrm{sign}(X_2^{x_n}(s)) - \mathrm{sign}(X_2^x(s))]^2\,ds \underset{n\to\infty}{\longrightarrow} 0\ . \tag{17}$$

For any $s \in [0,t]$ and $\varepsilon > 0$

$$\begin{aligned} \overline{E}[\text{sign}(X_2^{x_n}(s)) - \text{sign}(X_2^x(s))]^2 &= 4\overline{P}[X_2^{x_n}(s)\, X_2^x(s) < 0] \\ &\leq 4\overline{P}[|X_2^x(s)| < \varepsilon] + 4\overline{P}[|X_2^{x_n}(s) - X_2^x(s)| \geq \varepsilon] \\ &\rightarrow 4\overline{P}[|X_2^x(s)| < \varepsilon]\,, \quad \text{as } n \rightarrow \infty \text{ (using (14))}, \end{aligned}$$

so

$$\overline{E}[\text{sign}(X_2^{x_n}(s)) - \text{sign}(X_2^x(s))]^2 \underset{n\rightarrow\infty}{\longrightarrow} 0\,.$$

For (16), using (14) and the dominated convergence theorem, it is sufficient to state the following convergence in probability

$$\overline{P}\,(|u(X^{x_n}(s)) - u(X^x(s))| > \varepsilon) \underset{n\rightarrow 0}{\longrightarrow} 0\,, \quad \forall \varepsilon > 0\,, \quad \forall s \in [0,t]\,. \tag{18}$$

As $u \in \mathcal{U}$, for any $\delta > 0$ there exists a closed subset $D_\delta \subset \mathbb{R}^2$ and for any $\rho > 0$ there exists $C_\rho(\delta) \in [0,1]$ such that

$$\begin{aligned} &(i) \quad \overline{P}\,(X^x(s) \in D_\delta^c \cap B(0,\rho)) \leq C_\rho(\delta)\,, \quad \forall \rho, \delta > 0\,, \\ &(ii) \quad C_\rho(\delta) \longrightarrow 0\,, \quad \text{as } \delta \rightarrow 0\,, \quad \forall \rho > 0\,, \\ &(iii) \quad u \text{ is continuous on } D_\delta, \quad \forall \delta > 0\,, \end{aligned}$$

where $B(0,\rho) \triangleq \{x; |x| < \rho\}$. We have the following inequality

$$\begin{aligned} &\overline{P}\,(|u(X^{x_n}(s)) - u(X^x(s))| > \varepsilon) \\ &\leq \overline{P}\,(X^x(s) \in B(0,\rho)^c) \\ &\quad + \overline{P}\,(|u(X^{x_n}(s)) - u(X^x(s))| > \varepsilon\,;\; X^x(s) \in D_\delta \cap B(0,\rho)) \\ &\quad + \overline{P}\,(X^x(s) \in D_\delta^c \cap B(0,\rho))\,. \end{aligned} \tag{19}$$

Hence from (14) and because $u(x)$ is uniformly continuous on $D_\delta \cap B(0,\rho)$, we get

$$\begin{aligned} &\lim_{n\rightarrow\infty} \overline{P}\,(|u(X^{x_n}(s)) - u(X^x(s))| > \varepsilon) \\ &\qquad \leq \overline{P}\,(X^x(s) \in B(0,\rho)^c) + \overline{P}\,(X^x(s) \in D_\delta^c \cap B(0,\rho))\,. \end{aligned}$$

Let $\delta \rightarrow 0$ first and then $\rho \rightarrow \infty$, so we get (18), which proves the lemma. □

By means of usual techniques (e.g. [6] th. 9.3 ch. 4), lemmas 2.1, 2.2 yield

**Proposition 2.3** *For any $u \in \mathcal{U}$, the diffusion process (1) admits an invariant probability measure $\mu_u$.*

## 2.2 Uniqueness

In this section $\mu$ denotes a fixed invariant probability measure associated with system (1), and $X(t)$ is the solution of this system with $\mu$ as initial law (i.e. $X(0)$ has law $\mu$). We also define $Z(t)$ by (11) where $X^x$ is replaced by $X$.

**Lemma 2.4** *Under $\overline{P}$, for any $t > 0$, the law of $X(t)$ has a density $\overline{p}(t,x)$ such that*

$$\overline{p}(t,x) > 0\,, \quad \forall x\,.$$

**Proof** From now on we are working under $\overline{P}$. Consider the system (12) where $d\overline{W}$ is replaced by $v\,dt$ ($v \in L^2(\mathbb{R}^+)$), we get

$$\begin{pmatrix} \dot{x}_1 \\ \dot{x}_2 \end{pmatrix} = \begin{pmatrix} x_2 \\ -\beta\, x_1 \end{pmatrix} + \begin{pmatrix} 0 \\ \sigma \end{pmatrix} v\,, \quad x(0) = x\,. \tag{20}$$

Let $x^{x,v}(t)$ denote the solution of this last equation. We define the set of reachability

$$\mathcal{A}(t,x) \triangleq \left\{x^{x,v}(t)\ ;\ \forall v : \mathbb{R}^+ \to \mathbb{R},\ \dot{v} \in L^2(\mathbb{R}^+)\right\}\,.$$

(20) can be rewritten as $\dot{x} = A\,x + B\,v$ and the matrix $[B|A\,B]$ has full rank. Hence this system is controllable. So

$$\forall t > 0\,, \quad \forall x \in \mathbb{R}^2\,, \quad \mathcal{A}(t,x) = \mathbb{R}^2\,. \tag{21}$$

Using [12], we prove that — under $\overline{P}$ — the law of $X(t)$ is absolutely continuous with respect to Lebesgue measure and that its density $\overline{p}(t,x)$ is strictly positive for any $t > 0$ and $x$. □

**Lemma 2.5** *Let $\mu$ be an invariant measure for $X(t)$ under $P$. Then $\mu$ has a density $p(x)$ with respect to Lebesgue measure, and $p(x) > 0$ for any $x$ a.e. .*

**Proof** For any $\phi \in C_b(\mathbb{R}^2)$

$$\begin{aligned} \langle \mu\,, \phi \rangle &= \overline{E}[\phi(X(t))\,Z(t)]\,, \\ &= \overline{E}\left[\phi(X(t))\,\overline{E}[Z(t)|X(t)]\right]\,, \\ &= \int_{\mathbb{R}^2} \phi(x)\,\overline{E}[Z(t)|X(t) = x]\,\overline{p}(t,x)\,dx\,. \end{aligned}$$

Since $\overline{E}[Z(t)|X(t)] > 0$ $\overline{P}$-a.s. and under $\overline{P}$ the law of $X(t)$ is equivalent to Lebesgue measure, we get $\overline{E}[Z(t)|X(t) = x] > 0$ $\forall x$–a.e. . Using lemma 2.4 and the last inequality, we prove that $\mu$ has a density

$$q(x) \triangleq \overline{E}[Z(t)|X(t) = x]\,\overline{p}(t,x)\,,$$

and that this density is strictly positive for all $x \in \mathbb{R}^2$ a.e. . □

This lemma implies the following result: if there exists two invariant measures, they are equivalent. So there exits at most one extremal invariant measure. We can therefore state

**Proposition 2.6** *For any $u \in \mathcal{U}$, the diffusion process (1) admits a unique invariant measure $\mu_u$.*

# 3 Numerical Approximation

## 3.1 Approximation of the Control Problem

In a first step, the solution $X(t)$ of (1) is approximated by a controlled Markov process in continuous time and discrete (but infinite) state space. In a second step, it is approximated by a controlled Markov process in continuous time and finite state space.

a – *first step*

Let $h_i$ be the finite difference interval to be used to approximate the derivative w.r.t. the spatial direction $i$ $(i = 1,2)$. We define the grid

$$\mathbb{R}_h^2 \triangleq \{x \in \mathbb{R}^2 \,;\, x = (n_1 h_1 + h_1/2\,,\, n_2 h_2 + h_2/2)\,,\; n_1,\, n_2 \in \mathbf{Z}\}\;, \qquad h \triangleq (h_1\,,\, h_2)\;.$$

We will use the finite difference approximation

$$b_i^u(x)\,\frac{\partial \phi(x)}{\partial x_i} \simeq \begin{cases} b_i^u(x)\,\dfrac{\phi(x+e_i\,h_i)-\phi(x)}{h_i}\;, & \text{if } b_i^u(x) > 0, \\ b_i^u(x)\,\dfrac{\phi(x)-\phi(x-e_i\,h_i)}{h_i}\;, & \text{if } b_i^u(x) < 0, \end{cases} \qquad (i=1,2) \quad (22)$$

$$\frac{\sigma^2}{2}\,\frac{\partial^2 \phi(x)}{\partial x_2^2} \simeq \frac{\sigma^2}{2}\,\frac{\phi(x+e_2\,h_2)-2\,\phi(x)+\phi(x-e_2\,h_2)}{h_2^2}\;, \qquad (23)$$

where $e_i$ denotes the unit vector in the $i$th coordinate direction of $\mathbb{R}^2$ (the special choice for the finite difference approximation will be motivated in remark 3.2).

So $\mathcal{L}_u$ is approximated by a matrix $\mathcal{L}_u^h \in \mathbb{R}^{\mathbb{N}} \times \mathbb{R}^{\mathbb{N}}$

$$\mathcal{L}_u\phi(x) \simeq \mathcal{L}_u^h\phi(x) \triangleq \sum_{y\in\mathbb{R}_h^2} \mathcal{L}_u^h(x,y)\,\phi(y)\;, \quad \forall x \in \mathbb{R}_h^2\;.$$

Because of the finite difference approximations (22,23) we use, $\mathcal{L}_u^h$ can be regarded as the infinitesimal generator of a Markov process in continuous time and discrete state space $\mathbb{R}_h^2$ [7,10]. Using classical definition, a Markov chain $\{\xi_k^h\,;\, k \in \mathbb{N}\}$ is associated with $X^h(t)$ [6].

We then have an ergodic stochastic control problem for a Markov process with infinitesimal generator $\mathcal{L}_u^h$. The cost function is

$$J_h(u) \triangleq \lim_{T\to\infty} E\,\frac{1}{T}\int_0^T f^u(X^h(t))\,dt\;. \qquad (24)$$

$u$ is an element of the class $\tilde{\mathcal{U}}_h$ defined by

$$u \in \tilde{\mathcal{U}}_h \quad \Longleftrightarrow \quad u \text{ is an application from } \mathbb{R}_h^2 \text{ to } [\underline{u},\overline{u}].$$

b – *second step*

$X^h(t)$ has a discrete but infinite state space, and for numerical calculations we need to restrict ourselves to a *finite* state space. Let us then consider the following rectangular subset of $\mathbb{R}^2$

$$D \triangleq [-\overline{x}_1\,,\,\overline{x}_1] \times [-\overline{x}_2\,,\,\overline{x}_2]\,, \quad \overline{x}_i > 0 \quad (i = 1,2)\,, \tag{25}$$

from which we define the following new state space

$$\mathbb{R}^2_{h,D} \triangleq \mathbb{R}^2_h \cap D\,, \quad N \triangleq \mathrm{Card}\left(\mathbb{R}^2_{h,D}\right)\,. \tag{26}$$

Now we have to specify the boundary conditions. In practice, $D$ is chosen to be large enough so that the process will rarely reach the border. Hence, the choice of the boundary conditions is not crucial. Nevertheless, they have to insure that all the states communicate. Example of such conditions (usually reflecting conditions) will be given later.

So we obtain $\mathcal{L}^{h,D}_u$, an approximation of $\mathcal{L}^h_u$. $\mathcal{L}^{h,D}_u$ is a $N \times N$ matrix, it can be interpreted as the generator of a controlled Markov process $X^{h,D}(t)$ in continuous time and finite state space; $\{\xi^{h,D}_k\,;\, k \in \mathbb{N}\}$ denotes the corresponding Markov chain.

The cost function is of the form

$$J_{h,D}(u) \triangleq \lim_{T\to\infty} E\frac{1}{T}\int_0^T f^u(X^{h,D}(t))\,dt = \sum_{x\in\mathbb{R}^2_{h,D}} f^u(x)\,\mu^{h,D}_u(x)\,, \tag{27}$$

where $\mu^{h,D}_u$ is the invariant measure[1] of the process $X^{h,D}(t)$ (more details can be found in [1]). This measure is a solution of the following linear system

$$\sum_{y\in\mathbb{R}^2_{h,D}} \mathcal{L}^{h,D}_u(y,x)\,\mu^{h,D}_u(y) = 0\,, \quad \forall x \in \mathbb{R}^2_{h,D}\,, \quad \sum_{y\in\mathbb{R}^2_{h,D}} \mu^{h,D}_u(y) = 1\,. \tag{28}$$

where $u \in \tilde{\mathcal{U}}_{h,D}$, the class of policies which is defined by

$$u \in \tilde{\mathcal{U}}_{h,D} \iff u \text{ is a mapping from } \mathbb{R}^2_{h,D} \text{ into } [\underline{u},\overline{u}].$$

A Hamilton–Jacobi–Bellman equation can be stated for this ergodic control problem

$$\min_{u\in[\underline{u},\overline{u}]}\left(\sum_{y\in\mathbb{R}^2_{h,D}} \mathcal{L}^{h,D}_u(x,y)\,v(y) + f^u(x)\right) = \rho\,, \quad \forall x \in \mathbb{R}^2_{h,D}\,, \tag{29}$$

where $\rho$ is a positive constant and $v : \mathbb{R}^2_{h,D} \to \mathbb{R}$ (i.e. $v \in \mathbb{R}^N$) is defined up to an additive constant. In the first term of this equation, $u$ has to be considered as an element of $[\underline{u},\overline{u}]$ ($f^u(x) = f(u,x)$). Equation (29) has been studied in [1], for

[1]in the discrete case we also use the notation $\sum_x f^u(x)\,\mu^{h,D}_u(x) = \langle f^u\,,\,\mu^{h,D}_u\rangle$.

fixed $h$ and $D$. In particular, the existence and uniqueness property for a solution $(v, \rho) \in \mathbb{R}^N \times \mathbb{R}^+$ is established.

Equation (29) appears as an approximation for the Hamilton–Jacobi–Bellman equation (6), and leads to the solution of the ergodic control problem associated with the Markov process $X^{h,D}(t)$ in continuous time, finite state space, and infinitesimal generator $\mathcal{L}_u^{h,D}$.

## 3.2 The Policy Iteration Algorithm

In order to solve (29), we use the policy iteration algorithm [5,8]: suppose that $u^0 \in \mathcal{U}_{h,D}$ — the initial policy — is given. Starting with $u^0$ we generate a sequence $\{u^j; j \geq 1\}$. The iteration $u^j \to u^{j+1}$ proceeds in two steps

*compute* $(v^j, \rho^j)$ — we compute $(v^j, \rho^j) \in \mathbb{R}^N \times \mathbb{R}^+$ the solution of the linear system

$$\sum_{y \in \mathbb{R}^2_{h,D}} \mathcal{L}_{u^j}^{h,D}(x,y)\, v(y) + f^{u^j}(x) = \rho\,, \quad \forall x \in \mathbb{R}^2_{h,D}\,.$$

*compute* $u^{j+1}$ — we solve the $N$ following optimization problems: for any $x \in \mathbb{R}^2_{h,D}$

$$u^{j+1}(x) \in \operatorname{Arg} \min_{u \in [\underline{u}, \overline{u}]} \left( \sum_{y \in \mathbb{R}^2_{h,D}} \mathcal{L}_u^{h,D}(x,y)\, v^j(y) + f^u(x) \right).$$

The convergence of this algorithm is stated in [1] (where numerical results can be also found).

**Remark 3.1** The first step of this algorithm leads to a linear system of dimension $N$. Let $\mathbb{R}^2_{h,D} = \{x^i\,;\, i = 1, \ldots, N\}$, then the unknown parameters are

$$v(x^2)\,,\ v(x^3)\,, \ldots,\ v(x^N)\,,\ \rho\,,$$

and we take $v(x^1) = 0$.

**Remark 3.2** For the second step, the optimization problems are nonlinear and they are solved by means of iterative routines. The nonlinearity comes from the discretization technique we use. Indeed, the choice of finite difference approximation (22) depends on $u$. Instead of (22), we can use central difference approximation (so that it does not depend on $u$), in which case the second step becomes explicit because the functions to be optimized are now quadratic in $u$. On the other hand, with this kind of difference approximation, a certain condition on the parameter $h$ has to be fulfilled ($h$ must be small enough) for the matrix $\mathcal{L}_u^{h,D}$ to be the generator of a Markov process. See [10] p.175–179 for further considerations.

# 4 Convergence of the Cost Functions

We suppose that $h_1 = h_2$ and we denote it by $h$. In this section, we prove the convergence result

$$J_{h,D}(u) \to J(u) , \quad \forall u \in \mathcal{U}$$

as the discretization parameter $h$ tends to 0 and the set $D$ tends to $\mathbb{R}^2$. Let us fix $u \in \mathcal{U}$. All the results of this section — up to corollary 4.7 — are adapted from [7].

**a** – ***a sequence of discretization sets***

We consider two strictly increasing sequences $\{\overline{x}_1^h; h > 0\}$ and $\{\overline{x}_2^h; h > 0\}$, such that $\overline{x}_i^h > 0$ and $\overline{x}_i^h \to \infty$ as $h \to 0$. We define

$$D_h = [-\overline{x}_1^h, \overline{x}_1^h] \times [-\overline{x}_2^h, \overline{x}_2^h] .$$

We suppose that

$$\lim_{h \to 0} h\, \delta_h = 0 , \quad \text{where} \quad \delta_h \triangleq \text{radius}(D_h) . \tag{30}$$

Let $\Gamma_h$ be the boundary of $D_h$, we define $\overline{\Gamma}_h$, the discretization of $\Gamma_h$ as follows

$$x \in \overline{\Gamma}_h \iff \begin{cases} x \in \Gamma_h \cap \mathbb{R}_h^2 , \\ \text{or} \\ x \in D_h \cap \mathbb{R}_h^2 , \text{ and } \exists y \in V_h(x) \text{ such that } y \notin D_h \cap \mathbb{R}_h^2 , \end{cases}$$

where $V_h(x) \triangleq \{y \in \mathbb{R}_h^2 ; |x - y| \le \varepsilon \sqrt{2}\}$ is the set of points adjacent to $x$. We define

$$\overline{D}_h \triangleq D_h \cap \mathbb{R}_h^2 .$$

$\overline{D}_h$ is the state space for the discretized problem and $\overline{\Gamma}_h$ is the set of boundary points. We use the same set-up as in section 3 and we define

$$\begin{aligned} \overline{\mathcal{L}}_u^h &\triangleq \mathcal{L}^{h,D_h} , \\ \overline{X}^h(t) &\triangleq X^{h,D_h}(t) , \quad \text{with initial law } \nu^h , \\ \overline{X}^{x,h}(t) &\triangleq X^{h,D_h}(t) , \quad \text{with initial condition } \overline{X}^{x,h}(0) = x , \\ \overline{\xi}_k^h &\triangleq \xi_k^{h,D_h} . \end{aligned}$$

**b** – *the process* $\overline{X}^h(t)$

We can describe the process $\overline{X}^h(t)$ in the following way. We introduce

- a sequence $\{\Delta t_n^h ; n \ge 0\}$, where $\Delta t_n^h$ represents the elapsed time between the $n$–th and the $(n+1)$–th jump.

- a Markov chain $\{\overline{\xi}^h_n\,;\, n \geq 0\}$ with values in $\overline{D}^h$, where $\overline{\xi}^h_n$ represents the state of the process between the $n$–th and the $(n+1)$–th jump.

We consider

$$\begin{aligned}\lambda_h(x) &\triangleq -\overline{\mathcal{L}}^h_u(x,x) \geq 0\,, \qquad \forall x \in \overline{D}^h\,,\\ \pi^h(x,y) &\triangleq (\lambda_h(x))^{-1}\overline{\mathcal{L}}^h_u(x,y)\,,\ \forall x,y \in \overline{D}^h\,,\ x \neq y\,,\end{aligned}$$

(with $0/0 = 0$). We have the following properties

- the pair $(\Delta t^h_{n+1}, \overline{\xi}^h_{n+1})$ depends only on $\overline{\xi}^h_n$,
- under the conditional law $P(\,\cdot\,|\overline{\xi}^h_n = x)$, the random variables $\Delta t^h_{n+1}$ and $\overline{\xi}^h_{n+1}$ are independent.
- under the conditional law $P(\,\cdot\,|\overline{\xi}^h_n = x)$, $\Delta t^h_{n+1}$ is exponentially distributed with parameter $(\lambda_h(x))^{-1}$ ,
- $\pi^h(x,y)$ is the transition probability of the chain $\{\overline{\xi}^h_n\,;\, n \geq 0\}$,

Now we give a representation for the process $\overline{X}^h(t)$. For this purpose we must specify the boundary conditions. These are of Neumann type (reflected) in order to simplify the proof of the convergence result (lemma 4.8).

We define

$$a \triangleq \begin{pmatrix} 0 \\ \sigma \end{pmatrix} (0\ \sigma) = \begin{pmatrix} 0 & 0 \\ 0 & \sigma^2 \end{pmatrix}\,, \quad a^h(x) \triangleq a + h \begin{pmatrix} |b^u_1(x)| & 0 \\ 0 & |b^u_2(x)| \end{pmatrix}\,,$$

and the stopping time

$$\tau^h \triangleq \inf\left\{t \geq 0\,;\, \overline{X}^h(t) \in \Gamma^h\right\}\,.$$

It is easely seen that for any $x \in \overline{D}^h \setminus \Gamma^h$

$$\begin{aligned}\sum_{y\in\overline{D}^h} \overline{\mathcal{L}}^h_u(x,y)\,(y-x) &= b^u(x)\,,\\ \sum_{y\in\overline{D}^h} \overline{\mathcal{L}}^h_u(x,y)\,(y-x)\otimes(y-x) &= a^h(x)\,,\end{aligned}$$

which yield the

**Proposition 4.1** *The process $\{\overline{X}^h(t)\,;\, t \geq 0\}$ admits the following representation*[2]

$$\begin{aligned}\overline{X}^h(t) &= \overline{X}^h(0) + \int_0^t \begin{pmatrix} b^u_1(X(s))\,\mathbb{I}_{\{|X_1(s)|<\overline{x}^h_1\}} \\ b^u_2(X(s)) \end{pmatrix} ds + M^h(t) \qquad (31)\\ &\quad + \begin{pmatrix} h \\ 0 \end{pmatrix} N^{0-}(t) - \begin{pmatrix} h \\ 0 \end{pmatrix} N^{0+}(t) + \begin{pmatrix} 0 \\ h \end{pmatrix} N^{-}(t) - \begin{pmatrix} 0 \\ h \end{pmatrix} N^{+}(t)\,,\end{aligned}$$

[2] $\mathbb{I}_A$ denotes the indicator function of the event $A$.

*with*

$$N^{0-}(t) \triangleq \sum_{s\leq t} \mathbb{I}_{\{X_1(s^-)=-\bar{x}_1^h\}} \mathbb{I}_{\{X_2(s^-)=h/2\}} ,$$

$$N^{0+}(t) \triangleq \sum_{s\leq t} \mathbb{I}_{\{X_1(s^-)=\bar{x}_1^h\}} \mathbb{I}_{\{X_2(s^-)=-h/2\}} ,$$

$$N^{-}(t) \triangleq \sum_{s\leq t} \mathbb{I}_{\{X_2(s^-)=-\bar{x}_2^h\}} ,$$

$$N^{+}(t) \triangleq \sum_{s\leq t} \mathbb{I}_{\{X_2(s^-)=\bar{x}_2^h\}} ,$$

*and $M^h(t)$ is a square–integrable martingale, with increasing process*

$$\langle M^h , M^h \rangle_t = \int_0^t a^h(\overline{X}^h(s))\, ds .$$

c – *a priori estimations*

**Proposition 4.2**

$$E\left(\sup_{s\leq t} |M^h(s)|\right)^2 \leq 4\, K(h)\, t . \tag{32}$$

**Proof** Since $|b^u(x)| \leq C\,(1+|x|)$, we deduce that for any $x$ in $\overline{D}^h \setminus \Gamma^h$

$$|a^h(x)| = \mathrm{trace}(a^h(x)) \leq K(h) , \quad \text{with } K(h) \triangleq \sigma^2 + h\,\sqrt{2}\,C\,(1+\delta_h) \underset{h\to 0}{\longrightarrow} \sigma^2 .$$

And proposition 4.1 yields to

$$\mathrm{trace}\langle M^h , M^h \rangle_t = \int_0^t \mathrm{trace}(a^h(\overline{X}^h(s))\, ds \leq K(h)\, t .$$

From which (32) follows, using the Burkholder–Gungy inequality. □

**Remark 4.3** The jumps of $M^h$ and $\overline{X}^h$ coincide, and those of $\overline{X}^h$ are of amplitude less than $h$, so

$$\sup_{t\geq 0} |\Delta M_i^h(t)| \leq h , \quad i = 1, 2 .$$

**Proposition 4.4**

$$E\left(\sup_{s\leq t} |\overline{X}^{x,h}(s)|\right)^2 \leq 3\, \left(|x|^2 + (C\,t)^2 + 4\, K(h)\, t\right)\, \exp(2\, C\, t) . \tag{33}$$

**Proof** From (31)

$$|\overline{X}^h(t)| \leq \psi_h(t) + C\int_0^t |\overline{X}^h(s)|\,ds\,, \quad \text{where } \psi_h(t) \triangleq |\overline{X}^h(0)| + C\,t + |M^h(t)|\,.$$

Using Gronwall's lemma, we get

$$|\overline{X}^h(t)| \leq \psi_h(t) + C\int_0^t \psi_h(s)\,\exp(C\,(t-s))\,ds \leq \left(\sup_{s\leq t}\psi_h(s)\right)\,\exp(C\,t)\,.$$

Hence

$$\sup_{s\leq t}|\overline{X}^h(s)| \leq \left(|\overline{X}^h(0)| + C\,t + \sup_{s\leq t}|M^h(s)|\right)\exp(C\,t)\,.$$

And (32) leads to (33). □

**Remark 4.5** It follows from (33) that the sequence $\{\tau^h\,;\,h>0\}$ of random times tends to infinity in probability as $h\to 0$.

d – *convergence of invariant measures*

On the space $D^2 = \{\xi : \mathbb{R}^+ \to \mathbb{R}^2\,;$ right continuous and with a left limit$\}$ with the Borel $\sigma$-field $\mathcal{B}(D^2)$, we define the following probability laws

$$\begin{aligned}
\mathbb{P}^h_{\nu^h} &= \text{law of process } \overline{X}^h \text{ with } \overline{X}^h(0) \sim \nu^h\,,\\
\mathbb{P}_{\nu} &= \text{law of process } X \text{ with } X(0) \sim \nu\,,\\
\mathbb{P}^h_{x} &= \text{law of process } \overline{X}^h \text{ starting from } x\,,\\
\mathbb{P}_{x} &= \text{law of process } X \text{ starting from } x\,.
\end{aligned}$$

**Proposition 4.6** *Suppose that* $\nu^h \underset{h\to 0}{\Longrightarrow} \nu$, *then*

$$\mathbb{P}^h_{\nu^h} \underset{h\to 0}{\Longrightarrow} \mathbb{P}_{\nu} \quad \textit{on } (D^2, \mathcal{B}(D^2))\,.$$

**Proof** Fix $x$ in $\mathbb{R}^2$ and $\{x_h\,;\,h>0\}$ a sequence in $\mathbb{R}^2$ such that

$$\forall h>0\,,\quad x_h \in \overline{D}^h \quad \text{and } |x_h - x| \leq \frac{h}{2}\sqrt{2}\,,$$

(in particular $x_h \to x$ as $h\to 0$). Using the representation (31) in terms of semi-martingale for the process $\overline{X}^h$, the a priori estimation (33), the remark 4.3, and results from [14] (th. 5.8, ch. 2) we prove that

$$\mathbb{P}^h_{x_h} \underset{h\to 0}{\Longrightarrow} \mathbb{P}_x \quad \text{on } (D^2, \mathcal{B}(D^2))\,.$$

Furthermore, the convergence is uniform with respect to $x$ on any compact subset of $\mathbb{R}^2$, the proof of the proposition follows. □

**Corollary 4.7** *Let $\overline{\mu}_u^h \triangleq \mu_u^{h,D_h}$ be the invariant measure of the process $\overline{X}^h$ ($\forall h > 0$). If $\mu_u$ is a weak limit of some subsequence of $\{\overline{\mu}_u^h\}$, then $\mu_u$ is an invariant measure of the process $X$.*

We consider $\mu_u$ the invariant measure of $X$,

**Lemma 4.8**

$$\overline{\mu}_u^h \underset{h\to 0}{\Longrightarrow} \mu_u\,, \qquad \forall u \in \mathcal{U}\,.$$

**Proof** In view of corollary 4.7, it is enough to prove that

$$\text{the sequence } \{\overline{\mu}_u^h;\, h > 0\} \text{ is tight,} \tag{34}$$

and a sufficient condition for (34) is that there exists a constant $C$ independant of both $t$ and $h$, such that

$$E\left|X^{h,D_h}(t)\right|^2 \leq C\,. \tag{35}$$

For notational convenience, we denote $X = X^{h,D_h}$, $X_i = X_i^{h,D_h}$, $b_i^u(\cdot) = b_i(\cdot)$. Starting from the representation (31), the proof is identical to that of lemma 2.1. We are concerned with the behavior of the function

$$V(t) \triangleq E\left(\beta\, X_1^2(t) + \varepsilon\, X_1(t)\, X_2(t) + X_2^2(t)\right)\,.$$

Since $X$ is a pure jump process

$$\begin{aligned} V(t) - V(0) &= E\sum_{s\leq t}\big[\beta\,(X_1(s^-) + \Delta X_1(s))^2 - \beta\, X_1^2(s^-) \\ &\qquad +\varepsilon\,(X_1(s^-) + \Delta X_1(s))\,(X_2(s^-) + \Delta X_2(s)) \\ &\qquad -\varepsilon\, X_1(s^-)\, X_2(s^-) \\ &\qquad +(X_2(s^-) + \Delta X_2(s))^2 - X_2^2(s^-)\big] \\ &= E\sum_{s\leq t}\big[2\,\beta\, X_1(s^-)\,\Delta X_1(s) + \beta\,\Delta X_1(s)^2 + \varepsilon\, X_1(s^-)\,\Delta X_2(s) \\ &\qquad +\varepsilon\, X_2(s^-)\,\Delta X_1(s) + 2\, X_2(s^-)\,\Delta X_2(s) + \Delta X_2(s)^2\big]\,. \end{aligned}$$

This last equation and the representation (31), give

$$\begin{aligned} V(t) - V(0) &= 2\,\beta\, E\int_0^t X_1(s)\, X_2(s)\, \mathbb{I}_{\{|X_1(s)| < \overline{x}_1^h\}}\, ds \\ &-2\,\beta\, h\,\overline{x}_1^h \sum_{s\leq t}\Big[P(\{X_2(s^-) = -\frac{h}{2}\} \cap \{X_1(s^-) = -\overline{x}_1^h\}) \\ &\qquad\qquad + P(\{X_2(s^-) = \frac{h}{2}\} \cap \{X_1(s^-) = \overline{x}_1^h\})\Big] \\ &+\beta\, h^2 \sum_{s\leq t}\Big[P(\{X_2(s^-) = -\frac{h}{2}\} \cap \{X_1(s^-) = -\overline{x}_1^h\}) \\ &\qquad\qquad + P(\{X_2(s^-) = \frac{h}{2}\} \cap \{X_1(s^-) = \overline{x}_1^h\})\Big] \end{aligned}$$

$$
\begin{aligned}
&+\varepsilon \frac{h^2}{2} \sum_{s\le t} \Big[ P(\{X_2(s^-) = -\frac{h}{2}\} \cap \{X_1(s^-) = -\overline{x}_1^h\}) \\
&\qquad\qquad + P(\{X_2(s^-) = \frac{h}{2}\} \cap \{X_1(s^-) = \overline{x}_1^h\}) \Big] \\
&+\beta\, h\, E \int_0^t |b_1(X(s))|\, ds + \varepsilon\, E \int_0^t X_1(s)\, b_2(X(s))\, ds + \varepsilon\, E \int_0^t X_2(s)\, b_1(X(s))\, ds \\
&+\varepsilon\, h\, E \int_0^t X_1(s^-)\, dN^-(s) - \varepsilon\, h\, E \int_0^t X_1(s^-)\, dN^+(s) \\
&+2\, E \int_0^t X_2(s)\, b_2(X(s))\, ds + h\, E \int_0^t |b_2(X(s))|\, ds + \sigma^2 t \\
&+2\, h\, E \sum_{s\le t} X_2(s^-) \left[ \mathbb{I}_{\{X_2(s^-) = -\overline{x}_2^h\}} - \mathbb{I}_{\{X_2(s^-) = \overline{x}_2^h\}} \right] \\
&+h^2\, E \sum_{s\le t} \left[ \mathbb{I}_{\{X_2(s^-) = -\overline{x}_2^h\}} - \mathbb{I}_{\{X_2(s^-) = \overline{x}_2^h\}} \right] .
\end{aligned}
$$

the sum of terms 2, 3 and 4 is negative for $h$ and $\varepsilon$ small enough. The sum of the last two terms is negative as soon as $h < 2\,\overline{x}_2^h$.

We chose a law for $X(0)$ which is symmetrical with respect to 0, and the two axes. Then we get, using the symmetrical/dissymmetrical nature of the problem

$$
E \int_0^t X_1(s^-)\, dN^-(s) \le 0 , \qquad E \int_0^t X_1(s^-)\, dN^+(s) \ge 0 .
$$

So

$$
\begin{aligned}
V(t) - V(0) \;\le\; & 2\beta\, E \int_0^t X_1(s)\, X_2(s)\, \mathbb{I}_{\{|X_1(s)| < \overline{x}_1^h\}}\, ds + \varepsilon\, E \int_0^t X_1(s)\, b_2(X(s))\, ds \\
& + \varepsilon\, E \int_0^t X_2(s)\, b_1(X(s))\, ds + 2\, E \int_0^t X_2(s)\, b_2(X(s))\, ds + \sigma^2 t \\
& + h\, \beta\, E \int_0^t |b_1(X(s))|\, ds + h\, E \int_0^t |b_2(X(s))|\, ds ,
\end{aligned}
$$

which is the same expression as in lemma 2.1, except for the last two terms. But these terms are of linear growth with respect to $X_1$, $X_2$, with a multiplicative coefficient which tends to 0 as $h \to 0$. The first term is also different, but

$$
\begin{aligned}
2\beta\, E \int_0^t X_1(s)\, X_2(s)\, \mathbb{I}_{\{|X_1(s)| < \overline{x}_1^h\}}\, ds \;=\; & 2\beta\, E \int_0^t X_1(s)\, X_2(s)\, ds \\
& - 2\beta\, E \int_0^t X_1(s)\, X_2(s)\, \mathbb{I}_{\{|X_1(s)| = \overline{x}_1^h\}}\, ds ,
\end{aligned}
$$

and this last term is negative. This shows (35). □

Since $\mu_u$ has a density, the tools used for the proof of lemma 2.2 lead to

$$
\langle f^u , \overline{\mu}_u^h \rangle \underset{h\to 0}{\longrightarrow} \langle f^u , \mu_u \rangle ,
$$

which proves the

**Theorem 4.1**

$$
J_{h,D_h}(u) \underset{h\to 0}{\longrightarrow} J(u) , \quad \forall u \in \mathcal{U} .
$$

**Remark 4.9** In [1], we proved the existence of an optimal feedback control law for the discretized problem. With such a control, we can associate a feedback control law $\hat{u}_h$ for the continuous state space problem, where $\hat{u}_h$ is piecewise constant. Using theorem 4.1 we can easely conclude that

$$\limsup_{h\to 0} J_{h,D_h}(\hat{u}_h) \leq \inf_{u\in\mathcal{U}} J(u) .$$

We would like to prove the stronger result that the sequence $\{\hat{u}_h; h > 0\}$ is a minimizing sequence for the functional $J$, i.e.

$$J(\hat{u}_h) \to \inf_{u\in\mathcal{U}} J(u) , \quad \text{quand } h \to 0 .$$

# References

[1] S. BELLIZZI, R. BOUC, F. CAMPILLO, and E. PARDOUX. Contrôle optimal semi-actif de suspension de véhicule. In *Analysis and Optimization of Systems, A. Bensoussan and J.L. Lions (eds.)*, INRIA, Antibes, 1988. Lecture Notes in Control and Information Sciences 111, 1988.

[2] A. BENSOUSSAN. *Perturbation Methods in Optimal Control.* John Wiley & Sons, New-York, 1988.

[3] V.S. BORKAR and M.K. GHOSH. Ergodic control of multidimensional diffusions I: the existence results. *SIAM Journal of Control and Optimization*, **26**(1):112–126, January 1988.

[4] F. DELEBECQUE and J.P. QUADRAT. Contribution of stochastic control singular perturbation averaging and team theories to an example of large-scale systems: management of hydropower production. *IEEE Transactions on Automatic Control*, **AC–23**(2):209–221, April 1978.

[5] B.T. DOSHI. Continuous time control of Markov processes on an arbitrary state space: average return criterion. *Stochastic Processes and their Applications*, **4**:55–77, 1976.

[6] N. ETHIER and T.G. KURTZ. *Markov Processes - Characterization and Convergence.* J. Wiley & Sons, New-York, 1986.

[7] F. LE GLAND. *Estimation de Paramètres dans les Processus Stochastiques, en Observation Incomplète — Applications à un Problème de Radio-Astronomie.* Thèse de Docteur-Ingénieur, Université de Paris IX - Dauphine, 1981.

[8] R.A. HOWARD. *Dynamic Programming and Markov Processes.* J. Wiley, New-York, 1960.

[9] H.J. KUSHNER. Optimality conditions for the average cost per unit time problem with a diffusion model. *SIAM Journal of Control and Optimization*, **16**(2):330–346, March 1978.

[10] H.J. KUSHNER. *Probability Methods for Approximations in Stochastic Control and for Elliptic Equations.* Volume 129 of *Mathematics in Science and Engineering*, Academic Press, New-York, 1977.

[11] A. LEIZAROWITZ. Controlled diffusion processes on infinite horizon with the overtaking criterion. *Applied Mathematics and Optimization*, **17**:61–78, 1988.

[12] D. MICHEL and E. PARDOUX. An introduction to Malliavin's calculus and some of its applications. (to appear).

[13] J.P. QUADRAT. *Sur l'Identification et le Contrôle de Systèmes Dynamiques Stochastiques.* Thèse, Université de Paris IX - Dauphine, 1981.

[14] R. REBOLLEDO. *La méthode des martingales appliquée à l'étude de la convergence en loi de processus. Mémoire 62*, Bulletin SMF, 1979.

[15] M. ROBIN. Long-term average cost control problems for continuous time Markov processes: a survey. *Acta Applicandae Mathematicae*, **1**:281–299, 1983.

[16] R.H. STOCKBRIDGE. *Time-average control of martingale problems.* PhD thesis, University of Wisconsin-Madison, 1987.

# SUPPORT THEOREMS IN NON-LINEAR FILTERING

*M. CHALEYAT-MAUREL*

*Laboratoire de Probabilités - Université Paris VI*

*Tour 56, 4, Place Jussieu - 75252 PARIS CEDEX 05*

*D. MICHEL*

*Laboratoire de Statistiques et Probabilités*

*Université Paul Sabatier - 118 Route de Narbonne*

*31062 TOULOUSE CEDEX*

In this paper, we present two support theorems in filtering theory: the first one for the unnormalized filter and the other one for the density of the filter.

Support theorems for stochastic processes aim to describe the support of the law of the process - on its natural state space endowed by a suitable topology - in terms of the set of solutions of a deterministic controlled system.

Such studies were initiated by the celebrated theorem of D. Stroock and S. Varadhan [8] : they proved that the support of the law of a diffusion process, solution of the stochastic differential equation (written in Stratonovitch form) :

$$\begin{cases} dx_t = X_o(x_t)dt + X_i(x_t) \circ dw_t^i \\ x_o = x \end{cases}$$

coincides with the closure (for the natural Banach topology on $\mathcal{C}([0,1],\mathbb{R}^n)$) of the set of solutions of the following controlled system :

$$\begin{cases} \dfrac{dx_t^u}{dt} = X_o(x_t^u) + X_i(x_t^u)\, \dot{u}_t^i \\ x_o^u = x \end{cases}$$

where $u$ varies over $H^1([0,1],\mathbb{R}^n)$

In [2] and [3], we followed this route in the setting of filtering theory. A result similar to the Stroock-Varadhan theorem holds for the unnormalized filter in the correlated filtering theory of diffusions. More precisely, if $\varphi$ is a $C^{\infty}$-function from $\mathbb{R}^n$ to $\mathbb{R}$ with suitable growth conditions, the unnormalized filter $\rho_t\varphi$, taken as a function of the initial point $x$, is viewed as a stochastic process taking values in $C^{\infty}(\mathbb{R}^n,\mathbb{R})$. We describe the support of the law of $\rho_.\varphi$ on $\mathscr{C}([0,1],C^{\infty}(\mathbb{R}^n,\mathbb{R}))$ endowed with its natural Fréchet topology, as the closure of a set of solutions of a controlled partial differential system (in the weak sense).

In [4] we considered the uncorrelated case. Under hypoellipticity and boundedness hypotheses, the measures $\rho_t$ have densities $p_t$ w.r.t. the Lebesgue measure, and these densities belong to the Schwartz space $\mathscr{S}$ (the space of rapidly decaying functions). A support theorem then holds for $p_.$ in $\mathscr{S}$. This theorem is analogous to the previous one for $\rho_.\varphi$ but the proof is more direct as we work in a "robust situation".

The paper is organized as follows. In a first paragraph, we state the filtering problem. The second paragraph deals with the unnormalized filter and the third one is devoted to the density of the filter.

## 1. Statement of the problem.

Let $(\Omega,\mathscr{F},P)$ be a usual probability space and $(w,\tilde{w})$ two independent Wiener processes taking their values in $\mathbb{R}^n$ and $\mathbb{R}^p$ respectively.

Let $X_o,X_1,\dots,X_m,\tilde{X}_1,\dots,\tilde{X}_p$ be $m+p+1$ vector fields on $\mathbb{R}^n$, $h_1,\dots,h_p$ $p$ functions on $\mathbb{R}^n$. We denote by $h$ the vector $(h_1,\dots,h_p)$.

Denote by $d$ the Itô differential and by $\circ d$ the Stratonovitch differential.

The couple signal/observation $(x,y)$ with values in $\mathbb{R}^n\times\mathbb{R}^p$ is the solution of the system :

$$\begin{cases} dx_t = X_o(x_t)dt + X_i(x_t)\circ dw_t^i + \tilde{X}_i(x_t)\circ(d\tilde{w}_t^i + h_i(x_t)dt), \\ dy_t = h(x_t)dt + d\tilde{w}_t, \end{cases}$$

and $x_o$ has probability law $\pi_o$, $y_o = 0$.

We suppose that all the coefficients of this system are smooth with bounded derivatives of all orders and that $x_o$ is independent of $(w,\tilde{w})$.

For $\varphi$ on $\mathbb{R}^n$ such that $\forall t$, $E[\varphi^2(x_t)] < +\infty$, we define the unnormalized filter :

$$\rho_t\varphi = E_o^W[\varphi(x_t)L_t] ,$$

where

$$L_t = \exp\left\{\int_0^t h_i(x_s)dy_s^i - \frac{1}{2}\int_0^t \left(\sum_{i=1}^p h_i^2(x_s)\right) ds\right\} ,$$

$\mathcal{F}_t$ is the $\sigma$-field generated by $(w_s, \tilde{w}_s, 0 \le s \le t)$,

$P_o$ is the probability measure defined by:

$$\frac{dP_o}{dP}\Big|_{\mathcal{F}_t} = L_t^{-1} ,$$

and $E_o^W$ denotes the integration w.r.t to the law of $w$ under $P_o$, i.e. the Wiener measure.

It is well-known that $\rho_t\varphi$ solves the Zakaï equation :

$$\begin{cases} d\rho_t\varphi = \rho_t(L_o\varphi)dt + \rho_i(L_i\varphi) \circ dy_t^i \\ \\ \rho_o\varphi = \int\varphi(x)d\pi_o(x) \end{cases}$$

where $L_o = \frac{1}{2}\sum_{i=1}^m X_i^2 + X_o - \frac{1}{2}\tilde{X}_i h^i - \frac{1}{2}\sum_{i=1}^p h_i^2$ ,

$L_i = \tilde{X}_i + h_i$ .

## 2. A support theorem for the unnormalized filter.

In this section, we take $\pi_o = \delta_x$. The unnormalized filter $\rho_t\varphi$ is then a function of the starting point $x$.

It is proved in [3] that, provided $\varphi$ is in a suitable space (roughly speaking, smooth with exponential growth), the mapping $x \to \rho_.\varphi$ belongs to $E = \mathcal{C}([0,1], C^\infty(\mathbb{R}^n,\mathbb{R}))$. The space $E$ is equipped with its natural Fréchet topology.

Let $P^\varphi$ denote the law of $\rho_.\varphi$ and let $S(P^\varphi)$ denote the topological support of $P^\varphi$ (i.e. the smallest closed subset of $E$ that carries $P^{\varphi}$). The description of $S(P^{\varphi})$ involves the following "controlled Zakaï equation" :

$$\begin{cases} \dfrac{d\rho_t^u\varphi}{dt} = \rho_t^u(L_o\varphi) + \rho_t^u(L_i\varphi)\dot{u}_t^i, & u \in H^1([0,1],\mathbb{R}^p) \\ \rho_o^u\varphi = \varphi \end{cases}$$

*Theorem [3] : We have*

$$S(P^{\varphi}) = \overline{\{\rho_{.}^u\varphi,\ u \in H^1\}}^E$$

*where, for* $A \subset E$, $\overline{A}^E$ *denotes the closure of* $A$ *in the Fréchet topology of* $E$.

The proof is very technical (see [3] for details). It proceeds in two steps :

(i) $S(P^{\varphi}) \subset \overline{\{\rho_{.}^u\varphi,\ u \in H^1\}}^E = \overline{\mathcal{G}^{\varphi}}^E$ .

As in the diffusion case, this inclusion is proved by an approximation theorem.

We consider the usual polygonal approximations of the Brownian motion $y$, which we denote by $y^{\varepsilon}$, and we introduce the corresponding path by path Zakaï equation : for every $\varepsilon > 0$, $\rho_t^{\varepsilon}\varphi$ is defined as the solution of :

$$\begin{cases} \dfrac{d\rho_t^{\varepsilon}\varphi}{dt} = \rho_t^{\varepsilon}(L_o\varphi) + \rho_t^{\varepsilon}(L_i\varphi)\dot{y}_t^{i,\varepsilon} \\ \rho_o^{\varepsilon}\varphi = \varphi \end{cases}$$

The proof of the following proposition depends on a technique developed by J.M. Moulinier [7].

*Proposition [3] : Set* $\rho_{.}^{o}\varphi = \rho_{.}\varphi$ *; the mapping* $\varepsilon \to \rho_{.}^{\varepsilon}\varphi$ *is then a.s. continuous from* $[0,1]$ *into* $E$. *Moreover, as* $\varepsilon$ *tends to zero,* $\rho_{.}^{\varepsilon}\varphi$ *converges towards* $\rho_{.}\varphi$ *in the* $L^p$*-norm, for any* $p < +\infty$.

This result gives the a.s. convergence of the whole family ($\rho_{.}^{\varepsilon}\varphi$ , $\varepsilon>0$) and not only of a subsequence as in Wong-Zakaï type theorem. It leads to the desired inclusion since it shows that the laws $P^{\varepsilon,\varphi}$ of $\rho_{.}^{\varepsilon}\varphi$ are weakly convergent towards $P^{\varphi}$, which implies :

$$1 = \overline{\lim_{\varepsilon\to 0}} P^{\varepsilon,\varphi}(\mathcal{S}^{\varphi}) \le P^{\varphi}(\mathcal{S}^{\varphi})$$

(ii) $\overline{\mathcal{S}^{\varphi}}^{E} \subset S(P^{\varphi})$.

It would be sufficient to prove that, $\forall \varepsilon > 0$, $\forall u \in H^1$, $P_o[\|\rho_.\varphi - \rho^u_.\varphi\| \le \varepsilon] > 0$. Inspired by the diffusion case, we obtain the following stronger statement :

*<u>Proposition [3]</u> : For any* $\eta > 0$, $\varepsilon \in ]0,2[$, *for any bounded subset* $\mathcal{B}$ *of* $H^1$ *there exist two positive constants* $\delta_o$ *and* $C$ *such that :*

$$\forall u \in \mathcal{B},\ \forall \delta \le \delta_o,\quad P_o[\|\rho_.\varphi - \rho^u_.\varphi\|_E > \eta \mid \|y-u\| \le \delta] \le C \exp - \frac{1}{4\delta^{2-\varepsilon}}$$

*where the notation* $\|\cdot\|_E$ *stands for any semi-norm compatible with the topology of* $E$ *.*

Notice that, whenever the mapping $y \to \rho_.\varphi$ is continuous from the Wiener space (equipped with the Banach topology) into $E$, the left-hand side of the inequality of the proposition vanishes for $\delta$ small. However, this mapping is not continuous in the correlated case which is considered here.

## 3. <u>A support theorem for the density.</u>

We now consider the case when $\tilde{X}_i = 0$, $\forall i = 1,\dots,p$ and we do not assume that the initial law $\pi_o$ is a Dirac measure.

The existence of a density for the measure $\varphi \to \rho_t\varphi$ follows from the next result, due to J.M. Bismut - D. Michel [1] and S. Kusuoka - D. Stroock [6].

Let $\mathcal{B}$ be the Lie algebra generated by $(X_1,\dots,X_m)$ . Let $\mathcal{J}$ be the ideal generated by $\mathcal{B}$ in the Lie algebra generated by $(X_o,X_1,\dots,X_m)$, and $\mathcal{J}(x) = \{X(x), X \in \mathcal{J}\}$.

*Theorem [1],[6] : Suppose that* $\mathcal{J}(x) = \mathbb{R}^n$, *for every* $x$ *belonging to the topological support of* $\pi_o$. *Then, a.s., for all* $t>0$, $\rho_t$ *admits a density* $p_t$ *w.r.t. the Lebesgue measure on* $\mathbb{R}^n$. *Moreover* $p_t$ *belongs to* $\mathcal{S}$.

From now on, we make the following asumptions :

$(H_1)$ $\mathcal{J}(x) = \mathbb{R}^n \quad \forall x \in \text{supp}\cdot\pi_o$.

$(H_2)$ $\pi_o$ has a density $p_o$ w.r.t the Lebesgue measure, and $p_o$ is in $\mathcal{S}$.

The density $p_t$ of $\rho_t$, which exists by the previous theorem, solves the Zakaï equation :

$$\begin{cases} dp_t = L_o^* \, p_t \, dt + h_i \, p_t \circ dy_t^i \\ \\ p_o \end{cases}.$$

The controlled p.d.e. associated with this equation is :

$$(1) \qquad \begin{cases} \dfrac{dp_t^u}{dt} = L_o^* \, p_t^u + h_i \, p_t^u \, \dot{u}_t^i & (u \in H^1([0;1],\mathbb{R}^p)) \\ \\ p_o^u = p_o \end{cases}$$

We now recall the robust version of $\rho_t$ introduced by M.H.A.Davis [5].

**Definition** : Let $c \to p_t^c$ be the mapping from $\mathcal{C}([0,1],\mathbb{R}^p)$ into $\mathcal{M}(\mathbb{R}^n)$ (the set of bounded measures on $\mathbb{R}^n$) defined by

$$\forall \varphi \in C_b(\mathbb{R}^n) \qquad \rho_t^c = E_o^W(\varphi(x_t)L_t^c)$$

where $L_t^c = \exp\left[h(x_t)c_t - \int_0^t c_s \, d(h(x_s)) - \frac{1}{2}\int_0^t \left(\sum_{i=1}^p h_i^2(x_s)\right)ds\right]$.

Using Malliavin calculus and Fourier transforms, it is proved in [4] that, for every $c$ in $\mathcal{C}([0,1],\mathbb{R}^p)$ and for every $t>0$, $\rho_t^c$ has a density $p_t^c$ in $\mathcal{S}$. If $c$ belongs to $H^1$, $p_t^c$ coincides with the unique solution in $\mathcal{S}$ of (1) with $u = c$. Moreover, the map $\Phi : c \to p_.^c$ is continuous from $\mathcal{C}([0,1],\mathbb{R}^p)$, endowed with the Banach norm, into $F = \mathcal{C}(]0,1],\mathcal{S})$ equipped with the canonical Fréchet topology. In addition, $p_t = p_t^y$, where $y$ is the observation process.

Let us denote by $P^{p_o}$ the law of $p_.$ on F and by $S(P^{p_o})$ the support of $P^{p_o}$. Let $\mathcal{R}^{p_o}$ be the set of all functions $p^u_.$, $p^u$ solution of (1) for some u belonging to $H^1$.

*Theorem [4] : We have:*

$$S(P^{p_o}) = \overline{\mathcal{R}^{p_o}}^F$$

This result follows from the continuity of the mapping $y \longrightarrow p_.$ , from the Wiener space into F. We observe that, under $P_o$, y is a Brownian motion and so, if Q denotes the Wiener measure on $\mathcal{W}$ , we have:

$$S(P^{p_o}) = S(\text{law } \Phi(y)) = \overline{\Phi(S(Q))}^F = \overline{\Phi(\mathcal{W})}^F = \overline{\Phi(H^1)}^F ,$$

since $S(Q) = \mathcal{W}$ and $H^1$ is dense in $\mathcal{W}$.

**REFERENCES :**


[1] **J.M. Bismut-D. Michel** : *Diffusions conditionnelles.* *J. Funct. Anal. Part I, 44, 174-211 (1981), Part II, 4, 274-292 (1982).*

[2] **M. Chaleyat-Maurel - D. Michel** : *The support of the law of a filter in $C^\infty$-topology.* *Stochastic differential systems, stochastic control theory and applications. W. Fleming-P.L. Lions eds .I.M.A. volumes in mathematics and its applications, vol. 10, Springer (1988).*

[3] **M. Chaleyat-Maurel - D. Michel** : *A Stroock-Varadhan support theorem in non linear filtering theory.* *Submitted to Probab. Th. Rel. Fields.*

[4] **M. Chaleyat-Maurel - D. Michel** : *The support of the density of the filter in the uncorrelated case. Proceedings of the 1988 Trento Conference on Stochastic Partial Differential Equations. G. Da Prato, L. Tubaro, eds. To appear.*

[5] **M.H.A. Davis** : *Pathwise non linear filtering. In : Stochastic systems M. Hazewinkel, ed. Dordrecht Reidel (1981).*

[6] **S. Kusuoka - D. Stroock** : *The partial Malliavin calculus and its application to non linear filtering. Stochastics, vol. 12, p. 83-142 (1984).*

[7] **J.M. Moulinier** : *Théorèmes limites. To appear in Bull. Sc. Math.*

[8] **D. Stroock - S. Varadhan** : *On the support of diffussion processes with applications to the strong maximum principle. Proc.* $6^{th}$ *Berkeley Symp. Math. Stat. Probab. III, p. 333-359 (1972).*

ERGODICITE DES SYSTEMES STOCHASTIQUES
POLYNOMIAUX EN TEMPS DISCRET

- Abdelkader Mokkadem -
Université de Paris-Sud, Bât 425, Mathématiques,
91405 ORSAY - Cédex

Summary : We obtain geometric ergodicity and recurrence properties for the Markov chain $(X_t)$, defined by $X_{t+1} = \varphi(X_t, u_{t+1})$ where $\varphi$ is a rational application and $(u_t)$ is a sequence of independent identically distributed random variables.

## 1. Introduction

On considère un système stochastique d'équation d'état

$$(1) \quad X_{t+1} = \varphi(X_t, u_{t+1}) \quad , \quad t \in \mathbb{N} ,$$

où $X_t$ est à valeurs dans une variété algébrique W lisse (c'est à dire sans singularités), $(u_t)_{t\in\mathbb{N}}$ est une suite de variables aléatoires indépendantes et de même loi de probabilité dans un sous ensemble E d'une variété algébrique lisse V (les variétés algébriques que nous considérons sont des ensembles algébriques réels irréductibles comme les définit Bröcker [3]) ; $\varphi$ est un morphisme régulier de W x V dans V (i.e. ses composantes $\varphi_i$ sont des fractions rationnelles $P_i/Q_i$ où $Q_i$ n'a pas de zéros dans W x V, voir [2]).

De tels systèmes apparaissent souvent comme des représentations (ou réalisations) markoviennes de certains processus aléatoires (ARMA [1], bilinéaire [9], polynômial [11]...). La qualité des statistiques qu'on peut faire sur ces processus, identification, estimation, dépend des propriétés probabilistes de la chaîne de Markov $(X_t)$ définie par l'équation (1) ; ces propriétés ne sont pas toujours simples à dégager. Notons $\varphi_u$ le morphisme régulier $\varphi(.,u)$ de W dans W et soit $\mathcal{S}$ le semi-groupe d'applications engendré par la famille $\{\varphi_u\}_{u\in E}$ . Il est clair qu'en partant d'un état initial x, la chaîne $X_t$ reste dans l'orbite $\mathcal{S}x$ de x et n'en sort jamais ; on est donc amené à étudier la chaîne dans l'adhérence de $\mathcal{S}x$ et on se trouve alors confronté à certaines difficultés dont les deux qui suivent.

1.1. En général l'orbite $\mathcal{S}x$ est contenue dans une réunion dénombrable de variétés algébriques et il n'est pas exclu que l'adhérence de $\mathcal{S}x$ soit contenue dans une variété de dimension strictement plus grande. Une telle situation n'aura pas lieu si x appartient à $\mathcal{S}x$.

1.2. L'adhérence de $\mathcal{S}x$ ou $\mathcal{S}x$ elle même, peut contenir d'autres orbites et donc des ensembles absorbants pour la chaîne $(X_t)$. Cette difficulté sera surmontée en faisant une hypothèse de contraction sur un des morphismes $\varphi_u$ ∎

Enfin on notera que même dans le meilleur des cas, $W = \mathbb{R}^n$, $E = \mathbb{R}^p$, les orbites $\mathcal{S}x$ sont des réunions d'ensembles semi-algébriques qu'on ne peut pas en général plonger dans un espace affine de même dimension; celà signifie qu'on est inévitablement amené à considérer des variétés algébriques et donc on n'alourdit en rien le texte en supposant que $X_t$ et $u_t$ sont dans des variétés algébriques lisses W et V quelconques.

En faisant des hypothèses du même type que celles utilisées dans le cas des processus bilinéaires ([9], [12]) on établit dans la section 3 diverses propriétés probabilistes (ergodicité géométrique, Harris récurrence) pour la chaîne $(X_t)$. La section 2 est consacrée à des résultats préliminaires sur les chaînes de Markov et sur les mesures images.

## 2. Résultats préliminaires

2.1. Soit $(Y_t)$, $t \in \mathbb{N}$, une chaîne de Markov à valeurs dans un sous-espace S d'un espace topologique M séparable et dénombrable à l'infini. Soit $\mu$ une mesure positive régulière sur M telle que $\mu(S) > 0$ et soit $\mu_S$ la restriction de $\mu$ à S. Notons P(y,.) la probabilité de transition de la chaîne $(Y_t)$ ; on s'intéresse aux propriétés suivantes (voir [8], [10]) :

(P1) Existence d'une probabilité $\pi$ sur S telle que

$$(2) \quad \pi(A) = \int \pi(dy)\, P(y,A) \quad \text{pour tout borélien A dans S ;}$$

quand une telle probabilité existe, la chaîne $(Y_t)$ est dite positive récurrente (ou ergodique) et $\pi$ est appelée probabilité invariante de la chaîne.

(P2) La chaîne $(Y_t)$ de probabilité invariante $\pi$ est dite géométriquement ergodique si pour $\pi$-presque tout y on a

$$(3) \qquad \| P^n(y,.) - \pi \| = O(\rho^n)$$

où $0 < \rho < 1$ et $\| \, . \, \|$ désigne la norme en variation.

(P3) La chaîne $(Y_t)$ de probabilité invariante $\pi$ est dite Harris récurrente si pour tout A de $\pi$-mesure non nulle on a :

$$(4) \qquad P\left[ \bigcup_{t \geq 1} \{Y_t \in A\} \mid Y_0 = y \right] = 1 \text{ pour tout } y \text{ dans } S \ ;$$

en d'autres termes, presque sûrement toute trajectoire passe par A.

Quand $(Y_t)$ est Harris récurrente et géométriquement ergodique on a alors

$$(5) \qquad \int \pi(dy) \, \| P^n(y,.) - \pi \| = O(\rho^n) \quad [8] \ .$$

Ces propriétés sont essentielles pour évaluer les vitesses de convergence des décisions statistiques. Nous donnons maintenant un critère pour que la chaîne $(Y_t)$ possède ces propriétés ; on notera $Pg(y) = \int g(x) \, P(y,dx)$.

<u>Théorème 1 [6]</u>.

Si les conditions suivantes sont satisfaites :

(C1) pour tout compact K de M et tout borélien A de S de $\mu$-mesure nulle, il existe un entier $n_0$ tel que $P^{n_0}(y,A) = 0$ pour tout y dans $K \cap S$,

(C2) pour tout compact K de M et tout borélien A de S de $\mu$-mesure non nulle, il existe un entier $n_1$ tel que $\inf_{y \in K \cap S} P^{n_1}(y,A) > 0$ ,

(C3) il existe une fonction g positive sur S (fonction de Lyapounov), un compact K de M et des nombres réels $A > 0$, $B > 0$, $0 < \rho < 1$ tels que

(i) $\mu(K \cap S) > 0$

(ii) $P\,g(y) \leq g(y) - 1$ et $P\,g(y) \leq \rho\, g(y)$ si $y \notin K \cap S$

(iii) $B \leq P\,g(y) \leq A$ si $y \in K \cap S$ ,

alors la chaîne $(Y_t)$ est Harris récurrente et géométriquement ergodique ;

de plus sa probabilité invariante $\pi$ est équivalente à $\mu_S$ et le moment $\int g(y)\,\pi(dy)$ est fini ■

C'est ce théorème que nous appliquerons à la chaîne $X_t$ définie par (1) ; pour vérifier les conditions (C1) et (C2) on aura besoin de propriétés de continuité de la loi de probabilité $P^n(x,.)$ qui n'est rien d'autre que l'image de la loi de $(u_1, u_2,\dots,u_n)$ par un morphisme régulier. Nous présentons donc maintenant un résultat concernant l'image d'une mesure par un morphisme régulier.

2.2. Soient L, M, N des variétés algébriques contenues dans des variétés algébriques lisses L', M', N' ; on suppose que M est lisse. Soit $\psi(z,x)$ une application de classe $C^1$ de L' x M' dans N' telle que $\psi(L\times M) \subset N$ et que $\psi(z,.)$ soit un morphisme régulier.

On munit chaque variété algébrique X d'une mesure régulière $\mu_X$ qui est lebesguienne sur la partie non singulière de X [4] et nulle sur tout sous-ensemble algébrique de X de dimension inférieure ; une telle mesure est construite dans [5].

Soit $\theta$ une mesure positive régulière sur M, absolument continue par rapport à $\mu_M$ et E l'ensemble de positivité de sa densité f (i.e. E = {f > 0}). Pour tout z dans L on note $\theta_z$ la mesure image de $\theta$ par le morphisme $\psi(z,.)$. On a alors le résultat suivant :

<u>Théorème 2 [6].</u>

Soit $z_0$ un point de L tel que $\psi(z_0,.)$ soit un morphisme dominant de M dans N (celà signifie que $\psi(z_0,M)$ contient un ouvert de la partie non singulière de N), alors

(i) $\theta_{z_0}$ est absolument continue par rapport à $\mu_N$ et sa densité a pour ensemble de positivité $\psi(z_0,E)$

(ii) Pour tout borélien A de N on a : $\liminf_{z \longrightarrow z_0} \theta_z(A) \geq \theta_{z_0}(A)$ ;

si de plus $\theta$ est finie alors $\lim_{z \to z_0} \theta_z(A) = \theta_{z_0}(A)$ ∎

### 3. Propriétés de la chaîne $(X_t)$

3.1. On revient maintenant au système (1). On complète la suite $u_t$ en une suite de variables aléatoires indépendantes et de même loi, indexées par $\mathbb{Z}$, $(u_t)_{t\in\mathbb{Z}}$, et on fait les hypothèses suivantes :

($\mathcal{H}1$) La loi de probabilité de la variable aléatoire $u_t$ est absolument continue par rapport a $\mu_V$ ; et on notera E l'ensemble de posivité de sa densité f.

($\mathcal{H}2$) Il existe un point a dans E et un entier k tels que $\varphi_a^k$ soit lipschitzienne de rapport strictement inférieur à un ; on notera T le point fixe de $\varphi_a$.

($\mathcal{H}3$) La condition (C3) du théorème 1 est satisfaite par la chaîne $(X_t)$ ; de plus il existe un nombre réel positif s tel que la suite $U_k = \varphi_{u_t} \circ \varphi_{u_{t-1}} \circ \ldots \circ \varphi_{u_{t-k}}(T)$ soit convergente en moyenne d'ordre s.

L'hypothèse ($\mathcal{H}3$) est impliquée par l'hypothèse plus forte suivante

($\mathcal{H}3'$) Il existe un point $x_0$ dans W et deux nombres réels positifs s et $\rho$, $\rho < 1$, tels que :

$$E \left| \varphi_{u_t}(x_0) \right|^s < \infty \quad \text{et} \quad E \left|\varphi_{u_t}(x) - \varphi_{u_t}(x_0)\right|^s < \rho \, |x|^s \text{ pour tout } x \in W \;;$$

il est en effet facile de voir que la condition (C3) est vérifiée avec $g(x) = |x|^s + 1$ et que la suite $U_k = \varphi_{u_t} \circ \ldots \circ \varphi_{u_{t-k}}(x)$ admet une limite indépendante de x.

L'hypothèse ($\mathcal{H}3$) a une conséquence immédiate. Notons $Z_t$ la limite de $\varphi_{u_t} \circ \ldots \circ \varphi_{u_{t-k}}(T)$ ; le processus $(Z_t)_{t\in\mathbb{Z}}$ est un processus stationnaire en loi et markovien de même probabilité de transition que la chaîne $(X_t)$ ; la loi de probabilité $\pi$ de $Z_t$ est donc une probabilité invariante pour la chaîne $(X_t)$ et cette dernière est par conséquent positive récurrente.

Il est clair que le support de la probabilité $\pi$ est contenu dans l'adhérence de l'orbite $\mathcal{S}T$ de T et en particulier dans la fermeture de Zariski M de $\mathcal{S}T$. Comme nous allons le voir tout de suite, M est une variété algébrique dont la dimension peut être déterminée très simplement.

Notons $\varphi^k(.,.)$ l'application de $W \times V^k$ dans W définie par :

$$(6) \qquad \varphi^k(x, a_1, a_2,..,a_k) = \varphi_{a_k} \circ \varphi_{a_{k-1}} \circ .. \circ \varphi_{a_1}(x)$$

et $M_k$ la fermeture de Zariski de $D_k = \varphi^k(T,E^k)$.

Comme E est de $\mu_V$-mesure non nulle, sa fermeture de Zariski est V ; par conséquent $M_k$ est aussi la fermeture de Zariski de $\varphi^k(T,V^k)$ ; $V^k$ étant irréductible, il en est de même de $M_k$. D'autre part puisque $\varphi_a(T) = T$, il s'ensuit que la suite $M_k$ est une suite croissante de variétés algébriques contenues dans W. Pour des raisons de dimension cette suite est nécessairement constante à partir d'un certain rang $k_0$ ; et on a alors $M = M_{k_0}$. Plus précisément on a le résultat suivant :

<u>Proposition 1</u>. [11]

Si $M_k = M_{k+1}$ alors $M_k = M$ ∎

En particulier, si dim W = m, il découle de la proposition 1 que $M_m = M$ ; en d'autres termes la dimension de M est donnée par le rang du morphisme $\varphi^m(T,.)$ de $V^m$ dans W ∎

Nous allons maintenant montrer que la probabilité $\pi$ est portée par $\mathcal{S}T = UD_k$.

3.2. <u>Support de la probabilité invariante $\pi$.</u>

On sait que $\pi$ est portée par l'adhérence de $\mathcal{S}T$ ; en particulier $\pi(M) = 1$.

<u>Lemme 1</u>. Sous les hypothèses $(\mathcal{H}1)$, $(\mathcal{H}2)$, et $(\mathcal{H}3)$ on a $\pi(\mathcal{S}T) > 0$ ∎

<u>Preuve</u>.

Il est facile de voir que pour tout entier q

$$(7) \qquad \varphi^q(\mathcal{S}T, V^q) \subset M$$

et donc par continuité pour la topologie de Zariski

$$(8) \qquad \varphi^q(M, V^q) \subset M$$

On sait que $\varphi^m(T,.)$ est dominante et donc par continuité $\varphi^m(T',.)$ est aussi dominante pour T' dans un voisinage $M_1$ de T, contenu dans M.

D'autre part il est clair que $P^m(T, \mathcal{S}T) = 1$ et donc grâce à la deuxième partie du théorème 2 on a aussi

$$(9) \quad P^m(T', \mathcal{S}T) > 0$$

pour T' dans un voisinage $M_2$ de T, contenu dans $M_1$.

Maintenant, comme $\varphi^m(T',.)$ est dominante d'après la première partie du théorème 2, $P^m(T',.)$ est une mesure équivalente à la restriction de $\mu_M$ sur $\varphi(T',E^m)$, et donc (9) implique ;

$$(10) \quad \mu_M(\varphi^m(T',E^m) \cap \mathcal{S}T) > 0 \text{ pour tout } T' \text{ dans } M_2.$$

Soit K un compact de M tel que $\pi(K) > 0$ (celà est possible puisque $\pi(M) = 1$). Grâce à ($\mathcal{H}2$) et à (8) il existe un entier q tel que $x' = \varphi_a^q(x)$ soit dans $M_2$ pour tout x dans K ; comme $\varphi^m(x', E^m)$ est contenu dans $\varphi^{m+q}(x, E^{m+q})$ on a donc d'après (10)

$$(11) \quad \mu_M(\varphi^{m+q}(x, E^{m+q}) \cap \mathcal{S}T) > 0 \text{ pour tout } x \in K$$

et par conséquent d'après le théorème 2

$$(12) \quad P^{m+q}(x, \mathcal{S}T) > 0 \text{ pour tout } x \in K,$$

on a maintenant : $\pi(\mathcal{S}T) = \int \pi(dx)\, P^{m+q}(x, \mathcal{S}T) \geq \int_K \pi(dx)\, P^{m+q}(x, \mathcal{S}T)$

ce qui est strictement positif en vertu de (12).

Théorème 3 [6]

Sous les hypothèses ($\mathcal{H}1$), ($\mathcal{H}2$) et ($\mathcal{H}3$) on a $\pi(\mathcal{S}T) = 1$

■

La démonstration est donnée dans [6]; elle consiste, en utilisant le lemme 1, à appliquer convenablement le théorème ergodique au processus stationnaire $(Z_t)_{t\in\mathbb{Z}}$ défini par l'hypothèse ($\mathcal{H}3$).

### 3.3. Le théorème principal.

On a vu en (8) que $\varphi(M\times V) \subset M$; on se place donc dans M et on s'intéresse au comportement de la chaîne $(X_t)$ dans l'orbite $\mathcal{S}T$ de T. Notre résultat est le

Théorème 4

Sous les hypothèses ($\mathcal{H}1$), ($\mathcal{H}2$) et ($\mathcal{H}3$) le système (1) définit sur $\mathcal{S}T$ une chaîne de Markov $(X_t)$, Harris récurrente et géométriquement ergodique. De plus la probabilité invariante $\pi$ de $(X_t)$ est équivalente à la restriction de $\mu_M$ sur $\mathcal{S}T$ et $\int g(x)\,\pi(dx)$ est finie (g étant la fonction positive de (C3)) ∎

Démonstration.

Il s'agit de vérifier (C1) et (C2) du théorème 1 avec la mesure $\mu_M$.

Soit K un compact dans M et A un sous-ensemble de $\mathcal{S}T$ de $\mu_M$-mesure nulle ; comme dans le lemme 1 on peut trouver un entier r tel que $\varphi^r(x,.)$ soit dominant pour tout x dans K ; en utilisant la première partie du théorème 2 on conclut que : $P^r(x,A) = 0$ pour tout x dans K ; c'est la condition (C1).

Soit A un sous-ensemble de $\mathcal{S}T$ de $\mu_M$-mesure non nulle. En procédant pour A comme pour $\mathcal{S}T$ dans le lemme 1, on peut trouver un entier r tel que : $P^r(x,A) > 0$ et $\varphi^r(x,.)$ est dominante pour tout $x \in K$ ; d'après la deuxième partie du théorème 2 on a alors : $\inf_{x\in K} P^r(x,A) > 0$ ; c'est la condition (C2).

## 4. Exemples

### 4.1. Le système polynomiale affine.

C'est le système (1) avec $\varphi$ polynomiale de degré un en $X_t$ ; il se réécrit :

$$(13) \qquad X_{t+1} = \left[ A + \sum_\alpha B_\alpha u^\alpha_{t+1} \right] X_t + \sum_\alpha C_\alpha u^\alpha_{t+1} + D$$

où la sommation sur $\alpha$ est étendue à un ensemble fini dans $\mathbb{N}^d$ ; A, $B_\alpha$ sont des matrices ; $C_\alpha$, D sont des vecteurs et si $x = (x_1,\dots,x_d)$ et $\alpha = (\alpha_1,\dots,\alpha_d)$, $x^\alpha$ désigne $x_1^{\alpha_1}\dots x_d^{\alpha_d}$.

Le système (13) apparait dans les représentations markoviennes des processus polynomiaux [11] et bilinéaires [9]. La suite $(u_t)$ est supposée vérifier l'hypothèse ($\mathcal{H}$1) dans une variété algébrique lisse V contenant 0 ; on peut toujours supposer, que le point 0 est dans l'ensemble de positivité E de la densité de $u_t$ puisque cet ensemble est défini à un ensemble de $\mu_V$-mesure nulle près. ($\mathcal{H}$2) est vérifiée avec a=0 si

(H2) les valeurs propres de A sont en valeur absolue strictement inférieures à 1 ; (on a alors $T = (I-A)^{-1}D$).

L'hypothèse ($\mathcal{H}$'3), pour le système (13), est satisfaite avec $x_0 = 0$ , si :

(H3) $\exists$ s > 0 tel que $E\left|\sum_{\alpha} C_\alpha u_t^\alpha + D\right|^s < \infty$ et $E\left\|A + \sum_{\alpha} B_\alpha u_t^\alpha\right\|^s < 1$

où $\|\ .\ \|$ désigne ici la norme d'application linéaire.

Sous (H2) et (H3) le théorème 4 s'applique donc au système (13).

**4.2.** Le système linéaire.

C'est le système

$$(14) \qquad X_{t+1} = AX_t + Bu_{t+1}\ ,\ \text{où A et B sont des matrices.}$$

On suppose toujours que ($\mathcal{H}$1) est satisfaite comme en 4.1 ; on fait maintenant l'hypothèse :

(H) les valeurs propres de A sont en valeurs absolue strictement inférieures à un et la variable aléatoire $u_t$ est d'espérance nulle.

L'hypothèse (H) (dite de consalité) est souvent admise dans l'étude de (14). Sous (H) les hypothèses ($\mathcal{H}$2), avec a = 0, et ($\mathcal{H}$3), avec $x_0 = 0$, sont satisfaites (voir [7]).

**5. Conclusion**

L'étude statistique du système (1) peut-être abordée de deux manières différentes :

**5.1.** On part d'un état initial $x_0$ et on observe les sorties $X_1$, $X_2, \ldots, X_N$ ; on a donc dans ce cas un échantillon de la chaîne $(X_t)$ avec comme état initial $x_0$. Les résultats que nous avons obtenus ne peuvent s'appliquer que si $x_0$ est dans $\mathcal{ST}$ ; ce qui signifie que $\mathcal{ST}$ doit être connue à priori ou tout au moins identifiée.

**5.2.** Le système (1) est en régime stationnaire, c'est-à-dire qu'on observe un processus stationnaire en loi et vérifiant l'équation (1) ; dans ce cas en faisant l'hypothèse supplémentaire

$$(\mathcal{H}"3) \quad \lim_k E\left|\varphi_{u_t} \circ \ldots \circ \varphi_{u_{t-k}}(x) - \varphi_{u_t} \circ \ldots \circ \varphi_{u_{t-k}}(x')\right|^s = 0$$

pour tout x, x' dans W, (noter que ($\mathcal{H}$'3) implique ($\mathcal{H}$"3)), alors le système (1) n'admet qu'une solution stationnaire, le processus $(Z_t)_{t\in\mathbb{Z}}$

provenant de l'hypothèse ($\mathcal{H}3$). C'est donc ce processus qui est observé ; comme le théorème 4 s'applique, $(Z_t)_{t\in\mathbb{Z}}$ est Harris récurrent et géométriquement ergodique et donc géométriquement absolument régulier [6]. On notera que lorsque le système (1) est une représentation markovienne d'un processus stationnaire observable, $(Y_t)_{t\in\mathbb{Z}}$, (i.e. $Y_t = F(X_{t-s},\dots,X_{t+r})$ avec F mesurable), le processus $(Y_t)$ est géométriquement absolument régulier dès que $(Z_t)$ l'est ■

## Références.


[1] H. **AKAIKE**, *Markovian representation of stochastic processes*, Ann. Inst Stat. Math, 26, 1974, 363-387.

[2] J. **BOCHNAK**, M. **COSTE**, M.F. **ROY**, *Géométrie algèbrique réelle*, Springer Verlag, Berlin, 1987

[3] Th. **BRÖCKER**, *Differentiable germs and catastrophes.* Cambridge University Press, 1975.

[4] J. **DIEUDONNE**, *Eléments d'analyse, t.3*, Gauthier Villars, Paris 1979.

[5] S. **LOJASIEWICZ**, *Ensembles semi analytiques*, Multigraphie de l'I.H.E.S., Bures Sur Yvette, 1965.

[6] A. **MOKKADEM**, *Thèse de doctorat d'état*, Orsay, 1987.

[7] A. **MOKKADEM**, *Sur le mélange des processus ARMA vectoriel*, CRAS, t303, 519-521, 1986.

[8] E. **NUMMELIN**, P. **TUOMINEN**, *Geometric ergodicity of Harris recurrent Markov chain*, Stoch. Proc. Appl., 12, 187-202, 1982.

[9] T.D. **PHAM**, *Bilinear markovian representation and bilinear models*, Stoch. Proc. Appl., 20, 295-306, 1985.

[10] D. **REVUZ**, *Markov chains*, Nort Holland, Amsterdam, 1984.

[11] E.D. **SONTAG**, *Polynomial Response Maps*, Lecture Notes in Control and Information Sciences, 13, 1979.

[12] T. **SUBBA RAO**, M.M. **GABR**, *An introduction to bispectral analysis and bilinear time series models*, Lect. Notes in Statistics, 24, 1984.

# GENERALIZED INPUTS TO NON-LINEAR SYSTEMS

T. HUILLET[1], A. MONIN[1,2], G. MONTSENY[1,2], G. SALUT[1].

1. LAAS du CNRS, 7 Ave du Colonel Roche, 31077 TOULOUSE Cedex, FRANCE
2. DIGILOG, 21 rue F. Joliot, ZI, Les Milles Cedex, FRANCE

Abstract: The paper identifies the set of generalized inputs which are naturally associated to non-linear dynamical systems. This is done through completion of the class of functions with absolutely continuous primitives in a topology which is adapted to non-linear systems. One thus obtains a closed set of generalized inputs which allows an algebraic treatment of the input semi-group through coding by series of independant real numbers. Non trivial examples are given as an illustration.

Key-words: Non-linear systems, generalized inputs, Lie groups and algebras, Hall basis.

## 0. INTRODUCTION:

As is well known, generalized inputs to linear systems are defined through the coding of (causal) linear functionnals and are closely related to L.Schwartz's distributions [7]. They can be easily defined when being Lebesgue-Stieltjes measures, but also when they are distributions of higher order.

The situation is quite different for non-linear systems, due to their non-commutative properties, as first noticed by Sussmann [8] and Fliess [3]. We propose to define here a closed class of generalized inputs to non-linear systems which is "maximal" in the following sense:

We consider differential systems on a Banach space E, which are linear in the inputs, i.e.:

$$(0.1) \qquad \dot{x}_t = \sum_{i=1}^{\nu} f_i(x_t)\, u_i(t) \quad , \quad x_t \in E \ , \ u_i(t) \in \mathbb{R} \quad , \quad x_0 \text{ given}$$

One may associate to such systems the operator evolution equation:

$$(0.2) \qquad \dot{H}_t = H_t \sum_{i=1}^{\nu} u_i(t)\, A_i \ , \quad H_0 = I$$

where the non-commutative operators $A_i$ are defined as :

$$(A_i\, \varphi)(x) = \nabla\varphi(x).\, f_i(x) \ ,$$

and the transition operator $H_t$ satisfies:

$$(H_t\, \varphi)(x_0) = \varphi(x_t) \quad ,$$

$\varphi$ being an arbitrary function in the domain of $H_t$. In particular, when $\varphi$ is the identity function in E, one obtains $x_t$ as a function of $x_0$.

This is the classical way to obtain an abstract bilinear system from a system which is non-linear in x. There are two basic ways to define generalized inputs to a system such as (0.2). A first approach consists in using topological duality with a dense space of test functions, and weakening this space as much as possible, in order to get the widest possible dual space. A second approach is based on completion, with respect to a suitable topology, of a basic space of approximation elements.

It is the second approach which is used here. More precisely, we define from (0.2), a metric semi-group of basic inputs. Using algebraic representation results [5], we then proceed to the completion of this semi-group, and some of the typical new elements obtained are being studied. Direct application is made to the case when the functions f are linear ( bilinear systems ), and indications are given for the treatment of the more general anlytic case.

## 1. Definition of generalized inputs

As indicated above, the peculiar properties of generalized inputs to an abstract bilinear system such as (0.2) are going to be associated to the (non-commutative) properties of the A 's. We therefore set below a few algebraic preliminaries as well as notations.

### 1.1 Algebraic preliminaries

Let $X = (X_1, \ldots, X_\nu)$ be a set of non-commutative indeterminates, with the usual algebraic operations (+,.).

We consider:

A(X): the free associative algebra on R, with X as generator.
L(X): the corresponding free Lie algebra.
$\hat{A}(X)$: the algebra of non-commutative formal power series in X.
$\hat{L}(X)$: the associated algebra of Lie series.

Let $\mathfrak{J}$ be the set of all finite series of elements belonging to $\{1,2,\ldots\nu\}$, i.e.:
$\forall \sigma \in \mathfrak{J},\ \sigma = (i_1, \ldots, i_k)\ ,\ |\sigma| \triangleq k$

By convention: $\begin{cases} X_\sigma = X_{i_1} X_{i_2} \ldots X_{i_k} \\ X_\sigma = 1, \text{for } \sigma = \emptyset \end{cases}$

We may therefore write:

$$S \in A(X),\quad S = \sum_{\sigma \in \mathfrak{J}} a_\sigma X_\sigma$$

where all coefficients $a_\sigma \in R$ are null except a finite number of them.

For formal power series:

$$\forall S \in \hat{A}(X),\quad S = \sum_{\sigma \in \mathfrak{J}} a_\sigma X_\sigma$$

We define, on the other hand, the sets:

(1.1) $\mathcal{U}_T = \{ u = (u_i)_{i=1,\nu} \mid u_i \in L^1[0,T],\ u_i(t) = 0\ \forall t \notin [0,T] \ \}$

(1.2) $\mathcal{U} = \bigcup_{T \geq 0} \mathcal{U}_T$ ,

with a concatenation law $\vee$ on $\mathcal{U}$ defined as:

3) $\begin{cases} \forall u_1 \in \mathcal{U}_{T_1},\ \forall u_2 \in \mathcal{U}_{T_2}, u_1 \vee u_2 = u_3 \in \mathcal{U}_{T_1+T_2}, \\ \qquad \text{with: } u_3(t) = u_1(t), \text{ if } t < T_1, \\ \qquad u_3(t) = u_2(\ t-T\ )\ , \text{ if } t > T_1 \end{cases}$

It is easily checked that $(\mathcal{U}, \vee)$ has a semi-group structure, the unit element being $e(t) = 0$.

Let us define as well the set of primitives :

(1.4) $\mathcal{V} = \{ v = \int_0^t u(\tau)\, d\tau \ , \ u \in \mathcal{U} \}$

$\mathcal{V}$ is a semi-group which is isomorphic to $\mathcal{U}$ .Moreover,the elements of L of an element v is being defined as:

(1.5) $L(v) = \int \sup_{i=1,\ldots,\nu} |u_i(\tau)|\, d\tau$

We now consider the system:

$$(1.6) \quad \begin{cases} \dot{H}_t = H_t \sum_{i=1}^{\nu} X_i u_i(t), \ (u_i) \in \mathcal{U} \\ H_0 = I \ , I \text{ being the identity in } \hat{A}(X) \end{cases}$$

We may as well write:

$$(1.7) \quad \begin{cases} dH_t = H_t \sum_{i=1}^{\nu} X_i dv_i(t) \ , \quad (dv_i(t) = u_i(t)dt\ ) \\ H_0 = I \end{cases}$$

We now state a two-fold representation result whose proofs may be found in [4 , 5 , 6 ], as well as [ 8 ] for a part of it.

Proposition 1

The solution of system (1.6) in $\hat{A}(X)$, has the following representation:

$$H_t^u = \overleftarrow{\prod_{B \in \mathcal{B}}} \exp (S_B^u(t)\ B\ ) = \exp \Big( \sum_{B \in \mathcal{B}} r_B^u(t)\ B\ \Big)$$

$\overleftarrow{\prod}$ indicates the left product,and :

1-(B) = $\mathcal{B}$ is a Hall basis of L(X)
2- $(s_B^u(t))_{B \in \mathcal{B}}$ , $(r_B^u(t))_{B \in \mathcal{B}}$ belong to $R^{\mathcal{B}}$ and each set constitutes a minimal coding of u (cf. [ 5 , 6 ]) .

We shall be mainly interested here into the left infinite product representation ,through the $s_B^u$ coefficients.

Let us set:

(1.9) $H^u = \lim_{t \to \infty} H_t^u \quad , \quad s_B^u = \lim_{t \to \infty} s_B^u(t)$

As is easily seen,the map :

(1.10) $\psi : (\mathcal{U}, \vee) \dashrightarrow (\hat{A}(X), .)$
which associates $H^u$ to u ,is a semi-group homomorphism.

Moreover ([ 8 ]) $\psi(\mathcal{U})$ is a multiplicative group G:

(1.11) $H^{u^*}.H^u = H^u.H^{u^*} = I$ , with $u^*(t) = -u(T-t)$.

Definition 1:

Let us denote $(U,\vee)$ the quotient group:

(1.12) $U = \mathcal{U}/\psi^{-1}(0)$

$(U,\vee)$ is isomorphic to $(G,.)$.

If we denote $\mathfrak{R}$ the equivalence relation:

$u_1 \mathfrak{R} u_2 \Leftrightarrow H^{u_1} = H^{u_2}$ ,we have:

Proposition 2 [ 4 , 5 , 6 ]:

For all $u$ , $u \in \mathcal{U}$,

(1.13) $u_1 \mathfrak{R} u_2 \Leftrightarrow s_B^{u_1} = s_B^{u_2}$ , $\forall B \in \mathcal{B}$

Therefore, to any input $u \in (L^1(0,T))^\nu$ driving system (1.6), is associated a (denumerable) family of coefficients, this representation being faithful. It should be noted that this coding is a global picture of the map $I \longrightarrow H^u$ , for a given u, and not as a function of $t \in [0,T]$.

Definition 2: Let F be the map:

(1.14) $F: u \in U \longrightarrow (s_B^u)_{B \in \mathcal{B}} \in R^{\mathcal{B}}$ ,and:

(1.15) $S=F(U)$

S is therefore isomorphic to U (for the induced law).

Remark2: for notational convenience, we introduce the following subscripts

(1.16) $\mathcal{B} = (B)_{B \in \mathcal{B}} = (B_n)_{n \geq 1}$.

Let us now introduce the topological framework to this algebraic construction:

1.2. Generalized input semi-groups:

Let us define the following norm on A(X):

$$.17) \quad \| \sum_{\sigma \in \mathfrak{J}} a_\sigma X_\sigma \|_1 = \sum_{\sigma \in \mathfrak{J}} |a_\sigma| \, \Gamma^{|\sigma|} \quad ,0< \Gamma$$

As easily checked A(X) is a normed algebra , this norm being of multiplicative type. Indeed:

$$\|X_\sigma X_{\sigma'}\|_1 = \Gamma^{|\sigma|+|\sigma'|} = \|X_\sigma\|_1 . \|X_{\sigma'}\|_1$$

We then define:

$\forall\, s \in A(X)$ , $\|s\| = 2\, \|s\|_1$

One can easily show:

Proposition 3: Taking $\overline{A}(X)$ to be the completion of A(X) for this norm, $\overline{A}(X)$ is a complete normed Lie algebra.

Corollary 4: The set $\overline{A}^{*}(X)$ of invertible elements is a Lie group.

We now are in the position of defining the generalized inputs class admissible to (1.6).Let us first recall an existantial result:

Proposition 5[ 8 ]:

If $L(u) = \int \sup_i |u_i(\tau)| d\tau$ is small enough, then:

$$\sum_{n\geq 1} | s_n^u(t) | . \|B_n\|, \quad \overleftarrow{\prod_{n\geq 1}} \exp( s_n^u (t).B_n)$$

converge uniformly in t.

However,as shown by Sussmann,the infinite product does not converge necessarily.But,it is clear that any input $u \in U$ may be split into a finite concatenation of inputs u satisfying the above condition.Therefore:

Proposition 6: For $u \in U$,there exists $u_1,u_2,\ldots,u_N \in U$,such that:

(1.20) 1. $u = u_1 \vee u_2 \vee \ldots \vee u_N$

(1.21) 2. $H^u = \overleftarrow{\prod_{k=1}^{N}} \overleftarrow{\prod_{n\geq 1}} \exp (s_n^{u_k} B_n)$,

the infinite products being convergent in $\overline{A}(X)$.

The local property (1.8)may therefore be extended to the whole group in the above sense.

Let us now study the topological properties of G and U.Owing to the continuity of multiplication and inversion,it is clear that (G,.) is a topological group,the topology being defined by the metric induced by the norm; taking for U the initial topology,one gets:

Proposition 7 : U is a metric group,the distance being defined by:

(1.22) $d(u,u') = \|H^u - H^{u'}\|$

Definition 4: Let us denote: $\overline{U} = \psi^{-1}(\overline{G})$ (1.23)
$\overline{G}$ standing for the completion of G in $\overline{A}^{*}(X)$.

$\overline{U}$ is a group ( of generalized inputs) which is isometric to $\overline{G}$ (for the initial topology).Moreover:

Proposition 8: $\overline{U}$ is metrizable and complete.

Proof: $\overline{A}^{*}(X)$ is a Lie group,hence metrizable and complete [ 1 ]; one may therefore define the topology of $\overline{A}^{*}(X)$ from a metric for which $\overline{A}^{*}(X)$ is complete. $\overline{G}$ is closed within the complete set $\overline{A}^{*}(X)$ and is therefore complete. ∎

Remark: $\overline{G}$,together with $\overline{A}^{*}(X)$,are not necessarily complete for the metric induced by the norm.

We now define a subgroup of $\overline{U}$ together with a metric which is not as coarse and better suited to systems analysis.We first recall the following result:

<u>Lemma 9:</u> Taking the sequences $(s_n)_{n\geq 1}$ and $(s_{nk})_{n\geq 1}$, $k \in N$, such that:

$$\sum_{n\geq 1} |s_n|.\|B_n\| < +\infty \quad , \sum_{n\geq 1} |s_{n_k}|.\|B_n\| < +\infty$$

$$\text{if: } \sum_{n\geq 1} |s_{nk} - s_n|.\|B_n\| \xrightarrow[k\to\infty]{} 0, \tag{1.24}$$

then: $$\| \overleftarrow{\prod_{n\geq 1}} \exp(s_{nk}B_n) - \overleftarrow{\prod_{n\geq 1}} \exp(s_n B_n) \| \xrightarrow[k\to\infty]{} 0 \tag{1.25}$$

The proof consists in a trivial extension of a result to be found in [ 6 ,p92]. ■

Let us now consider:

(1.26) $$U_o = \{ u \in U, \sum_{n\geq 1} |s_n^u|.\|B_n\| < +\infty \}$$

(1.27) $$\mathcal{U}_o = \{ u \in \mathcal{U}, \sum_{n\geq 1} |s_n^u(t)|.\|B_n\| < +\infty, \text{ uniformly w.r. to t } \}$$

Let us define on $U_o$ the distance:

(1.28) $$\rho(u,u') = \sum_{n\geq 1} |s_n - s_n'|.\|B_n\|$$

Using (1.24),(1.25), the metric thus defined on $U_o$ is less coarse than the metric induced by the norm.We then define:

<u>Definition 6:</u>

Let $\tilde{U}_o$ be the completion of $U_o$ for the metric defined by (1.28).

Obviously, $\tilde{U}_o \subset \overline{U}$.

We also define:

<u>Definition 7:</u>

(1.29) $$\tilde{U} = \{ u \in \overline{U}, u = u_1 \vee u_2 \vee \ldots \vee u_N , u_i \in \tilde{U}_o , N \in \mathbb{N}^* \}$$

We endow $\tilde{U}$ with the topology:

$$u_n \longrightarrow u \text{ within } \tilde{U} \overset{\Delta}{\Leftrightarrow} u_n \vee u^{-1} \longrightarrow e \text{ within } \tilde{U}_o , \tag{1.30}$$
e being the neutral element for $\vee$ and $u^{-1}(t) = -u(T-t)$.

<u>Propositon 9:</u> $\widetilde{U}$ is a separated topological group ( of generalized inputs). Moreover $\widetilde{U}_0$ is closed within $\widetilde{U}$ .

See [9] for the proof. ■

<u>Remark 3:</u> The one to one mapping $\widetilde{U} \hookrightarrow \overline{U}$ is continuous.

Finally, let us define two semi-groups of generalized inputs associated to $\overline{U}$ and $\widetilde{U}$ . .

Let $\tau_1$ , $\tau_2$ be the following topologies over $\mathcal{U}$:

1. $u_n \xrightarrow{\tau_1} u \Leftrightarrow \mathbb{1}[0,t] . u_n \longrightarrow \mathbb{1}[0,t].u$ within $\overline{U}$ uniformly w.r. to t (1.31)

( $\mathbb{1}[0,t]$ being the indicatrix of $[0,t]$ ).

2. Taking: $u = u_1 \vee u_2 \vee \ldots \vee u_N$ , $u_n = u_{n_1} \vee u_{n_2} \vee \ldots \vee u_{n_N}$ ,
$u_n \xrightarrow{\tau_2} u \Leftrightarrow \mathbb{1}[0,t]\, u_{n_i} \longrightarrow \mathbb{1}[0,t].u_i$ within $\widetilde{U}_0$ , uniformly w.r. to t (and i) (1.32)

<u>Definition 8:</u> Let $\overline{\mathcal{U}}$ and $\widetilde{\mathcal{U}}$ be respectively the completions of $\mathcal{U}$ for the topologies $\tau_1$ and $\tau_2$ defined by (1.31),(1.32).

One may easily check that :

<u>Proposition 10:</u>

$$\begin{array}{ccccc} u_n \xrightarrow{\widetilde{\mathcal{U}}} u & \Rightarrow & u_n \xrightarrow{\overline{\mathcal{U}}} u & \Leftrightarrow & H_t^{u_n} \xrightarrow{\overline{A}(X)} H_t^u \text{ uniformly w.r. to } t \\ \Downarrow & & \Downarrow & & \Downarrow \\ u_n \xrightarrow{\widetilde{U}} u & \Rightarrow & u_n \xrightarrow{\overline{U}} u & \Leftrightarrow & H^{u_n} \xrightarrow{\overline{A}(X)} H^u \end{array} \qquad (1.33)$$

<u>Remark 4:</u> The (separated topological) semi-groups $\overline{\mathcal{U}}$ and $\widetilde{\mathcal{U}}$ include (generalized) inputs for all $t > 0$, by the topology which has been used for uniform convergence. To the contrary, the groups $\overline{U}$ , $\widetilde{U}$ are made of inputs which are considered globally as :

$$I \longrightarrow H^u \in \overline{A}(X)$$

$\overline{\mathcal{U}}$ and $\widetilde{\mathcal{U}}$ will be useful if one wishes to consider the system's trajectory. $\overline{U}$ and $\widetilde{U}$ may be used for example in the study of system's controllability.

Let finally $\overline{\mathcal{V}}$ be the completed set of $\mathcal{V}$ (cf.(1.4)) for the image metric induced by $\overline{\mathcal{U}}$.

$\overline{\mathcal{V}}$ is then isometric to $\overline{\mathcal{U}}$. We therefore have:

$$v_n \in \mathcal{V} \longrightarrow v \in \overline{\mathcal{V}} \Leftrightarrow u_n = dv_n/dt \in \mathcal{U} \longrightarrow u \in \overline{\mathcal{U}} \qquad (1.34)$$

We shall denote, by extension :

$$(1.35) \quad dv/dt \stackrel{\Delta}{=} u$$

Let us remark however that v is generally no longer a path of R .

We are now in the position to give a definition applicable to systems (1.6),(1.7).

For all $u_n \in \mathcal{U}$, we set:

$$(1.36) \quad C.u_n = \sum_{i=1}^{\nu} X_i u_{n_i}$$

(1.6) may then be written as:

$$(1.37)\quad \begin{cases} \dot{H}_t = H_t\, C.u_n(t) \\ H_o = I \end{cases}$$

whose solution is denoted by $H_t^{u_n}$.

Definition 9: For all u in $\overline{\mathcal{U}}$, such as ;

(1.38) $u = \lim u_n$ , $u_n \in \mathcal{U}$, and the formal equation:

$$(1.39)\quad \begin{cases} \dot{H} = H_t C.u(t) \\ H = I \end{cases} \qquad \text{admits as a solution:}$$

$$(1.40)\quad H_t^u = \lim H_t^{u_n}$$

Besides,(1.39) may be written as the differential form:

$$(1.41)\quad \begin{cases} dH_t = H_t C.dv(t) \\ H_o = I \end{cases}$$

Of course dv is not a measure.It should be also remarked that,by construction, (1.35) is a reversible system.

## 2.Properties of generalized inputs

The aim of this paragraph is to provide the semi-groups introduced above with additional properties,and to exhibit the new elements obtained through the completion operation.In particular,one gets that the set of coefficients $(s_n)_{n\geq 1}$ coding the input-paths may be chosen independantly when taking u within $\overline{U}$ or $\tilde{U}$ (this being not the case for U).

This appears of utmost importance if one wishes to study systems controllability from an algebraic point of view ([5,6]) .

### 2.1.Description of some non-trivial generalized inputs

We start with a fundamental result:

Theorem 11 [1] Let $\mathcal{G}$ be a Lie group on R.There is a neighbourhood W of 0 within the Lie algebra of $\mathcal{G}$, $L(\mathcal{G})$, such that, for all $x,y \in W$:

$$(2.1)\quad \exp(t(x+y)) = \lim_{q\to\infty} (\exp(tx/q)\exp(ty/q))^q$$

$$(2.2)\quad \exp(t^2[x,y]) = \lim (\exp(tx/q)\exp(ty/q)\exp(-tx/q)\exp(-ty/q))^{q^2}$$

uniformly w.r. to $t \in [0,T]$.

Remark 5: From corollary 4 ,$\mathcal{G}$ and $L(\mathcal{G})$ may be identified with $\overline{A^*}(X)$ and $\overline{A}(X)$ ([1]), respectively.

Let us now establish a result which will be useful in the sequel:

Lemma 12: for all $x,y \in W$ , with T small enough, one has:

$$(2.3)\quad 1.\ \exp(t(x+y)) = \lim (\exp(x/q).\exp(y/q))^{E(tq)},$$

E standing for integer part.

$$(2.4)\quad 2.\ \exp(t[x,y]) = \lim (\exp(x/q)\exp(y/q)\exp(-x/q)\exp(-y/q))^{E(tq)}$$

uniformly w.r. to t on [0,T].

see [9] for the proof. ∎

This result allows a construction of generalized inputs which have no equivalent in the linear theory.

<u>Definition 10</u> We denote Br(X) the set of all formal Lie brackets in the indeterminates X .

<u>Remark 6</u> Br(X) may be identified to the free magma in the indeterminates X .

We show below that any element of Br(X) may be activated by some input of $\bar{U}$, i.e.:

<u>Proposition 13</u> For all $B \in Br(X)$ there is a $u \in \bar{\mathcal{U}}$ such that:

(2.7) $H_t^u = \exp(tB)$ , $0<t<T$ small enough.

<u>Proof</u>: By recurrence on the length of B and with lemma 12 (see[9] for more details).■

<u>Corollary 13</u> For all $n \in N^*$ ,there exists an input $u \in \bar{\mathcal{U}}$ such that:

$$(s_n^u(t)) = (0,0,\ldots,\underset{\substack{\uparrow\\ n\text{-place}}}{t},0,\ldots) \quad , t<T .$$

<u>Proof</u>: this comes from the fact that any element $B \in \mathcal{B}$ is an element of Br(X).■

<u>Remark 7:</u> 1. The elements of $\bar{\mathcal{U}}$ thus exhibited are no more functions.We give below a technique of approximation by functions from $\mathcal{U}$.

2. The generalized inputs within $\bar{U}$ are easily derived using proposition 10.

We now wish to show that any sequence $(s_n)_{n\geq 1}$ ,rapidly decreasing to zero, is the coding of an input of $\bar{U}$.

Let us define the following set of sequences:

<u>Definition 11</u> We denote by $l^1_{\mathcal{B}}$ :

$$l^1_{\mathcal{B}} = \{ \quad (s_n)_{n\geq 1} \in R^{N^*} \quad , \sum_{n\geq 1} |s_n|.\|B_n\| \quad < +\infty \} \qquad (2.13)$$

<u>Remark 8</u> $l^1_{\mathcal{B}}$ may be isometrically identified to $l^1$ by the mapping:

$$l^1_{\mathcal{B}} \longrightarrow l^1$$
$$(s_n) \longrightarrow (\ s_n\|B_n\| \ )$$

From lemma 8, if $(s_n) \in l^1_{\mathcal{B}}$, $\overleftarrow{\prod_{n\geq 1}} \exp(s_n B_n)$ converges.

<u>Proposition 14:</u> For any sequence $(s_n) \in l^1_{\mathcal{B}}$, there exists $u \in \bar{U}$ satisfying: $s_n^u = s_n$ , $\forall n\geq 1$.

<u>Proof</u>: From corollary 13, there is a sequence $(u_q)_{q\geq 1}$, such that: $(s_n^{u_q}) = (0,0,\ldots,\underset{\substack{\uparrow\\ q\text{-place}}}{s},0,\ldots)$

On the other hand, for all q,q' , q'>q, one has:

$H^{u_q \vee u_{q'}} = H^{u_q} . H^{u_{q'}} = \exp(s_{q'} B_{q'}).\exp(s_q B_q)$.

Since q'>q,

$$s^{u_q \vee u_{q'}} = (0,0,\ldots,\underset{\substack{\uparrow\\ q\text{-place}}}{s_q},0,\ldots,0,\underset{\substack{\uparrow\\ q'\text{-place}}}{s_{q'}},0,\ldots)$$

Let then: $\bar{u}_N = u_1 \vee u_2 \vee \ldots \vee u_N$

It can easily be checked that:

$$1.\quad H^{\bar{u}_N} = \overleftarrow{\prod_{n=1}^{N}} \exp(s_n B_n)$$

$$2.\quad H^{\bar{u}_N} \xrightarrow[N\to\infty]{} \overleftarrow{\prod_{n\geq 1}} \exp(s_n B_n) = H^{u} \quad ,\text{with} \quad u = \bigvee_{q=1}^{\infty} u_q$$

Therefore : $u = \lim u_q \Rightarrow u \in \bar{U}$. ■

We may then conclude that each term of the sequence $(s_n^u)$ may be chosen independantly provided that $(s_n^u) \in l^1_{\mathfrak{B}}$. This does not hold within the (uncomplete) group U.

Let us now recall an interesting result on exponentials of Lie series:

**Proposition 15** For any Lie series S converging within $\bar{A}(X)$, there exists $u \in U$ satisfying :

$$H^u = \exp(S).$$

**Proof:** 1. Any Lie polynomial P may be written as:

$$P = \sum_{B \in Br(X)} \alpha_B B$$ , with vanishing $\alpha_B$ except for a finite number of them.

From Proposition 13 and (2.1), exp ( P ) is the limit of a sequence $H^{u_q}$, $u_q \in U$. Therefore exp ( P ) = $H^u$ .

2. S being convergent, one has:

$S = \lim P_q$, where $P_q$ is a Lie polynomial.

Therefore : $\exp(S) = \exp(\lim P_q) = \lim H^{u_q} = H^u$ , where $u \in \bar{U}$. ■

## 2.2 Examples of generalized inputs:

2.2.1 Take $\nu=2$ , so that $X = (X_1, X_2)$. The simplest (although not trivial) among the generalized inputs is the one represented by:

$(s_n^u(t)) = (0,0,t,0,0,\dots)$ , $0 \leq t \leq T$,

with: $H^u = \exp(t[X_1, X_2])$.

From (2.9) and lemma 12, one easily shows that such an input is the limit of the series of inputs $(u_q)_{q\in N} = (u_{q1}, u_{q2})_{q\in N}$ depicted as below:

(2.14)

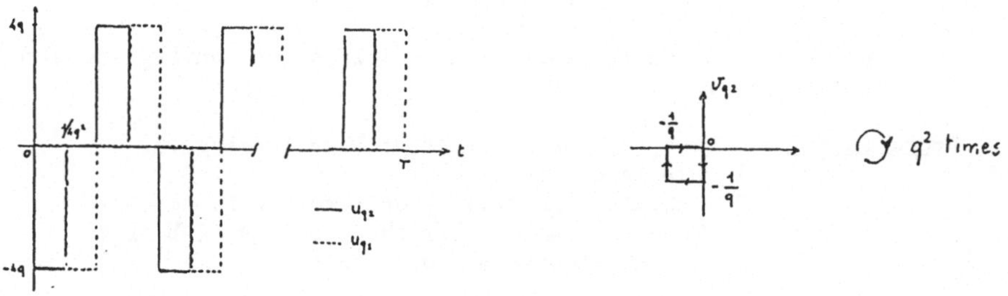


This type of oscillating input has no significant effect in linear system's theory. More precisely, it may be shown that u vanishes in the space of distributions $\mathfrak{D}'$.

To the contrary, in the case of bilinear systems or analytic non-linear systems ( linear in the control ), this input activates the first Lie bracket of $\mathfrak{B}$.

We are now going to show that generalized inputs characterized by the series $(s_n) \in l^1_{\mathcal{B}}$ are also in $\widetilde{U}$, that is are limit of elements of U for the topology defined by (1.28).

2.3 Generalized inputs of $\mathcal{U}$ and U : a description:

The main result of this subsection consists in establishing that, with a condition on the norm, the series in $\mathcal{U}$ (resp. U ) which are convergent within $\bar{\mathcal{U}}$ ( resp. $\bar{U}$ ) to elements coded by (0,0,...,0,t,0,...) are also converging within $\widetilde{\mathcal{U}}$ ( resp $\widetilde{U}$ ), convergence within $\widetilde{\mathcal{U}}$ ( resp. $\widetilde{U}$ ) being defined from the coding $(s^u_n)_{n\geq 1}$.

Recalling that $\|\sum_{\sigma\in\mathcal{J}} a_\sigma X_\sigma\| = \sum_{\sigma\in\mathcal{J}} |a_\sigma|\Gamma^{|\sigma|}$, $\Gamma>0$ , we now give the essential result:

Propositon 16 Assume that $0<\Gamma<1/2$. Then:
For any sequence $(s(t)) = (0,0,\ldots,0,t,0\ldots)$, $t\in[0,T]$, T small enough, there exists $u\in\bar{\mathcal{U}}$ such that:

$$(s^u_n(t)) = (s_n(t))$$

The proof may be founded in [9] . ∎

One then easily show, from the definition of $\widetilde{\mathcal{U}}$ , in a similar to that of proposition 14:

Corollary 17: For any sequence $(s_n) \in l^1_{\mathcal{B}}$ ,there exists $u \in \widetilde{U}$ such that $s^u_n = s_n$ , $n\geq 1$.

It follows that the terms of the $(s^u_n)$ coding of an input $u \in \widetilde{U}$ are independant. Moreover, such an input may be approximated by a sequence of inputs in $\mathcal{U}$ for the topology defined by (1.28), which is simpler to manipulate.

2.4 Some additional results and remarks:

The topology of $\bar{\mathcal{U}}$ being that of uniform convergence defined by (1.32), and $H^u_t$ being continuous w.r. to t, for $u \in \mathcal{U}$ , we get:

Proposition 18: For all $u \in \bar{\mathcal{U}}$,
$H^u_t$ is continuous w.r. to t.

Let us give as well a sufficient criterion of convergence within $\mathcal{U}$ for the topology induced by $\bar{\mathcal{U}}$.

Proposition 19 Let $(u_n)$ be a series of $\mathcal{U}$ satisfying :
1. $u_n$ is continuous.
2. $u_n(t) \longrightarrow u(t)$ , uniformly w.r. to t.

Then $u_n \longrightarrow u$, for the topology of $\bar{\mathcal{U}}$, i.e. :
$H^{u_n}_t \longrightarrow H^u_t$ , uniformly w.r. to t.

See [9] for the proof. ∎

Let us make a final remark concerning the differential equation of the form (1.41):

(2.55) $dH_t = H_t C dv(t)$ , where $v \in \mathcal{V}$.

Equation (2.55) is independant of the parametrization of H and v , i.e. for any function $\varphi$: $R\rightarrow R$ increasing and regular enough, (2.55) implies :

$dH_{\varphi(t)} = H_{\varphi(t)}\, C\, dv(\varphi(t))$ or, setting $t' = \varphi(t)$ :

$dH_{t'} = H_{t'}\, C\, dv(t')$

This property extends to $v \in \bar{\mathcal{V}}$ by continuity. It is then possible to define a type of generalized inputs which is analogous to Dirac measures for linear systems. Thus, consider the differential system:

(2.56) $$dH_t = H_t(\, X_1 dv_1(t) + X_2 dv_2(t))$$

and suppose that $v = (v_1, v_2)$ is discontinuous at $t = t_0$. It is then impossible to integrate such a system unless , as in [ 5 , 6 ] , one does consider the "cocheminement" of the components of v.

This amounts to set a fictitious time parametrization $t = t(\tau)$ so that one can describe the path followed by v(t) within the plane of discontinuity.

(2.57)

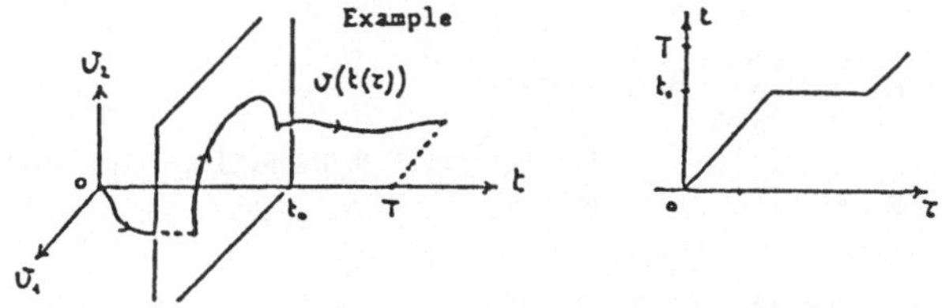


We are then led back to the case $v \in \mathcal{V}$ (v being absolutely continuous w.r. to $\tau$).

Obviously, in a similar fashion, the trajectory $H_{t(\tau)}$ is continuous w.r. to $\tau$, but not w.r. to t. This method allows to account in a simple way for impulsive inputs to the system (1.6) and naturally implies to consider the "cocheminement" of the components.

Remark that, for linear systems of the type: $\dot{H} = AH + Bu$, the "cocheminement" of an input when a vector impulse occurs, has no effect, which makes such considerations irrelevant.

By extension and continuity, the above remarks apply to all integral inputs of $\bar{\mathcal{V}}$.

Let us finally note that this type of impulsive inputs preserves reversibility of the system, by construction.

## 3. Application to dynamical systems

The preceding study concerns abstract differential equations (1.6),(1.7). This is a necessary condition to define generalized inputs in the least restrictive generality. There remains to show, of course, that the domain of applicability is not empty. The next step consists in applying the obtained results to systems of the announced type:

(3.1) $$\dot{x}_t = \sum_{i=1}^{\nu} f_i(x_t).u_i(t), \quad x_t \in E \text{ Banach space},$$

where the $f_i$ are sufficiently regular.

A first simple particular case is that where the $f_i$ are linear (bilinear systems) :

$$(3.2)\quad \begin{cases} \dot{x}_t = \sum_{i=1}^{\nu} A_i . x_t . u_i(t) \ , \ A_i \in \mathcal{L}(E) \\ x_0 \text{ being fixed.} \end{cases}$$

Let us now consider the associated opeartor equation:

$$\begin{cases} \dot{K}_t = \sum_{i=1}^{\nu} A_i K_t u_i(t), \qquad K_t \in \mathcal{L}(E) \\ K_0 = I \end{cases}$$

or in a dual way:

$$(3.3)\quad \begin{cases} \dot{K}_t^* = K_t^* \sum_{i=1}^{\nu} A_i^* u_i(t) \\ K_0^* = I \end{cases}$$

$*$ denoting the dual operation

Let us take (in (1.17)): $\Gamma = \sup_i \|A_i\|$ (3.4)

Let $\omega$ be the map :

$$\omega: \quad A(X) \longrightarrow \mathcal{L}(E)^*$$

$$S = \sum_{\sigma\in\mathfrak{J}} a_\sigma X_\sigma \longrightarrow \sum_{\sigma\in\mathfrak{J}} a_\sigma A_\sigma^* \triangleq \tilde{S} \qquad (3.5)$$

where $A_\sigma^* = A_{i_1}^* A_{i_2}^* \dots A_{i_k}^*$

One has from (1.17),(1.18) :

$$(3.6)\qquad \|\tilde{S}\| = \|\sum_\sigma a_\sigma A_\sigma^*\| < \sum_\sigma |a_\sigma| . \|A_\sigma^*\| < \sum_\sigma |a_\sigma| . \|X_\sigma\|_{\bar{A}(X)} = \|S\|_{\bar{A}(X)}$$

$\omega$ is thus continuous and may be extended through continuity/density to $\bar{A}(X)$. Let us denote $A^* u = \sum_{i=1}^{\nu} A_i^* u_i$ (3.7)

We may then write:

<u>Proposition 20</u> The system:

$$\begin{cases} \dot{K}_t^* = K_t^* A^* u(t), \quad u \in \bar{\mathcal{U}} \\ K_0^* = I \end{cases} \qquad (3.8)$$

admits as a solution:

$K_t^{u*} = \tilde{H}_t^u$, as defined by (3.5),

where $H_t^u$ is solution of :

$$(3.9)\quad \begin{cases} \dot{H}_t = H_t C\, u(t) \\ H_0 = I \end{cases}$$

Moreover :

if $u_n \to u$ within $\bar{\mathcal{U}}$, $K_t^{u_n *} \to K_t^{u *}$ within $\mathcal{L}(E)^*$ (3.10)

<u>Corollary 21</u> The system

(3.11)
$$\begin{cases} \dot{x}_t = \sum_{i=1}^{\nu} A_i x_t u_i(t) , & u \in \overline{\mathcal{U}} \\ x_0 \text{ fixed} \end{cases}$$

admits as a solution :

(3.12)
$$x_t^u = K_t^u \, x_0$$

Moreover:

(3.13) if $u_n \to u$ within $\overline{\mathcal{U}}$ , $x_t^{u_n} \to x_t^u$ within E, unif. wr to t

<u>Example:</u>

We consider the bilinear scalar system:

(3.14)
$$\begin{cases} \dot{x} = x.u_1 + u_2 \\ x_0 = 0, \end{cases}$$

where $u_1$ and $u_2$ are defined by the limit depicted below when n goes to infinity:

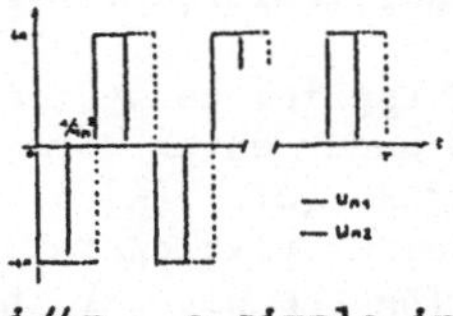

Denoting t = 4i/4n ,a simple integration of (3.14) leads to the following recurence:

(3.15)
$$\begin{cases} x^n_{t+1} = ( x^n_t \, e^{1/n} + 1/n )e^{-1/n} - 1/n = x^n_t + 1/n \, ( e^{-1/n} - 1 ) \\ x_0 = 0, \text{ hence:} \end{cases}$$

(3.16)
$$x^n_t = x_0 + t \, n^2/n \, ( e^{-1/n} - 1) = t \, n^2 ( e^{-1/n} - 1 )/n$$

We obtain at the limit:

(3.17)
$$x_t = -t$$

We now set $x_2 = x$, and $x_1 = 1$. We get from (3.14):

(3.18)
$$\begin{cases} \dot{x}_1 = 0 \\ \dot{x}_2 = x_2 . u_1 + x_1 . u_2 \\ x_1(0) = 1 \\ x_2(0) = 0 \end{cases}$$

This also reads:

(3.19)
$$\begin{cases} X = \begin{pmatrix} 0 & 0 \\ 0 & 1 \end{pmatrix} X.u_1 + \begin{pmatrix} 0 & 0 \\ 0 & 1 \end{pmatrix} X.u_2 \\ X_0 = \begin{pmatrix} 1 \\ 0 \end{pmatrix} \end{cases}$$

We deduce from the preceding results that the input $(u_1, u_2)$ is being coded by (0,0,t,0,...) and (3.19) has for a solution:

$$X_t = \exp\left( t\left[\begin{pmatrix}0 & 0\\ 0 & 1\end{pmatrix}, \begin{pmatrix}0 & 0\\ 1 & 0\end{pmatrix}\right]\right) X_0$$

$$= \exp\left( t \begin{pmatrix}0 & 0\\ 1 & 0\end{pmatrix}\right) X_0 = \begin{pmatrix}1 & 0\\ t & 1\end{pmatrix}\cdot\begin{pmatrix}1\\ 0\end{pmatrix} = \begin{pmatrix}1\\ -t\end{pmatrix} \tag{3.20}$$

One can check that $x(t) = -t$.

As was shown above, application of the general result to the case of real bilinear systems ( linear $f_i$'s ) is straightforeward using the usual matrix norm for the $f_i$'s. A more general application holds when the $f_i$'s are analytic . However, in that case, the substitution of analytic vector fields to the $A_i$ operators introduces a domain of convergence on the initial conditions $x_0$, for the series' development to hold globally.

Conclusion:

As was stressed in the beginning, generalized inputs to non-linear systems are not of the same nature as those to linear systems, due to non-commutative aspects that arise between their components, a situation which does not occur in linear theory. This makes the topology of such generalized inputs non trivial and, in particular, different from that of tributions.

The case which has been studied applies to systems' representations which are linear in the inputs, a form under which more general situations can be handled, using marked measures (cf. [6] ). The procedure here consists in considering abstract non-commutative operators as coefficients of the inputs. They are used for deriving the natural input topology which defines the state space solution of such systems in a non ambiguous way.

Such a procedure was already used by Fliess in [3], with the help of the generating series in the associative algebra of $A_i$ operators. The further and crucial benefit obtained here, with the help of exponential representations is to identify in a precise manner the generalized inputs to a *Lie* serie of independant coefficients ($s_n$) which are a one to one mapping of those inputs.

The non-commutative effects are being coded in the ($s_n$) coefficients in a way which has no equivalent in linear functionnal analysis. In particular, the Fourier coefficients of usual commutative harmonic analysis appear as a corser topology which is in fact a quotient topology to the one exhibited here.

REFERENCES

[1] N.BOURBAKI : Groupes et algèbres de Lie Ch.2,3, Hermann, 1972

[2] H.CARTAN : Cours de calcul différentiel, Hermann, 1979.

[3] M.FLIESS : Fonctionnelles causales non-linéaires et indéterminées non-commutatives. Bull. Soc. Math. de France, N° 109, 1981, p 3-40.

[4] T.HUILLET,A.MONIN,G.SALUT : Lie algebraic representation results for non-stationnary evolution operators. Math. Syst. Theory, N°19, 1987, p 205,226.

[5] T.HUILLET,A.MONIN,G.SALUT : Lie algebraic canonical representations in non-linear control systems. Math.Syst. Theory, N°20, 1987, p 193,213.

[6] A.MONIN: Contributions algébriques et topologiques à la représentation des systèmes non-linéaires. Thèse Toulouse 1987.

[7] L.SCHWARTZ : Théorie des distributions, Hermann , 1978.

[8] H.J.SUSSMANN : A product expansion for the Chen series. Theory and applications of non-linear control systems. North-Holland, 1986.

[9] T.HUILLET,A.MONIN,G.MONTSENY,G.SALUT : Generalized inputs to non-linear systems. Rapport LAAS .

# BIFURCATION DYNAMIQUE avec BRUIT MULTIPLICATIF

E. BENOIT
B. CANDELPERGHER
C. LOBRY

## 1 Introduction .

On considère l'équation différentielle :

$$dx/dt = f(x,\mu) \quad , \; x \in \mathbb{R}$$

telle que 0 soit une position d'équilibre stable pour $t < t_0$ , instable pour $t > t_0$ . Des études théoriques récentes [5, 7, 9-14, 16 , 17, 19, 20] ont montré que lorsqu'on fait croitre lentement le paramètre de part et d'autre de la valeur de bifurcation l'instabilité sera effectivement observée après un délai ; précisément on a, entre autres, le résultat suivant :

Relation Entrée - Sortie [1, 2] :

Soit $\varepsilon > 0$ fixé *infiniment* petit . Soit le problème de cauchy :

$$(1) \qquad \begin{cases} dx/dt = f(x,\mu) \\ d\mu/dt = \varepsilon \\ x(0) = x_0 \;,\; \mu(0) = \mu_0 < t_0 \end{cases}$$

Sous des hypothèses de régularité très générales sur f, la solution $(x(t), \mu(t))$ *pénètre* dans le *halo* de la droite $x = 0$ en un point *infiniment* proche de $(0,\mu_0)$ puis *longe* cette droite jusqu'en un point $\underline{\mu} > \mu_0$ défini par :

$$\underline{\mu} = \mu_0 + \varepsilon \underline{t}$$

$$\int_0^{\underline{t}} f_x(0, \mu_0+\varepsilon s)ds = 0$$

où elle la *quitte* .
(Ce théorème est illustré par la figure 1)

On peut trouver dans [1], [2] et [8] des explications sur une formalisation possible, dans le langage de l'Analyse Non Standard, des mots écrits en italique . Toute cette note peut être comprise en leur attribuant leur sens naïf évident . L'objet des notes [1] et [2] n'était pas l'étude des bifurcations dynamiques, dont la problèmatique revient principalement aux auteurs de [5] , [6] et [7], mais celle des équations "lentes rapides" de $\mathbb{R}^2$, c'est à dire de la forme :

$$(2) \qquad \begin{cases} \varepsilon\, dx/dt = f(x,y) \\ dy/dt = g(x,y) \end{cases}$$

dont les systèmes de la forme (1) ci dessus constituent ( après changement de temps ) un cas particulier . C'est pourquoi la relation entrée sortie n'y figure pas explicitement sous la forme indiquée ; une démonstration directe en est donnée dans [8].

Des études expérimentales (par exemple [6] et [15]) ont montré que ce phénomène est observable dans divers contextes . Les aspects théoriques de la prise en considération du bruit commencent à être envisagés ([4,17] et leur bibliographie ). Dans [17] on aborde l'étude de l'influence du bruit sur ce retard à la bifurcation dans le cas particulier où $f(x, \mu)$ est de la forme $xF(\mu)$ . Les cas du bruit additif et multiplicatif y sont abordés .

C'est sur le cas du **bruit multiplicatif** que nous revenons dans la présente note .

## 2 - Représentation du bruit .

soit $\mu \longmapsto F(\mu)$ une fonction continue *standard* telle que :

$$\mu < 0 \Rightarrow F(\mu) < 0$$
$$\mu > 0 \Rightarrow F(\mu) > 0$$

**(Dire qu'une fonction est *standard* c'est dire que dans son expression elle ne contient aucun paramètre *non standard* , *infiniment petit* ou *infiniment grand* ).**

Nous considérons la suite récurrente :

$$(3)\quad \begin{cases} x_{t+dt} = x_t + x_t F(\mu_t)dt \pm a x_t \sqrt{dt} & t = ndt \text{ , dt } \textit{infiniment} \text{ petit .} \\ \mu_{t+dt} = \mu_t + \varepsilon dt & \varepsilon \textit{ infiniment} \text{ petit} \\ x_0 \text{ est la variable aléatoire certaine prenant la valeur } \textit{standard } x_0 > 0 & \\ \mu_0 \text{ est donné } \textit{standard} \text{ négatif .} & \end{cases}$$

où ± symbolise une suite de variables aléatoires indépendantes prenant les valeurs + 1 et − 1 avec la probabilité 1/2 . Il est montré dans [3] qu'un tel processus (avec dt *infiniment* petit) est l'équivalent du processus de diffusion classique défini par l'équation différentielle :

$$(4)\quad \begin{cases} dx_t = x_t F(\mu_t)dt + a\, x_t dW_t \\ d\mu_t = \varepsilon \\ x_0 \text{ est la variable aléatoire certaine prenant la valeur } x_0 > 0 \\ \mu_0 \text{ est donné négatif .} \end{cases}$$

considérée au sens de Ito . Nous insistons sur le fait que dt étant *infiniment* petit , l'objet défini par (3) n'est pas une approximation de la diffusion définie par (4) , mais une **idéalisation** ayant le même statut .

Dans [17] les auteurs considèrent la quantité $E[x_t^2]$ et définissent le temps caractéristique $t^*$ du processus bruité, comme le premier instant où cette espérance est supérieure à $x_0^2$ , l'idée étant qu'à partir de cet instant les solutions du processus seront "en moyenne" plus éloignées de l'état d'équilibre 0 que ne l'était la condition initiale puisque la "dispersion" mesurée par le moment d'ordre 2 sera plus grande que $x_0^2$ . Un calcul élémentaire montre que :

$$E[x_{t+dt}^2] = E[x_t^2] + E[x_t^2](\, 2F(\mu_0 + \varepsilon t) + a^2 \,)dt + o(dt)$$

donc que ( à un *infiniment* petit près ) $s^2(t) = E[x_t^2]$ est solution de l'équation différentielle:

$$ds^2/dt = s^2(t)\,(\,2F(\mu_0 + \varepsilon t) + a^2\,)$$

ce qui donne un temps caractéristique $t^*$ défini par la relation :

$$\int_0^{t^*} (F(\mu_0 + \varepsilon t) + a^2/2\,)dt = 0$$

dont il est aisé de voir qu'il est inférieur au temps $\underline{t}$ de *sortie* du *halo* de la droite $x = 0$ défini dans l'introduction . Ceci conduit les auteurs de [17] à déclarer que le retard à la bifurcation est **réduit** par la présence d'un bruit multiplicatif .

Nous allons voir que concrètement ( par exemple sur des simulations numériques) il n'en est rien et que, au contraire, la présence d'un bruit multiplicatif **accroit** le retard à la bifurcation . Le bruit multiplicatif **augmente la stabilité** ! Nous fournissons une explication théorique simple de ce fait à première vue paradoxal .

## 3- Changement de variable

Comme le laisse prévoir la forme multiplicative de l'équation le passage au logarithme fournit des information utiles . Nous posons :

$$t = \tau/\varepsilon \qquad y_\tau = \varepsilon\,\mathrm{Log}(x_t)$$

Ce changement de variable est légitime car nous avons supposé $x_0 > 0$ ce qui entraine que $x_{ndt}$ est également strictement positif pour tout n .

$$y_{\tau+d\tau} = \varepsilon\,\mathrm{Log}(x_{t+dt})$$

$$y_{\tau+d\tau} = \varepsilon\,\mathrm{Log}(x_t + x_t\,F(\mu_t)dt \pm a\,x_t\sqrt{dt})$$

$$y_{\tau+d\tau} = \varepsilon\,\mathrm{Log}(x_t) + \varepsilon\,\mathrm{Log}(\,1 + F(\mu_t)dt \pm a\,\sqrt{dt}))$$

$$y_{\tau+d\tau} = y_\tau + \varepsilon F(\mu_t)dt - \varepsilon (a^2/2)\, dt \pm \varepsilon a \sqrt{dt} + \varepsilon O(dt^{3/2})$$

$$y_{\tau+d\tau} = y_\tau + ( F(\mu_\circ+\tau) - (a^2/2))\, d\tau \pm a\sqrt{\varepsilon}\sqrt{d\tau} + o(d\tau)$$

Si nous notons

$$m(\tau) = E[y_\tau] \quad \text{et} \qquad \sigma^2(\tau) = E[(y_\tau - m(\tau))^2]$$

un calcul élémentaire nous donne

$$m(\tau+d\tau) = m(\tau) + (F(\mu_0+\tau) - (a^2/2))\, d\tau + o(d\tau)$$
$$\sigma^2(\tau+d\tau) = \sigma^2(\tau) + a^2 \varepsilon\, d\tau + o(d\tau)$$

d'où nous déduisons que pour tout $\tau$ *non infiniment* grand $m(\tau)$ et $\sigma^2(\tau)$ sont *infiniment* proches des solutions des équations différentielles :

$$(5) \quad \begin{cases} dm/d\tau = F(\mu_0 + \tau) - a^2/2 \\ m(0) = \varepsilon \operatorname{Log}(x_0) \\ \\ d\sigma^2/d\tau = a^2 \varepsilon \\ \sigma^2(0) = 0 \end{cases}$$

donc

$$m(\tau) \simeq \varepsilon \operatorname{Log}(x_\circ) + \int_0^\tau (F(\mu_0 + s) - a^2/2)\, ds$$

$$m(\tau) \simeq \int_0^\tau (F(\mu_0 + s) - a^2/2)\, ds$$

$$\sigma^2(\tau) \simeq a^2 \varepsilon \tau$$

Définissons l'instant $\tau^{**}$ par la relation :

$$\int_0^{\tau^{**}} (F(\mu_o + s) - a^2/2)\, ds = 0$$

On a évidemment l'inégalité $\tau^{**} > \underline{t}$ . Pour tout $\tau$ *nettement* plus grand que 0 et *nettement* plus petit que $\tau^{**}$ , $m(\tau)$ est *nettement* plus petit que 0 . Comme d'autre part $\sigma^2(\tau)$ est *infiniment petit* nous déduisons , par l'inégalité des martingales, appliquée au processus :

$$y_\tau - m(\tau)$$

que , presque certainement, $y_\tau$ est *nettement* plus petit que 0 pour $\tau$ *nettement* positif et *nettement* plus petit que $\tau^{**}$ . Alors $x_t$ , qui est égal à $\exp( y_\tau/\varepsilon)$ , est presque certainement *infiniment* proche de 0 pour toutes les valeurs du paramètre $\mu$ *nettement* plus grandes que $\mu_o$ et *nettement* plus petites que $\mu_o + \tau^{**}$. Il y a donc , **presque certainement**, un retard à la bifurcation supérieur au retard qu'on observerait en l'absence de bruit multiplicatif .

## 4 - Quelques simulations

Nous prenons pour fonction F la fonction $F(\mu) = \mu$ , pour conditions initiales $x_o = 1$ et $\mu_o = -1$ . Dans ce cas les valeurs caractéristiques de $\mu$ sont :

1 - Bifurcation statique

définie par le changement de signe de $F(\mu)$, soit :

$$\mu = 0$$

2 - Bifurcation retardée, en l'absence de bruit

définie par les relations :

$$\underline{\mu} = \varepsilon \underline{t} - 1$$

$$\int_0^{\underline{t}} (-1 + \varepsilon t)\, dt = 0$$

soit $\underline{\mu} = 1$ .

3 - Bifurcation retardée en présence de bruit multiplicatif :

définie par les relations :

$$\mu^{**} = \varepsilon \tau^{**} - 1$$

$$\int_0^{\tau^{**}} (-1 + t - a^2/2)\, dt = 0$$

soit $\mu^{**} = 1 + a^2$

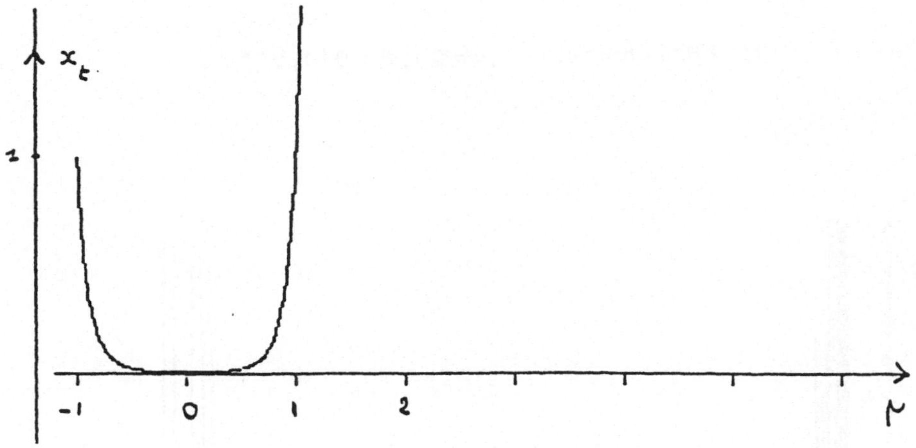


fig- 1 Bruit nul a = 0 , $\varepsilon$ = 0.1, dt = 0.01

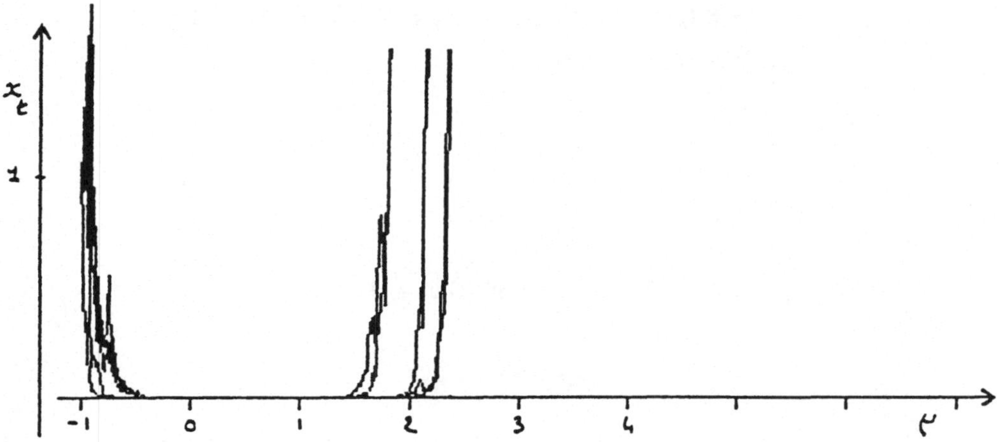


fig- 2 Bruit non nul $a = 1, \varepsilon = 0.1, dt = 0.01, \mu^{**} = 2$

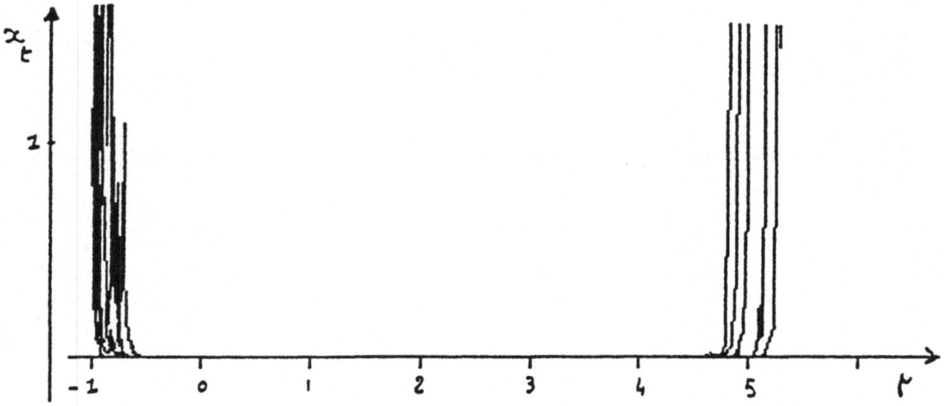


fig- 3 Bruit non nul $a = 2, \varepsilon = 0.1$ , $dt = 0.01, \mu^{**} = 5$

La dispersion observée provient de ce que ni $\varepsilon$ ni dt ne sont *infiniment petits* . Il n'était pas possible, avec la précision dont nous disposions de prendre $\varepsilon$ et dt plus petits . En effet les valeurs prises par $x_t$ sont extrèmement petites. Pour les exemples traités on observe $10^{-30}$ . Pour des valeurs plus petites de $\varepsilon$ et dt il arrive que $x_t$ s'annule définitivement .

## 5 - Conclusion

1) L'effet stabilisateur du bruit multiplicatif provient du fait que le processus :

$$z_{t+dt} = z_t \pm z_t\sqrt{dt}$$

$$z_0 = 1$$

a la curieuse propriété que **presque toutes ses trajectoires** tendent vers 0 quand t tend vers l'infini . (Ceci se déduit de la loi du logarithme itéré après avoir fait le même changement de variables que ci dessus) . Pourtant on constate que l'espérance de $z_t$ reste égale à 1 et sa variance croit exponentiellement ; ceci est dû à ce que quelques trajectoires inobservables, parce que de probabilité très petite, croissent exceptionnellement vite compensant ainsi une majorité de trajectoires tendant vers 0 . Il n'est donc pas légitime de conclure à la dispersion d'une variable aléatoire sur la simple observation que sa variance est grande . Ceci n'est vrai que pour les variables aléatoires Gaussiennes et celles qui en sont proches . Lorsque ce n'est pas le cas, comme dans l'exemple étudié ici, la connaissance des deux premiers moments n'apporte pas necessairement une information valable .

2) Les phénomènes de retard à la bifurcation, avec ou sans bruit, doivent être interprètés avec précautions . En effet, les effets prévus théoriquement ne peuvent être reproduits en machine que si on dispose d'une précision très grande ( au moins $10^{-20}$ ) ce qui peut faire douter de leur réalité dans des dispositifs où le bruit est présent . A cela nous pouvons répondre que : Un bruit additif, comme le montrent les études théoriques de [17] , réduit le retard à la bifurcation, sans nécessairement le supprimer, comme on le constate dans certaines situations concrètes ; voir [4 , 6 , 10-16] . L'effet du bruit multiplicatif du type décrit ici est plus curieux ; il reste à savoir dans quelles conditions il représente effectivement une situation réelle .

3) Le retard à la bifurcation pourrait avoir des conséquences importantes sur certains types de capteurs comme nous allons l'expliquer sur un exemple . Imaginons que nous désirions déceler la rotation d'une barre à l'aide d'un dispositif du type du régulateur de Watt dessiné ci contre . Une analyse élémentaire des forces en présence montre que pour une vitesse inférieure à une certaine valeur $\omega_0$ l'angle $\emptyset$ reste nul et, au delà, $\emptyset = 0$ devient une position d'équilibre instable . Actuellement les théories évoquées dans cet article ne permettent pas de prévoir le comportement d'un tel dispositif mais des simulations numériques ont montré que le phénomène de retard était effectivement présent, avertissant que la valeur $\omega_0$ est dépassée quand la vitesse réelle est déja nettement supérieure, donc peut être trop tard .

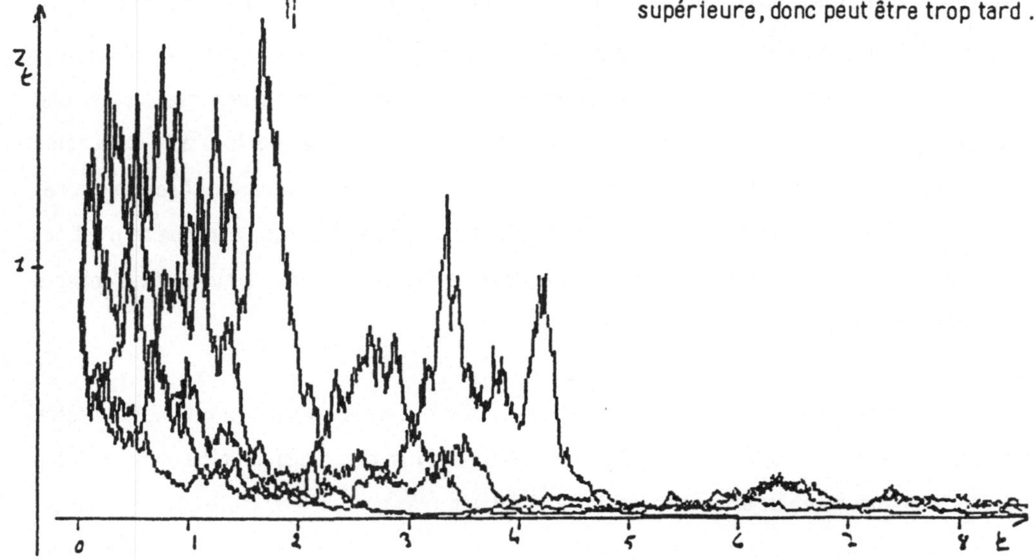


fig – 4 Cinq réalisations de : $z_{t+dt} = z_t \pm z_t\sqrt{dt}$

$z_0 = 1$ , dt = 0,001 , t variant de 0 à 8

**Remerciements** : Les auteurs remercient P . MANDEL pour de nombreuses conversations stimulantes sur le thème des bifurcations retardées .

## Références


[1] BENOIT E. " Tunnels et entonnoirs " C.R. Acad. Sci. Paris Ser. I 292 (1981) 283-286.

[2] ——— " Relation entrée - sortie " C.R. Acad. Sci. Paris Ser. I 293 (1981) 293- 296 .

[3] BENOIT E. , B. CANDELPERGHER , C. LOBRY : "Diffusions discrètes et méchanique stochastique" . Première partie : " Diffusions " par E. BENOIT - Prépublication Université de Nice, Ecole des Mines, Octobre 1987 .

[4] BROGGI G. , A. COLOMBO , L.A. LUGIATO ET P. MANDEL "Influence of White noise on delayed bifurcations " Physical Rew. A , VOL 33, n+ 5 3635-3637.

[5] ERNEUX T. and P. MANDEL " Imperfect bifurcation with slowly-varying control parameter " **SIAM J. on Appl. MATH , VOL 46 n° 1 (1986) 1-15.**

[6] GLORIEUX P. et D. DANGOISSE " Dynamical behavior of a laser containing a saturable absorber " IEEE Journal of QUANTUM ELECTRONICS , VOL QE-21 n° 9 (1985) 1486-1490 .

[7] KAPRAL R. and P. MANDEL " Bifurcation structure of the nonautonomous quadratic map " Physical Rew. A , VOL 32, n° 2 1076-1081.

[8] LOBRY C. " Analyse Non Standard et Théorie des Bifurcations " Université de Nice, Prébublications Mathématiques , n° 139 Octobre 1987 .

[9] ——— et G. WALLET "La traversée de l'axe imaginaire n'a pas toujours lieu là où l'on croit l'observer " Actes du colloque de "Mathématiques finitaires" de Lumigny , préprint Université de Poitier 1987 à paraitre ...

[10] MANDEL P. "Dynamic versus static stability "dans "Frontier in quantum optics" ; Adam Hilger, Bristol and Boston (1986) 430-452 .

[11] -------- and T.ERNEUX " Laser Lorenz equation with a time dependent parameter " Physical Rev. Letters Vol 53, n° 19 (1984) 1818-1820 .

[12] ———————— " Nonlinear control in optical bistability " IEEE Journal of QUANTUM ELECTRONICS , VOL QE-21 n° 9 (1985) 1352 - 1355.

[13] ———————— " The slow passage through a steady bifurcation : delay and memory effects " J. of Stat Phys. 48 (1987) 36-38 .

[14] MANDEL P. and XIAO-GUAN WU " Second harmonic generation in a laser cavity " J. of Optical Society of America B, VOL 3 p 940 (1986) 940-948.

[15] MORRIS B. and F. MOSS " Postponed bifurcation of a quadratic map with a swept parameter " , PHYSICS Let. A VOL 118, n° 3 (1986) 117 - 120.

[16] RUSCHIN S. and S.H. BAUER "Bistability , Hysteresis and critical behavior of a $CO_2$ laser, with $SF_6$ interactivity as a structurable absorber " CHEMICAL PHYSICS Let. VOL 66 n° 1 (1979) 100-103 .

[17] Van den BROECK and P. MANDEL "Delayed bifurcations in the presence of noise " Phys. Letters A122 (1987) 1059-1070 ;

[18] WALLET G. " Entrée sortie dans un tourbillon " Annales de l'Institut Fourier , Tome 36 Fasc. 4 (1986) 157-184 .

[19] ——— " Bifurcation dynamique foyer stable/foyer instable " Preprint (1987) .

[20] -------- " Dérive lente du champ de Lienard " Actes du colloque de "Mathématiques finitaires" de Luminy , préprint Université de Poitiers 1987 à paraitre ...

Adresse des auteurs .

E. BENOIT : Centre de Mathématiques Appliquées, Ecole des Mines, Sophia-Antipolis, 06565 VALNONNE Cedex FRANCE.

B. CANDELPERGHER et C. LOBRY Département de Mathématiques, Université de NICE, Parc Valrose, 06084 NICE Cedex FRANCE.

# VI - APPLICATIONS

# ROBOTICS

# REMARKS ON SOME WORKED OUT APPLICATIONS OF NONLINEAR CONTROL THEORY

J. LEVINE
Centre d'Automatique et Informatique
Section Automatique
Ecole Nationale Supérieure des Mines de Paris
35, rue St. Honoré, 77305 FONTAINEBLEAU, FRANCE.

**Abstract**

The improvements of nonlinear control theory are discussed in regard to real life applications. Some of the interactions between modeling and control are displayed. The importance of identification, of adapted choice of coordinates for inputs, state and outputs is described in view of noises analysis and of control objectives. Applications to robot control, tracking filters and distillation columns are presented.

## 1 Introduction : when are nonlinear control techniques useful?

During the last decade, considerable developments of nonlinear control theory have been produced. The purpose of most of these works consists in obtaining nonlinear controllers and observers such that the closed-loop system's design reduces to a simple application of linear techniques, with, in particular, easy stability criteria(see for ex. [21], [22], [25]).

In applications, however, these improvements are not always clear for many reasons that we shall try to demonstrate, and it turns out that the use of nonlinear control techniques requires a careful analysis of the system and its control objectives.

At this occasion, let us try to give a precise meaning to *applications of nonlinear control theory*. It is not original to remark that every system is nonlinear though system's linearity is often used as a convenient assumption. Nevertheless, this does not prove at all that linear techniques fail in the presence of nonlinearities or that nonlinear control design tools have a natural superiority. This would be in contradiction with the spirit of modern automatic control, namely to work with a simplified model and design a robust controller that will automatically compensate noises, modeling errors, parasitics or neglected dynamics, etc. It results that we shall only be concerned with applications of nonlinear control theory for which linear techniques, or easy adaptations of them, do not satisfactorily work or impose unacceptable limitations. The classical tangent linearization technique, for instance, becomes quite hazardous if the control objective consists in following reference trajectories with large amplitudes and fast dynamics, in the presence of bifurcations, or of singularities such as loss of controllability or observability of the tangent model.

The aim of this paper is, on the one hand, to give an idea of the kind of improvements that one can expect from the nonlinear approach and more specifically from exact state or input-output feedback linearization techniques, and on the other hand to demonstrate how important is the choice of the basic modeling ingredients to succeed in the control design : choice of coordinates, of outputs, of discrete or continuous time, of time scales, account of noises, disturbances and errors. Several real life applications, such as observers for target tracking, control of robots and of distillation columns, will be presented to illustrate our discussion. Note that we do not aim at surveying applications of nonlinear control and that many other examples have been treated by many authors in various fields :

- Aerospace : control of aircrafts or missiles [37], [3], of helicopters [33], of satellites [2], [34] ...
- Mechanical Systems : control of robots [15] (and many others!), active compensation of vibrations or external forces [16], [7] ...
- Electrical, Electronic Systems and Networks : DC drives [20], [35], [11], circuits analysis [4], [19], [38] ...
- Chemical and Biochemical Processes : distillation [17], [30], reactors [18], fermentations [13], [23] ...
- Medical Problems : drugs injection [6] ...
- Observation Problems : target tracking [29], [27], intelligent sensors [28] ...

This list is far from being exhaustive.

The paper is organized as follows : to initialize the discussion on the relationships between modeling and control, we begin with a short presentation of the actual status of nonlinear identification (Section 2). Then, in Section 3, we show some of the basic couplings that have to be taken into account between modeling, control and observer design in a real life application : the importance of well adapted coordinates of inputs, state and outputs, of time-scales, in view of control objectives, noises and errors.

## 2 Some remarks about nonlinear identification

The literature on this difficult subject is extremely poor compared to the volume of papers dealing with nonlinear control. However, the applicability of nonlinear control techniques strongly depends on the efficiency of identification algorithms.

Recall that the identification problem consists in finding a state-space representation, or the associated Volterra or generating series, of the input-output map known only through a finite number of experiments. More precisely, if we focus attention on the state-space approach, we are looking for a model :

$$\left\{ \begin{array}{rcl} \dot{x} & = & f(x,u) \\ y & = & h(x) \end{array} \right. \tag{1}$$

where the smooth manifold $X$ associated to the state vector $x$ and, of course, its dimension $n$, are unknown, where $f$ is a smooth unknown vector field on this manifold $X$ and $h$ a smooth unknown function (without restriction on their algebraic structure), such that if the known inputs $u_k(t), t \in [0,T], k = 1,\ldots,K$ ($u_k(t) \in \mathbf{R}^m$ for every $t$) produce the measured outputs $\bar{y}_k(t), t \in [0,T], k = 1,\ldots,K$, then the responses $y_k$ of (1) for each $u_k$ are arbitrarily close (in a sense to be defined) to the measured responses $\bar{y}_k$ for every $k$. We assume that $y_k(t) \in \mathbf{R}^p$ for every $t$.

Analogous definitions can also be stated involving Volterra series or generating series in place of system (1). Clearly, it is always possible to associate to system (1) a Volterra series, namely a functional developement of the form :

$$y(t) = \sum_{k\geq 0} \sum_{i_1,\ldots,i_k=1}^{m} \int_{\mathbf{R}^k} w_{i_1,\ldots,i_k}(t_1,\ldots,t_k) u_{i_1}(t-t_1)\cdots u_{i_k}(t-t_k) dt_1 \cdots dt_k \tag{2}$$

valid for small $t$ and $u$, or for every $t$ in the neighborhood of a stable equilibrium. The input-output identification problem consists thus in finding the Volterra kernels $w_{i_1,\ldots,i_k}$ up to a given order such that, as before, the outputs corresponding to these kernels and to the known inputs are arbitrarily close to the measured outputs. The same definition can be stated in terms of generating series.

Note that one can obtain, under mild regularity assumptions, a local state-space realization from these series and that the (local) state-space identification problem includes the two others.

It should be emphasized that identification is mainly used to design control strategies and therefore is assumed to produce acceptable answers to a wide range of control functions, whereas the model itself is generally obtained from a small number of experiments. Moreover, the advantage of building a nonlinear model becomes only significant if its validity range strictly includes the one of a tangent linear approximation. This point is extensively discussed below.

Roughly speaking, two kinds of identification algorithms are developed :

- purely quantitative approaches, or more precisely, ignoring qualitative informations that might be available on the behavior of the process to be identified, in particular on asymptotics, (Volterra-Wiener techniques [1], [36], small time bilinear approximations [12], nonlinear ARMA models [24], on-line parameter adaptation [13], etc.) ; all these techniques have in common the fact that the algebraic structure of the system or of the input-output map is assumed to be given a priori.
- approaches including additional informations on the qualitative behavior of the process [10].

For both approaches, current methods display serious drawbacks : In addition to the fact that series representations yield a generally untractable over-parametrization, more often requiring a truncation at the order 2, quantitative approaches can produce erroneous informations when used on an infinite time interval or with too large control variations. Moreover, the assumption that the system or the input-output map has a global given structure is, in practice, almost impossible to check and can only work locally. The gain with respect to linear identification is thus questionable.

In the second category of methods, one basically finds bilinear approximation techniques at a stable equilibrium point or at a collection of stable equilibria, using preidentified tangent linearizations at these equilibria. This last approach seems to be currently the most popular and we shall discuss it in greater details.

For this purpose, we need some definitions and notations. Let us introduce the asymptotic manifold which partly describes the system's asymptotic behavior. The asymptotic manifold $A$ is the submanifold of $X \times \mathbf{R}^m$ defined by :

$$A = \{(x,u) \in X \times \mathbf{R}^m | f(x,u) = 0\} \tag{3}$$

Recall that, since $f$ is smooth by assumption, $A$ is an at least $m$ dimensional smooth manifold. The set of stable equilibrium points $A_s$ is clearly a closed subset of $A$ and is generally a manifold with boundary. In a practical point of view, apart from exceptional systems, only structurally stable equilibrium points can be observed and we assume that enough experiments can be made on the process to provide a precise description of $A_s$. Now we can describe the bilinear identification technique of ([10]).

Assuming that we are given a collection of linear systems described by the matrices $F(x,u)$, $G(x,u)$ and $H(x)$ for every $(x,u)$ in $A_s$, we want to find a minimal bilinear model

$$\left\{\begin{array}{rcl} \dot{x} & = & (A + \sum_{i=1}^{m'} u'_i B_i)x \\ y & = & Cx \end{array}\right. \tag{4}$$

with $m' \geq m$ and $u'_i = u_i$ for every $i \leq m$, namely with possibly more inputs than in the given linear models, such that the tangent linearization of (4) on $A_s$ is the best approximation, in the least square sense, of the family of linear systems $(F, G, H)$.

The major advantages of this method are firstly that a bilinear model can be easily obtained by using known linear identification algorithms and secondly that the volume of computations is significantly smaller than with the other approaches. However, we shall try to prove that the validity range of the identified bilinear model is not clearly improved compared to the linear ones and more information is needed about transients. It should also be noted that the independance of the extra inputs $u'_i$, $m+1 \leq i \leq m'$ is not clear. Finally the assumption that a collection of linear systems $(F, G, H)$ is given may cause problems in specific, but not rare, situations.

Let us first address a more general problem, namely to find a vector field $f$ and an output function $h$ such that

$$\left\{ \begin{array}{lcl} \frac{\partial f}{\partial x}(x,u) & = & F(x,u) \\ \frac{\partial f}{\partial u}(x,u) & = & G(x,u) \\ \frac{\partial h}{\partial x}(x) & = & H(x) \end{array} \right. \tag{5}$$

for every $(x, u)$ in $A_s$.

Assuming that the dependance of $(F, G, H)$ with respect to $(x, u)$ is analytic and that $F$ (resp. $G$, resp. $H$) is an $n \times n$ (resp. $n \times m$, resp. $p \times n$) matrix for every $(x, u)$, it is easy to check that the system of partial differential equations (5) has at least one local solution if and only if

$$\left\{ \begin{array}{ll} \frac{\partial F_{i,j}}{\partial x_k} = \frac{\partial F_{i,k}}{\partial x_j} & i,j,k = 1,\ldots,n \\ \frac{\partial F_{i,j}}{\partial u_k} = \frac{\partial G_{i,k}}{\partial x_j} & i,j = 1,\ldots,n, \ k = 1,\ldots,m \\ \frac{\partial G_{i,j}}{\partial u_k} = \frac{\partial G_{i,k}}{\partial u_j} & i = 1,\ldots,n, \ j,k = 1,\ldots,m \\ \frac{\partial H_{i,j}}{\partial x_k} = \frac{\partial H_{i,k}}{\partial x_j} & i = 1,\ldots,p, \ ,j,k = 1,\ldots,n \end{array} \right. \tag{6}$$

But since $F$, $G$ and $H$ are only known on the submanifold $A_s$ of $X \times \mathbf{R}^m$, it results that the partial derivatives appearing in (6) corresponding to transversal directions to $A_s$ cannot be computed and that it is generally impossible to check if such a family of tangent models effectively corresponds to a system of the form (1). Moreover, if we assume that the compatibility conditions (6) hold, an infinite number of vector fields $f$ and output functions $h$ are such that their restriction to $A_s$ satisfies (5) and this proves that the validity of such an identified model, without further informations outside $A_s$, is comparable to the one of the given linear approximations, unless particular controls are used to move slowly on $A_s$. In this case, however, the dynamic properties of the system are not really exploited.

On the other hand, the assumption that linear approximations can be identified on $A_s$ is not as natural as it seems to be. Assume for instance that, in a given neighborhood, the tangent approximation to the true model, if it exists, is not controllable whereas the true nonlinear model is. Small perturbations on the control and the state, after some time, will produce trajectories that do not remain in the controllability subspace of the tangent model and the identification of this tangent model from such signals will of course produce a controllable linear system. It results that a nonlinear model produced by the previous method will be first order controllable and, even if the associated asymptotic responses are arbitrarily close to the true ones, the identified transients will be faster, or more precisely, the controllability indices of the identified model will be strictly smaller than the true model ones. Therefore, if we try to follow a reference model having the same controllability indices as our identified model, an exploding control may result.

To conclude, we have displayed some of the intrinsic problems that might appear when using known identification methods. All of them suffer for lack of information on the qualitative behavior of the process, the difficulty relying on the way to include such informations in the identification algorithms. This is certainly the main reason why successful applications concern only nonlinear systems whose model is derived from physical considerations, with input, state and output variables having precise physical interpretation and where the use of identification techniques is not required. All the examples presented below have this specificity.

# 3 What is a good model for observation and control?

It has already been remarked (see for ex. [21]) that the choice of coordinates for the inputs, state and outputs are crucial for nonlinear systems. In several non generic (but not so rare) situations it may happen that physical variables give rise to complicated nonlinear models which, once expressed in more adapted coordinates are significantly simplified. In particular, when there exists a feedback change of coordinates of the inputs, a change of coordinates in the state space and possibly a choice of coordinates for the outputs such that the system becomes linear controllable, with possibly a stable unobservable part, then the design of a robust controller becomes easier. However, in the forthcoming discussion, we want to point out that this analysis is not complete if noises, disturbances, errors, neglected dynamics, etc., are not taken into account and that unexpected behaviors may result from their ignorance.

## 3.1 Choosing good coordinates

We begin with an example of robot with pneumatic actuators where a non trivial choice of coordinates suffices to design an efficient observer - controller, whereas more traditional approaches encounter serious difficulties coming in particular from the pneumatic complicated nonlinear behavior.

### 3.1.1 An observer-controller for a robot arm with pneumatic actuators

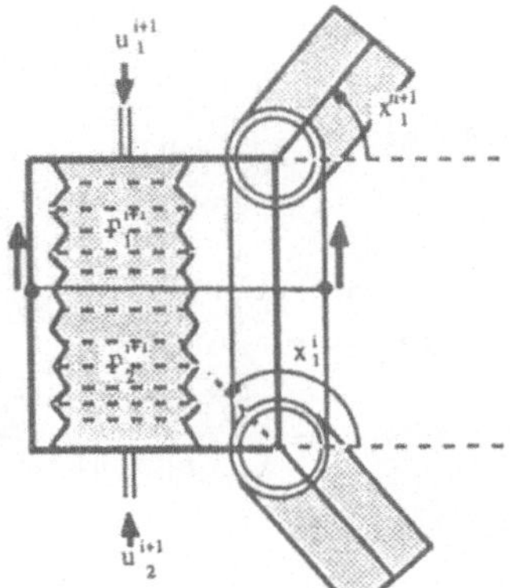

Figure 1: the pneumatic actuators

We consider a robot arm made of an arbitrary number $n_1$ of segments. The actuators (servo-valves) inject air under constant pressure, with air flows $u_1$ (resp. $u_2$), at the top (resp. the bottom) of each balloons (see Figure 1). Denoting $p_1$ the $n_1$ dimensional vector of pressures in the superior half part of the balloons and $p_2$ the $n_1$ dimensional vector of pressures in the inferior half part of the balloons, the inflation of the superior or inferior part of each balloon induce the rotation of the corresponding joint in the positive or negative sense. $x_1$ is the $n_1$ dimensional vector of relative angular positions, $x_2$ the $n_1$ dimensional vector of angular velocities. A model of the robot can be written as :

$$\left\{ \begin{array}{lcl} \dot{x}_1 & = & x_2 \\ \dot{x}_2 & = & \Gamma_0^{-1}(x_1)(\Gamma(x_1,x_2)+Q(x_1,x_2,p_1,p_2)) \\ \dot{p}_1 & = & a_1(x_1,x_2,p_1,p_2)+b_1(x_1,x_2,p_1,p_2)u_1 \\ \dot{p}_2 & = & a_2(x_1,x_2,p_1,p_2)+b_2(x_1,x_2,p_1,p_2)u_2 \end{array} \right. \tag{7}$$

where the $x$ part of the system corresponds to the mechanical model $\Gamma_0(x_1)\ddot{x}_1 = \Gamma(x_1,\dot{x}_1) + Q(x_1,\dot{x}_1,p_1,p_2)$ and where the $p$ part corresponds to the pressure model which is not detailed here since the linearization result is independent of the precise form of $a$ and $b$. We only assume that one can find two real numbers $\lambda_1$ and $\lambda_2$, at least one being non zero, such that

$$\text{rank}(\lambda_2\frac{\partial Q}{\partial p_1} - \lambda_1\frac{\partial Q}{\partial p_2}) = n_1 \ \forall(x,p) \tag{8}$$

Remark that the $p$ part can be considered as a nonlinear dynamic feedback if $p_1$ and $p_2$ were the control variables in the $x$ part. Such a nonlinear integrator is generally the source of persistent state dependent oscillations. It results that classical control techniques are extremely difficult to apply here. This difficulty can in fact be easily circumvented as shown in the following proposition :

**Proposition 3.1** *the change of coordinates $\xi = \varphi(x_1,x_2,p_1,p_2)$ and the feedback change of inputs $u = \alpha(x_1,x_2,p_1,p_2)+\beta(x_1,x_2,p_1,p_2)v$, are such that $\xi(t) = \varphi(x(t),p(t))$ satisfies :*

$$\left\{ \begin{array}{lcl} \dot{\xi}_1 & = & \xi_2 \\ \dot{\xi}_2 & = & \xi_3 \\ \dot{\xi}_3 & = & v_1 \\ \dot{\xi}_4 & = & v_2 \end{array} \right. \tag{9}$$

*with $\varphi$, $\alpha$, $\beta$ given by (10), (11), (12) and (13) :*

$$\varphi(x_1,x_2,p_1,p_2) = \begin{pmatrix} x_1 \\ x_2 \\ \Gamma_0^{-1}(x_1)(\Gamma(x_1,x_2)+Q(x_1,x_2,p_1,p_2)) \\ \lambda_1 p_1 + \lambda_2 p_2 \end{pmatrix} \tag{10}$$

$$\alpha(x,p) = M^{-1}(x,p)\begin{pmatrix} \gamma(x,p) \\ -\lambda_1 a_1(x,p) - \lambda_2 a_2(x,p) \end{pmatrix}, \ \beta(x,p) = M^{-1}(x,p)\begin{pmatrix} \Gamma_0(x) & 0 \\ 0 & I \end{pmatrix} \tag{11}$$

$$M(x,p) = \begin{pmatrix} \frac{\partial Q}{\partial p_1}b_1(x,p) & \frac{\partial Q}{\partial p_2}b_2(x,p) \\ \lambda_1 b_1(x,p) & \lambda_2 b_2(x,p) \end{pmatrix} \tag{12}$$

*and*

$$\gamma(x,p) = -\Gamma_0(\frac{\partial}{\partial x_1}(\Gamma_0^{-1}(\Gamma+Q))x_2 + (\frac{\partial}{\partial x_2}(\Gamma_0^{-1}(\Gamma+Q)))(\Gamma_0^{-1}(\Gamma+Q))) - \frac{\partial Q}{\partial p_1}a_1 - \frac{\partial Q}{\partial p_2}a_2 \tag{13}$$

Consequently, the controller design can be easily done by means of the well-known linear observer-controller approach even if the state is not entirely measured. It suffices, for example, to implement sensors on $x_1$ and $p$. Furthermore, this controller is robust in the sense that small modeling errors or small perturbations will not affect the stability of a stable linear system. However, for large errors, the linear structure can be destroyed since important nonlinearities are reinjected in (9) by

$\varphi$ and by the feedback. As far as modeling erors are concerned, it means that the present model needs to be refined. But if exogeneous perturbations or noises are involved, the proposed observer design may fail and techniques taking into account probabilistic characteristics of the noise have to be considered. We shall examine this point on the next example where influence of noise can be more easily analysed.

### 3.1.2 Observers : the role of noises and of continuous/discrete time design

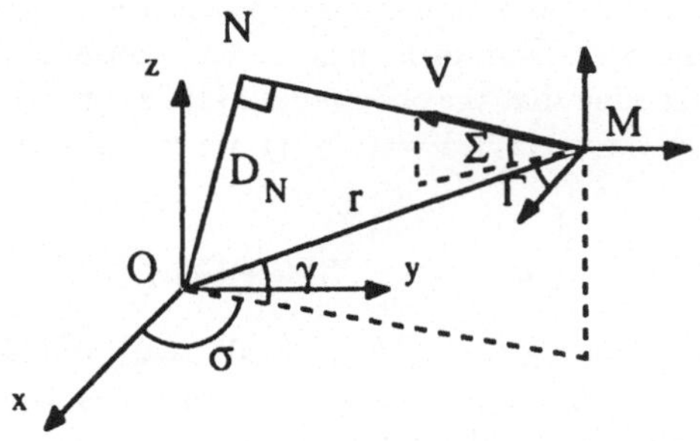


Figure 2: Tracking coordinates

We are now interested in tracking for a moving object with constant velocity ([27]), which can be observed by means of sensors providing the angles $\sigma$ and $\gamma$ of Figure 2. We assume in addition that the distance $r$ may not be always available. The state and observation equations are given by

$$\left\{\begin{array}{lcl} \dot{x} & = & v\cos\Sigma\cos\Gamma \\ \dot{y} & = & v\cos\Sigma\sin\Gamma \\ \dot{z} & = & v\sin\Sigma \\ \dot{\Sigma} & = & 0 \\ \dot{\Gamma} & = & 0 \\ \dot{v} & = & 0 \end{array}\right. \tag{14}$$

$$\left\{\begin{array}{lcl} r & = & \sqrt{x^2+y^2+z^2} \\ \sigma & = & \arcsin(\frac{z}{\sqrt{x^2+y^2+z^2}}) \\ \gamma & = & \arctan(\frac{y}{x}) \end{array}\right. \tag{15}$$

When $r$ is measured, the system is observable and, by inversion of the trigonometric relations (15) together with their time derivatives, one easily obtains :

$$\left\{\begin{array}{lcl} x & = & r\cos\sigma\cos\gamma \\ y & = & r\cos\sigma\sin\gamma \\ z & = & r\sin\sigma \\ v_x & = & \dot{r}\cos\sigma\cos\gamma \;-\; r\dot{\sigma}\sin\sigma\cos\gamma \;-\; r\dot{\gamma}\cos\sigma\sin\gamma \\ v_y & = & \dot{r}\cos\sigma\sin\gamma \;-\; r\dot{\sigma}\sin\sigma\sin\gamma \;+\; r\dot{\gamma}\cos\sigma\cos\gamma \\ v_z & = & \dot{r}\sin\sigma \;+\; r\dot{\sigma}\cos\sigma \end{array}\right. \tag{16}$$

with $v_x = v\cos\Sigma\cos\Gamma$, $v_y = v\cos\Sigma\sin\Gamma$, $v_z = v\sin\Sigma$.

Otherwise stated, system (14) becomes :

$$\begin{cases} \dot{x} &= v_x \\ \dot{v}_x &= 0 \\ \dot{y} &= v_y \\ \dot{v}_y &= 0 \\ \dot{z} &= v_z \\ \dot{v}_z &= 0 \end{cases} \tag{17}$$

and the problem is transformed into a trivial linear problem with new observation functions $x$, $y$ and $z$. However, this approach requires a small enough noise level to compute the required transformations in (16). Assuming that the observations (15) are now corrupted by white gaussian noise $w_t$, denoting $X = (x, y, z, v_x, v_y, v_z)$, $Y = (r, \sigma, \gamma)$, we can rewrite (17) and (15) as follows :

$$\begin{cases} \dot{X} &= AX \\ dY_t &= h(X_t)dt + Gdw_t \end{cases} \tag{18}$$

where $h$ is obtained by differentiating the right-hand-side $H(X)$ of (15) with respect to time. To evaluate the effect of noise through the preceding transformations, let us denote $K = H^{-1}$ (given by the right-hand-side of (16)) and $Z_t \stackrel{\text{def}}{=} K(Y_t) = H^{-1}(H(X_t) + Gw_t)$. Then, by Itô formula, we have :

$$dZ_t = (\frac{\partial K}{\partial Y}(Y_t)h(X_t) + \frac{1}{2}\text{tr}(GG'\frac{\partial^2 K}{\partial Y^2}(Y_t)))dt + \frac{\partial K}{\partial Y}(Y_t)Gdw_t \tag{19}$$

Comparing to $dX_t = \frac{\partial K}{\partial Y}(H(X_t))h(X_t)dt$, it results that :

$$\mathsf{E}(Z_t - X_t) = \int_0^t \mathsf{E}((\frac{\partial K}{\partial Y}(Y_s) - \frac{\partial K}{\partial Y}(H(X_s)))h(X_s))ds + \frac{1}{2}\int_0^t \mathsf{E}(\text{tr}(GG'\frac{\partial^2 K}{\partial Y^2}(Y_s)))ds \tag{20}$$

which is small when the noise variance $GG'$ is small, but becomes significantly different than zero when $GG'$ grows. It follows that for a large enough noise variance, $Z_t$ cannot be represented as the sum of $X_t$ and a gaussian white noise. Moreover, the mean error (20) may tend to infinity with t.

The situation is more complicated when the distance $r$ is not anymore available. Firstly, a one dimensional unobservable foliation results. It can be easily checked that without distance measurement, the equations of motion and of observation are invariant by similarity and the similarity's ratio is unobservable along each trajectory. Consequently, inversion formulas can be obtained for 5 over the 6 components of the state, and can be obtained independently of the unobservable state variable at the condition that the first and second time derivatives of $\sigma$ and $\gamma$ are available (see [27]). It results that the preceding observer technique can be adapted, but that the error analysis of the preceding paragraph is now much worse since differentiations of signals up to the second order are needed. For too large noise level, exact or approximated nonlinear filtering techniques may be more efficient (see [27]).

To conclude this paragraph, it should be mentionned that such differentiations can be avoided by using an "exact" discrete-time counterpart of the problem, taking the linearity of the state equation into account. Furthermore, the choice of sampling time plays the role of a (nonlinear) low-pass filter and can be used to attenuate the noise. Consequently, the choice of sampling time has to be considered as important as choices of state or outputs coordinates and the noise analysis may deliver criteria to choose between an observer design and the filtering approach. Note that filtering means in particular that the system's algebraic or geometric structure is balanced by probabilistic considerations or more precisely, that a too accurate nonlinear model is meaningless when noises strongly corrupt the state and the observations.

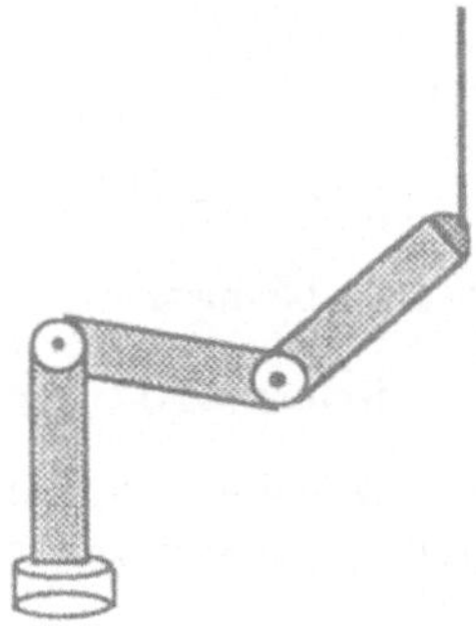

Figure 3: the inverted pendulum at the end body of the robot arm

## 3.2 Choosing good output functions : the equilibrist robot arm

For linear systems, the well-known separation principle provides an easy tool to construct output feedback laws. Under observability and controllability assumptions, an estimate of the state is obtained from the outputs by means of an observer and a stabilizing controller can be designed on the basis of such state estimates.

Unfortunately, this result is no longer true for nonlinear systems. Moreover, nonlinear observers only work in very specific cases and the theory needs further developments. It results that the role played by outputs in nonlinear systems theory is very different than the linear case and we generally assume that the whole state is measured. In fact, the output functions can be considered as tools for the control design, in particular they can be used to parametrize the controller. As for linear systems, the choice of outputs may result, in the decoupling problem or the input-output linearization, in stable or unstable feedback systems. More precisely, the decoupling feedback generally makes part of the system unobservable and the stability of this part is not anymore affected by control. Consequently, the choice of outputs may produce a stable observable and controllable part and simultaneously, an unstable controller. This suggests that the outputs should be used as an additional degree of freedom to obtain stabilizing controls and particularly around reference trajectories whose stability is difficult to analyze. We shall discuss this aspect on a robotic problem.

We consider a robot arm topped by an inverted pendulum (Figure 3) and we aim at stabilizing the ruler at its upper (unstable) equilibrium position while the end body simultaneously follows a reference trajectory.

A model of this mechanical system is easily delivered by Lagrangian methods :

$$\Gamma_0(z)\ddot{z} = \Gamma(z,\dot{z}) + Qu \tag{21}$$

where $z$ is the $n_1$ dimensional vector of relative angular variables between two successive links, $\dot{z}$ being the vector of angular velocities, $u$ is the $m$ dimensional vector of motor torques (inputs), with $m \leq n_1$. Note that the (constant) matrix $Q$ is assumed to have full rank, namely $\text{rk}Q = m$, and the inequality $m \leq n_1$ results from the fact that the pendulum turns freely and thus is not directly controlled by actuators. It can be easily checked that feedback linearization results do not hold in this case, unless $m = n_1$, which is contrary to our assumption.

Let us denote $x = (z, \dot{z})$. Let $\{x_r(t) | t \in \mathbf{R}_+\}$ be a trajectory satisfying (21) for the reference input trajectory $\{u_r(t) | t \in \mathbf{R}_+\}$.

The problem consists in choosing $m$ output functions $h_1(x,t), \ldots, h_m(x,t)$ and state feedback $u$ such that, denoting $y = h(x,t)$ the $m$ dimensional vector of outputs, we have :

(i) the static state feedback $u(x,t)$ is such that the output $y = h(x,t)$ satisfies the first order linear decoupled differential system :

$$\dot{y} = -A_0 y + w \tag{22}$$

with $A_0$ a constant diagonal $m \times m$ stable matrix.

(ii) $u(x,t)$ is such that the reference trajectory $x_r(.)$ is asymptotically stable.

Precisely, we are looking for a feedback law $u$ such that the closed-loop system is the stable system (22) around $y_r = \lim_{t\to\infty} h(x_r(t),t)$, and such that the unobservable part remains locally stable around $x_r(.)$.

To state the result, more notations are needed.

Assuming that $\Gamma_0$ is everywhere invertible, equation (21) can be rewritten as follows :

$$\dot{x} = f(x) + g(x)u \tag{23}$$

with $x = (x_1, x_2)$, $f(x) = \begin{pmatrix} x_2 \\ \Gamma_0^{-1}(x_1)\Gamma(x_1,x_2) \end{pmatrix}$ and $g(x) = \begin{pmatrix} 0 \\ \Gamma_0^{-1}(x_1)Q \end{pmatrix}$. We also denote $\Delta$ the decoupling matrix associated to (23) with outputs $h$. Since the output must satisfy (22), it is trivially seen that all the characteristic numbers must be 0 and that $\Delta = \frac{\partial h}{\partial x_2}\Gamma_0^{-1}Q$ (see [21], [8]).

We also denote $F(t) = (\frac{\partial f}{\partial x}(x_r(t)) + \frac{\partial g}{\partial x}(x_r(t))u_r(t))$ , $G(t) = g(x_r(t))$ and $H(t) = \frac{\partial h}{\partial x}(x_r(t),t)$ the time-varying matrices of the tangent linear approximation of (23) with outputs $h$.

The idea consists simply in finding h such that the state feedback law which simultaneously decouples and linearizes the input-output map $u \to y$ :

$$u(x,t) = \Delta^{-1}(x,t)(-A_0 h(x,t) - L_f h(x,t) - \frac{\partial h}{\partial t}(x,t) + w) \tag{24}$$

guarantees that the first order approximation of (23) is stabilized at the reference trajectory, precisely :

**Proposition 3.2** *Assume that there exists a time varying $m \times n$ matrix $C(t)$ such that $(F(t) - G(t)C(t))$ is stable. Let also $H(t)$ be a time-varying $m \times n$ matrix satisfying :*

$$\dot{H}(t) = -H(t)(F(t) - G(t)C(t)) - A_0 H(t) \tag{25}$$

*then $u$ and $h$ solution of (i) and (ii) are locally defined by :*

$$\frac{\partial u}{\partial x}(x_r(t),t) = -C(t) \quad , \quad \frac{\partial h}{\partial x}(x_r(t),t) = H(t) \tag{26}$$

The proof can be found in [8] and is an easy adaptation of [9]. Note that the desired matrix $C(t)$ can be computed by linear quadratic optimization techniques which means solving a matrix Riccatti equation.

In our context, this result can be successfully applied to follow particular periodic orbits such as, for example, the robot arm turning around its vertical axis with constant velocity, with the inverted pendulum slightly bending to compensate the centrifugal torque. It should be remarked firstly that , even if $u$ and $h$ are designed by means of their linear approximations, the fact that $u$ satisfies (24) guarantees that the nonlinearities of the system (23) are taken into account and locally compensated, providing a larger stability domain than with a linear first order stabilizing control. Secondly, this approach produces a locally robust controller since small errors on the model do not affect the stability of $F - GC$. See [8] for a detailed discussion and simulation results.

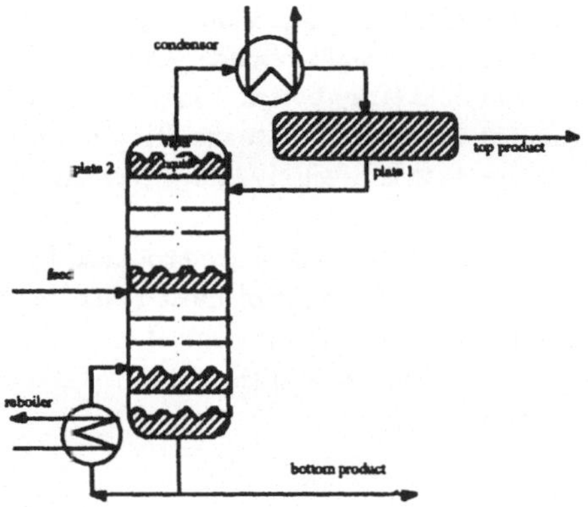


Figure 4: Distillation column

## 3.3 Choosing good time-scales : distillation columns

We have already pointed out the difficulties to derive reliable models by the actual input-output methods. In fact, an intermediary technique exists to reduce large knowledge based models with different time-scales and obtain a simplified model for control purpose. Another aspect of time scales relies on the links between modeling and the definition of the control objectives : to control slowly varying processes, a simplified model suffices, whereas control of slow trends coupled with fast transients may lead to impossibilities unless the control is designed according to the natural scaling of the process, with a choice of coordinates for which the slow process does not depend on the fast one (see [32]). Distillation columns provide a convincing example of such a situation.

One of the main distillation control problems consists in obtaining distillated products with fixed composition whereas the feed composition varies significantly and cannot be measured. The problem of rejection of perturbations is thus immediately invoked.

Note that, since high purity is wanted for the top and bottom products and since nonlinear phenomena cannot be neglegted in this case, linear control techniques produce mediocre performances.

On the other hand, depending on the nonlinear model used to design the controller, very different behaviors concerning stability and robustness can be observed, whereas the models are almost equivalent. More precisely, if a tray-by-tray model is used, one obtains a blowing up control at the steady-state, whereas with a suitable aggregation of trays, the associated controller is smooth and robust. Let us comment this point.

In the process of distillation, the law of mass conservation is strongly coupled with thermodynamical equilibria between liquid and vapor on each plate of the column (see Figure 4). Actually, the resulting process may have two different time scales, one related to the rest time of one unit of each product on each tray and the second related to the time needed to achieve the liquid-vapor equilibrium. However, even slower aspects are of interest in a column since the only things that can generally be observed are the products of distillation for which the rest time inside the column is very large compared to the one on each tray. On the other hand, modern columns are designed in such a way that the steady state on each plate is fastly reached whatever the load. This stability consideration suggests that the separations realized on each tray are comparable and that one can deal with groups of trays for which an average behavior can be defined. Finally, since the first and last trays are much bigger than the others, it seems natural to group the small trays such that their sum is comparable to the two big ones. Following the techniques described in [30], one obtains a slow (reduced) model for which the rejection of perturbations of the feed composition is achieved by slow controls varying inside tight bounds. Comparing to the controller solving the same problem but with a more complete (tray-by-tray) model (see [17]), one can easily check that the decoupling matrix is singular at any steady-state resulting in an exploding control. Note that this singularity can be passed if the state is precisely known, which seems to be related to the time scales considerations.

To summarize, certain singularities of the control are produced by a wrong choice of time scales that can be interpreted by the fact that the control must reach long term objectives, while the stable short term behavior of the system damps the control and prevents it from enough exciting the system. Such a singularity clearly disappears in the coordinates adapted to the slow time scale. A more detailed discussion can be found in [30].

# References


[1] S. A. BILLINGS, *Identification of nonlinear systems - a survey.* IEE PROC., 127, D, 6, 272-285, 1980.

[2] B. BONNARD, *Contrôle de l'attitude d'un satellite rigide.* In Développements et Utilisation d'Outils et Modèles Mathématiques en Automatique, Analyse des Systèmes et Traitement du Signal. I. Landau coordonnateur, Editions du CNRS, Vol. 3, 1983.

[3] B. CHARLET, J. LEVINE and R. MARINO, *Dynamic feedback linearization with application to aircraft control.* Proc. of the 27th IEEE Conf. on Decision and Control, Austin, Texas, 1988.

[4] L. O. CHUA, *Dynamic nonlinear networks : state-of-the-art.* IEEE. Trans. Circuits Systems, 27, 1980, 1059-1087.

[5] D. CLAUDE, *Everything you always wanted to know about linearization.* In Algebraic and Geometric Methods in Nonlinear Control Theory, M. Fliess, M. Hazewinkel eds., North-Holland, 181-226, 1986.

[6] D. CLAUDE *Automatique et régulation biologique.* This volume.

[7] D. CLAUDE, A. GLUMINEAU and C. MOOG, *Nonlinear decoupling and immersion techniques applied to a single point mooring of a tanker.* Proc. of the 24th CDC, Fort Lauderdale, 1985.

[8] B. D'ANDREA and J. LEVINE *Synthesis of nonlinear state feedback for the stabilization of a class of manipulators.* Large Scale Systems and Applications, 1988.

[9] B. D'ANDREA and L. PRALY *About finite nonlinear zeros for decouplable systems.* Systems and Control Letters, 10, 1988, 103-109.

[10] H. DANG VAN MIEN and D. NORMAND-CYROT, *Nonlinear state affine identification methods : applications to electrical power plants.* Automatica, 20, 2, 1984, 175-188.

[11] A. DE LUCA and G. ULIVI, *Dynamic decoupling of voltage frequency controlled induction motors.* In Analysis and Optimization of Systems, A. Bensoussan, J. L. Lions eds., Lect. Notes in Control and Inf. Sci., Springer, 111, 127-137, 1988.

[12] H. DIAZ and A. A. DESROCHERS, *Modeling of nonlinear discrete-time systems from input-output data.* Proc. of the 10th IFAC World Congress, Munich, 1987.

[13] D. DOCHAIN and G. BASTIN, *Adaptive identification and control algorithms for nonlinear bacterial growth systems.* Automatica, 20, 5, 1984, 621-634.

[14] M. FLIESS, *Un outil algébrique : les séries formelles non commutatives.* in Mathematical Systems Theory, G. Marchesini, S. K. Mitter eds., Lect. Notes in Economics, Mathematics and Systems, Vol 131, Springer, 1976, 122-148.

[15] E. FREUND, *Fast nonlinear control with arbitrary pole placement for industrial robots and manipulators.* Int. J. Robotics Research, 1 (1), 1982.

[16] A. FROMENT, *Commande digitale d'un amortisseur actif.* Thèse de Docteur-Ingénieur, ENSMP, 1984.

[17] J. P. GAUTHIER, G. BORNARD, S. BACHA and M. IDIR, *Rejet de perturbations pour un modèle non linéaire de colonne à distiller.* In Développements et Utilisation d'Outils et Modèles Mathématiques en Automatique, Analyse des Systèmes et Traitement du Signal. I. Landau coordonnateur, Editions du CNRS, Vol. 3, 1983.

[18] E. D. GILLES, *Model based techniques for controlling processes in chemical engineering.* Proc. of the 10th IFAC World Congress, Munich, 1987.

[19] M. HASLER and J. NEIRYNCK, *Circuits Non Linéaires.* Presses Polytechniques Romandes, Lausanne, 1985.

[20] M. ILIC-SPONG, R. MARINO, S. PERESADA and D. G. TAYLOR *Feedback linearizing control of switched reluctance motors.* IEEE Trans. Autom. Control, AC-32, 5, 1987, 371-379.

[21] A. ISIDORI, *Nonlinear Control Systems : An Introduction.* Lecture Notes in Control and Information Sciences, Vol. 72, Springer 1985.

[22] B. JAKUBCZYK and W. RESPONDEK, *On linearization of control systems.* Bull. Acad. Polonaise Sci. Ser. Sci. Math. 28, 1980, 517-522.

[23] A. JOHNSON, *The control of fed-batch fermentation processes - a survey.* Automatica, 23, 6, 1987, 691-705.

[24] M. KORTMANN and H. UNBEHAUEN, *Identification methods for nonlinear MISO systems.* Proc. of the 10th IFAC World Congress, Munich, 1987.

[25] A. KRENER and W. RESPONDEK, *Nonlinear observers with linearizable error dynamics.* SIAM J. Control & Optimiz. 23, 2, 1985, 197-216.

[26] J. LEVINE and R. MARINO, *Nonlinear systems immersion, observers and finite-dimensional filters.* Systems & Control Letters, 5, 1986, 403-412.

[27] J. LEVINE and R. MARINO, *Constant-speed target tracking via bearings-only measurements.* Submitted.

[28] J. LEVINE and R. MARINO, *On fault-tolerant observers.* Submitted.

[29] J. LEVINE and G. PIGNIE, *Exact finite dimensional filters for a class of nonlinear discrete-time systems.* Stochastics, 18, 1986, 97-132.

[30] J. LEVINE and P. ROUCHON, *Disturbances rejection and integral control of aggregated nonlinear distillation models.* In Analysis and Optimization of Systems, A. Bensoussan, J.L. Lions eds., Lecture Notes in Control and Information Sciences, Springer, 1986.

[31] R. MARINO, *On the largest feedback linearizable subsystem.*Systems and Control Letters, 6, 1986, 345-351.

[32] R. MARINO and P. KOKOTOVIC, *A geometric approach to nonlinear singularly perturbed control systems.* Automatica, 24, 1, 1988, 31-41.

[33] G. MEYER, R. SU and L. HUNT, *Applications to aeronautics of the theory of transformations of nonlinear systems.* In Développements et Utilisation d'Outils et Modèles Mathématiques en Automatique, Analyse des Systèmes et Traitement du Signal. I. Landau coordonnateur, Editions du CNRS, Vol. 3, 1983.

[34] S. MONACO and S. STORNELLI, *A nonlinear feedback control law for attitude control.* In Algebraic and Geometric Methods in Nonlinear Control Theory, M. Fliess, M. Hazewinkel eds., North-Holland, 573-595, 1986.

[35] C. REBOULET, P. MOUYON and C. CHAMPETIER, *About the local linearization of nonlinear systems.* In Algebraic and Geometric Methods in Nonlinear Control Theory, M. Fliess, M. Hazewinkel eds., North-Holland, 311-322, 1986.

[36] W. RUGH, *Nonlinear System Theory. The Volterra-Wiener Approach.* The Johns Hopkins University Press, Baltimore, 1981.

[37] S. N. SINGH and A. SCHY, *Output feedback nonlinear decoupled control synthesis and observer design for maneuvering aircraft.* Int. J. Control, 31, 1980, 781-806.

[38] J. WOOD, *Power conversion in electrical networks.* PHD Dissertation, Harvard University, 1974.

# PUTTING PHYSICS BACK IN CONTROL

**Jean-Jacques E. Slotine**
Nonlinear Systems Laboratory
Massachusetts Institute of Technology
Cambridge MA 02139, USA

## ABSTRACT

*In designing feedback controllers for complex nonlinear systems, the physical properties of the system plants are often overlooked. In this paper, we argue that scalar summarizing properties, such as energy conservation or entropy production, may often be used effectively in the design of controllers for multi-input nonlinear physical systems. For instance, the conservation of total mechanical energy allows one to show simply the stability of simple proportional-derivative position controllers for robot manipulators or attitude controllers for rigid spacecraft, and can also be systematically exploited to design adaptive tracking controllers for these systems. Similar approaches may be used in the design of controllers for complex chemical processes. The development points towards a more "hand-crafted", physically motivated approach to nonlinear control system design.*

### 1. INTRODUCTION

Feedback control is well understood and widely used for systems with predominantly linear dynamics. It is also well developed for large classes of nonlinear sytems with single inputs or uncoupled multiple inputs. For general multi-input nonlinear systems, however, feedback control is still very much a research topic, whose urgency has been rendered more acute by the recent development of machines with challenging nonlinear dynamics, such as robot manipulators, themselves made economically viable by the availability of cheap computation.

In this paper, we argue that the new, richer problems posed by nonlinear systems control, may require a departure from the purely mathematical approach of most traditional design methodologies, and especially may benefit from a closer look at the physics of the systems considered. In particular, we show that scalar summarizing properties such as energy conservation hold significant potential in the design of position and tracking controllers for nonlinear systems. In a sense, the discussion may be viewed as a renewed emphasis on the motivation of early fundamental work in nonlinear system analysis [Lyapunov, 1907], which evolved from physical considerations.

We motivate the approach by detailing specific examples, namely, the trajectory control of robot manipulators [Slotine and Li, 1986] in Section 2, the attitude tracking control of spacecraft [Slotine and Di Benedetto, 1988] in Section 3, and the control of chemical stirred tanks [Ydtsie and Slotine, 1988] in Section 4. Generalizations are discussed in Section 5.

### 2. ROBOTICS AS A PROTOTYPE

Robot manipulators are familiar examples of trajectory-controllable mechanical systems. Their strongly nonlinear dynamics present, however, a challenging control problem, to which traditional control approaches, most of which developed for linear systems, do not easily apply. The difficulty was for a while mitigated by the fact that manipulators were highly geared, thereby strongly reducing the interactive dynamic effects between links. However, as designs evolved in recent years towards "cleaner" approaches, such as gear-free, direct-drive arms, featuring reduced friction and avoiding backlash altogether, explicit account of the nonlinear dynamic effects became critical in order to exploit the full dynamic potential of the new high-performance manipulator arms.

Consider, for instance, a planar, two-link, articulated manipulator, whose position can be described by a 2-vector $\mathbf{q}$ of joint angles, and whose actuator inputs consist of a 2-vector $\tau$ of torques applied at the manipulator joints. The dynamics of this simple manipulator (detailed in the Appendix) is strongly nonlinear. It can be written in the general form (see e.g., [Asada and Slotine, 1988])

$$\mathbf{H}(\mathbf{q})\ddot{\mathbf{q}} + \mathbf{C}(\mathbf{q},\dot{\mathbf{q}})\dot{\mathbf{q}} + \mathbf{g}(\mathbf{q}) = \tau \tag{1}$$

where $\mathbf{H}(\mathbf{q})$ is the 2×2 manipulator inertia matrix (which is symmetric positive definite), $\mathbf{C}(\mathbf{q},\dot{\mathbf{q}})\dot{\mathbf{q}}$ is a 2-vector of centripetal and Coriolis torques (with $\mathbf{C}(\mathbf{q},\dot{\mathbf{q}})$ a 2×2 matrix), and $\mathbf{g}(\mathbf{q})$ is the 2-vector of gravitational torques. The feedback control problem for such a system is to compute the required actuator inputs to perform desired tasks (e.g., follow a desired trajectory), given the measured system state, namely the vector $\mathbf{q}$ of joint angles, and the vector $\dot{\mathbf{q}}$ of joint velocities.

### 2.1 Position Control

Let us first assume, for simplicity, that the manipulator is in the horizontal plane ( $\mathbf{G}(\mathbf{q}) \equiv \mathbf{0}$ ), and that the task is to move it to a given final position, as specified by a constant vector $\mathbf{q}_d$ of desired joint angles $q_{dj}$. It is physically clear that a so-called joint P.D. (proportional-derivative) controller, namely a feedback control law that selects each actuator input independently, based on the local measurements of position errors $\tilde{q}_j = q_j - q_{dj}$ and joint velocities $\dot{q}_j$

$$\tau_j = - k_{Pj}\tilde{q}_j - k_{Dj}\dot{q}_j \qquad\qquad j = 1,2 \tag{2}$$

will achieve the desired position control task. Indeed, the control law (2) , where $k_{Pj}$ and $k_{Dj}$ are strictly positive constants, is simply mimicking the effect of equiping each of the manipulator's joints with a passive mechanical device composed of a coil-spring and a damper, and having the desired $q_{dj}$ as its rest position; the resulting passive physical system would simply exhibit damped oscillations towards the rest position $\mathbf{q}_d$.

Yet, writing the system dynamics in the " $\mathbf{f} = m\,\mathbf{a}$ " (i.e., Newtonian) form (1) does not easily capture this simple fact. In order to formalize the above discussion, a more appropriate approach is to rewrite the system's dynamics in terms of energy transfers (i.e., in a Lagrangian or Hamiltonian form). This remark, which should sound familiar to physicists, dates back to Kelvin [Thomson and Tait, 1886], and was more recently rediscovered in the literature under various forms [Takegaki and Arimoto, 1981; Koditschek, 1984; Van der Schaft, 1986], as well as exploited implicitly in robotic task-oriented methodologies such as impedance control [Hogan, 1981, 1985a,b; Salisbury, 1980]. We believe that the physical reasoning can be most easily formalized by writing conservation of energy in the form

$$\frac{1}{2}\frac{d}{dt}[\,\dot{\mathbf{q}}^T\mathbf{H}\dot{\mathbf{q}}\,] = \dot{\mathbf{q}}^T.\tau \tag{3}$$

where the left-hand side is the derivative of the manipulator's kinetic energy, and the right-hand side represents the power input from the actuators. Expression (3) does not mean, of course, that the Coriolis and centripetal terms of (1) have disappeared, but simply that they are now accounted for implicitly, since they stem from the time-variation of the inertia matrix **H**. The stability and convergence proof for the P.D. controller above can then be derived very simply. Let us actually take the control input in a form slightly more general than (2), namely

$$\tau = -\mathbf{K}_P \tilde{\mathbf{q}} - \mathbf{K}_D \dot{\mathbf{q}} \tag{4}$$

where $\mathbf{K}_P$ and $\mathbf{K}_D$ are constant symmetric positive definite matrices (expression (2) corresponds to having $\mathbf{K}_P$ and $\mathbf{K}_D$ diagonal), and let us consider the total mechanical energy $V$ that would be associated with the system if control law (4) were implemented using physical springs and dampers, namely

$$V = \frac{1}{2}[\, \dot{\mathbf{q}}^T \mathbf{H} \dot{\mathbf{q}} + \tilde{\mathbf{q}}^T \mathbf{K}_P \tilde{\mathbf{q}} \,]$$

To analyze the closed-loop behavior of the controlled system, *we shall use this virtual mechanical energy V as our Lyapunov function*. The time-derivative of $V$ can be written, given (3), as

$$\dot{V} = \dot{\mathbf{q}}^T (\tau + \mathbf{K}_P \tilde{\mathbf{q}}\,) \tag{5}$$

which, using control law (4), simplifies as

$$\dot{V} = -\dot{\mathbf{q}}^T \mathbf{K}_D \dot{\mathbf{q}} \leq 0$$

Not surprisingly, we recognize $\dot{V}$ as the power dissipated by the virtual dampers. We now only need to check that the system cannot get "stuck" at a stage where $\dot{V} = 0$ while $\mathbf{q} \neq \mathbf{q}_d$, or, to put it more technically, invoke the invariant-set theorem [La Salle and Lefschetz, 1961]. Since

$$\dot{V} = 0 \quad \Rightarrow \quad \dot{\mathbf{q}} = \mathbf{0} \quad \Rightarrow \quad \ddot{\mathbf{q}} = -\mathbf{H}^{-1}\mathbf{K}_P \tilde{\mathbf{q}}$$

one has $\dot{V} \equiv 0$ only if $\tilde{\mathbf{q}} = \mathbf{0}$, and therefore the system does converge to $\mathbf{q} \equiv \mathbf{q}_d$, as the physical reasoning suggested.

### 2.2 Adaptive Trajectory Control

In this section, we extend the previous discussion to the case where the manipulator is actually required to follow a desired trajectory, rather than merely reach a desired position. While the simple P.D. controller above cannot effectively handle the dynamic demands of trajectory tracking (although it does preserve overall stability), the new control problem can still be addressed *without any a priori knowledge of the system's mass properties*, by having the control system be *adaptive*, as we now discuss. The development is based on our earlier work [Slotine and Li, 1986], and applies to general *n*-degree-of-freedom manipulators.

In order to easily address trajectory control problems, our previous discussion of energy conservation needs to be formalized a step further. In the presence of external gravity torques $\mathbf{G}(\mathbf{q})$, energy conservation can be written (generalizing expression (3) ) as

$$\frac{1}{2}\frac{d}{dt}[\, \dot{\mathbf{q}}^T \mathbf{H} \dot{\mathbf{q}} \,] = \dot{\mathbf{q}}^T (\tau - \mathbf{G}) \tag{6}$$

Differentiating the left-hand side explicitely

$$\frac{1}{2}\frac{d}{dt}[\dot{\mathbf{q}}^T \mathbf{H} \dot{\mathbf{q}}] = \dot{\mathbf{q}}^T \mathbf{H} \ddot{\mathbf{q}} + \frac{1}{2}\dot{\mathbf{q}}^T \dot{\mathbf{H}} \dot{\mathbf{q}}$$

and expanding the term $\mathbf{H}\,\ddot{\mathbf{q}}$ using the system dynamics

$$\mathbf{H(q)}\ddot{\mathbf{q}} + \mathbf{C}(\mathbf{q},\dot{\mathbf{q}})\dot{\mathbf{q}} + \mathbf{G(q)} = \tau$$

we get

$$\dot{\mathbf{q}}^T(\tau - \mathbf{G}) = \dot{\mathbf{q}}^T[\tau - (\mathbf{C}(\mathbf{q},\dot{\mathbf{q}})\dot{\mathbf{q}} + \mathbf{G(q)})] + \frac{1}{2}\dot{\mathbf{q}}^T \dot{\mathbf{H}} \dot{\mathbf{q}}$$

and therefore conclude that for all $\dot{\mathbf{q}}$

$$\dot{\mathbf{q}}^T(\dot{\mathbf{H}} - 2\mathbf{C})\dot{\mathbf{q}} = 0 \tag{7}$$

More specifically, one can easily show [Slotine and Li, 1987a] that with a proper definition of $\mathbf{C}$ (infinitely many definitions of the matrix $\mathbf{C}$ yield the same vector $\mathbf{C}\dot{\mathbf{q}}$ of Coriolis and centripetal torques), the matrix $(\dot{\mathbf{H}} - 2\mathbf{C})$ is *skew-symmetric* – this is not quite a direct consequence of (7), since the matrix itself depends on $\dot{\mathbf{q}}$. With this further formalization of energy conservation, we are now ready to address the trajectory control problem.

Given the desired trajectory $\mathbf{q}_d(t)$ (we shall assume that $\mathbf{q}_d(t)$, $\dot{\mathbf{q}}_d(t)$, and $\ddot{\mathbf{q}}_d(t)$ are all bounded), and with some or all the manipulator parameters being unknown, the controller design problem is to derive a control law for the actuator torques, and an estimation law for the unknown parameters, such that the manipulator output $\mathbf{q}(t)$ closely tracks the desired trajectory. To this effect, we consider the Lyapunov function candidate

$$V(t) = \frac{1}{2}[\mathbf{s}^T\mathbf{H}\mathbf{s} + \tilde{\mathbf{a}}^T\Gamma\tilde{\mathbf{a}}] \tag{8}$$

where $\tilde{\mathbf{a}} = \hat{\mathbf{a}} - \mathbf{a}$ is the parameter estimation error, with $\mathbf{a}$ being a constant vector of unknown manipulator parameters (which we shall detail later), and $\hat{\mathbf{a}}$ its estimate; $\Gamma$ is a positive definite matrix; and the vector $\mathbf{s}$, a measure of tracking accuracy, is defined as

$$\mathbf{s} = \dot{\tilde{\mathbf{q}}} + \Lambda\tilde{\mathbf{q}} = \dot{\mathbf{q}} - \dot{\mathbf{q}}_r \tag{9}$$

with

$$\dot{\mathbf{q}}_r = \dot{\mathbf{q}}_d - \Lambda\tilde{\mathbf{q}} \tag{10}$$

where $\tilde{\mathbf{q}} = \mathbf{q} - \mathbf{q}_d$ is the tracking error, and $\Lambda$ is a symmetric positive definite matrix.

The "reference velocity" vector $\dot{\mathbf{q}}_r$ of (10), formed by modifying the desired velocities $\dot{\mathbf{q}}_d$ using the position error $\tilde{\mathbf{q}}$, is introduced to guarantee the convergence of the tracking error to zero, as we shall see below. It simply represents a notational manipulation which allows one to easily translate energy-related properties (expressed in terms of the actual joint velocity vector $\dot{\mathbf{q}}$ ) into trajectory control properties (expressed in terms of the virtual velocity error vector $\mathbf{s}$ ). The vector $\mathbf{s}$ conveys information about boundedness and convergence of $\mathbf{q}$ and $\dot{\mathbf{q}}$ , since *the definition* (9) *of* $\mathbf{s}$ *can also be viewed as a stable first-order differential equation in* $\tilde{\mathbf{q}}$ , with $\mathbf{s}$ as an input. Thus, assuming bounded initial conditions, showing the boundedness of $\mathbf{s}$ also shows the boundedness of $\tilde{\mathbf{q}}$ and $\dot{\tilde{\mathbf{q}}}$ , and therefore of $\mathbf{q}$ and $\dot{\mathbf{q}}$ ; similarly, if $\mathbf{s}$ tends to $\mathbf{0}$ as $t \rightarrow \infty$, so do $\tilde{\mathbf{q}}$ and $\dot{\tilde{\mathbf{q}}}$ .

Let us now differentiate $V(t)$ in (8)

$$\dot{V}(t) = \mathbf{s}^T(\mathbf{H}\ddot{\mathbf{q}} - \mathbf{H}\ddot{\mathbf{q}}_r) + \tilde{\mathbf{a}}^T\Gamma\dot{\tilde{\mathbf{a}}} + \frac{1}{2}\,\mathbf{s}^T\dot{\mathbf{H}}\mathbf{s}$$

that is, substituting $\mathbf{H}\ddot{\mathbf{q}}$ from the system dynamics,

$$\dot{V}(t) = \mathbf{s}^T(\tau - \mathbf{H}\ddot{\mathbf{q}}_r - \mathbf{C}\dot{\mathbf{q}}_r - \mathbf{G}) + \tilde{\mathbf{a}}^T\Gamma\dot{\tilde{\mathbf{a}}}$$

where the skew-symmetry of $(\dot{\mathbf{H}} - 2\mathbf{C})$ has been used to eliminate the term $1/2\,\mathbf{s}^T\dot{\mathbf{H}}\mathbf{s}$ . Now, given a proper definition of the unknown parameter vector $\mathbf{a}$, we can define a matrix $\mathbf{Y} = \mathbf{Y}(\mathbf{q}, \dot{\mathbf{q}}, \dot{\mathbf{q}}_r, \ddot{\mathbf{q}}_r)$ such that

$$\mathbf{H}(\mathbf{q})\ddot{\mathbf{q}}_r + \mathbf{C}(\mathbf{q},\dot{\mathbf{q}})\dot{\mathbf{q}}_r + \mathbf{G}(\mathbf{q}) = \mathbf{Y}(\mathbf{q}, \dot{\mathbf{q}}, \dot{\mathbf{q}}_r, \ddot{\mathbf{q}}_r)\,\mathbf{a}$$

and therefore

$$\tilde{\mathbf{H}}(\mathbf{q})\ddot{\mathbf{q}}_r + \tilde{\mathbf{C}}(\mathbf{q},\dot{\mathbf{q}})\dot{\mathbf{q}}_r + \tilde{\mathbf{G}}(\mathbf{q}) = \mathbf{Y}(\mathbf{q}, \dot{\mathbf{q}}, \dot{\mathbf{q}}_r, \ddot{\mathbf{q}}_r)\,\tilde{\mathbf{a}}$$

where

$$\tilde{\mathbf{H}} = \hat{\mathbf{H}} - \mathbf{H} \qquad \tilde{\mathbf{C}} = \hat{\mathbf{C}} - \mathbf{C} \qquad \tilde{\mathbf{G}} = \hat{\mathbf{G}} - \mathbf{G}$$

While such linear parametrization of the dynamics is obvious for the gravitational torques, it can obtained more generally by referring all mass properties to the center of mass of each link (e.g., [Khosla and Kanade, 1985; Atkeson, *et al.*, 1985]).

Taking the control law to be

$$\tau = \hat{\mathbf{H}}\ddot{\mathbf{q}}_r + \hat{\mathbf{C}}\dot{\mathbf{q}}_r + \hat{\mathbf{G}} - \mathbf{K}_D\mathbf{s} = \mathbf{Y}\hat{\mathbf{a}} - \mathbf{K}_D\mathbf{s} \tag{11a}$$

so as to now include a "feedforward" term $\mathbf{Y}\hat{\mathbf{a}}$ , in addition to the P.D. term $\mathbf{K}_D\mathbf{s}$ ; and updating the parameter estimates $\hat{\mathbf{a}}$ according to the correlation integrals

$$\dot{\hat{\mathbf{a}}} = -\Gamma^{-1}\mathbf{Y}^T\mathbf{s} \tag{11b}$$

then yields

$$\dot{V}(t) = -\,\mathbf{s}^T \mathbf{K}_D \mathbf{s} \;\leq\; 0 \tag{12}$$

This implies (as intuition suggests, and as detailed below mathematically) that the output error converges to the surface

$$\mathbf{s} = \dot{\tilde{\mathbf{q}}} + \Lambda \tilde{\mathbf{q}} = \mathbf{0}$$

which in turn shows that $\tilde{\mathbf{q}} \rightarrow \mathbf{0}$ as $t \rightarrow \infty$. Therefore, both global stability of the system, and convergence of the tracking error, are guaranteed by the above adaptive controller.

Let us detail the mathematics. The function $V$ is not formally a Lyapunov function, but simply a positive non-increasing function of time. Since $\dot{V}$ is negative or zero, and further since $V$ is lower bounded (by zero), $V$ tends towards a constant as $t \rightarrow \infty$, and therefore remains bounded for $t \in [0, \infty]$. Given the definition (8) of $V$, this in turn implies, since $\mathbf{H}$ is uniformly positive definite (i.e., $\mathbf{H} \geq h\mathbf{I}$ for some strictly positive $h$), that $\mathbf{s}$ is bounded, and therefore that $\mathbf{q}$ and $\dot{\mathbf{q}}$ are bounded; it also implies that $\tilde{\mathbf{a}}$ is bounded, and therefore that $\hat{\mathbf{a}}$ is bounded. From the system dynamics, this then makes $\dot{\mathbf{s}}$ bounded, and thus $\mathbf{s}$ is *uniformly continuous* on $t \in [0, \infty]$. Assuming that the (perhaps time-varying) matrix $\mathbf{K}_D$ is chosen to be uniformly continuous (for instance, to be constant), $\dot{V}$ is then uniformly continuous on $t \in [0, \infty]$; and therefore, since $V$ is bounded on that time interval, and $\dot{V}$ is of constant sign ($\dot{V} \leq 0$), $\dot{V}$ *goes to zero as* $t \rightarrow \infty$. Assuming that $\mathbf{K}_D$ is uniformly positive definite (as is again the case if $\mathbf{K}_D$ is chosen to be constant), this implies from (12) that $\mathbf{s} \rightarrow \mathbf{0}$ as $t \rightarrow \infty$, and therefore that $\tilde{\mathbf{q}} \rightarrow \mathbf{0}$ as $t \rightarrow \infty$.

Note that the scheme does not necessarily estimate the unknown parameters exactly, but simply generates values that allow the the desired task to be achieved. Conditions on the desired trajectory for this to imply that the parameter estimates do converge to the exact values can be found in [Slotine and Li, 1987b]. These "sufficient richness" conditions [Morgan and Narendra, 1978] indicate how demanding the desired trajectory should be for tracking convergence to necessarily require parameter convergence. Also, of course, the tracking error does not merely tend "asymptotically" to zero, but for all practical purposes, converges within finite time constants determined for a given trajectory by the values of the gain matrices $\Lambda$, $\mathbf{K}_D$, and $\Gamma^{-1}$, themselves limited by the presence of high-frequency unmodelled dynamics and measurement noise. More precisely, the algorithm can be modified so as to guarantee exponential convergence of the tracking, with predictable convergence rates depending on the richness of the desired trajectory [Slotine and Li, 1988b].

While it may seem somewhat superfluous to keep estimating the parameter vector $\mathbf{a}$ in order to drive the very same manipulator, the practical relevance of the above approach is that the possibly large unknown loads that the manipulator may carry can also be directly accounted for, simply by considering the (securely) grasped load as part of the second link. Furthermore, as shown in [Slotine and Li, 1987a], the methodology can be easily extended to the case where only a few significant parameters have to be adapted upon (in order to simplify the computations, for instance), while the controller is robust to residual errors resulting from inaccuracies on the *a priori* estimates of the other parameters (as well as, perhaps, to bounded time-varying additive disturbances such as stiction). In particular, one may adapt only on the mass properties of the load (which can be described by at most ten parameters, namely load mass, three parameters describing the position of the center of mass, and six independent parameters describing the symmetric load inertia matrix).

Exprimental illustrations of the above development can be found in [Slotine and Li, 1988a]. They show that the dynamic parameters of the manipulator, assumed to be initially unknown, can be estimated within the first half second of a typical run, and that accordingly the manipulator trajectory can be precisely controlled. Furthermore, these experimental results demonstrate that the adaptive controller enjoys essentially the same level of robustness to unmodelled dynamics as a PD controller, yet achieves much better tracking accuracy than either PD or computed-torque schemes. This superior performance for high speed operations, in the presence of parametric and non-parametric uncertainties, and its relative computational simplicity, make it an attractive option both to address complex industrial tasks, and to simplify high-level programming of more standard operations. It is interesting to notice that the design, although designed assuming full state feedback, achieves its performance in the presence of significant measurement inaccuracies, and in particular of large sensor noise and phase-lag in the velocity signals. Also, this performance is obtained along smooth and rather "unexciting" desired trajectories, which represent a clear challenge to an adaptive controller.

Note that a trajectory control problem may arise even when the task is merely to move a load from its initial position to a final desired position. This may be due to requirements on the maximum time to complete the task, or may be necessary to avoid overshooting or otherwise bumping into obstacles during the task. These needs in turn specify the required "tracking" accuracy, and therefore which of the algorithm simplifications mentioned earlier are allowable.

Let us now discuss another example.

## 3. ATTITUDE TRACKING CONTROL OF SPACECRAFT

Robotic spacecraft potentially represent a safe and economical alternative or complement to man in the construction, maintenance, and operation of the space structures to be deployed in the next decade. They present, however, specific and difficult control problems, largely due to their nonlinear dynamics. Furthermore, while robotic spacecraft can potentially be expected to easily handle objects of masses and sizes comparable to or larger than their own (as, e.g., in releasing a payload from a shuttle orbiter, retrieving a sattelite, or performing docking or construction operations), thanks to weightlessness, such tasks involve by nature large dynamic uncertainties.

In order to deal with the inherent nonlinearity of spacecraft dynamics, many existing control schemes, e.g., [D'Amario and Stubbs, 1979; Bryson, 1985; Redding and Adams, 1987; Bergmann, *et al.*, 1987], have to rely on various aproximations (such as assuming that the attitude changes are very slow or that the applied torques are much larger than the Euler coupling torques) so that linear controller designs be applicable. Others methods, e.g. [Crouch, 1984; Dwyer, *et al.*, 1984, 1985; Monaco and Stornelli, 1985; Wie and Barba, 1985; Byrnes and Isidori, 1986], do account for (or "invert") the full nonlinear spacecraft dynamics, but require precise knowledge of all dynamic parameters.

This section, based on [Slotine and Di Benedetto, 1988], shows that the approach discussed above in a robotic context can be translated in a straightforward fashion to the accurate attitude tracking control of rigid spacecraft handling large loads of unknown dynamic properties. The method fully accounts for the nonlinear dynamics of the spacecraft, and, as in the robotics case, also presents advantages over techniques based on inverse dynamics, in terms of simplicity, easier handling of robustness issues, and capability of adaptation to the unknown mass properties of the spacecraft or the loads.

After briefly reviewing, in section 3.1, the dynamics and kinematics of spacecraft, the control algorithm is derived in Section 3.2. Extensions are discussed in Section 3.3.

### 3.1 The Spacecraft Model

We consider the attitude control of a spacecraft driven by reaction wheels. In practice, the spacecraft may also be equipped with gas-jet systems (used e.g. to control translational motion of the system, to compensate for non-zero translational momentum imparted by the loads, or to desaturate the reaction wheels), the control of which shall be commented upon in Section 3.3.

The spacecraft is treated as a rigid body whose attitude can be described by two sets of equations, namely, *kinematic* equations, which relate the time derivatives of the angular position coordinates to the angular velocity vector, and *dynamic* equations, which describe the evolution of the angular velocity vector. The development can be directly applied to the case of a spacecraft having rigidly secured a (possibly) large load of unknown dynamic properties. The results are also directly applicable to a spacecraft having itself inadequately known mass properties, due to e.g., reconfiguration, fuel variations in the gas-jet systems, thermal deformation, and so on.

#### Dynamic Equations

Let us first define the (classical) reference frames in which our attitude control problem shall be described. We assume that the control torques are applied through a set of three reaction wheels along orthogonal axes. Based on these axes, we define an arbitrary orthonormal reference frame linked to the spacecraft, which we shall refer to as the *spacecraft frame*. The origin of this frame is not necessarily the center of mass of the system, nor are the axes necessarily the principal axes of the spacecraft. We also assume that an arbitrary *inertial frame* has been defined, with respect to either fixed stars or to a reference that can be considered inertial for the duration of the attitude maneuver (e.g., a space station).

Let $\omega$ denote the angular velocity vector of the spacecraft, expressed in the spacecraft frame. The equations describing the evolution of $\omega$ in time may be written as (e.g., [Crouch, 1984])

$$\mathbf{H}\dot{\omega} = \mathbf{p} \times \omega + \tau$$

The "inertia" matrix $\mathbf{H}$ is symmetric positive definite, and can be written

$$\mathbf{H} = \mathbf{H}^0 - \mathbf{H}^A$$

where $\mathbf{H}^0$ is the total (spacecraft with reaction wheels) central inertia matrix, $\mathbf{H}^A$ is the (diagonal) matrix of axial wheels' inertias, and $\mathbf{p}$ is the total spacecraft angular momentum, all expressed in spacecraft coordinates. Note that since $\tau$ is the torque vector applied to the spacecraft by the reaction wheels, $-\tau$ is the vector of control torques actually applied by the reaction wheels motors. The $\times$ operator denotes the vector product operation, and the notation $[\mathbf{p} \times]$ refers to the skew-symmetric matrix defining the vector product by $\mathbf{p}$

$$[\mathbf{p} \times] = \begin{bmatrix} 0 & -p_3 & p_2 \\ p_3 & 0 & -p_1 \\ -p_2 & p_1 & 0 \end{bmatrix}$$

**Kinematic Equations**

The angular position of the body may be described in various ways. For example, one can consider the so-called Gibbs vector

$$\mathbf{q} = tan(\rho/2)\,\mathbf{e}$$

which derives from a transformation of the quaternion parametrization [Dwyer and Batten, 1985]. The vector **q** represents the result of a virtual rotation of ρ radians about a virtual unit axis **e**, with reference to the inertial reference frame. In that case, one can write

$$\dot{\mathbf{q}} = \mathbf{J}(\mathbf{q})\,\omega$$

where

$$\mathbf{J}[\mathbf{q}] = \frac{1}{2}\,[\mathbf{I} + \mathbf{q}\mathbf{q}^T + \mathbf{q}\times]$$

and **I** is the $3 \times 3$ identity matrix. This description is valid for $-\pi < \rho < \pi$. Using this representation, the momentum **p** can be expressed as a function of **q** by noting that

$$\mathbf{p} = \mathbf{R}(\mathbf{q})\,\mathbf{p}^I$$

where $\mathbf{p}^I$ is the (constant) inertial angular momentum, and the matrix **R(q)** represents the coordinate transformation from the inertial frame to the spacecraft frame:

$$\mathbf{R}(\mathbf{q}) = 2\,(1 + \mathbf{q}^T\mathbf{q})^{-1}\,[\mathbf{I} + \mathbf{q}\mathbf{q}^T - \mathbf{q}\times] - \mathbf{I}$$

Note that $\mathbf{R}^{-1}(\mathbf{q}) = \mathbf{R}^T(\mathbf{q}) = \mathbf{R}(-\,\mathbf{q})$.

Note that control-moment-gyroscopes (CMGs) may be used in place of reaction wheels while keeping a similar formalism. Instead of controlling angular momentum by varying the angular speeds of rotors of constant orientation, as in the case of reaction wheels, CMGs obtain the same effect by using simple or double gimbal mechanisms to vary the orientation of constant-speed rotors.

**3.2 Adaptive Attitude Tracking Control**

Let us choose state-space coordinates as the components of the vectors **x** and $\dot{\mathbf{x}}$ defined by

$$\mathbf{x} = \mathbf{q}$$

$$\dot{\mathbf{x}} = \mathbf{J}(\mathbf{q})\omega$$

This set of coordinates is well-defined, since the matrix **J** remains invertible in the domain of validity of the kinematic representation. By differentiating the expression of $\dot{\mathbf{x}}$, the equations of motion can be written, in the new coordinates, as

$$\mathbf{H}^*(\mathbf{x})\ddot{\mathbf{x}} + \mathbf{C}^*(\mathbf{x},\dot{\mathbf{x}})\dot{\mathbf{x}} = \mathbf{F}$$

with

$$\tau = \mathbf{J}^T\mathbf{F}$$

$$\mathbf{H}^*(\mathbf{x}) = \mathbf{J}^{-T}\mathbf{H}(\mathbf{x})\mathbf{J}^{-1}$$

$$\mathbf{C}^*(\mathbf{x},\dot{\mathbf{x}}) = -\mathbf{J}^{-T}\mathbf{H}\mathbf{J}^{-1}\dot{\mathbf{J}}\mathbf{J}^{-1} - \mathbf{J}^{-T}[\mathbf{p}\times]\mathbf{J}^{-1}$$

As in the robotic case, two properties of the above dynamics are exploited in the adaptive controller design. First, the matrix $(\dot{\mathbf{H}}^* - 2\mathbf{C}^*)$ is again *skew-symmetric*. Indeed

$$\begin{aligned}\dot{\mathbf{H}}^* - 2\,\mathbf{C}^* &= \frac{d}{dt}(\mathbf{J}^{-T})\,\mathbf{H}\,\mathbf{J}^{-1} + \mathbf{J}^{-T}\mathbf{H}\frac{d}{dt}(\mathbf{J}^{-1}) + 2\,\mathbf{J}^{-T}[\mathbf{p}\times]\mathbf{J}^{-1} + 2\,\mathbf{J}^{-T}\mathbf{H}\,\mathbf{J}^{-1}\dot{\mathbf{J}}\,\mathbf{J}^{-1} \\ &= \frac{d}{dt}(\mathbf{J}^{-T})\,\mathbf{H}\,\mathbf{J}^{-1} - \mathbf{J}^{-T}\mathbf{H}\frac{d}{dt}(\mathbf{J}^{-1}) + 2\,\mathbf{J}^{-T}[\mathbf{p}\times]\mathbf{J}^{-1}\end{aligned}$$

which, given the skew-symmetry of the matrix $[\mathbf{p}\times]$, implies the skew-symmetry of $(\dot{\mathbf{H}}^* - 2\,\mathbf{C}^*)$. Second, the dynamics is *linear* in terms of a properly defined constant parameter vector $\mathbf{a}$. While many such parametrizations are possible, we choose $\mathbf{a}$ to consist of the 6 independent components of the symmetric central inertia matrix $\mathbf{H}$, and of the 3 components of the constant inertial angular momentum $\mathbf{p}^I$. Given the above expressions of $\mathbf{H}^*$ and $\mathbf{C}^*$ and the relation $\mathbf{p} = \mathbf{R}(\mathbf{q})\,\mathbf{p}^I$, the matrices $\mathbf{H}^*$ and $\mathbf{C}^*$ are indeed linear in the constant parameter vector $\mathbf{a}$.

Based on the above dynamic formulation, it is now straightforward to derive an adaptive attitude tracking controller for the spacecraft, using the same proof as in the robot manipulator case. We assume that the system's state vector, namely $\mathbf{x}$ and $\dot{\mathbf{x}}$, is available (or computable) from measurements, and that the desired $\mathbf{x}_d$, $\dot{\mathbf{x}}_d$, and $\ddot{\mathbf{x}}_d$ are all bounded. Then, using the Lyapunov function candidate

$$V(t) = \frac{1}{2}\,[\,\mathbf{s}^T\mathbf{H}^*\mathbf{s} + \tilde{\mathbf{a}}^T\Gamma\tilde{\mathbf{a}}\,]$$

with the vector $\mathbf{s}$ defined as

$$\mathbf{s} = \dot{\tilde{\mathbf{x}}} + \Lambda\tilde{\mathbf{x}} = \dot{\mathbf{x}} - \dot{\mathbf{x}}_r$$

the same proof as in Section 2.2 shows that the control law and adaptation law

$$\mathbf{F} = \hat{\mathbf{H}}^*(\mathbf{x})\ddot{\mathbf{x}}_r + \hat{\mathbf{C}}^*(\mathbf{x},\dot{\mathbf{x}})\dot{\mathbf{x}}_r - \mathbf{K}_D\mathbf{s}$$

$$\dot{\hat{\mathbf{a}}} = -\Gamma^{-1}(\mathbf{Y}^*)^T\mathbf{s}$$

yield

$$\dot{V}(t) = -\mathbf{s}^T\mathbf{K}_D\mathbf{s} \le 0$$

and therefore guarantee tracking convergence. The actual control action, expressed in terms of the torques applied by the reaction wheels, can be written

$$\tau = \mathbf{J}^T(\hat{\mathbf{H}}^*\ddot{\mathbf{x}}_r + \hat{\mathbf{C}}^*\dot{\mathbf{x}}_r - \mathbf{K}_D\mathbf{s}) = \mathbf{J}^T(\mathbf{Y}^*\hat{\mathbf{a}} - \mathbf{K}_D\mathbf{s})$$

As before, the controller can be modified easily to be robust to bounded time-varying disturbances, such as those that may be created by vibrations of flexible appendages (e.g., solar panels).

Note that, since the state $(\mathbf{x}, \dot{\mathbf{x}})$ of the system is bounded (given the above Lyapunov proof), the singularity of our quaternion-based representation is not reached as long as the rotation corresponding to the desired maneuver does not exceed 180 degrees. In the case that the desired maneuver does not correspond to this requirement, one can simply decompose the control problem in subtasks, and change reference frames between tasks [Dwyer and Batten, 1985]. Also, note that while we are estimating the *central* inertia matrix **H** as part of the adaptation process, the actual position of the center of mass of the system is not assumed to be known. Finally, note that we only assumed that **x** and $\dot{\mathbf{x}}$ (i.e., the state vector) are available from measurements. If the initial angular velocity $\Omega(0)$ of the reaction wheels about their axes can also be measured, then, using the fact that

$$\mathbf{p}^I = \mathbf{R}^T(\mathbf{x}(0))\,[\mathbf{H}^0\omega(0) + \mathbf{H}^A\Omega(0)] = \mathbf{R}^T(\mathbf{x}(0))\,\mathbf{H}^A\,[\omega(0) + \Omega(0)] + \mathbf{R}^T(\mathbf{x}(0))\,\mathbf{H}\,\omega(0)$$

the above adaptive controller can be expressed (under the mild assumption that the matrix $\mathbf{H}^A$ of axial wheels inertias is known) in terms of only the 6 independent components of the inertia matrix **H**. Of course, if $\mathbf{p}^I$ itself is known (e.g., to be zero), then again the adaptive controller need only estimate the 6 independent components of the inertia matrix **H**.

Simulations in [Slotine and Di Benedetto, 1988] illustrate the approach. Also, note that, as in the robotic case, a simple P.D. controller in **F**, namely

$$\tau = \mathbf{J}^T\mathbf{F} = -\mathbf{J}^T[\mathbf{K}_P\tilde{\mathbf{x}} + \mathbf{K}_D\dot{\mathbf{x}}]$$

guarantees stable *position* control (i.e., to have $\tilde{\mathbf{x}}$ converge to **0** if $\mathbf{x}_d$ is *constant*), as can be shown easily using the Lyapunov function

$$V(t) = \frac{1}{2}[\dot{\mathbf{x}}^T\mathbf{H}^*\dot{\mathbf{x}} + \tilde{\mathbf{x}}^T\mathbf{K}_P\tilde{\mathbf{x}}]$$

### 3.3 Discussion

We believe that advanced control algorithms present exceptional potential in the dynamically clean, weightless environment of space. The above development can in principle be extended easily to include translational control of the spacecraft using gas-jets, by adding the system's mass and position of the center of mass to the list of components of the unknown parameter vector **a**. Again, if we assume the reaction wheels' axial angular velocities to be measured, and their axial inertias to be known, then only a total of 10 parameters need to be estimated as part of the adaptation process. Of course, the necessity of averaging on-off gas-jet action presents well-known problems of its own (e.g., [Bergmann, *et al.*, 1987]). In addition, the lightness requirements in space components may also present difficulties linked to the presence of low-frequency structural modes (see e.g., [Junkins and Turner, 1986]). In particular, while the previous discussion can be extended easily to control the rigid dynamics of manipulators mounted on the spacecraft, by applying the results of [Slotine and Li, 1987] and this paper using the "virtual manipulator" formalism of [Vafa and Dubowsky, 1987], practical implementation will require flexibility issues to be explicitly addressed.

## 4. CHEMICAL STIRRED TANK

This section, based on [Slotine and Ydtsie, 1988], illustrates the application of a similar approach to a non-Hamiltonian problem, namely the control of chemical stirred tanks. While the example we discuss here is particularly simple, it can be easily extended to complex multi-input applications, as detailed in [Slotine and Ydtsie, 1988].

Consider the problem of controlling temperature in a typical continuous stirred tank reactor $A \rightarrow B$ (see e.g., [Stephanopoulos, 1984]). Our approach starts by focusing attention on the *enthalpy balance* of the system, rather than the nonlinear state equations. We have

$$\dot{H} = -F\,[c_p\,\rho\,(T - T_d) + c_A\,\tilde{H}_A + c_B\,\tilde{H}_B] \;+\; F_i\,h_i\,\rho_i \;-\; Q$$

where the standard notations of [Stephanopoulos, 1984] are used, with $H$ being the enthalpy of the system, $F$ the flow, $Q$ the heat removed, $T_d$ the desired temperature, and the subscript $i$ indicating inflow variables. The process is to be controlled in the presence of possibly large uncertainties on the parameter vector

$$\mathbf{a} = [\rho c_p \;\; h_i \;\; \tilde{H}_A \;\; \tilde{H}_B]^T$$

Let us define the known (state-dependent) vector $\mathbf{y}$ as

$$\mathbf{y} = [(T - T_d)\,F + \dot{T}_d\,,\, F\,,\, \dot{c}_A - F\,c_A\,,\; \dot{c}_B - F\,c_B\,]^T$$

Then, using the known sign property

$$\partial H / \partial T \;=\; \rho\, c_p \;>\; 0$$

the Lyapunov function

$$V = 1/2\,\rho\,c_p\,(T - T_d)^2 \;+\; 1/2\,\tilde{\mathbf{a}}^T\,\Gamma\,\tilde{\mathbf{a}}$$

with the corresponding control and adaptation laws

$$Q = -\,\alpha\;(T - T_d) \;+\; \mathbf{y}^T\,\hat{\mathbf{a}}$$

$$\dot{\hat{\mathbf{a}}} = -\,\Gamma^{-1}\,\mathbf{y}\;(T - T_d)$$

where $\alpha$ is a positive gain, yields after some algebraic manipulation

$$\dot{V} = -\,\alpha\,(T - T_d)^2$$

which in turn can be shown easily to guarantee tracking convergence, using uniform continuity arguments as before.

As in the robotic case, exploiting scalar balance equations for the system and known sign properties allowed us to solve the control problem in a straightforward manner. The approach can be easily extended to complex multi-input process control applications, as detailed in [Slotine and Ydtsie, 1988].

## 5. TOWARDS PHYSICAL CONTROL OF PHYSICAL SYSTEMS

Extensions of the approach to large classes of control problems seem likely. Indeed, physical systems verify energy conservation equations of the form

$$\frac{d}{dt}[\text{ Stored Energy }] = [\text{ External Power Input }] + [\text{ Internal Power Generation}]$$

of which e.g. (6) is a particular instance. The external power input term can be written $\mathbf{y}^T\mathbf{u}$, where $\mathbf{u}$ is the input vector ("effort" or "flow"in the vocabulary of physical system modelling techniques, such as bond-graphs) and $\mathbf{y}$ is the output vector (flow or effort). By properly defining which of the physical inputs are "internal", we can always assume that the full vector $\mathbf{u}$ can be used for control purposes. Furthermore, we are not necessarily limited to dissipative or passive systems (i.e., such that their internal power generation be negative or zero), as long as the internal power generated can be compensated for by the external power input, i.e. by a proper choice of $\mathbf{u}$ (as in the case of gravitational forces in the robotic example). The major step is then to design a feedback controller structure (e.g. equation (11a) in the robotic example) that translates the energy conservation equation in terms of a dissipative mapping between parametric uncertainty and a vector $\mathbf{s} = \mathbf{y} - \mathbf{y}_r$ (rather than between $\mathbf{u}$ and $\mathbf{y}$), such that $\mathbf{s} = \mathbf{0}$ represent an exponentially stable differential equation in the the tracking error – this implies, predictably, that the number of components of s (and, accordingly, the number of independent variables to be tracked) must be equal to (or smaller than) the dimension of the control input vector $\mathbf{u}$. Furthermore, while an explicit *linear* parametrization of the dynamics still remains a key element in adaptive controller designs, bounds on the parametric uncertainty (which, in the robotic example, would correspond to bounds on the components of the vector $\mathbf{Y}\,\tilde{\mathbf{a}}$ and on disturbances) may be sufficient for simple robust designs [Slotine and Li, 1987a].

Furthermore, besides basic forms of energy conservation, the scalar summarizing property actually exploited for control system design may depend on the problem considered. In underwater vehicle dynamics, for instance, conservative effects such as added-mass can be directly accounted for by using the fluid's velocity potential flow. In systems with implicit interaction ports, available energy (in the case of interactions with quasi-ideal sources of temperature, pressure, or chemical potential) or entropy production [Glandsdorff and Prigogine, 19871] may be most suitable.

Extensions of the approach to other aspects of the control problem, in particular robustness to high-frequency unmodelled dynamics, and the exploitation of dissipative terms in the system's dynamics, are discussed in [Slotine, 1988].

Clearly, this study only represents a step towards developing systematic control methodologies for nonlinear physical systems. Actually, the very nature of the approach that we advocate implies a more "hand-crafted" , and therefore less general, view than the $\dot{\mathbf{x}} = \mathbf{f}(\mathbf{x},\mathbf{u})$ paradigm usually investigated in nonlinear systems analysis and control. This may be a reasonable price to pay for the accrued simplicity that physical insight is likely to provide. Further, this more intuitive approach could also facilitate constructive interactions between machine design and controller development.

# REFERENCES

**Asada, H., and Slotine, J.J.E.**, 1986. Robot Analysis and Control, *John Wiley & Sons.*

**Atkeson, C.G., An, C.G., and Hollerbach, J.M.**, 1985. *Int. Symp. Robotics Res.*, Gouvieux.

**Bergmann, E.V., Walker, B.K., and Levy, D.R.**, 1987. Mass Property Estimation for Control of Asymmetrical Satellites, *A.I.A.A. Journal of Guidance, Control, and Dynamics*, vol. 10, n° 5.

**Bryson, A.E.**, 1985. Control of Spacecraft and Aircraft, Stanford University, Department of Aeronautics and Astronautics Report.

**Byrnes, C.I., and Isidori, A.**, 1986. On the Attitude Stabilization of Rigid Spacecraft, *preprint.*

**Craig, J.J., Hsu, P. and Sastry, S.**, 1986. Adaptive Control of Mechanical Manipulators, *I.E.E.E. Int. Conf. Robotics and Automation*, San Francisco.

**Crouch, P.**, 1984. Spacecraft Attitude Control and Stabilization: Application of Geometric Control Theory to Rigid Body Models, *I.E.E.E. Trans. Autom. Control*, AC-29 (4).

**D'Amario, L.A., and Stubbs, G.S.**, 1979. A New Single Rotation Axis Autopilot for Rapid Spacecraft Attitude Maneuvers, *A.I.A.A. Journal of Guidance, Control, and Dynamics*, vol. 2.

**Dwyer, T.A.W.**, 1984. Exact Nonlinear Control of Large Angle Rotational Maneuvers, *I.E.E.E. Trans. Autom. Control*, AC-29.

**Dwyer, T.W.A., and Batten, A.L.**, 1985. Exact Spacecraft Detumbling and Reorientation Maneuvers with Gimbaled Thrusters and Reaction Wheels, *A.A.S. J. of the Astronautical Sciences*, vol.3.

**Dwyer, T.A.W., Fadali, M.S., and Ning Chen**, 1985. Single Step Optimization of Feedback - Decoupled Spacecraft Attitude Maneuvers, *I.E.E.E. 24th Conf. Dec. Control*, Fort Lauderdale, FL.

**Glansdorff P., and Prigogine, I.**, 1971. Thermodynamics of Structure, Stability, and Fluctuation, *John Wiley and Sons.*

**Hogan, N.**, 1981. Impedance Control of a Robotic Manipulator, *A.S.M.E. Winter Annual Meeting*, Washington.

**Hogan, N.**, 1985a. Impedance Control: An Approach to Manipulation, *A.S.M.E. J. Dynamic Systems, Measurement, and Control*, 107.

**Hogan, N.**, 1985b. Control Strategies from Computed Movements Derived from Physical Systems Theory, *Int. Symp. Synergetics*, Bavaria.

**Junkins, J.L., and Turner, J.D.**, 1986. Optimal Spacecraft Rotational Maneuvers, *Elsevier.*

**Khosla, P.,and Kanade, T.**, 1985. Parameter Identification of Robot Dynamics, *I.E.E.E. Conf. Decision and Control*, Fort Lauderdale.

**Koditschek, D.** 1984. Natural Motion of Robot Arms, *I.E.E.E. Conf. Decision and Control*, Las Vegas.

**La Salle, J., and Lefschetz, S.**, 1961. Stability by Lyapunov's Direct Method, *Academic Press.*

**Lyapunov**, 1907. Probleme General de la Stabilite du Mouvement, *Ann. Fac. Sci. Toulouse*, 9.

**Monaco, S., and Stornelli, S.**, 1985. A Nonlinear Attitude Control Law for a Satellite with Flexible Appendages, *I.E.E.E. 24th Conf. Dec. Control*, Fort Lauderdale, FL.

**Morgan, A.P., and Narendra, K.S.**, 1977. On the Uniform Asymptotic Stability of Certain Linear Non-Autonomous Differential Equations, *S.I.A.M. J. Control and Optimization*, 15.

**Nicolis, G., and Prigogine, I.**, 1978. Self-Organization in Non-Equilibrium Systems, *John Wiley & Sons*

**Redding, D.C., and Adams, N.J.**, 1987, *A.I.A.A. Journal of Guidance, Control, and Dynamics*, vol. 10, n° 1.

**Salisbury, J.K.**, 1980. Active Stiffness Control of a Manipulator in Cartesian Coordinates, *I.E.E.E. Conf. Decision and Control*, Albuquerque, NM

**Slotine, J.J.E.**, 1984. Sliding Controller Design for Nonlinear Systems, *Int. J. Control*, 40(2).

**Slotine, J.J.E.**, 1988. Putting Physics in Control, I.E.E.E. Control Systems Magazine, 8-4.

**Slotine, J.J.E., and Di Benedetto, M.D.**, 1988. Hamiltonian Adaptive Control of Spacecraft, NSL-880603 Report, submitted to I.E.E.E. Trans. Autom. Control.

**Slotine, J.J.E., and Li, W.**, 1986. On The Adaptive Control of Robot Manipulators, *A.S.M.E. Winter Annual Meeting*, Anaheim, CA.

**Slotine, J.J.E., and Li, W.**, 1987a. On The Adaptive Control of Robot Manipulators, *Int. J. Robotics Research*, 6(3).
**Slotine, J.J.E., and Li, W.**, 1987b. Theoretical Issues in Adaptive Manipulator Control, *Fifth Yale Workshop on Applications of Adaptive Systems Theory*, Yale.
**Slotine, J.J.E., and Li, W.**, 1988a. Adaptive Manipulator Control: A Case Study, *I.E.E.E. trans. Autom. Control*, AC-33 (11).
**Slotine, J.J.E., and Li, W.**, 1988b. Composite Adaptive Robot Control, NSL-880201 Report, submitted to *Automatica*.
**Slotine, J.J.E., and Ydtsie, B.E.**, 1988. Control of Nonlinear Chemical Processes: A Physically-Motivated Approach, submitted to *Automatica*.
**Stephanopoulos, G.**, 1984. Chemical Process Control, Prentice-Hall.
**Takegaki, M., and Arimoto, S.**, 1981. A New Feedback Method for Dynamic Control of Manipulators, *A.S.M.E. J. Dynamic Systems, Measurement, and Control*, 102.
**Thomson, W., and Tait, P.**,1886. Treatise on Natural Philosophy, *University of Cambridge Press*.
**Van der Schaft**, 1986. Stabilization of Hamiltonian Systems, *Non. An. Th. Meth. Appl.*, Vol. 10.
**Vafa, Z., and Dubowsky, S.**, 1987. On the Dynamics of the Manipulators in Space Using the Virtual Manipulator Approach, *Proc. of the 1987 I.E.E.E. Int. Conf. on Robotics and Automation*, Raleigh, NC.
**Wie, B., and Barba, P.M.**, 1985. Quaternion Feedback for Spacecraft Large Angle Maneuvers, *A.I.A.A. Journal of Guidance, Control, and Dynamics*, vol. 8, n° 3.

# FEEDBACK LINEARIZATION OF A ONE-LINK FLEXIBLE ROBOT ARM MODELLED BY PARTIAL DIFFERENTIAL-INTEGRAL EQUATIONS


Tzyh-Jong Tarn
Dept. of Systems Science
Washington University
St. Louis, MO 63130
U.S.A.

Xuru Ding
Dept. of Systems Science
Washington University
St. Louis, MO 63130
U.S.A.

Antal K. Bejczy
Jet Propulsion Lab.
Pasadena, CA 91109
U.S.A.


## I. INTRODUCTION

A popular approach for modelling of robot arms with flexible links is to use finitely truncated modal expansions to approximate the distributed elastic coordinates (For listing of appropriate references, see [2]). These approximated dynamic models may serve certain engineering purposes to some degree of satisfaction. However, this approach of modelling brings up the problem of spill-over, and defies revealing some important features of the original distributed parameter system.

In order to solve this problem and give more flexibility to analysis and control of the system, we have derived the dynamic model for a large class of flexible robot arms as infinite-dimensional dynamic system, which may be reduced to various simplified dynamic models, when justified assumptions are made. The dynamic model we obtained is a set of partial differential-integral equations and a set of dynamic boundary conditions associated with it. Some observations on properties of the dynamic model have been made. Also, a nonlinear control is proposed for a one-link flexible robot arm example to achieve partial linearization and input-output decoupling.

## II. GENERAL DYNAMIC MODEL

The basic assumptions we make for the general dynamic model discussed here are:

(a) Joints of the flexible robot arm can be either rotary or prismatic. The joint elasticity is considered as the effect of a linear torsional spring. Also, the motors satisfy the assumptions (A1) and (A2) made in [8].

(b) Link deformation is small, and longitudinal deformation is negligible. Each flexible link consists of hyper-elastical material, and they can be of extended shape.

The general dynamic model was derived using Hamilton's principle. The Denavit-Hartenberg four-parameter representation and the frame structure by Richard Paul [1] for undeformed robot arms are adopted, and the effect of the link deformation is incorporated into the model through the distributed elastic coordinates. For the complete model and its derivation, see [2].

If we assume that the flexible links of a robot arm are Euler-Bernoulli-beams, then the general dynamic model can be further reduced to a simpler one [2]. Also, we compared our dynamic model with various dynamic models derived by other investigators, and showed that our model can be reduced to most existing dynamic models for flexible robot arms [2].

Basically, the dynamic model we derived is a system of partial differential-integral equations together with the dynamic boundary conditions. The reason that we have dynamic boundary conditions rather than static ones is that the boundary of the elastic structure is moving due to the action of the actuators. This is clear from [2, 7]. Also, as we demonstrated in [7], these dynamic boundary conditions can degenerate to lower-order differential equations or even algebraic constraints. This degeneration may happen under certain arm configurations, or regardless of the arm configuration. We can define "psuedo-inertia operator matrix" for the dynamic model as an analogue to the "psuedo-inertia matrix" for dynamic model of rigid-body robot arm. It is our observation that the psuedo-inertia operator matrix is symmetric, nonnegative semi-definite, etc [7]. These properties can be used to simplify the calculation and help us to gain a better understanding of the dynamic behavior of the original system.

## III. FEEDBACK LINEARIZATION AND STABILITY ANALYSIS OF A ONE-LINK FLEXIBLE ARM

In this section we shall propose a methodology of input-output decoupling and partial linearization using diffemorphic state transformation and nonlinear feedback, and apply it to a one-link flexible robot arm. The local stability of the resultant system under the control will then be discussed.

The dynamic model of a flexible robot arm with one Euler-Bernoulli-beam link and rigid joint (see Figure 1) is the following

$$\frac{\partial^2 w}{\partial t^2} + x\ddot{\theta} - w\dot{\theta}^2 - g\cos\theta = -\frac{\partial^2}{\partial x^2}\left(\frac{EI}{\rho}\frac{\partial^2 w}{\partial x^2}\right),$$

$$[J_m + \int_0^L (x^2 + w^2)\rho\, dx]\ddot{\theta} + \int_0^L x\rho\frac{\partial^2 w}{\partial t^2}dx + 2\int_0^L w\frac{\partial w}{\partial t}\rho\, dx\,\dot{\theta}$$

$$- mg(\bar{x}\, \text{cps}\theta - \bar{w}\sin\theta) = \tau,$$

$$\frac{\partial}{\partial x}\left(EI\frac{\partial^2 w}{\partial x^2}\right)\Big|_{x=L} = 0,$$

$$\frac{\partial^2 w}{\partial x^2}\Big|_{x=L} = 0,$$

$$\frac{\partial w}{\partial x}\Big|_{x=0} = 0,$$

$$w\Big|_{x=0} = 0,$$

where $w(x,t)$ is the displacement of a point on the beam in the $y$ direction, $\theta$ is the joint angle, $EI$ is the stiffness of the beam (E: Young's modulus; I: inertia), $\rho$ is the material density of the link, $g$ is the gravity constant, $\tau$ is the reflected motor torque, $L$ is the length of the beam, $m$ is the mass of the beam, and

$$\bar{x} = \frac{1}{m}\int_0^L x\,\rho\,dx$$

$$\bar{w} = \frac{1}{m}\int_0^L w(x,t)\,\rho dx$$

are the X-Y coordinates of the center of mass of the beam. For simplicity, we assume that EI and $\rho$ are constant across the beam.

The above dynamic model can be rewritten in its state space expression. Let $q_1 = \theta(t)$, $q_2 = w(x,t)$, $q_3 = \dot{\theta}(t)$, $q_4 = \frac{\partial w}{\partial t}(x,t)$, and $u = \tau$. We have

$$\begin{bmatrix} \frac{dq_1}{dt} \\ \frac{\partial q_2}{\partial t} \\ \frac{dq_3}{dt} \\ \frac{\partial q_4}{\partial t} \end{bmatrix} = \begin{bmatrix} q_2 \\ q_4 \\ f_3(q) \\ f_4(q_1,x) \end{bmatrix} + \begin{bmatrix} 0 \\ 0 \\ g_3(q_2) \\ q_4(q_2,x) \end{bmatrix} u \quad := f + gu \qquad (1a)$$

where $q = [q_1\ q_2\ q_3\ q_4]'$. $f_3$, $f_4$, $g_3$ and $g_4$ involve operators, and are specified in [6].

Now we choose the output of the system to be the $X_0$ - coordinate of the end-effector, which is at the tip point of the beam. That is [7]:

$$y = L\cos\theta - w(L,t)\sin\theta$$
$$= L\cos q_1 - q_2(L,t)\sin q_1 \quad := h(q) \ . \qquad (1b)$$

Choosing suitable Sobolev space as the state space, the operators involved in (1) will be bounded, and the Lie derivatives of those operators can be found [7]. We now let the state transformation $\Phi$: $q \to \xi$ be given by

$$\begin{aligned} \xi_1 &= h(q) \\ \xi_2 &= \mathcal{L}_f\, h(q) \\ \xi_3 &= x\, q_1 + q_2(x,t) \\ \xi_4 &= \frac{\partial \xi_3}{\partial t} \ . \end{aligned} \qquad (2)$$

$\xi_3$ now has the physical meaning of the arc length of a point on the beam.

One can check that this state transformation is diffeomorphic generically. Moreover, we have

$$\mathcal{L}_g h = 0,$$
$$\mathcal{L}_g \mathcal{L}_f h \neq 0 \quad \text{generically.}$$

Therefore, in the new state variable $\xi$, the system can be written as

$$\dot{\xi}_1 = \xi_2$$

$$\dot{\xi}_2 = \mathcal{L}_f^2 h + \mathcal{L}_g \mathcal{L}_f h \quad u \,, \tag{3a}$$

$$\frac{\partial \xi_4}{\partial t} = \xi_4$$

$$\frac{\partial \xi_4}{\partial t} = -\frac{EI}{\rho}\frac{\partial^4 \xi_3}{\partial x^4} + \bar{f}_4(\xi, x) \,, \tag{3b}$$

$$y = \xi_1 \,. \tag{3c}$$

The input $u$ does not appear in (3b) because of the particular state transformation we have chosen. Now we let the control $u$ be given by

$$u = \alpha(q) + \beta(q)\, v \tag{4}$$

where $\alpha(q) = -(\mathcal{L}_g \mathcal{L}_f h)^{-1}\, \mathcal{L}_f^2 h,$

$$\beta(q) = (\mathcal{L}_g \mathcal{L}_f h)^{-1}$$

and $v$ is a new input. Then we have

$$\begin{bmatrix} \dot{\xi}_1 \\ \dot{\xi}_2 \end{bmatrix} = \begin{bmatrix} 0 & 1 \\ 0 & 0 \end{bmatrix} \begin{bmatrix} \xi_1 \\ \xi_2 \end{bmatrix} + \begin{bmatrix} 0 \\ 1 \end{bmatrix} v \tag{5a}$$

$$y = [1 \quad 0] \begin{bmatrix} \xi_1 \\ \xi_2 \end{bmatrix} \tag{5b}$$

$$\begin{bmatrix} \frac{\partial \xi_3}{\partial t} \\ \frac{\partial \xi_4}{\partial t} \end{bmatrix} = \begin{bmatrix} \xi_4 \\ -\frac{EI}{\rho}\frac{\partial^4 \xi_3}{\partial x^4} + \bar{f}_4(\xi, x) \end{bmatrix} \tag{5c}$$

and $\frac{\partial}{\partial x}\left(EI \frac{\partial^2 \xi_3}{\partial x^2}\right)\Big|_{x=L} = 0,$

$$\frac{\partial^2 \xi_3}{\partial x^2}\Big|_{x=L} = 0,$$

$$\frac{\partial \xi_3}{\partial x}\Big|_{x=0} = 0,$$

$$\xi_3\Big|_{x=0} = 0 \,. \tag{5d}$$

We can see that (5a) and (5b) give a two-dimensional controllable linear subsystem which reflects the input-output relation of the full system, while (5c) is an infinite-dimensional subsystem with boundary conditions given by (5d). the new input $v$ can be designed to control the $X_0$-coordinate of the end-effector using linear dynamic system theory. This control does not enter the PDE given by (5c), but it does affect the boundary condition (5d). One question arises here: Is the vibration of the link under the control stable? To answer this question, we first define the "zero dynamics" for this system.

Let the output be kept at zero all the time. That means we have $\xi_1(t) = \xi_2(t) = 0$ and $v = 0$, $\forall t$. The dynamics of the system now left is

$$\frac{\partial \xi_3}{\partial t} = \xi_4$$

$$\frac{\partial \xi_4}{\partial t} = -\frac{EI}{\rho}\frac{\partial^4 \xi_3}{\partial x^4} + \bar{f}_4(\xi,x)\Big|_{\xi_1=\xi_2=0} \tag{6a}$$

with boundary conditions

$$\xi_3\Big|_{x=0} = 0 \ ,$$

$$\frac{\partial^3 \xi_3}{\partial x^2}\Big|_{x=L} = 0 \ ,$$

$$\frac{\partial^3 \xi_3}{\partial x^3}\Big|_{x=L} = 0 \ ,$$

and $\xi_3\Big|_{x=L} = 0$ . (6b)

The last boundary condition in (6b) is due to the assumption that the output $y$ is kept at zero all the time. The dynamics described by (6) is defined as the "zero-dynamics" of the system.

If we include the structure damping and the viscous damping in the dynamic model, then (6) can be rewritten as the following:

$$\begin{bmatrix} \frac{\partial \xi_3}{\partial t} \\ \frac{\partial \xi_4}{\partial t} \end{bmatrix} = \begin{bmatrix} \xi_4 \\ -\frac{EI}{\rho}\frac{\partial^4 \xi_3}{\partial x^4} + \gamma(\frac{EI}{\rho})^{1/2}\frac{\partial^2 \xi_4}{\partial x^2} + \delta\xi_4 \end{bmatrix} + \begin{bmatrix} 0 \\ \bar{f}_4(\xi,t)\Big|_{\xi_1=\xi_2=0} \end{bmatrix}$$

$$:= A\begin{bmatrix} \xi_3 \\ \xi_4 \end{bmatrix} + B \tag{7a}$$

where

$$A = \begin{bmatrix} 0 & 1 \\ -\frac{EI}{\rho}\frac{\partial^4}{\partial x^4} & \delta(\frac{EI}{\rho})^{1/2}\frac{\partial^2}{\partial x^2} + \delta \end{bmatrix} \tag{7b}$$

$$B = \begin{bmatrix} 0 \\ \bar{f}_4(\xi,t)\Big|_{\xi_1=\xi_2=0} \end{bmatrix} ,$$

where $\gamma$ is the structural damping constant, and $\delta$ is the viscous damping constant [7].

<u>Claim</u>: The zero dynamics given by (7a) and (6b) is locally asymptotically stable.

A sketch of the proof is given in the following.

Step 1. Find a suitable Banach space $\bar{X}$ as state space. In our case, we choose $\bar{X}$ to be a Sobolev space $H_{-n}(A)$ with $n > 1/2$ [7].

Step 2. Solve the PDE

$$\begin{bmatrix} \frac{\partial \xi_3}{\partial t} \\ \frac{\partial \xi_4}{\partial t} \end{bmatrix} = A \begin{bmatrix} \xi_3 \\ \xi_4 \end{bmatrix} \tag{8}$$

with boundary condition (6b), where $A$ is given by (7b).

Identify $S(t)$ generated by $A$ and verify that $A(t)$ is a strongly continuous semigroup using the norm of $H_{-n}$. Also, notice that there exists some positive number $\lambda$ such that $(\lambda I + A)$ is sectorial.

Step 3. Compare the eigenvalues of (8) with the eigenvalues of the following system

$$\begin{bmatrix} \frac{\partial \xi_3}{\partial t} \\ \frac{\partial \xi_4}{\partial t} \end{bmatrix} = (A + \bar{B}) \begin{bmatrix} \xi_3 \\ \xi_4 \end{bmatrix}$$

where $\bar{B}$ is the linear part of the power series expansion of $B$ around the equilibrium point. Prove that if $(\lambda I + A)$ is sectorial, then there exists some $\eta > 0$, s.t. $Re(\sigma(A + \bar{B})) < - \eta < 0$.

Step 4. Define the $\alpha$-norm $(0 < \alpha < 1)$ $\|\cdot\|_\alpha$ on $\bar{X}$ [11], and denote $\bar{X}$ equipped with $\|\cdot\|_\alpha$ by $\bar{X}^\alpha$. Write

$$B = \bar{B} + F(\xi_3, \xi_4)$$

where $F$ is the remainder of the power series expansion of $B$ around the equilibrium point. Then prove the following:

i) $B$ is locally Lipschitz in x;

ii) $\bar{B} = \bar{X}^\alpha \to \bar{X}$ is bounded;

iii) $\|B\|_{-n} = 0 (\|\begin{bmatrix} \xi_3 \\ \xi_4 \end{bmatrix}\|_\alpha) \to 0$ uniformly in $t \geq 0$.

Now, by Theorem 4.4.1 in Bank's book [5], the zero dynamics is locally asymptotically stable.

The asymptotical stability of the zero dynamics then leads to the asymptotic stability of the full system. This can be proved using perturbation method (perturbing the initial state and the 4th boundary condition in (6b)) [4].

Therefore we conclude that every small motion around the equilibrium point will be attracted to the asymptotically stable center manifold described by the zero dynamics.

## IV. CONCLUSIONS

In this paper, we briefly discussed the general dynamic model for flexible robot arms. A one-link Euler-Bernoulli-beam robot arm is modeled as an distributed parameter system, and the method of input-output linearization by state transformation and nonlinear feedback for finite-dimensional nonlinear system [3] is generalized to this infinite-dimensional system with finite number of input and output. Moreover, using the existing theory of distributed parameter systems, we are able to determine the local stability of the controlled system.

## Acknowledgement

This research was supported in part by the National Science Foundation Grant Numbers DMC-8615963, DMC-8505843 and ECS-8515899.

## REFERENCES

[1] R.P. Paul, "Robot Manipulators: Mathematics, Programming and Control," MIT Press 1981

[2] T.J. Tarn, A.K. Bejczy and X. Ding, "On the Modelling of Flexible Robot Arms, (revised)," Robotics Laboratory Report SSM-RL-88-11, Department of Systems Science and Mathematics, Washington University, St. Louis, MO 63130, U.S.A.

[3] A. Isidori, "Nonlinear Control Systems: An Introduction," Lecture Notes in Control and Information Sciences Vol. 72, Springer-Verlag, 1985

[4] J. Carr, "Applications of Centre Manifold Theory," Springer-Verlag, 1981

[5] S. Banks, "State-space and frequency-domain methods in the control of distributed parameter systems," Peter Peregrinus LTD, 1983

[6] T.J. Tarn, A.K. Bejczy and X. Ding, "On the Modelling of Flexible Robot Arms for Control," to appear in Proceedings of the 8th Internatinal Symposium on Mathematical theory of Networks and Systems, Phoenix, Arizona, U.S.A., June 15-19, 1987

[7] X. Ding, T.J. Tarn and A.K. Bejczy, "A Novel Approach to the Modelling and Control of Flexible Robot Arms," to appear in Proceedings of the 1988 IEEE Conference on Decision and Control, Austin, Texas, December 1988

[8] M. Spong, "Modeling and Control of Elastic Joint Robots," 1986 ASME Winter Annual Meeting, Anaheim, California, December 1986

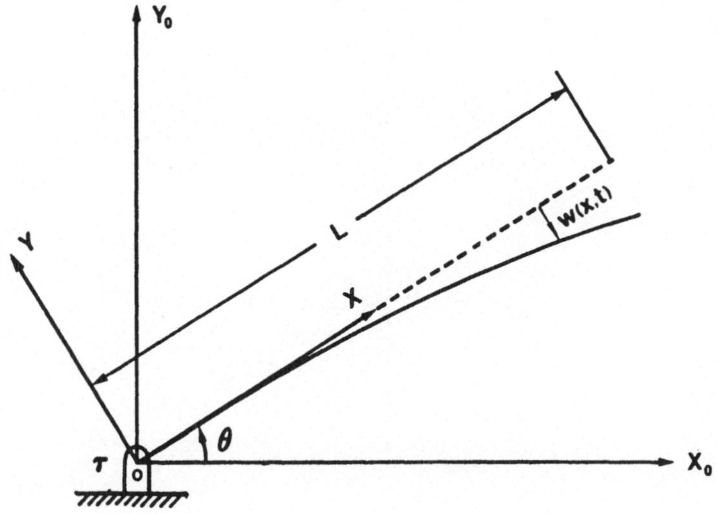


Figure 1. A One-Joint Robot Arm

# MISCELLANEOUS

# Hybrid Dynamical Systems theory

# and nonlinear dynamical systems over finite fields.

Albert Benveniste, Paul Le Guernic
IRISA/INRIA, Campus de Rennes Beaulieu
35042 RENNES CEDEX, FRANCE

*Summary: we study the logic and synchronization characteristics of general dynamical systems called Hybrid Dynamical Systems. Our theory generalizes the notion of Discrete Event Dynamical Systems by handling numerics as well as symbolics. Our theory is supported by the programming language SIGNAL and a mathematical model of relational style. This framework allows us to formulate in the same way HDS programming or specification and HDS control. The core of the theory is the notion of HDS resolution which is based on a reduction technique mapping any HDS specification program into a polynomial dynamical system on the finite field of integers modulo 3; all the algorithms are then based on the study of this dynamical system. This paper is devoted to an informal introduction to this approach.*

## 1 Introduction.

### 1.1 Requirements from applications: Hybrid Dynamical Systems (HDS).

Discrete Event Dynamical Systems (DEDS) have been introduced as a theoretical framework for the study of flexible manufacturing and related systems by Wonham and Ramadge [Ramadge and Wonham 1987, a−b], and have been widely studied since their introduction. Roughly speaking, DEDS are finite state transition systems which are observed and can be controlled by the language generated by the labels that are attached to each transition, regardless of the precise meaning of these labels. However, in most of the complex signal processing and control applications, some actions involving possibly complex numerics can influence transitions (e.g. firing actions based on the behaviour of some internally generated signals). This restriction makes the use of the DEDS approach not suitable to the global study of complex dynamical systems of mixed symbolic/numeric nature. In the sequel, ***Hybrid Dynamical Systems (HDS)*** theory will refer to a theory handling synchronization, logic, and their interconnections to numerics in dynamical systems. As the reader will understand while reading this paper, the mixed nature of HDS make them definitely more difficult to study than DEDS, which justifies the development of a new theory and paradigm.

### 1.2 A new paradigm.

Our first remark was about the highly combinatorial complexity of HDS. Such a complexity faces us with a new problem which was not considered before in the control community, namely the difficulty of simply *describing* or *constructing* HDS. This has been for a long time recognized by computer scientists as a sufficient reason for introducing programming languages, i.e. *concrete syntax* with the usual hierarchical constructs. The core of our approach is the kernel of the language SIGNAL we shall present later, which is composed of only 5 instructions to describe any HDS.

A second claim is that a HDS should be described via a set of *relations* or *constraints*, rather than as a complicated input–output map as usually in control science. In fact, this argument has been already recognized by J.C. Willems [Willems 1987] for the theory of linear dynamical systems. An immediate consequence of this choice is that such HDS specifications cannot be effective, i.e. it is not immediately possible to compute the outputs of a so-specified HDS in response to some sequences of inputs. The control scientist will recognize a standard situation when handling descriptor or implicit linear dynamical systems. By ***HDS resolution***, we have in mind a procedure to transform any relational HDS specification into a machine which can execute the desired behaviours, and thus represents the desired equivalent input–output map.

Another consequence of this point of view is that *exact model following control problems* (such as considered by Ramadge and Wonham) are just particular cases of HDS resolution, since requiring an exact model following is just achieved by adding further constraints described within the same framework as the HDS itself.

What is the nature of time for HDS? Complex applications such as mentioned above are inherently distributed in nature. Hence every subsystem possesses his own time reference, namely the ordered collection of all the communications or actions this subsystem performs. Hence the nature of time in HDS is by no means universal, but rather local to each subsystem, and consequently multiform. A fundamental consequence is that communications between subsystems impose constraints on the timing of these subsystems. Hence handling these multiform time references and reasoning about them is one of the fundamental tasks we have to perform. In the sequel, signals possessing identical time references will be said to have the same *clock*.

# 2 The programming language SIGNAL–kernel; some small examples.

As we have indicated before, HDS should be specified via programming languages. We shall now present such a language: a subset of the language SIGNAL[1]. To be concise, we shall introduce only the primitives of the language, and drop any reference to typing and various declarations; the interested reader is referred to [Gautier 1987].

## 2.1 SIGNAL-kernel as a specification language for HDS.

SIGNAL handles (possibly infinite) sequences of data with time implicit: such sequences will be referred to as ***signals***. For example, x denotes the infinite sequence $\{x_t\}_{t\geq 1}$ where the time index $t$ is attached to this signal; signals possessing the same time index are said to have the same ***clock***, so that clocks are equivalence classes of simultaneous signals (a formal definition is

---

1 SIGNAL is a joint trademark of CNET and INRIA

discussed in the companion paper [Benveniste & al. 1988]). Instructions of the language SIGNAL are intended to relate clocks as well as values of the various signals involved in a given HDS.

A basic principle in SIGNAL is that a single name is assigned to every signal, so that in the sequel (and unless explicitely stated), identical name refers to identical signals. The kernel-language SIGNAL possesses 5 instructions, the first of them being a generic one.

| | |
|---|---|
| p(x(1),...,x(n)) | (i) |
| y := x$ *init* $x_0$ | (ii) |
| y := x *when* b | (iii) |
| y := u *default* v | (iv) |
| P \| Q | (v) |

Their intuitive meaning is the following:

(i): direct extension of instantaneous relations into relations acting on flows:

$$p(x(1),\ldots,x(n)) \Leftrightarrow \forall t{:}p(x_t(1),\ldots,x_t(n))$$

Examples are functions such as $z = x + y$ $(\forall t{:}z_t = x_t + y_t)$ or statements such as $(a \text{ and } b) \text{ or } c = true$ $(\forall t{:}\ (a_t \text{ and } b_t) \text{ or } c_t = true)$. A byproduct of this instruction is that *all referred signals must have the same time index, i.e. must be present simultaneously.* This is a generic instruction, i.e. we assume a family of relations is at hand. If one chooses an instantaneous relation accepting any n-uple, the resulting SIGNAL instruction only constrains the invloved signals to have the same clock: the so-obtained instruction *synchro*x,y,... has thus as only effect to force the two signals x,y,... to have the same clock.

(ii): logical delay (the usual operator $z^{-1}$)

$$y = x\$ \ \textit{init}\ x_0 \Leftrightarrow \forall t > 1{:}y_t = x_{t-1}, y_1 = x_0$$

(iii): condition (b is boolean): y equals x when the signal x and the boolean b are available and b is true; otherwise, y is not emitted; the result is an event−based undersampling of signals.

(iv): y merges u and v, with priority to u when both signals are simultaneously present; this instruction is the key to oversampling as we shall see later.

The instructions (i−iv) specify the elementary processes, we shall call *generators.* The objects named x, y, u, v, b, will be called *signals.*

(v): communication of already defined processes: P and Q communicate through their signals with common names; for example

$$y = zy + a \mid zy = y\$ \textit{ init } x_0$$

denotes the system of recurrent equations for $t \geq 1$

$$y_t = zy_t + a_t$$
$$zy_t = y_{t-1}, zy_1 = x_0$$

which is equivalent to $y_t = y_{t-1} + a_t, y_0 = x_0$.

## 2.2 Discussing some examples.

We present here only short classroom examples; the reader interested to more realistic ones should refer to [Benveniste & al. 1988]. Here we shall only present a toolbox of elementary mechanisms which will be used similarly to basic instructions in the sequel (something like «macros»). To avoid the need for explicit typing, we shall use the following generic notations

- u,x,y,z...: signal of any type
- a,b,c...: boolean signals
- h,k,l: signals of type *event*, i.e boolean signals which take only the value *true*

Here follows a first example.

**Access to the clock of a signal.**

h := *event* x

=

h := (x = x)

The pure clock h is delivered when x is present (since x = x always holds). In this example, the notation

processname = body of the process

is used to give a name to a process defined by a set of formulas. Here, «processname» is of the form

*list of output signals* := NAME (*list of input signals*)

Using this notation, we present further basic mechanisms.

**Extraction of the occurrences *true* of a boolean signal.**

```
h := when(b)
      =
    h := b when b
```

**A synchronized memory.**

```
y := x cell b init y0
      =
    zy := y$ init y0
  | y := x default zy
  | synchro y, (x default when(b))
```

The output y returns either the present value of x (when x is received), or the last received value of x when b is present and true.

# 3 HDS resolution: an informal presentation.

Here follows the intuitive description of our method. Recall that SIGNAL is obtained by extending a given «instantaneous language» (the set of data types and corresponding relations known by the instruction (i)) with a small set of primitives to obtain a model of dynamical systems. Denoting by IL (Instantaneous Language) this starting point, any homomorphism $\psi$ from IL into itself is easily extended to an homomorphism $\Psi$ from SIGNAL into itself.

In our case, the homomorphism $\psi$ is constructed as follows

STEP 1: among the relations of IL, select the subfamily of relations and corresponding data types for which you accept to solve systems of equations (and are supposed to can!). Here we shall select the boolean variables together with the boolean relations generated by {:=, ***and***, ***or***, ***not***} and the constants *true*, *false*; this choice is motivated by the particular role played by the booleans in the instruction ***when***.

STEP 2: other instantaneous relations must be *functions*, and are encoded into their dependence graph:

$$y = f(x_1, \ldots, x_n) \Rightarrow \{x_1 \rightarrow y, \ldots, x_n \rightarrow y\}$$

## 3.1 Encoding SIGNAL programs.

The image of SIGNAL programs we shall obtain will be referred to as ***synchro-processes***. Synchro-processes are defined via constraints involving clocks, booleans, and labels. We shall provide an algebra with a convenient calculus where the pairs {clocks, booleans} can be represented. All we need to encode are the following status: *absent, present, true, false*. These are encoded onto the finite field $F_3 = Z/3Z$ of integers modulo 3 as follows

$$\begin{gathered} \textit{true}: +1 \\ \textit{false}: -1 \\ \textit{absent}: 0 \\ \textit{present}: \pm 1 \end{gathered}$$

where $\pm 1$ denotes a non determinate choice of $+1$ or $-1$; i.e. we handle in the same way clocks and boolean of nondeterminate value. Let us apply this idea to encode SIGNAL programs. The following notation will be used to present this coding:

$$\textit{synch}\,(\text{program}) :: \left( \begin{array}{c} \textit{clock calculus} \\ \hline \textit{conditional dependence graph} \end{array} \right)$$

Here,

- program denotes the program to be encoded, and ***synch*** is the encoding map;
- *clock calculus* denotes the set of algebraic equations encoding the constraints on synchronisation or logic; these equations can represent either static or dynamical systems as we shall see later;
- *conditional dependence graph* denotes the set of possibly occurring dependencies together with the clocks where these dependencies are in force.

**Instruction (i): relation or function.**
**Boolean relation.**
The coding of all boolean relations is easily derived from the coding of the following instructions and the coding of the composition we shall see below:

$$\textit{synch}\,(a := \textit{true}) :: \left( \begin{array}{c} a^2 - a = 0 \\ \hline \varnothing \end{array} \right)$$

$$\textit{synch}\,(b := \textit{not}\, a) :: \left( \begin{array}{c} b = -a \\ \hline \varnothing \end{array} \right)$$

$$\textit{synch}\,(c := a\,\textit{and}\,b) :: \left( \begin{array}{c} a^2 = b^2 \\ c = a^2 - (ab + a + b) \\ \hline \varnothing \end{array} \right) \qquad (3\text{-}1)$$

The algebraic equation of the first formula possesses $a = 1, a = 0$ as only solutions, which means that $a$ is either absent or true. The second equation is obvious. To derive the last one, remark that its first component encodes the fact that both signals $a$ and $b$ must have the same clock (they are either both present or absent, which is encoded as $a^2 = 1$ or $a^2 = 0$); then it is straightforward to verify that the last equation maps the pairs (0,0), (1,1), (−1,1), (1,−1), (−1,−1) respectively onto 0, 1, −1, −1, −1. Since only boolean are involved, no coding of dependencies is required hence the symbol ∅ in the second field.

**Non boolean function.**

$$\textit{synch}\,(y := f(x_1, \ldots, x_n)) :: \left( \begin{array}{c} y^2 = x_1^2 = \ldots = x_n^2 \\ \hline y^2 : x_1 \to y \;\ldots\; x_n \to y \end{array} \right) \qquad (3\text{-}2)$$

The first field encodes the constraints on clocks (equality), while the second one encodes the data dependencies. The meaning of the second field is «the listed dependencies hold when $y^2 = 1$». Notice that a := (u<v) produces a boolean, but is a nonboolean function.

**Instruction (ii): the register.**
**Boolean register.**
This is the key case where dynamical systems in $F_3$ come out.

$$\textit{synch}\,(b := a\$\,\textit{init}\,u) :: \left( \begin{array}{c} \xi = (1 - a^2)\xi' + a; \;\; \textit{initial cond} = u \\ b = a^2 \xi' \\ \hline \varnothing \end{array} \right) \qquad (3\text{-}3)$$

Here, $\xi$ is the current state of the dynamical system, $\xi'$ its previous state, and $u$ its initial condition (±1 valued). The corresponding explicit form of this dynamical system is

$$\xi_t = (1 - a_t^2)\xi_{t-1} + a_t; \quad \xi_0 = u$$
$$b_t = a_t^2 \xi_{t-1}$$

where $t$ is any time index fast enough to capture every presence of signal. Notice that the state takes $+1$ or $-1$ as only values, i.e. states are persistent. The state is modified when a new input is received, and at the same instant the old state is delivered at the output. Again is no dependence graph necessary.

**Non boolean register.**

$$\textit{synch}\,(y := x\,\$\,\textit{init}\,u) :: \left(\begin{array}{c} y^2 = x^2 \\ \hline \emptyset \end{array}\right) \qquad (3\text{–}4)$$

The first field expresses that clocks must be identical; the second field is empty even if we are considering non boolean types, since the current value of $y$ does not depend on the current value of $x$, but on the content of the memory (which has been lost in the coding via ***synch***).

**Instruction (iii): the filtering.**
**Filtering with boolean output.**

$$\textit{synch}\,(c := b\ \textit{when}\ a) :: \left(\begin{array}{c} c = b(-a - a^2) \\ \hline \emptyset \end{array}\right) \qquad (3\text{–}5)$$

**Filtering with non boolean output.**

$$\textit{synch}\,(y := x\ \textit{when}\ a) :: \left(\begin{array}{c} y^2 = x^2(-a - a^2) \\ \hline y^2 : x \to y \end{array}\right) \qquad (3\text{–}6)$$

The second field expresses that x influences y when y is produced.

**Instruction (iv): the merge.**
**Merge with boolean output.**

$$\textit{synch}\,(c := a\ \textit{default}\ b) :: \left(\begin{array}{c} c = a + b(1 - a^2) \\ \hline \emptyset \end{array}\right) \qquad (3\text{–}7)$$

**Merge with non boolean output.**

$$\mathbf{synch}\,(y := u\ \mathit{default}\ v) :: \left( \begin{array}{c} y^2 = u^2 + v^2(1 - u^2) \\ \hline u^2 : u \rightarrow y \\ v^2(1 - u^2) : v \rightarrow y \end{array} \right) \quad (3\text{-}8)$$

The second field expresses the fact that u influences y when it is present, while v influences y when it is present and u is absent.

**Instruction (v): the composition.**

$$\mathbf{synch}\,(P\,|\,Q) = \mathbf{synch}\,(P) \cup \mathbf{synch}\,(Q) \quad (3\text{-}9)$$

Here the symbol U means that the fields of P and Q have to be merged to produce a single clock field and a single conditional dependence graph field. This U is in fact a composition in our sense.

***Definition***: *In the coding above, the first field will be called the* ***clock calculus*** *of the program, while the second one will be called the* ***conditional dependence graph*** *of the program.*

### 3.1.1 Examples.

To allow drawing of graphs, conditional dependence graphs will be depicted using the following notation:

$$x \xrightarrow{h} y \quad \text{instead of} \quad h : x \rightarrow y$$

**The instruction *cell*.**

This instruction has been used as a macro in the «shared track» example. Recall the corresponding program:

y := x *cell* b *init* $y_0$

=

| zy := y\$ *init* $y_0$
| y := x *default* zy
| *synchro* y, (x *default* *when*(b))

We shall distinguish two cases: x boolean, and x non boolean.

**Encoding the boolean *cell* into its clock calculus.**

$$\xi = (1 - y^2)\xi' + y, \textit{init} = y_0$$
$$zy = y^2\xi'$$

$$y = x + zy(1 - x^2)$$
$$y^2 = x^2 + (-b - b^2)(1 - x^2) \tag{3-10}$$

Eliminating $zy$ yields

$$\xi = (1 - x^2)\xi' + x, \textit{init} = y_0$$
$$y = x + (-b - b^2)(1 - x^2)\xi'$$

which reflects exactly the meaning of the instruction ***cell***: the memory is refreshed when x is received, and y delivers the current or last value of x when x is received or b is received and true.

**Encoding the non boolean *cell* into its clock calculus and conditional dependence graph.**

*Clock calculus:*

$$zy^2 = y^2 = x^2 + (1 - x^2)zy^2$$
$$y^2 = x^2 + (1 - x^2)(-b - b^2)$$

which yields

$$zy^2 = y^2 = x^2 + (1 - x^2)(-b - b^2)$$

*Conditional Dependence Graph:*

$$zy \xrightarrow{\;y^2(1 - x^2)\;} y \xleftarrow{\;x^2\;} x$$

Here the dynamics has been lost; the clock calculus expresses only how clocks of signals are related.

# 4 Conclusion.

We have presented a brief account of HDS theory. Our theory is supported by the programming language SIGNAL. Our approach is relational, which results in a simple treatment of exact model following control problems, as well as proof systems such as usually done using temporal logic [Pnueli 1977] (proving a property is in fact an exact model following control problem). The main problem is then HDS resolution, which can be considered as a realization theory (program → state machine executing the program). Our theory captures both notions of observability and deadlock, the latter notion being related to controllability. Much remains to be done to fully exploit non linear system theory over the finite field $F_3$. Calculating the orbits of such dynamical systems is the key to observability and deadlock checking. Finally, let us mention the works around the other declarative (but functional) language LUSTRE [Caspi & al. 1987] and the imperative language ESTEREL [Berry & Cosserat 1984][Gonthier 1988] the syntax of which possesses some flavour of ADA.

# REFERENCES.

[Benveniste & Le Guernic 1987]: A. Benveniste, P. Le Guernic, «A denotational theory of synchronous communicating systems», INRIA research report No 685.

[Benveniste & al. 1988]: A. Benveniste, B. Le Goff, P. Le Guernic, «Hybrid Dynamical Systems theory and the language SIGNAL.», INRIA research report No 838.

[Berry & Cosserat 1984]: G. Berry, L. Cosserat, «The ESTEREL Programming Language and its Mathematical Semantics», INRIA Res. Rep. No 327, Rocquencourt, France, to appear in *Science of Computer Programming*.

[Caspi et al. 1987]: P. Caspi, D. Pilaud, N. Halbwachs, J.A. Plaice, «LUSTRE: a declarative language for programming synchronous systems», Proc of the 14th ACM symp. on Principles of Programming Languages, 1987.

[Gautier 1987]: T. Gautier, P. Le Guernic, L. Besnard, «SIGNAL, a Declarative Language for Synchronous Programming of Real–time Systems», proc. of the third Conference on Functional Programming Languages and Computer Architecture, G. Kahn Ed, Lect Notes in Computer Science, vol 274, Springer V.

[Gonthier 1988]: G. Gonthier, thesis, Univ. de Nice and Ecole des Mines, 1988.

[Le Guernic & al. 1986]: P. Le Guernic, A. Benveniste, P. Bournai, T. Gautier, «SIGNAL: a Data–Flow oriented Language for Signal Processing», IEEE Trans. on ASSP, ASSP–34 No 2, 362–374.

[Pnueli 1977]: A. Pnueli, «The Temporal Logic of Programs», Proc. of the IEEE Symposium on the Foundations of Computer Science, Providence, Rhode Island.

[Ramadge and Wonham 1987–a]: P.J. Ramadge, W.M. Wonham, «Supervisory Control of a class of Discrete Event Processes», SIAM J. Control and Opt., 25, 1, 1987, 206–230.

[Ramadge and Wonham 1987–b]: P.J. Ramadge, W.M. Wonham, «On the Supremal Controllable Sublanguage of a Given Language», SIAM J. Control and Opt., 25, 3, 1987, 637–659.

[Willems 1987]: J.C. Willems «From time series to linear systems», Automatica

# AUTOMATIQUE ET REGULATION BIOLOGIQUE

Daniel CLAUDE

Laboratoire des Signaux et Systèmes,
C.N.R.S.- E.S.E.,
Plateau du Moulon, 91190 Gif-sur-Yvette, France.

*Résumé :* En biologie, de nombreuses régulations font appel à plusieurs agents hormonaux aux actions couplées et non linéaires. Ainsi, toute action thérapeutique mesurable doit reposer sur une modélisation non linéaire multivariable. C'est l'objet des contrôles bipolaires.

*Abstract :* In biology, there are numerous regulations which call upon several hormonal agents with coupling and nonlinear actions. All measurable therapeutic action should thus be based on a nonlinear and multivariable model policy. This is the object of bipolar controls.

## I. INTRODUCTION

Depuis maintenant plusieurs décennies, de nombreux chercheurs ont pensé à créer un lien entre les mathématiques et la médecine ( cf. les livres récents de Winfree [ 26 ] et de Swan [ 25 ] ), en particulier par les essais de modélisation de certains phénomènes biologiques et par exemple, en cancérologie , par la recherche de procédures médicamenteuses (chimiothérapie) ou par la mise en place de protocoles d'émission de particules actives spécifiques (radiothérapie). Ils souhaitaient ainsi réunir la théorie mathématique et la pratique médicale. L'automatique, appliquée à certaines régulations biologiques, répond à cette exigence et à cette espérance.

En biologie, de nombreuses régulations font appel à plusieurs agents aux actions couplées. Il en est ainsi de la régulation de l'hydratation cellulaire ou du contrôle de la mitose dans lesquels interviennent les corticoïdes d'une part et la vasopressine d'autre part, de même que l'insuline et le glucagon régulent l'activité glycémique. La faillite dans certaines pathologies des thérapeutiques consistant à administrer une seule hormone trouve son explication dans le fait que l'on a négligé les réactions de l'autre hormone qui intervient à cause d'un jeu subtil de feedbacks croisés. En outre, la biologie est un domaine fortement non linéaire où le principe de superposition des actions n'a pas cours.

Ainsi, toute action thérapeutique mesurable doit passer par une modélisation non linéaire multivariable, suffisamment riche pour prendre en compte les aspects prépondérants des phénomènes étudiés, et assez simple pour envisager d'une manière raisonnable les possibilités de commande de ces systèmes et en déduire les actions thérapeutiques. A cause des couplages, les solutions proposées, par leur caractère faussement paradoxal, peuvent surprendre, déranger, voire provoquer des hostilités. Pourtant, les résultats cliniques sont là, authentifiés par les radiographies et les scanners, et on doit espérer que les deux exemples que nous allons traiter, permettent de convaincre de la nécessité de développer rapidement le champ d'action des thérapeutiques bipolaires dont Bernard-Weil est à l'origine.

## II DE L'UTILISATION DE L'AUTOMATIQUE EN BIOLOGIE

On connaît la complexité du monde biologique et on peut naturellement s'interroger sur l'attitude à adopter en face de cette réalité inéluctable. Doit-on se contenter d'admirer et de décrire les merveilleuses régulations que la nature a mises en place, ou bien, dans un brillant accès d'indépendance, toutefois tempéré et raisonné, se satisfaire d'une formalisation descriptive, ou alors, par une audace prométhéenne, vouloir comprendre et commander ?

Cette dernière position me semble assez proche de la démarche de l'automaticien. Face à un système complexe, son rôle est de concevoir une modélisation qui soit bien entendu suffisamment précise, vis-à-vis du comportement qu'elle entend décrire, mais aussi, qui soit assez simple pour que l'on puisse envisager des possibilités raisonnables de contrôle, après avoir dégagé les entrées et les sorties du système à asservir ( cf. Fliess [ 16 ] ). Le contrôle s'effectuant alors avec les méthodes connues ou à venir !

On voit ainsi que l'on doit, dans cette optique, écarter en biologie les modèles dits de connaissance qui par essence sont de grande dimension et contiennent un grand nombre de paramètres ; l'utilité de ces modèles résidant dans la grande quantité d'informations qu'ils sont à même de fournir. A l'opposé, une modélisation de type boîte noire, dans un emploi systématique, est certainement dangereuse pour l'étude des systèmes non linéaires, en raison de l'existence d'équations différentielles universelles, telles que celles données par Rubel [ 22 ] et Boshernitzan [ 2 ], dont les solutions sont à même d'approcher des éléments quelconques de vastes classes de fonctions ; ce fait devant inciter à la prudence.

Une solution intermédiaire est alors, dans une approche phénoménologique, de ne considérer que les aspects prépondérants de certaines classes de comportements non linéaires et de fixer un cadre formel conduisant à des systèmes non linéaires aux dimensions et au nombre de paramètres les plus faibles possibles, seule voie qui autorise la conception de commandes robustes.

En biologie, les systèmes de régulations ago-antagonistes définis par Bernard-Weil se prêtent à cette méthodologie.

## III LE SYSTEME SURRENO-POSTHYPOPHYSAIRE ET LA VASOPRESSINO-CORTICOTHERAPIE

Dans le cadre de l'application de l'automatique aux traitements chimiothérapiques en cancérologie, Sundareshan et Fundakowski [ 24 ], s'interrogent sur le caractère dual de l'objet de ces thérapeutiques et souhaitent trouver des agents qui soient capables de détruire les cellules malignes tout en préservant les cellules saines. En fait, au sein de l'organisme existe un important système qui assure la régulation du développement cellulaire tant au point de vue de la mitose que de l'hydratation de la cellule, c'est le système hormonal surréno-posthypophysaire.

Le système surréno-posthypophysaire, formé par les cortico-surrénales d'une part et par la neuro-posthypophyse d'autre part, intervient ainsi au premier chef dans les manifestations cliniques observées chez le malade neuro-chirurgical. Ce système est responsable de manifestations aussi diverses que certains oedèmes du cerveau, certains collapsus cérébraux aggravant les suites d'intervention pour hématome sous-dural, et intervient dans l'évolution des tumeurs cérébrales malignes.

La reconnaissance du couplage entre ces deux glandes date des années 30 ( cf.[ 23 ] ), et ce système, aux actions ago-antagonistes ( cf.[ 4, 6, 7 ] ), assure des régulations majeures. Ainsi, la cortisone, secrétée par les cortico-surrénales, est un merveilleux agent, non seulement contre l'hyperhydratation cellulaire mais aussi comme produit anti-mitotique, comme cela a été démontré *in vitro* aussi bien dans le cas de tumeurs cérébrales malignes en culture que dans celui de toute autre lignée cancéreuse en culture de tissu. Quant à la vasopressine, secrétée par la neuro-posthypophyse, elle est responsable de la réabsorption de l'eau par le tube rénal et est un facteur de croissance tout à fait important. Ce premier facteur de croissance polypetidique a été découvert en 1968 par Bernard-Weil, Dalage, Olivier et Piette [ 9 ] et leur résultat a été confirmé ultérieurement par les auteurs américains, Rozengurt et all. [ 20 ], en 1979, et Monaco et all. [ 18 ] en 1982. Nous renvoyons à Pawlikowski [ 19 ] pour avoir un rappel récent des actions mitogéniques des neuropeptides. Le déséquilibre entre les corticoïdes et la vasopressine, avec un excès de vasopressine favorisant le développement tumoral, a été de nouveau mesuré récemment en cancérologie digestive ( cf. [ 11 ] ), mais il a été constaté dans bien d'autres cas. De plus, à cause du couplage entre ces deux hormones, certains oedèmes cérébraux résistent à la cortisone et les tumeurs cancéreuses ne sont vraiment influencées par les corticoïdes que pour un court laps de temps et avec des doses très élevées de ces hormones. Pire encore, l'organisme malade se place dans une position " d'homéostasie pathologique" ( cf. Bernard-Weil [ 4 ] ) et ce déséquilibre régulé bénéficie de sauvegardes biologiques puissantes qui tendent à le maintenir en l'état comme s'il s'agissait du fonctionnement physiologique " normal".

C'est ainsi qu'est préservé le déséquilibre vasopressine-corticoïdes chez le malade cancéreux, l'administration de corticoïdes ayant pour effet d'augmenter le taux de vasopressine pourtant déjà anormalement élevé ( cf. [ 3 ] ).

La solution consiste donc à envisager l'administration simultanée de vasopressine et de corticoïdes ( cf.[ 5 ] ), un modèle multivariable non linéaire venant conforter les intuitions premières du médecin ( cf.[ 4, 6, 7 ] ).

Ce modèle, représenté par un système différentiel non linéaire à deux entrées, $e_1$ et $e_2$, deux perturbations, p et q, et deux sorties, $z_1$ et $z_2$, peut s'écrire sous la forme suivante ( cf. [ 15 ] ) :

$$\dot{\mathcal{X}} = \sum_{i=1}^{3} [\, k_i ( u+p )^i + c_i ( v+q )^i \,] + e_1$$

$$\dot{\mathcal{Y}} = \sum_{i=1}^{3} [\, k_i' ( u+p )^i + c_i' ( v+q )^i \,] + e_2$$

$$\dot{X} = e_1 \qquad (3.1)$$

$$\dot{Y} = e_2$$

$$z_1 = \mathcal{X} - \mathcal{Y}$$

$$z_2 = \mathcal{X} + \mathcal{Y} - m$$

avec $\mathcal{X} = x+X$ ; $\mathcal{Y} = y+Y$, où x et y désignent respectivement les actions des corticoïdes et de la vasopressine endogènes et X et Y les actions des hormones exogènes ( thérapeutique).
Il s'agit d'un développement en série dans lequel apparaissent des expressions antagonistes $u = \mathcal{X} - \mathcal{Y}$ et des expressions agonistes $v = m \operatorname{Log} [\, 1 + ( \mathcal{X} + \mathcal{Y} - m )/m \,] + \theta( t )$, avec $\theta( t ) = A + B \sin( \omega t ) + C \cos( \omega t )$, où les constantes A, B, C et $\omega$ ( $\omega = 2\pi / 24$ dans un rythme circadien ) déterminent le synchroniseur $\theta( t )$ lié aux rythmes biologiques. L'introduction de la puissance cubique se justife par les conditions de stabilité du système ( cf. [ 4 ] ) ; p( t ) représente un possible stimulus osmotique ; q( t ) correspond à un éventuel stimulus volémique ( hémorragie par exemple) ou un stress ; les paramètres, $k_i$, $c_i$, $k_i'$, $c_i'$ (i =1,2,3) sont

constants ; le paramètre m est pris en général constant ( m = 0,8 ) mais peut aussi être considéré comme variable dans le temps. Ainsi, lorsque q a des valeurs positives, par forte augmentation de la volémie par exemple, et telles que x et y deviendraient négatifs, on prévoit la possibilité de faire quitter à m la valeur 0,8 pendant la période transitoire nécessaire.
Le système est écrit dans un système d'unité commune ( u.c. ) pour lequel :
0,4 u.c. = 77 ng/ml de cortisol ( F ) = 1,1 µU/ml de vasopressine ( VP ),
valeurs qui correspondent à la moyenne des valeurs expérimentales des rythmes circadiens de ces hormones.
Les valeurs x, y, X, Y peuvent être assimilées à des concentrations hormonales et sont ainsi sujettes à des contraintes de positivité. Dans le cas physiologique ( X = 0, Y = 0 ; p = 0, q = 0 ), l'équilibration est simulée avec un champ paramétrique de ( 3.1 ) donnant un cycle-limite tel que le couple ( u,v ) admette l'origine ( 0, 0 ) comme point critique. L'équilibration ( X = 0, Y = 0 ) devient pathologique si une modification du champ ( 3.1 ) permet à un nouveau cycle-limite d'apparaître.
Les paramètres $k_i$, $c_i$, $k_i'$, $c_i'$ ( i = 1, 2, 3 ) pour le système simulant la pathologie, et $\bar{k}_i$, $\bar{c}_i$, $\bar{k}_i'$, $\bar{c}_i'$ pour le

système simulant le cycle physiologique, sont identifiés à partir des données cliniques et physiologiques, à l'aide de la méthode d'intégration numérique de Davidon-Fletcher-Powell avec contraintes ( cf. [ 1 ] ). Le critère à minimiser, $\mathfrak{J}$ ( $k_i$, $c_i$, $k_i'$, $c_i'$, T ), est donné par :

$$\mathfrak{J} ( k_i, c_i, k_i', c_i', T ) = \sum_j [ ( \bar{x}_j - x_j )^2 + ( \bar{y}_j - y_j )^2 ] \tag{3.2}$$

où $\bar{x}$ et $\bar{y}$ désignent des valeurs expérimentales et x et y les solutions "endogènes" du système ( 3.1 ) dans lequel on a pris X = 0,Y = 0, p = 0 et q = 0. La quantité T correspond à trois cycles, soit ici, à 72 heures. Dans le cas d'une homéostasie pathologique, la "simulation thérapeutique" consiste à déterminer les hormones exogènes X et Y de façon à ramener le système dans une position d'homéostasie physiologique. Une première méthode ( cf. [ 6, 7 ] ) consiste à écrire les entrées $e_1$ et $e_2$ dans une forme semblable à celle des hormones endogènes, soit :

$$\begin{aligned} e_1 &= \sum_{i=1}^{3} [ k_{3+i} ( u + p )^i + c_{3+i} ( v + q )^i ] + \sum_{i=1}^{3} \lambda_i ( X - \alpha_1 )^i \\ e_2 &= \sum_{i=1}^{3} [ k_{3+i}' ( u + p )^i + c_{3+i}' ( v + q )^i ] + \sum_{i=1}^{3} \lambda_i' ( X - \alpha_1' )^i \end{aligned} \tag{3.3}$$

avec $\lambda_1$, $\lambda_2$, $\lambda_3$, $\lambda_1'$, $\lambda_2'$, $\lambda_3'$, $\alpha_1$, $\alpha_1'$ des paramètres constants ayant pour rôle d'éviter la dérive du cycle-limite de dimension 4 que suivent les quatre états du système. On identifie alors les paramètres des relations ( 3.3 ) à l'aide de la méthode de Davidon-Fletcher-Powell.

*Remarque* 1: La tentation de prendre pour les entrées $e_1$ et $e_2$ la différence entre les équations de l'état physiologique et de l'état pathologique conduit à un contrôle qui peut ne pas satisfaire les conditions de positivité des variables x, y, X, Y, ni assurer l'existence d'un cycle-limite ( cf. [ 6 ] ).

Une seconde méthode, basée en premier ( cf. [ 15 ] ) sur le découplage et la linéarisation des systèmes non linéaires ( cf. [ 12, 13, 14 ] et les bibliographies afférentes ), consiste en fait à inverser le système ( 3.1 ) ( cf. [ 16, 17 ] ).

A partir du système ( 3.1 ), on considère alors les relations suivantes :

$$\mathcal{X} = 1/2\,( z_1 + z_2 + m )$$

$$\mathcal{Y} = 1/2\,( - z_1 + z_2 + m )$$

$$X = \mathcal{X} - x \qquad (3.4)$$

$$Y = \mathcal{Y} - y$$

x et y étant solutions des équations différentielles:

$$\dot{x} = \sum_{i=1}^{3} [\, k_i ( z_1 + p )^i + c_i ( m \,\mathrm{Log}( 1 + z_2/m ) + \theta + q )^i \,]$$

$$\dot{y} = \sum_{i=1}^{3} [\, k_i' ( z_1 + p )^i + c_i' ( m \,\mathrm{Log}( 1 + z_2/m ) + \theta + q )^i \,] \qquad (3.5)$$

Il s'agit donc de permettre aux sorties $z_1$ et $z_2$ du système ( 3.1 ) de passer de la position pathologique, donnée par les équations différentielles :

$$\dot{\psi}_1 = \sum_{i=1}^{3} [\, ( k_i - k_i' ) ( \psi_1 + p )^i + ( c_i - c_i' ) ( v + q )^i \,]$$

$$\dot{\psi}_2 = \sum_{i=1}^{3} [\, ( k_i + k_i' ) ( \psi_1 + p )^i + ( c_i + c_i' ) ( v + q )^i \,] ) \qquad (3.6)$$

avec $v = m \,\mathrm{Log}\, ( 1 + \psi_2/m ) + A + B \sin( \omega t ) + C \cos( \omega t )$ et $\omega = 2\pi / 24$,

à l'équilibre physiologique décrit par les équations différentielles obtenues à partir des données expérimentales :

$$\dot{\varphi}_1 = \sum_{i=1}^{3} [\, ( \bar{k}_i - \bar{k}_i' ) \varphi_1^i + ( \bar{c}_i - \bar{c}_i' ) v^i \,]$$

$$\dot{\varphi}_2 = \sum_{i=1}^{3} [\, ( \bar{k}_i + \bar{k}_i' ) \varphi_1^i + ( \bar{c}_i + \bar{c}_i' ) v^i \,] ) \qquad (3.7)$$

avec $v = m \,\mathrm{Log}\, ( 1 + \varphi_2/m ) + A + B \sin( \omega t ) + C \cos( \omega t )$ et $\omega = 2\pi / 24$. .

On écrit $z_1$ et $z_2$ sous la forme :

$$z_1 = \delta_1 + \varphi_1$$
$$z_2 = \delta_2 + \varphi_2 \qquad (3.8)$$

Le souhait du thérapeute est alors de trouver des fonctions $\delta_1$ et $\delta_2$ qui permettent en premier de définir un transitoire amenant les courbes pathologiques initiales, représentées par x et y, vers les courbes physiologiques que doivent suivre les variables $\mathcal{X}$ et $\mathcal{Y}$, somme des actions des hormones endogènes et exogènes. En second, après la période transitoire ( deux à trois jours ), le thérapeute souhaite à la fois, voir s'installer un régime permanent aussi proche que possible du rythme circadien physiologique pour les variables $\mathcal{X}$ et $\mathcal{Y}$, et mettre en place, pour de nombreuses raisons faciles à deviner, une action thérapeutique - représentée par X et Y - périodique de période égale ici à 24 heures.

Cependant, l'analyse immédiate des équations ( 3.5 ) montre qu'avec les coefficients du pathologique, il n'y a aucune raison pour que l'introduction dans ces équations des rythmes physiologiques entraîne l'apparition d'un cycle-limite. Bien au contraire, comme le confirment les simulations numériques, on assiste à une dérive affine du cycle. La démonstration de ce phénomène étant évidente.

Ainsi, la seule possibilité, en régime permanent, est de déformer aussi peu que possible le rythme physiologique pour assurer la périodicité de la thérapeutique représentée par X et Y, les fonctions $\delta_1$ et $\delta_2$ étant alors elles aussi périodiques. On est conduit ainsi à réaliser une optimisation sous les contraintes $x \geq 0$, $y \geq 0$, $X \geq 0$, $Y \geq 0$. Enfin, il faut s'assurer que le cycle-limite obtenu est stable et que le système est en plus structurellement stable.

Il est à noter que le principe de l'utilisation de l'optimisation est judicieux au regard de la notion de rythme physiologique moyen qui est utilisée et aussi vis à vis des incertitudes qu'amène l'utilisation d'un modèle.

Ainsi les fonctions $\delta_1$ et $\delta_2$ doivent permettre de satisfaire les conditions de positivité des variables x, y, X, Y, et, après un transitoire, doivent assurer l'existence d'un régime permanent cyclique et basé sur le rythme circadien. Il s'agit alors de trouver une classe de fonctions suffisamment riche pour pouvoir contenir les solutions cherchées. La famille de fonctions à quatre paramètres, dense dans l'ensemble des fonctions continues sur tout intervalle compact, et définie par :

$$f(x) = \int_0^{x+a} \frac{bd}{1 + d^2 - \cos(bt)} \cos(e^t)\, dt + c \quad \text{avec } d > 0 \qquad (3.9)$$

donne une idée du nombre minimal de paramètres qui pourraient être nécessaires.

Cette classe de fonctions est utilisée par Boshernitzan [ 2 ] dans la recherche des équations différentielles universelles ( cf. [ 2, 22 ] ).

*Remarque 2* : Le système ( 3.8 ) admet toujours au moins la solution mathématique $\delta_1 = \psi_1 - \varphi_1$ ; $\delta_2 = \psi_2 - \varphi_2$, mais cette solution ne peut bien sûr être en aucun cas la solution thérapeutique!

A l'heure actuelle, nous envisageons un contrôle non linéaire de la forme suivante :

$$\dot{z}_1 = \sum_{i=1}^{3} [\, k_{3+i} ( z_1 + p )^i + c_{3+i} ( m \,\mathrm{Log} ( 1 + z_2/m ) + \theta + q )^i \,] + \sum_{i=1}^{3} \lambda_i ( z_1 - \alpha_1 )^i$$

$$\dot{z}_2 = \sum_{i=1}^{3} [\, k'_{3+i} ( z_1 + p)^i + c'_{3+i} ( m \,\mathrm{Log} ( 1 + z_2/m ) + \theta + q )^i \,] + \sum_{i=1}^{3} \lambda'_i ( z_2 - \alpha'_1 )^i \qquad (3.10)$$

Les paramètres des équations ( 3.10 ) étant alors déterminés par la minimisation de l'écart entre les solutions des équations ( 3.7 ) et ( 3.10 ).
Cette seconde méthode est en cours d'étude.

*Remarque 3* : On pourrait s'inquiéter de l'impossibilité de trouver un contrôle thérapeutique capable de rétablir les rythmes physiologiques, mais on ne doit pas oublier que dans la réalité les paramètres qui déterminent le comportement du système sont variables et si ils sont passés de la position physiologique à la situation pathologique, le thérapeute, dans les cas de réversibilité, postule qu'un maintien forcé d'un rythme proche du rythme physiologique, pendant une période suffisante, permettra aux paramètres de se recaler sur l'homéostasie physiologique.

## IV. LE COUPLAGE INSULINE-GLUCAGON ET LE DIABETE.

L'activité glycémique peut être considérée comme la résultante des actions antagonistes du glucagon hyperglycémiant et de l'insuline hypoglycémiante, ces deux hormones agissant d'une façon couplée. Ce système présente, par rapport au système surréno-posthypophysaire, une particularité remarquable au plan anatomique. Dans le cas de la réponse glycémique, la nature a installé le mécanisme de commande de la régulation dans un même endroit - les îlots de Langerhans - au sein du pancréas. On trouve dans ces amas cellulaires la fabrication simultanée de l'insuline et du glucagon sous l'action coordonnée de la somatostatine. Devant les résultats cliniques obtenus à l'aide de la vasopressino-corticothérapie, il semblait, au regard des enjeux en diabétologie, intéressant de proposer une modélisation du système insuline-glucagon sous l'angle de la vision bipolaire des systèmes ago-antagonistes définis par Bernard-Weil.
La modélisation proposée prend la forme d'un système différentiel non linéaire, à trois entrées, $e_1$, $e_2$ et p, et trois sorties, $z_1$, $z_2$ et $z_3$, ainsi défini ( cf. [ 8 ] ) :

$$\dot{\mathcal{X}} = \sum_{i=1}^{3} [\, k_i (\mathcal{X} - \mathcal{Y} + p))^i + c_i (\mathcal{X} + \mathcal{Y} - m)^i \,] + e_1$$

$$\dot{\mathcal{Y}} = \sum_{i=1}^{3} [\, k_i' (\mathcal{X} - \mathcal{Y} + p)^i + c_i' (\mathcal{X} + \mathcal{Y} - m)^i \,] + e_2$$

$$\dot{X} = e_1$$

$$\dot{Y} = e_2 \qquad (4.1)$$

$$\dot{G} = g_1 (G_0 - G) + g_2 [\, g_3 [\, th( g_4 (\mathcal{X} - \mathcal{Y} + Y - X + g_5)) + th( g_4 (X - Y + g_5)) - 2th( g_4\, g_5)] + p \,]$$

$$z_1 = \mathcal{X} - \mathcal{Y}$$

$$z_2 = \mathcal{X} + \mathcal{Y} - m$$

$$z_3 = G$$

avec $\mathcal{X} = x + X$, $\mathcal{Y} = y + Y$, où x et y désignent respectivement les actions du glucagon et de l'insuline endogènes et X et Y les actions des hormones exogènes ( thérapeutique ).
$G_0 = 0{,}78$, désigne le taux de base physiologique de la réponse glycémique G( t ) et m = 2,1.
Le système est écrit dans un système d'unité commune pour lequel une unité commune vérifie :
10 μU/ml d'insuline = 100 pg/ml de glucagon.
Dans le cas de l'étude du test de tolérance au glucose, l'entrée p, qui est liée à la prise orale de 100 g de glucose, est représentée par la fonction :

$$p(t) = ( p_1 / ( p_1 - p_2 ) ) \,.\, 100 \,.\, p_3 [\, \exp(- p_2 t) - \exp( - p_1 t) \,] \qquad (4.2)$$

Comme pour le modèle du système surréno-pothypophysaire, les paramètres des équations ( 4.1 ) et ( 4.2 ) ont été identifiés, à l'aide de la méthode d'optimisation non linéaire de Davidon-Fletcher-Powell, à partir des courbes expérimentales. Les paramètres définissant la fonction p( t ) ont été ajustés une seule fois car les conditions d'absorption intestinale du glucose sont moins influencées par les anomalies hormonales que les autres processus du métabolisme glucidique. Par contre, bien entendu, les paramètres de l'équation ( 4.1 ) sont à identifier dans le cas physiologique et dans le cas pathologique.
La recherche du contrôle ( thérapeutique ) visant à corriger les anomalies de la réponse glycémique chez le diabétique a été obtenue dans un premier temps ( cf. [ 8 ] ) en prenant les entrées $e_1$ et $e_2$ sous la forme :

$$e_1 = \sum_{i=1}^{3} \left[ k_{3+i} ( \mathbf{H} - \mathbf{Y} + p )^i + c_{3+i} ( \mathbf{H} + \mathbf{Y} - m )^i \right]$$

$$e_2 = \sum_{i=1}^{3} \left[ k'_{3+i} ( \mathbf{H} - \mathbf{Y} + p )^i + c'_{3+i} ( \mathbf{H} + \mathbf{Y} - m )^i \right] \qquad (4.3)$$

Elles permettent de mettre en place un contrôle asymptotique tendant à ramener la position limite pathologique à la valeur physiologique moyenne de la glycémie ( 1 g/l ), le déséquilibre initial glucagon-insuline avant la charge en glucose, comme l'équilibre physiologique, étant des points critiques stables des modèles pathologique et physiologique.

On peut aussi opérer comme pour le système surréno-posthypophysaire et considérer les relations :

$$\mathbf{H} = 1/2 ( z_1 + z_2 + m )$$

$$\mathbf{Y} = 1/2 ( - z_1 + z_2 + m )$$

$$X = \mathbf{H} - x \qquad (4.4)$$

$$Y = \mathbf{Y} - y$$

x et y étant solutions des équations différentielles :

$$\dot{x} = \sum_{i=1}^{3} \left[ k_i ( z_1 + p )^i + c_i ( z_2 )^i \right]$$

$$\dot{y} = \sum_{i=1}^{3} \left[ k'_i ( z_1 + p )^i + c'_i ( z_2 )^i \right] \qquad (4.5)$$

Il s'agit ici de permettre aux sorties $z_1$, $z_2$ et $z_3$ du système ( 4.1 ) de passer de la position pathologique :

$$z_1 ( 0 ) ; z_2 ( 0 ) ; G ( 0 ) \qquad (4.6)$$

à l'équilibre physiologique " asymptotique" :

$$z_1 = 0 ; z_2 = 0 ; G = 1 \qquad (4.7)$$

L'équilibre physiologique devant bien entendu être atteint avant l'ingestion suivante, soit dans un délai d'environ 5 heures.

Pour déterminer la "thérapeutique" - X, Y - à appliquer au système "pathologique" ( 4.1 ) on peut alors, par exemple, utiliser de nouveau la méthode donnée par les relations ( 3.10 ) et effectuer une optimisation, sous les contraintes $x \geq 0$, $y \geq 0$, $X \geq 0$, $Y \geq 0$, en minimisant l'écart entre les trois sorties $z_1$, $z_2$ et $z_3$ du système "pathologique" contrôlé ( 4.1 ) et les trois sorties $\varphi_1$, $\varphi_2$ et $\varphi_3$ du système "physiologique" ( 4.1 ) soumis aux entrées $e_1 = 0$, $e_2 = 0$ et $p(t)$.

Ceci fera l'objet d'une prochaine étude, mais les simulations effectuées avec les entrées $e_1$ et $e_2$ sous la forme ( 4.3 ) ( cf. [ 8 ] ) montrent déjà qu'une meilleure approche de la courbe glycémique est obtenue avec l'intervention simultanée des deux actions X et Y (insuline et glucagon ) plutôt qu'avec l'insuline seule.

## V. CONCLUSION

Nous avons présenté et illustré par deux exemples une nouvelle méthode de recherche liant étroitement l'automatique et la biologie. Cette voie dont Bernard-Weil est l'initiateur, ouvre un champ d'investigation immense en permettant, par un procédé de modélisation original qui s'apparente aux "dynamical metaphors" de Rosen [ 21 ] , de prendre en compte l'aspect ago-antagoniste qui intervient dans un grand nombre de régulations biologiques. Cette modélisation, à même de simuler aussi bien le pathologique que le physiologique, propose des contrôles bipolaires aux incidences thérapeutiques parfois surprenantes. Il n'est pas question que l'automaticien rentre dans les précisions médicales dont il n'a pas la compétence, mais il peut tout de même indiquer, comme le montre déjà un certain nombre de publications médicales ( cf. [ 5, 10, 11 ] ), que la pratique des thérapeutiques bipolaires étend pas à pas son champ d'application. Il n'est pas douteux que dans un avenir que l'on doit rendre aussi proche que possible, ces thérapeutiques conduisent à supprimer l'état de souffrance d'un grand nombre d'êtres humains.

BIBLIOGRAPHIE


[ 1 ] M.S. BAZARAA et C.M. SHETTY, Nonlinear Programming, Theory and Algorithms, Wiley, New York, 1979.

[ 2 ] M. BOSHERNITZAN, Universal formulae and universal differential equations, Annals of Mathematics, 124, 1986, pp. 273-291.

[ 3 ] E. BERNARD-WEIL, Effects of a week of ACTH or corticosteroid treatment on the neuropostpituitary response to corticosteroid load, Steroids Lip. Res., 3 , 1972, pp.24-29.

[ 4 ] E. BERNARD-WEIL, Formalisation et contrôle du système endocrinien surréno-posthypophysaire par le modèle mathématique de la régulation des couples ago-antagonistes, Thèse d'Etat, Universté Paris VI, France, 1979.

[ 5 ] E. BERNARD-WEIL, Lack of response to a drug : a system theory approach, Kybernetes, 14, 1985,pp. 25-30.

[ 6 ] E. BERNARD-WEIL, Interactions entre les modèles empirique et mathématique dans la vasopressino-corticothérapie de certaines affections cancéreuses dans "Régulations physiologiques : Modèles récents", G. Chauvet et J.A. Jacquez, éd., Masson, Paris, 1986, pp. 133-155.

[ 7 ] E. BERNARD-WEIL, A general model for the simulation of balance, imbalance and control by agonistic-antagonistic biological couples, Mathem. Modelling, 7, 1986, pp. 1587-1600.

[ 8 ] E. BERNARD-WEIL et D. CLAUDE, Simulation du test de tolérance au glucose par le modèle de la régulation des couples ago-antagonistes. Contrôle bipolaire, C.R. Acad. Sci. Paris, 305, série I, 1987, pp. 303-306.

[ 9 ] E. BERNARD-WEIL, C. DALAGE, C. PIETTE et L. OLIVIER, Action of lysine-vasopressin on the protein content of Hela cell cultures and on the RNA and DNA concentrations of tissue incubation, Experta Med. Internat. Congr. Ser. : Protein and Polypeptide Hormones, 161, 1968, pp. 547-548.

[ 10 ] E. BERNARD-WEIL et B. PERTUISET, Mathematical model for hormonal therapy ( vasopressin, corticoids ) in cerebral collapse and malignant tumors of the brain ( 36 cases ), Neurol. Res., 5, 1983, pp. 19-35.

[ 11 ] E. BERNARD-WEIL, J.L. JOST et P. VAYRE, Nouvel aspect des relations hôte-tumeur. Etude du système surréno-post-hypophysaire chez le cancéreux digestif, Chirurgie, 113, 1987, pp. 293-298.

[ 12 ] D. CLAUDE, Découplage des systèmes non linéaires, séries génératrices non commutatives et algèbres de Lie, SIAM J. Control Optimiz., 24, 1986, pp. 562-578.

[ 13 ] D. CLAUDE, Everything you always wanted to know about linearization but were afraid to ask, in "Algebraic and geometric methods in nonlinear control theory", M. Fliess and M. Hazewinkel, ed., D. Reidel Publishing Company, 1986, pp. 181-226.

[ 14 ] D. CLAUDE, Découplage et linéarisation des systèmes non linéaires par bouclages statiques, Thèse d'Etat, Université Paris-Sud, France, 1986.

[ 15 ] D. CLAUDE et E. BERNARD-WEIL, Découplage et immersion d'un modèle neuro-endocrinien, C.R. Acad. Sci. Paris, 299, série I, 1984, pp.129-132.

[ 16 ] M. FLIESS, Automatique et corps différentiels, Forum mathématiques, 1, 1989.

[ 17 ] U. KOTTA, Application of inverse system for linearization and decoupling, Systems Control Lett., 8, 1987, pp. 453-457.

[ 18 ] M. MONACO, P.H. KOHN, W.R. KIDWELL, J.S. STROBL et M.E. LIPPMAN, Vasopressin action on WRK-1 rat mammory tumor cells, J.N.C.I., 68, 1982, pp. 267-270.

[ 19 ] M. PAWLIKOWSKI, The effect of neuropeptides on cellular proliferation, Materia Medica Polona,Fasc. 1, 61, 1987, pp. 17-20.

[ 20 ] E. ROZENGURT, A. LEGG et P. PETTICAN, Vasopressin stimulation of mouse 3T3 cell growth, Proc. Natl. Acad. Sci. USA, 76, 1979, pp. 1284-1287.

[ 21 ] R. ROSEN, Dynamical System Theory in Biology, Wiley-Interscience, New York, 1970.

[ 22 ] A. RUBEL, A universal differential equation, Bull. Amer. Math. Soc. ( N.S. ), 4, 1981, pp. 345-349.

[ 23 ] A. SILVETTE et S. BRITTON, A theory of corticoadrenal and postpituitary influences on the kidney, Science, 88, 1938, pp. 150-151.

[ 24 ] M.K. SUNDARESHAN et R.A. FUNDAKOWSKI, Stability and control of a class of compartmental systems with application to cell proliferation and cancer therapy, IEEE Trans. Automat. Contr., AC-31, 1986, pp. 1022-1032.

[ 25 ] G.W. SWAN, Application of Optimal Control Theory in Biomedicine, Dekker, New York, 1984.

[ 26 ] A.T. WINFREE, The Geometry of Biological Time, Springer-Verlag, New York, 1980.

# INTERNAL MODEL CONTROL OF A STEAM GENERATOR WITH VARYING DELAYS


S. Aksas*,**,*** and D. Meizel*

* Laboratoire d'Automatique et d'Informatique Industrielle (L.A.I.I.D.N)
INSTITUT INDUSTRIEL DU NORD (I.D.N.) - B.P. 48
59651 VILLENEUVE D'ASCQ CEDEX - FRANCE

** Laboratoire de Génie Chimique et d'Automatique (L.G.C.A)
same adress

*** CERCHAR
Rue A. Dubost - B.P. 19
62670 MAZINGARBE - FRANCE



Abstract

We present the internal model control (IMC) of a steam generator described by a discrete time linear model with delays and dynamical parameters depending upon the steam demand. The simultaneous consideration of IMC structure together with standard design objectives such as tracking and regulation reference models, or generalised minimum variance brings a very simple expression of the control law even in the linear stationnary case. In this context, the consideration of typically multivariable structures such as the interactor is of great interest. The obtained linear IMC requires no model inversion nor diophantine technique , and can easily be extended to the varying-load case into a scheduled-gain IMC .


## 1 Introduction

In this paper, we present a scheduled-gain control law of a pulverized-coal steam generator. The process under consideration has two outputs, two control inputs and a scalar measured disturbance input (the steam demand) .

One major feature of this study stems from its practical context. The modelization is performed with the aim of control and the identification is constrained by the process instrumentation and operation.

This leads to integrate discrete time linear identifications into a global model that can be termed as "linear with time varying coefficients". Its dynamical behaviour as well as its "static operating points" depend greatly on the variations of the supplied steam , especially the delays and interactors of the local linear model .

The control objectives are regulation around operating constant set points and disturbance rejection. Considering the non stationnary behaviour of the plant, Internal Model Control (IMC) stands as an appealing technique . The fact that the IMC design of a controller is an open-loop design is particularly interesting to take into account the variable delays of the process.

The paper is organized as follows . After the physical description of the process we present its dynamical behaviour under the form of a "linear with varying coefficients" model .

Some structural aspects, centered around the linear notion of interactor, considered here as a multivariable extension of the pure delay, are then pointed out and will be used for the synthesis of an IMC control-law.

The IMC structure is first recalled. By use of this control structure together with design objectives such as reference-models-matching or generalized minimum variance , we exhibit a particularly simple expression of the control law.

It is obtained first in the linear case with a constant dynamic, then extended to the non stationnary case .

## 2 Process description

The process under consideration is a pulverized coal steam generator.It is represented as follows

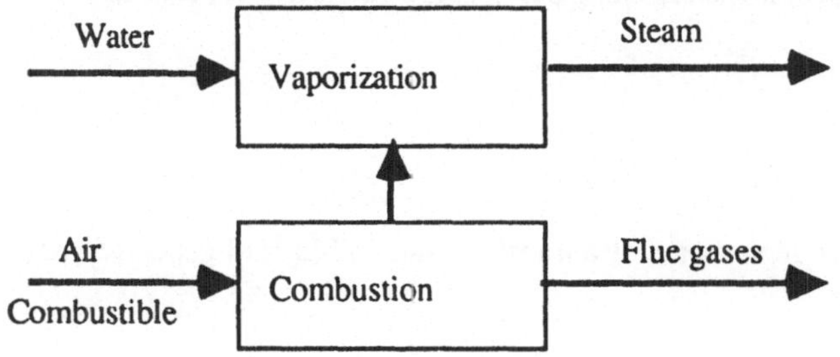


Figure 1 : Steam-generator principle

| | |
|---|---|
| $U'_1(t)$ | states the coal flow set point defined in an adimensional sacale [0,100%]. |
| $U_1(t)$ | " the set point of a local analog steam pressure regulation " ". |
| $U_2(t)$ | " the air flow set point " ". |
| $Y_1(t)$ | " the steam pressure measure " ". |
| $Y_2(t)$ | represents the oxygen concentration measure " ". |
| $d(t)$ | represents the steam flow measure " ". |

All informations are sampled with a convenient $\Delta$=5s sampling period.
Vector U (resp. Y) is the control-input (resp controlled-output . ) ; d is the disturbance input.
The supplied steam flow can be considered as a varying load parameter depending on the steam-users need ; its quality is here essentially given by its pressure which has to be kept approximately constant despite any variation of the steam flow . Energetical considerations leads the producer to realize a trade-off between two requirements:

-have a good combustion rate (i.e burn everything that is possible to burn) .

-have a good steam production rate (i.e transfer the maximum available energy to the steam ) .

From chemical engineering considerations, a set of "optimal" trade-off has been computed, for any admissible d, in the form of set-points $Y_{2S}(d)$ for the oxygen concentration in the flue gases and $U_S$ for the control input .

The investigated controller should then keep the controlled signals around their optimal set-points for any constant value of d and, moreover for any admissible variation of d(t).

The question of a variable d has two aspects (i)&(ii):

(i) the tracking of the variations of the operating-point dependent signals $U'_1(d)$, $U_2(d)$, $Y_2(d)$,

(ii) the modification of the local dynamical behaviours identified for d constant.

For brievety'sake, we shall not discuss here the first question (i), focusing our attention on the second point.

In the sequel, $y = (y_1, y_2)^t = (Y-Y_S)$ and $u = (u_1, u_2)^t = (U-U_S)$ (1)

will denote the deviations with respect to the operating point, assumed piece-wise constant

3 Dynamic model

The transfer from u to y can be represented by a transfer function $F(q^{-1})$ of general form (2) .

$$y(k)=F(q^{-1}).u(k) \qquad (2)$$

$$F(q^{-1}) = \begin{bmatrix} q^{-RT11}H_{11}(q^{-1}) & 0 \\ q^{-RT21}H_{21}(q^{-1})(1-q^{-1}) & q^{-RT22}H_{22}(q^{-1}) \end{bmatrix}$$

$H_{11}$,$H_{21}$,$H_{22}$ are second order transfer functions . Their coefficients together with the delays $RT_{11}$, $RT_{21}$ and $RT_{22}$ vary with d. Static gain and delay matrices are given for 3 characteristic loads on table 1 below.

| d = 4 t/h | | d = 6 t/h | | d = 8 t/h | | d = 4 t/h | | d = 6 t/h | | d=8t/h | |
|---|---|---|---|---|---|---|---|---|---|---|---|
| 13 | ∞ | 8 | ∞ | 5 | ∞ | 1 | 0 | 1 | 0 | 1 | 0 |
| 13 | 17 | 13 | 19 | 9 | 13 | 0 | 1.033 | 0 | 1.167 | 0 | 0.827 |
| Delays | | | | | | Static gains | | | | | |

Table 1 : Delay and Static Gain matrices for 3 characteristic loads

Some comments are made on the structure of the linearized process model :

i) The static gain matrix is invertible for each d .

ii) The structure of the delay matrix ( $T_{11} < T_{21} < T_{22}$ ) implies that the system is not decouplable unless adding artificial delays on $u_1$ .

Following those considerations, It is interesting to factorize out all the non invertible part of a transfer matrix F as in (3) [1,2,3] (F is here a full rank n*n matrix) where $q^T J(q^{-1})$ is termed ***interactor*** of the transfer matrix F in [3] . :

$$F(q^{-1})=q^{-T}J^{-1}(q^{-1})A^{-1}(q^{-1})B(q^{-1}) \quad (3)$$
$$A(q^{-1})=I+A_1q^{-1}+.......A_{nA}q^{-nA}$$
$$B(q^{-1})=B_0+B_1q^{-1}+....B_{nB}q^{-nB}\det(B_0)\neq 0.$$

This factorization is an important tool for the so-called "model matching problem" [1] , particularly considering the following result [3] :

*Let $F_1(q^{-1})$ and $F_2(q^{-1})$ be two (m*m) transfer matrices with interactors $q^{T1}J_1(q^{-1})$ and $q^{T2}J_2(q^{-1})$ . There exist a matrix $F(q^{-1})$ such that (4) is verified if and only if $(q^{T1}J_1(q^{-1}))*(q^{T2}J_2(q^{-1}))^{-1}$ is proper .*

$$F_1(q^{-1})*F(q^{-1})=F_2(q^{-1}) \quad (4)$$

## 4 Control

### 4.1 Internal model control structure

The basic IMC structure is shown on fig.2 . First developped for linear systems [4,5] this structure can conveniently be used for non-stationnary and/or non-linear systems [6,7] .

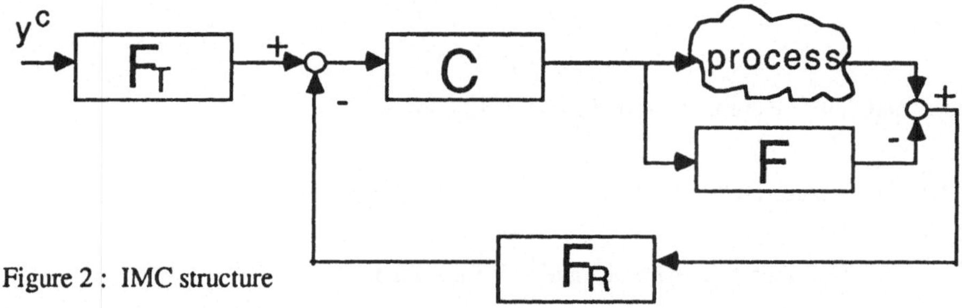


Figure 2 : IMC structure

yc represents a piece-wise constant output set-point,C is a dynamical controller, F is an input-output model of the process to be controlled . $F_T$ specifies the desired tracking trajectories and $F_R$ is concerned with robustness [4,5] and regulation .

The essential hypothese is that both the process and its model are supposed input-output stable .

One can make two simple remarks on that control scheme :

-in case of an exact model the process is open loop .

-the control is directly affected by the model/plant mismatch .

The three following and essential properties of IMC rely on those remarks :

P1-Stability : In case of an exact model stability of both controller and process ensures stability of the control .

P2-Perfect controller : If the controller C is chosen as the right inverse of the model F. It achieves perfect set point tracking for any plant/model mismatch that preserves stability .

P3-Zero-offset : any controller C that satisfies the static gain properties (5) (here expressed in discrete-time transfer function terms) and produces a stable closed loop yields zero-offset for any piece-wise constant set-point and/or output-additive perturbation

$$\{ \quad F(1)\, C(1) F_T(1)=I \quad , \quad F_R(1)=I \quad \} \tag{5}$$

The IMC design can then be structured in three steps :

-Modeling : find a model F of the plant (plant and model are assumed stable)

-Open-loop design of a control law ( C ) for this model that satisfies P1 & P3

-Robustness : design a filter $F_R$ that ensures conservation of stability and precision properties for a set of given modelling errors.

The studied process is primarily stabilized by a zero-level regulator (§2.1) and, due to the varying delays feature (table 1), the design of a closed-loop control law is tedious whereas an open-loop design is quite simple. For all these reasons, Internal Model Control (IMC) stands as an appealing control method.

In this context we present a particular design technique based upon reference models [8] and generalized minimum variance [9] control techniques. This structure is first introduced in the linear case (constant load) and then generalised to the varying load case.

4.2 Constant operating point : open loop synthesis

In this paragraph, we first consider the following process model (6) in which for sake of simplicity we omit the relationship between the load and the dynamical parameters.

$$y(k+T)=J^{-1}(q^{-1}).A^{-1}(q^{-1}).B(q^{-1}).u(k) \tag{6}$$

The design synthesis philosophy is based upon the model reference approach [8] enlarged later by the generalized minimum variance control [9].

A desired trajectory is generated by an admissible tracking reference model (7), chosen with the same interactor as the controlled process (§ 2.4)

$$y^d(k+T)=J^{-1}(q^{-1})A_T^{-1}(q^{-1})B_T(q^{-1})y_c(k) \tag{7}$$

$J^{-1}(1)A_T^{-1}(1)B_T(1)=I$

$J^{-1}(1)$ is normalized i.e $J^{-1}(1)=I$

Consider now that regulation objectives are specified into the following form [1] :

$$C_y(q^{-1})J(q^{-1})\,(y(k+T)-y^d(k+T))=0 \tag{8}$$

$C_y(q^{-1})=I+C_{y1}q^{-1}+.....+C_{ynCy}q^{-nCy}$ is a stable regulation reference-model

We get then following open loop control law :

$$u(k)=B^{-1}(q^{-1})A(q^{-1})(J(q^{-1})y^d(k+T)) \tag{9}$$

One possible IMC scheme is then displayed in fig.4. This scheme corresponds to the primary concept of open-loop control synthesis by model inversion [4,5,6].

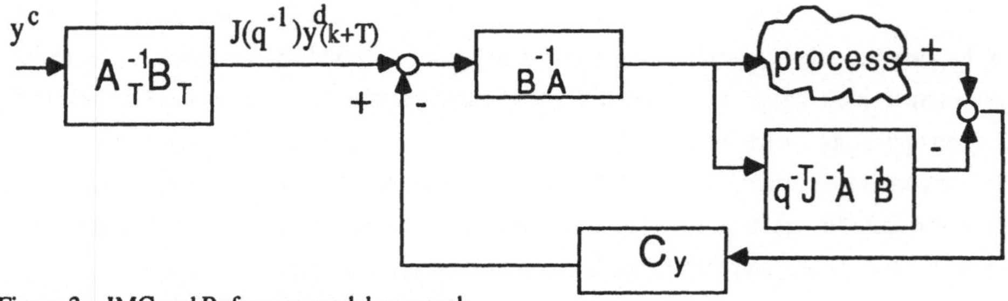


Figure 3 : IMC and Reference model approach

A major restriction of this method lies in the fact that the system should be minimum phase . IMC researchers have thus defined methods to find approximate inverses [4,5].
Neglecting here the idea of "approximate inverse" we only consider standard design objectives such as the minimization of a generalized variance since it is a natural extension of the model reference concept that handles stable non-minimum-phase processes .

Instead of the previous synthesis objective (7,8) we propose to select the control input by the minimization of a quadratic criterion $R_1$ (10) or, more generally $R_2$ (11).

$$R_1(u(k)) = \| C_y(q^{-1})J(q^{-1})(y(k+T)-y^d(k+T)) \|^2 + \lambda\|u(k)\|^2 \quad \lambda>0 \tag{10}$$

$$R_2(u(k)) = \| C_y(q^{-1})J(q^{-1})(y(k+T)-y^d(k+T)) \|^2 + \| C_u(q^{-1})(u(k)) \|^2 \tag{11}$$

$C_u(q^{-1})$ is a polynomial matrix and $C_y(q^{-1})$ is as in ( 8 ) . In particular, integral action will be exhibited with a u(k) derived from the minimization of a criterion $R_2$ with $C_u(1)=0$ and $B_0{}^tC_y(1)$ invertible .Zeroing the gradient of $R_2(u(k))$ (11) is easy due to the simple expression of the following gradient (12).It yields together with the process model ( 6 ) the open loop control law (13):

$$\nabla_{u(k)}(J(q^{-1})y(k+T))=B_0 \tag{12}$$

$$u(k)=[\, B_0{}^t\, C_y(q^{-1})\, A^{-1}(q^{-1})B(q^{-1}) + C_{u0}{}^t\, C_u(q^{-1})\, ]^{-1}\, B_0{}^t\, C_y(q^{-1})\, J(q^{-1})\, y^d(k+T) \tag{13}$$

The display of that control law in fig.4 emphasizes the role of the bicausal part of the process-model into the proposed open loop control law (13) .

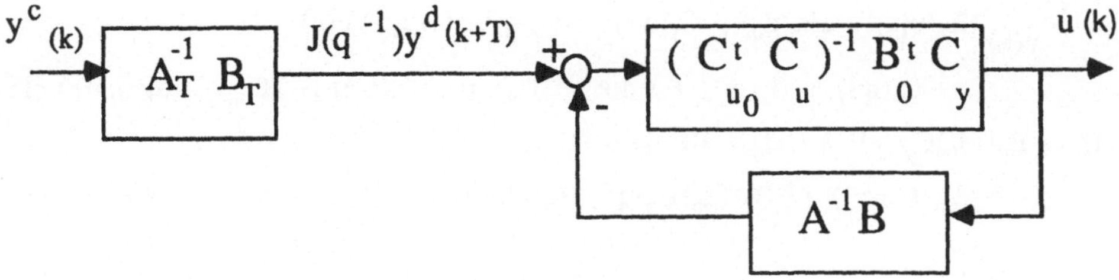


Figure 4 : IMC and Generalized Minimimun Variance .
Open Loop control law

4.3 Varying operating point

We consider now the load parameter variations effects on the output and on the dynamics of the "linear with varying parameters" model, and try to exhibit a scheduled-gain control law . According to (2&3), a prediction model could be the following (14) :

$$y(k+T(d(k)))=J^{-1}(d(k),q^{-1})A^{-1}(d(k),q^{-1})B(d(k),q^{-1})u(k) \tag{14}$$

To obtain a simple scheduled-gain version, a possible approach would be to replace (13) by its "varying-load version" . In such a case, the stability analysis of the non-stationnary controller is cumbersome, even in the case of a perfect modeling: we loose here the property P1 of IMC.

Another method is thus chosen here . We propose to center the open-loop synthesis of the IMC control law for the fixed value $d_0$ , in order to obtain a fixed controller C (see fig.2) .With this aim in mind , consider the following definition (15) of the bicausal part of the model, centered around a particular value $d_0$.

$$A^{-1}(d(k),q^{-1})B(d(k),q^{-1})=F_n^{-1}(d(k),q^{-1})A^{-1}(d_0,q^{-1})B(d_0,q^{-1})$$
$$F_n(d_0,q^{-1})=I\,,\ F_n(d,q^{-1}) \text{ is invertible and bistable for any d} \tag{15}$$

According to this definition, we propose to replace the tracking reference-model (7) (resp the design criterion (11)) by the follwing ones (16) (resp (17)).

$$y^d(k+T(d(k))=J^{-1}(d(k),q^{-1}).A_T^{-1}(q^{-1}).B_T(q^{-1}).u(k) \tag{16}$$

$$R_2(u(k)) = \| C_y(q^{-1})\, J(\, d(k),q^{-1})\, F_n(d(k),q^{-1})\, (y(k+T(d(k)))-y^d(k+T\,(d(k))))\, \|^2 \ldots$$
$$\ldots+\| C_u(q^{-1})(u(k))\, \|^2 \tag{17}$$

As $J(d(k),q^{-1}).\ F_n(d(k),q^{-1}).y(k+T(d(k)))=A^{-1}(d_0,q^{-1})B(d_0,q^{-1}).u(k)$, we get:

$$\nabla_{u(k)} (R_2(u(k)))=2B_0(d_0)^t[\ C_y(q^{-1})\ A^{-1}(d_0,q^{-1})B(d_0,q^{-1})\ u(k)\ldots$$
$$+ C_y(q^{-1}).J(d(k),q^{-1}).\ F_n(\ d(k),q^{-1})\ .y^d\ (k+T(d(k)))\ ]+ 2C_{u0}^t\ C_u\ (q^{-1}).u(k) \quad (18)$$
$$u(k)=[\ B_0^t\ (d_0)C_y(q^{-1})\ A^{-1}(\ d_0,q^{-1})B(\ d_0\ ,q^{-1})\ \ldots$$
$$+ C_{u0}^t\ C_u(q^{-1})\ \ ]^{-1} \cdot B_0^t\ (d_0)\ C_y(q^{-1}).J(d(k),q^{-1}).\ F_n(\ d(k),q^{-1})\ .y^d\ (k+T\ (d(k))\ ) \quad (19)$$

An I.M.C closed-loop realisation of this controller is displayed in fig.5.Notice that the open-part of the control-law(19) is similar to the one of the constant load's case (fig.4).Due to the convenient choice of the non-stationnary reference-model (14) and design-criterion, the non-stationnary part of the expression(19) is rejected in the open-loop part of this control-law. In the case of a correct modeling, the stability-analysis of this control-system is then reduced to a linear-system analysis at the single-point $d_0$.

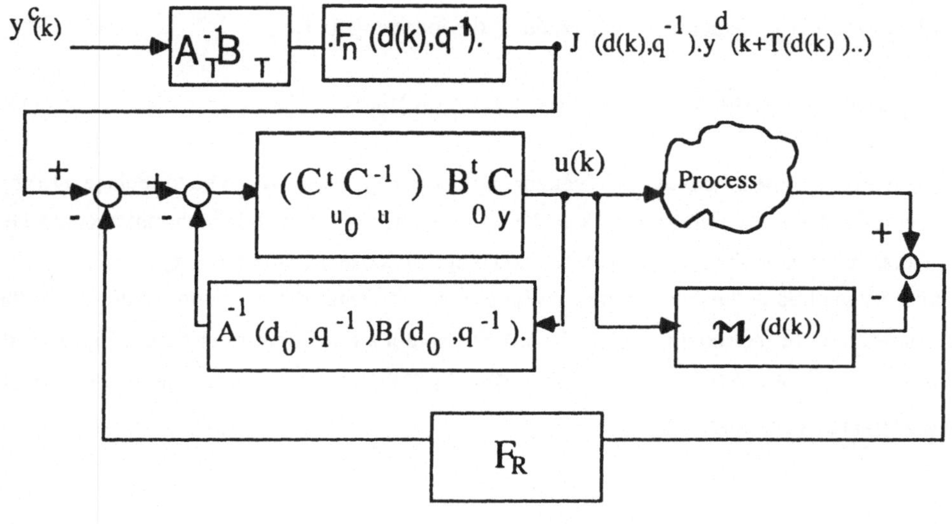


Figure 5 : Variable load ,

closed-loop control-law ( $\mathcal{M}(d(k)) = J^{-1}(d(k),q^{-1})A^{-1}(d(k),q^{-1})B(d(k),q^{-1})$ )

This contoller has been successfully implemented on the actual plant [11].

5. Conclusion

The use of Internal Model Control structure transforms the nominal control problem of a stable plant into an open-loop design question . This fact explains the ability of IMC to handle nonlinear or nonstationnary processes as in our practical application . Using both IMC stucture together with conventional design criteria and structural comments on linear multivariable systems, we obtain a simple controller that requires no model inversion nor diophantine technique in the linear case . The use of open-loop control enables to adress directly the non-stationnary case by throwing the non-stationnary part of the controller outside any feedback-loop thus getting rid of the problems caused by non-constant delays .

## REFERENCES

[1] DION J.M. , DUGARD L. (1986)
"Panorama de la commande adaptative multivariable"
in "Commande adaptative : aspects pratiques et theoriques" pp 229-305
coordinateurs I.D. LANDAU et L. DUGARD
Ed. Masson . Paris

[2] COMMAULT C. (1986)
"Structure des systèmes linéaires : approche transfert"
in "Colloque $\Sigma_\infty$ : Propriétés structurelles des systèmes linéaires
multivariables. Application à des problèmes de commandes ."
CNRS , 1986 , PARIS

[3] WOLOWITCH W.A. and FALB P.L. (1976)
"Invariants and canonical forms under dynamic compensation"
SIAM J. Control and Optimization, vol. 14, N°. 6, pp 996-1008

[4] GARCIA C.E. , MORARI M. (1982)
"Internal model control . 1 . A unifying review and some new results"
Ind. Eng. Chem. Process Des. Dev., 1982, 21 , pp 308-323

[5] GARCIA C.E. , MORARI M. (1985)
"Internal model control. 2 . Design procedure for multivariable systems"
Ind. Eng. Chem. Process Des. Dev., 1985, 24 , pp 472-484

[6] ECONOMOU C.G. , MORARI M. (1986)
"Internal model control. 5 . Extension to nonlinear systems"
Ind. Eng. Chem. Process Des. Dev., 1986, 25 , pp 403-411

[7] DUFOUR J. , GREHANT B. , LOTTIN J. (1987)
"Robustness : internal model control approach . Extension to state affine representation"
Elsevier-Science Publisher B.V pp 575-580

[8] LANDAU I.D. (1986)
"La commande adaptative : un tour guidé"
in "Commande adaptative : aspects pratiques et theoriques" pp 1-82
coordinateurs I.D. LANDAU et L. DUGARD
Ed. Masson . Paris

[9] CLARKE D.W. , GAWTHROP P.J. (1975)
"Self-tuning controller"
Proc. IEE, 1975, 122 (9), pp 929-934

[10] MEIZEL.D, AKSAS.S
"An I.M.C version of generalized minimum variance control for multivariable processes"
Proc of the "SMS'88" IMACS Congress. 18-21 sept 1988 - Cetraro Italy

[11] AKSAS.S
" Commande d'un générateur de vapeur soumis à une charge variable"
D.I.Thesis n°19, 4 Nov 1988, I.D.N, BP.48, 59651 Vn d'Ascq Cedex, FRANCE

# NON-LINEAR MODELING AND CONTROL OF SOME CLASSES OF INDUCTION MACHINE

G.A. CAPOLINO
ESIM/IMT Department of Electrical Engineering
28, rue des Electriciens 13012 Marseille
FRANCE

ABSTRACT.

The aim of this paper is to analyze and to compare different methods of modeling, by means of non-linear state space approach, some classes of induction machine. From an electrical engineer point of view, the problem is not very new but it has never been related to automatic control since the methods applied to AC machine are still very empirical. What we are going to present is related to previous works from PARK and KRON which are involved in mathematical modeling of AC machines in order to study transient performances.

In the first part of the paper, we present different methods to elaborate the mathematical model of three phase induction machine with the assumption of linear flux analysis. Our purpose is to give explicit state space models which are very useful in order to perform identification or control of the machine associated with a drive. In the second part of the paper, a non-linear flux analysis extension is given in order to develop a more accurate model. The third part of the paper is dedicated to some remarks about non-linear control of the induction machine derived from the previous models.

NOMENCLATURE.

| | |
|---|---|
| s,r | : subscripts referring to stator and rotor windings. |
| d,q | : subscripts referring to synchronous rotating or fixed axes. |
| Rj, Lj | : resistance and self-inductance of winding j. |
| lj | : leakage inductance of winding j. |
| M | : mutual inductance between stator and rotor windings. |
| theta | : angular position of the rotor in stationary reference frame. |
| sig | : dispersion factor. |
| Tauj | : time constant of winding j. |
| Tl | : load torque. |
| f | : friction coefficient. |
| J | : moment of inertia. |

I. INTRODUCTION.

The elaborate knowledge of electrical machines is about one hundred years old but their transient performances have been investigated only thirty years later. The first papers discussing methods of obtaining mathematical models of electrical machines have been published by STEINMETZ (1909) and LYON (1923) and they have elaborated a general theory of transient in electrical circuits and rotating machines. Nevertheless, the most significant works about mathematical models which have been the basic theory already used in recent investigations are, on our opinion, derived from PARK (1923) for synchronous machines and KRON (1938) for the tensor analysis of the generalized machine. It is not necessary to give there a complete bibliography for transient analysis of electrical machines, but it is clear that about a thousand papers have been published till today.

Unfortunately, all these works have been only exploited at the beginning of the sixties with numerical computers and the electrical engineer till founds the mathematical models too sophisticated in order to design controller for machine drives. Today, most of electrical machine drives are still designed with constant parameter PID controllers even if it is well known that electrical machines have nonlinear state space model. It seems that these designers, even with numerical methods, are far from modern control tools and that they prefer classical ways. On the other hand, the electromagnetic knowledge of electrical machines does not permit the use of "black box" model and the real models are often too difficult to exploit for actual development of both automatic control theory and digital signal processor technology

Today, only one class of machines is well adapted for automatic control theory because of the structure of its state space model. For a DC machine, the state vector (armature current, field current, speed or position) is completely measurable and the command is quite simple. So, the use of modern control tools has permitted to design electrical drives for DC machines with high performances and a lot of papers have been published on the subject.

The AC drives are more complicated both for theoretical and technological point of view. So, their development is submitted to several improvements, and if a standard exists for DC machine, it has not yet appeared for synchronous and induction machines. The main difference between these two classes of AC machines resided in the fact that the synchronous machine seems to be more closed for DC structure for the control (brushless drives) when induction machines are not so easy to transform into a decoupled structure (field oriented control drives).

In the first part of the paper, we propose a classical way in order to determine the explicit state space model of a three phase squirrel cage induction machine (I.M.) for linear fluxes approach. The method is derived from classical papers on the subject (KRON 1938, JORDAN 1965) but several models can be performed and it is necessary to investigate their own performances from automatic control point of view. An other part of the paper related to modelling is dedicated to nonlinear fluxes approach which is an important improvement in state space model of AC machines (VAS 1981, KOVACS 1984, CAPOLINO 1987). Some additional remarks about the control of the machine have been set in order to expose the principle of field orientation (BLASCHKE 1972) and the use of decoupling for induction machine drives.

## II. MODELING I.M. WITH LINEAR FLUXES.

### II.1. Basic modeling of an I.M. in static reference frame.

The idea developed by PARK and KRON is to transform a three-phase AC machine in a two-phase one which is identical both from magnetic and electrical point of view. The observer looking to the voltage drop at the stator terminals is supposed to be fixed and it measures rotating emf from rotor influence. Then, the classical equation of the machine can be derived both from OHM generalized law and mechanical equation of the motion.

$$V = R * I + L * \frac{dI}{dt} \qquad (1)$$

$V = ( Vds ; Vqs ; 0 ; 0 ; -Tl )^t$

$I = ( ids ; iqs ; idr ; iqr ; w )^t$

$R11 = R22 = Rs$

$R33 = R44 = Rr$

$R32 = -R41 = M * w$

$R34 = -R43 = Lr * w$

$R55 = f \; ; \; R51 = M * iqr \; ; \; R52 = -M * idr$

$L11 = L22 = Ls$

$L33 = L44 = Lr$

$L55 = J$

$L13 = L24 = L31 = L42 = M$

It is not an usual form of the state equation but it can be derived from (1) by inverting the inductance matrix and giving the expression of the derivative of the state vector.

The state vector contains the dq components of the stator and rotor currents and the speeds of the shaft. It is also possible to add the position as a component of the state vector without having to modify deeply the equation. So, the state has both linear and a non linear parts which appear just as product of state variables. The control vector contains both the inputs (Vds, Vqs) and the perturbation (Tl).

For magnetizing convenience the inputs have the following form :

$$Vds = V_m * \sqrt{\frac{3}{2}} * \sin(ws*t)$$

$$Vqs = -V_m * \sqrt{\frac{3}{2}} * \cos(ws*t)$$

where $V_m$ is the maximum magnitude of the voltage at stator terminals and ws its pulsation with a ratio that remains constant. The perturbation (Tl) comes from the load on the shaft and cannot be measured by a simple way. The different coefficients correspond with the classical notations already defined with a machine supposed to be bipolar.

$$\frac{dI}{dt} = A(w) * I + B * V \qquad (2)$$

$$A11 = A22 = \frac{-1}{(sig*Taus)}$$
$$A33 = A44 = \frac{-1}{(sig*Taur)}$$
$$A55 = \frac{-f}{J}$$
$$A12 = -A21 = (1-sig) * \frac{w}{sig}$$
$$A13 = A24 = \frac{M}{sig*Ls*Taur} \; ; \; A31 = A42 = \frac{M}{sig*Lr*Taus}$$
$$A14 = -A23 = M * \frac{w}{sig*Ls} \; ; \; A41 = -A32 = M * \frac{w}{sig*Lr}$$
$$A43 = -A34 = \frac{w}{sig}$$
$$A51 = \frac{-M}{J} * iqr \; ; \; A52 = \frac{M}{J} * idr$$
$$B11 = B22 = \frac{1}{sig*Ls}$$
$$B31 = B42 = \frac{-M}{sig*Ls*Lr}$$
$$B53 = \frac{-1}{J}$$

II.2. Basic modeling of an I.M. in rotating reference frame.

The transformation used in order to develop a two-phase model is performed by means of rotation matrix product. Then, it is also possible to define an other observer measuring the voltage drops of the two-phase equivalent machine. Let us suppose that this observer rotates at the pulsation of the stator supply (wa=ws), then there is a simple way (LEONHARD 1985) to derive from equation (2) a state space model in a rotating reference frame without having to overcome calculations.

$$A11 = A22 = \frac{-1}{sig*Taus} \qquad (3)$$
$$A33 = A44 = \frac{-1}{sig*Taur}$$
$$A55 = \frac{-f}{J}$$
$$A12 = -A21 = wa + (1-sig) * \frac{w}{sig}$$

$$A13 = A24 = \frac{M}{sig*Ls*Taur} \; ; \; A31 = A42 = \frac{M}{sig*Lr*Taus}$$

$$A14 = -A23 = M * \frac{w}{sig*Ls} \; ; \; A41 = -A32 = M * \frac{w}{sig*Lr}$$

$$A43 = -A34 = -wa + \frac{w}{sig}$$

$$A51 = \frac{-M}{J} * iqr \; ; \; A52 = \frac{M}{J} * idr$$

$$B11 = \frac{1}{sig*Ls}$$

$$B31 = \frac{-M}{sig*Ls*Lr}$$

$$B52 = \frac{-1}{J}$$

The main difference appears in the control vector which has only two components :

$$Vds = V_m * \sqrt{\frac{3}{2}}$$

$$wa = ws$$

where $V_m$ is also the maximum magnitude of the voltage at the stator terminals and ws its pulsation. In this case, the component Vqs becomes zero and the total dimension of the control vector is 2 instead of 3 in the case of the static reference frame.

The model seems to be more complicated but the response of the system to the same output is quite different from equivalent state vector point of view. For a rotating reference frame, we shall see that the response is closed for DC equivalent machine and this approach is very interesting for field oriented

## III. MODELING I.M. WITH NONLINEAR FLUXES.

### III.1. Basic assumptions for saturated I.M..

The principle of saturation in electrical machines is not new because it has been one of the preoccupation of POTIER at the end of the nineteenth century. What is new in modeling AC machines is the introduction of saturation in the state space model of the machine.

In order to define a physical model of the saturated AC machine related to the previous state vector, it is necessary to compute the fluxes in the air-gap and to give a matrix expression with dq components. The results are developed for static reference frame but it is obvious that the same procedure can be used for rotating reference frame. Every component of the flux both for stator and rotor has to be separated in magnetizing and leakage parts and a new component of the current has been defined.

$$phi = phi0 + l * i \qquad (4)$$

$$im = id + iq$$

$$id = ids + idr$$

$$iq = iqs + iqr$$

Then, it is possible to define also a static mutual inductance (Mst) as the ratio of the magnetizing flux to its current and a dynamic mutual inductance (Mdy) as its derivative. Some other inductances have to be derived in order to simplify the final expression of the generalized OHM law (KOVACS 1984, CAPOLINO 1987). The basic equation is the same as the presentation in (1) with only different terms as

$$V = ( Vds ; Vqs ; 0 ; 0 ; -Tl )^t \qquad (5)$$
$$I = ( ids ; iqs ; idr ; iqr ; w )^t$$
$$R11 = R22 = Rs$$
$$R33 = R44 = Rr$$
$$R32 = -R41 = Mst * w$$
$$R34 = -R43 = Lr * w$$
$$R55 = f \; ; \; R51 = Mst * iqr \; ; \; R52 = -Mst * idr$$
$$L11 = Lds \; ; \; L22 = Lqs$$
$$L33 = Ldr \; ; \; L44 = Lqr$$
$$L55 = J$$
$$L12 = L14 = L21 = L23 = L32 = L34 = L41 = L43 = L2s$$
$$L13 = L31 = L0 + L2c \; ; \; L24 = L42 = L0 - L2c$$

III.2. State space model for saturated I.M..

The derivation of an explicit form of the state space model is more difficult than for linear fluxes approach. For the equation, both Mst and Mdy come from the magnetizing characteristic of the machine but the values L2s and L2c depend on the rotor position and are of course time dependent.

The coefficients bj depend on the different inductances and mutual inductances defined above and are also nonlinear functions of time. The main remark is that saturation introduces two non linearities in the state space model : the first one is related to the magnetizing curve (phim as a function of im) and the second one comes from time dependance of inductances.

$$A11 = -Rs*b1 + b4*Mst*w \; ; \; A12 = -Rs*b2 - b3*Mst*w \qquad (6)$$
$$A13 = -Rr*b3 + b4*Lr*w \; ; \; A14 = -Rr*b4 - b3*Lr*w$$
$$A21 = -Rs*b5 + b8*Mst*w \; ; \; A22 = -Rs*b6 - b7*Mst*w$$
$$A23 = -Rr*b7 + b8*Lr*w \; ; \; A24 = -Rr*b8 - b7*Lr*w$$
$$A31 = -Rs*b9 + b12*Mst*w \; ; \; A32 = -Rs*b10 - b11*Mst*w$$
$$A33 = -R*b11 + b12*Lr*w \; ; \; A34 = -Rr*b12 - b11*Lr*w$$
$$A41 = -Rs*b13 + b16*Mst*w \; ; \; A42 = -Rs*b14 - b15*Mst*w$$

$A43 = -Rr*b15 + b16*Lr*w$ ; $A44 = -Rr*b16 - b15*Lr*w$

$A51 = -Mst * iqr$ ; $A52 = Mst * idr$

$A55 = \frac{-f}{J}$

$B11 = b1$ ; $B12 = b2$ ; $B21 = b5$ ; $B22 = b6$

$B31 = b9$ ; $B32 = b10$ ; $B41 = b13$ ; $B42 = b14$

$B53 = \frac{-1}{J}$

## IV. NONLINEAR CONTROL OF I.M..

### IV.1. Scalar control.

The most popular way to control the speed of an I.M. is to implement a scalar regulator based upon the correction of speed error with a condition on constant flux in the air-gap. If the performances of the model are examined (figures 1 a and b), it is obvious that the flux cannot be constant if it is not controlled. So, if the control law has to be based upon a constant ratio (V/ws) or the same corrected by a term proportional to the slip frequency (ws - w), it is well known that the dynamic performances of the drive are rather poor and that it is not equivalent to a DC drive.

### IV.2. Decoupling control.

The complex nonlinear structure of the I.M. can be decoupled by controlling the machine with a suitable vector (BLASCHKE 1972). This can be done by feeding the machine with unique set of voltages whose magnitude and frequency are defined by decoupling control scheme.

The decoupling control scheme can be implemented in order to perform perfectly independent command of torque, which imposes the speed, and flux in the air-gap. Then, the I.M. drive is equivalent to a DC machine with both armature and field control.

The decoupling controllers can be implemented in three different ways for stator flux magnetizing flux or rotor flux regulation (figures 2 a, b and c). The implementation is given for rotating reference frame, so it is necessary to perform inverse transformation of the current by means of rotor position and matrix computation.

## V. CONCLUSION.

This paper has tried to present some investigations in modeling and control of some classes of induction machines and to open perspectives in using modern control tool for electrical machine drives.

Several graduations in modeling I.M. have been presented and simulation results are given in order to show the validity of the different models. All these models are derived from knowledge of the device and are simplified in order to minimize both the order of the system and the nonlinear aspect.

The development in nonlinear modeling of electrical machines should help the designer of drives to implement more and more powerful control structure. In the last part of the paper, we have tried to present some remarks about control of I.M. drives in order to show that improvement in this way has to be performed when general control tools for nonlinear system shall be available.

REFERENCES.

(KRON 1938) : The application of tensor to the analysis of rotating electrical machinery, General Electric Ed., New-York.

(JORDAN 1965) : Analysis of induction machines in dynamic systems, Proc. IEEE Power Meeting, Paper 31.

(BLASCHKE 1972) : The principle of field orientation as applied to new transvector closed loop control system for rotating field machines, Siemens Review, vol. 34, pp.217-220.

(VAS 1981) : Generalized analysis of saturated AC machines, Archiv.f.Elektrotech., vol.64, pp.57-62.

(KOVACS 1984) : On the theory of cylindrical rotor AC machines including main flux saturation, IEEE Trans.PAS, vol 103, n°4, pp.754-761.

(LEONHARD 1985) : Control of electrical drives, Springer Verlag, Berlin.

(BOSE 1986) : Power electronics and AC drives, Prentice Hall, Englewood Cliffs.

(CAPOLINO, BOUSSAK, GAUTIER 1987) : Modélisation et simulation du comportement dynamique des machines à induction en régime saturé, Journées SEE Club 11, Développement récents des méthodes numériques appliquées aux machines électriques, Gif-sur-Yvette, pp.101-112.

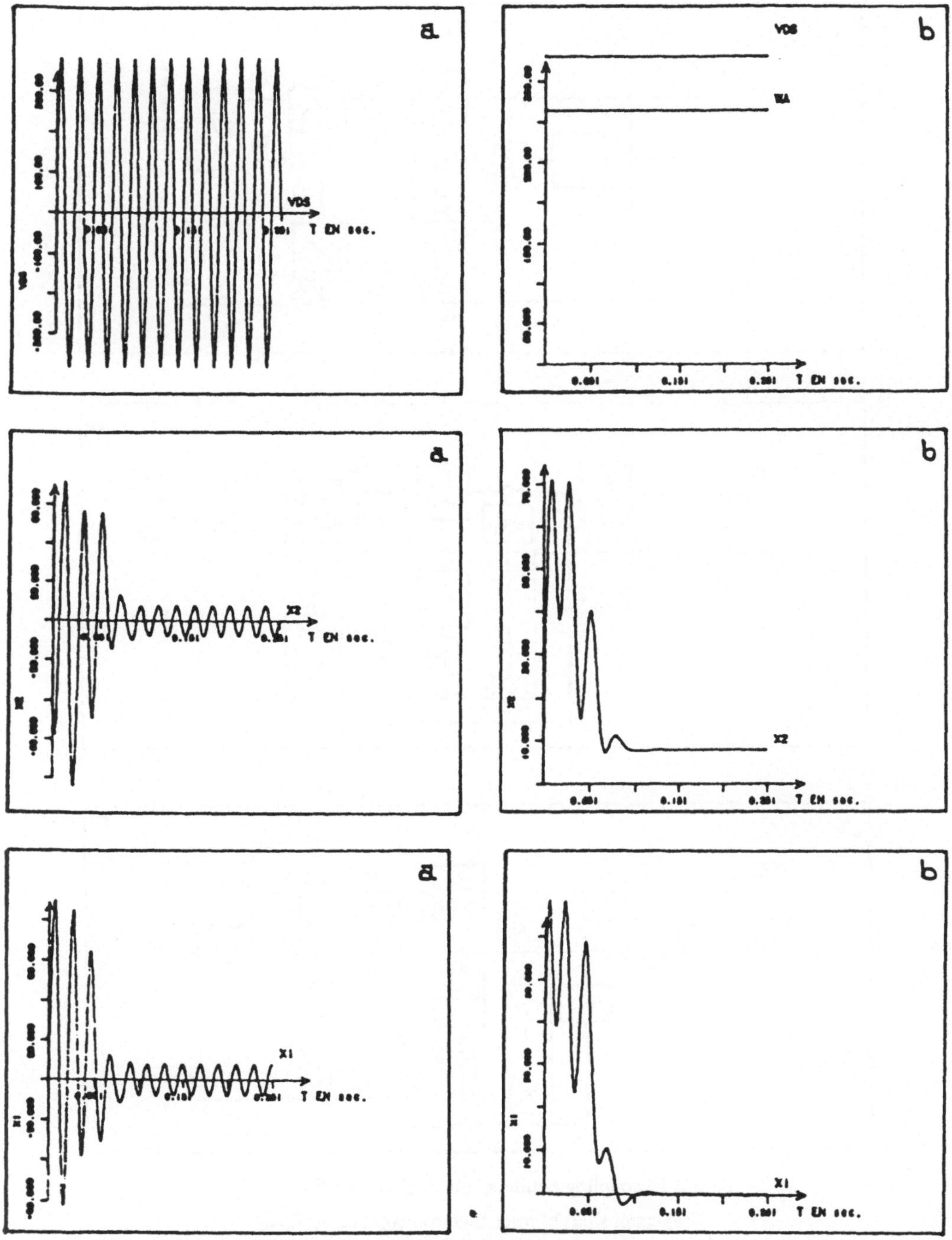


<u>Figure 1.</u> Simulation results (command and partition of state vector) for linear flux machine:
a) static reference frame; b) rotating reference frame.

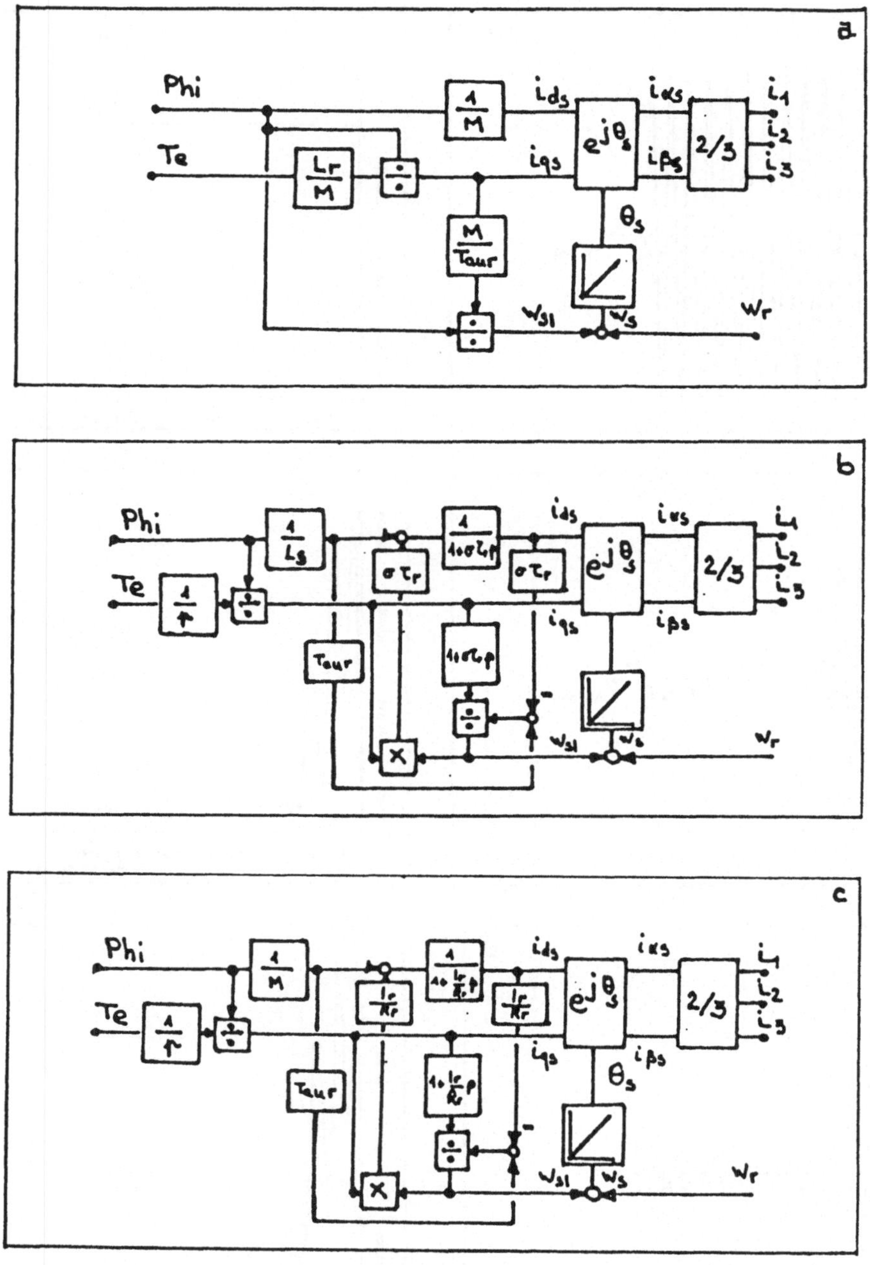


Figure 2. Decoupling nonlinear controllers with flux regulation:
a) stator flux; b) rotor flux; c) magnetizing flux.

# Lecture Notes in Control and Information Sciences

Edited by M. Thoma and A. Wyner

Vol. 81: Stochastic Optimization
Proceedings of the International Conference, Kiew, 1984
Edited by I. Arkin, A. Shiraev, R. Wets
X, 754 pages, 1986.

Vol. 82: Analysis and Algorithms of Optimization Problems
Edited by K. Malanowski, K. Mizukami
VIII, 240 pages, 1986.

Vol. 83: Analysis and Optimization of Systems
Proceedings of the Seventh International Conference of Analysis and Optimization of Systems
Antiba, June 26-27, 1986
Edited by A. Bensoussan, J. L. Lions
XVI, 901 pages, 1986.

Vol. 84: System Modelling and Optimization
Proceedings of the 12th IFIP Conference
Budapest, Hungary, September 2–6, 1985
Edited by A. Prékopa, J. Szelezsán, B. Strazicky
XII, 1046 pages, 1986.

Vol. 85: Stochastic Processes in Underwater Acoustics
Edited by Charles R. Baker
V, 205 pages, 1986.

Vol. 86: Time Series and Linear Systems
Edited by Sergio Bittanti
XVII, 243 pages, 1986.

Vol. 87: Recent Advances in System Modelling and Optimization
Proceedings of the IFIP-WG 7/1 Working Conference
Santiago, Chile, August 27-31, 1984
Edited by L. Contesse, R. Correa, A. Weintraub
IV, 199 pages, 1987.

Vol. 88: Bruce A. Francis
A Course in $H_\infty$ Control Theory
XI, 156 pages, 1987.

Vol. 88: Bruce A. Francis
A Course in $H_\infty$ Control Theory
X, 150 pages, 1987.
Corrected - 1st printing 1987

Vol. 89: G. K. H. Pang/A. G. J. McFarlane
An Expert System Approach to Computer-Aided Design of Multivariable Systems
XII, 223 pages, 1987.

Vol. 90: Singular Perturbations and Asymptotic Analysis in Control Systems
Edited by P. Kokotovic, A. Bensoussan, G. Blankenship
VI, 419 pages, 1987.

Vol. 91 Stochastic Modelling and Filtering
Proceedings of the IFIP-WG 7/1 Working Conference
Rome, Italy, Decembre 10-14, 1984
Edited by A. Germani
IV, 209 pages, 1987.

Vol. 92: L. T. Grujić, A. A. Martynyuk, M. Ribbens-Pavella
Large-Scale Systems Stability Under Structural and Singular Perturbations
XV, 366 pages, 1987.

Vol. 93: K. Malanowski
Stability of Solutions to Convex Problems of Optimization
IX, 137 pages, 1987.

Vol. 94: H. Krishna
Computational Complexity of Bilinear Forms
Algebraic Coding Theory and Applications to Digital Communication Systems
XVIII, 166 pages, 1987.

Vol. 95: Optimal Control
Proceedings of the Conference on Optimal Control and Variational Calculus
Oberwolfach, West-Germany, June 15-21, 1986
Edited by R. Bulirsch, A. Miele, J. Stoer and K. H. Well
XII, 321 pages, 1987.

Vol. 96: H. J. Engelbert/W. Schmidt
Stochastic Differential Systems
Proceedings of the IFIP-WG 7/1 Working Conference
Eisenach, GDR, April 6-13, 1986
XII, 381 pages, 1987.

# Lecture Notes in Control and Information Sciences

Edited by M. Thoma and A. Wyner

Vol. 97: I. Lasiecka/R. Triggiani (Eds.)
Control Problems for Systems
Described by Partial Differential Equations
and Applications
Proceedings of the IFIP-WG 7.2
Working Conference
Gainesville, Florida, February 3-6, 1986
VIII, 400 pages, 1987.

Vol. 98: A. Aloneftis
Stochastic Adaptive Control
Results and Simulation
XII, 120 pages, 1987.

Vol. 99: S. P. Bhattacharyya
Robust Stabilization Against
Structured Perturbations
IX, 172 pages, 1987.

Vol. 100: J. P. Zolésio (Editor)
Boundary Control and Boundary Variations
Proceedings of the IFIP WG 7.2 Conference
Nice, France, June 10-13, 1987
IV, 398 pages, 1988.

Vol. 101: P. E. Crouch,
A. J. van der Schaft
Variational and Hamiltonian
Control Systems
IV, 121 pages, 1987.

Vol. 102: F. Kappel, K. Kunisch,
W. Schappacher (Eds.)
Distributed Parameter Systems
Proceedings of the 3rd International Conference
Vorau, Styria, July 6–12, 1986
VII, 343 pages, 1987.

Vol. 103: P. Varaiya, A. B. Kurzhanski (Eds.)
Discrete Event Systems:
Models and Applications
IIASA Conference
Sopron, Hungary, August 3-7, 1987
IX, 282 pages, 1988.

Vol. 104: J. S. Freudenberg/D. P. Looze
Frequency Domain Properties of Scalar
and Multivariable Feedback Systems
VIII, 281 pages, 1988.

Vol. 105: Ch. I. Byrnes/A. Kurzhanski (Eds.)
Modelling and Adaptive Control
Proceedings of the IIASA Conference
Sopron, Hungary, July 1986
V, 379 pages, 1988.

Vol. 106: R. R. Mohler (Editor)
Nonlinear Time Series and
Signal Processing
V, 143 pages. 1988.

Vol. 107: Y. T. Tsay, L.-S. Shieh, St. Barnett
Structural Analysis and Design
of Multivariable Systems
An Algebraic Approach
VIII, 208 pages, 1988.

Vol. 108: K. J. Reinschke
Multivariable Control
A Graph-theoretic Approach
274 pages, 1988.

Vol. 109: M. Vukobratović/R. Stojić
Modern Aircraft Flight Control
VI, 288 pages, 1988.

Vol. 110: In preparation

Vol. 111: A. Bensoussan, J. L. Lions (Eds.)
Analysis and Optimization
of Systems
XIV, 1175 pages, 1988.

Vol. 112: Vojislav Kecman
State-Space Models of Lumped
and Distributed Systems
IX, 280 pages, 1988

Vol. 113: M. Iri, K. Yajima (Eds.)
System Modelling and Optimization
Proceedings of the 13th IFIP Conference
Tokyo, Japan, Aug. 31 – Sept. 4, 1987
IX, 787 pages, 1988.

Vol. 114: A. Bermúdez (Editor)
Control of Partial Differential Equations
Proceedings of the IFIP WG 7.2
Working Conference
Santiago de Compostela, Spain, July 6–9, 1987
IX, 318 pages, 1989

Vol. 115: H.J. Zwart
Geometric Theory for Infinite
Dimensional Systems
VIII, 156 pages, 1989.

Vol. 116: M.D. Mesarovic, Y. Takahara
Abstract Systems Theory
VIII, 439 pages, 1989

# Lecture Notes in Control and Information Sciences

Edited by M. Thoma and A. Wyner

Vol. 117: K.J. Hunt
Stochastic Optimal Control Theory
with Application in Self-Tuning Control
X, 308 pages, 1989.

Vol. 118: L. Dai
Singular Control Systems
IX, 332 pages, 1989

Vol. 119: T. Başar, P. Bernhard
Differential Games and Applications
VII, 201 pages, 1989

Vol. 120: L. Trave, A. Titli, A. M. Tarras
Large Scale Systems:
Decentralization, Structure Constraints
and Fixed Modes
XIV, 384 pages, 1989

Vol. 121: A. Blaquière (Editor)
Modeling and Control of Systems
in Engineering, Quantum Mechanics,
Economics and Biosciences
Proceedings of the Bellman Continuum
Workshop 1988, June 13–14, Sophia Antipolis, France
XXVI, 519 pages, 1989

Vol. 122: J. Descusse, M. Fliess, A. Isidori,
D. Leborgne (Eds.)
New Trends in Nonlinear Control Theory
Proceedings of an International
Conference on Nonlinear Systems,
Nantes, France, June 13–17, 1988
VIII, 528 pages, 1989